Fourth International Workshop on Environment and Geoscience

IOP Conference Series: Earth and Environmental Science
Volume 865

Hangzhou, China
17-19 July 2021

Editor:

Christophe Claramunt

ISBN: 978-1-7138-4847-9
ISSN: 1755-1307

Printed from e-media with permission by:

Curran Associates, Inc.
57 Morehouse Lane
Red Hook, NY 12571

Some format issues inherent in the e-media version may also appear in this print version.

Institute of Physics
Dirac House, Temple Back
Bristol BS1 6BE UK

Phone: 44 1 17 929 7481
Fax: 44 1 17 920 0979

techtracking@iop.org

Additional copies of this publication are available from:

Curran Associates, Inc.
57 Morehouse Lane
Red Hook, NY 12571 USA
Phone: 845-758-0400
Fax: 845-758-2633
Email: curran@proceedings.com
Web: www.proceedings.com

TABLE OF CONTENTS

Preface
Christophe Claramunt

Peer review declaration

Spatial-temporal variation patterns of groundwater tables in the middle section of the Hexi Corridor 1
Lingqi Li, Kai Wu, Huijuan Yin, Bo Qu, Liyong Jiang, Fei Yang

Attention-based DenseNet network for multi-source remote sensing classification 8
Lishuo Zhang, Hong Lin, Fanyang Zeng

Design and application of sewage treatment technology in fieldbus control system... 14
Kai Guo, Fuxiao Li

Monitoring and analysis of green sustainable development in impoverished counties based on nighttime light and RSEI 22
Zelin Jiang, Jian Deng, Haijun Luan, Zhangjin Qiu

Analysis of cyanobacteria outbreak effects in southern offshore area of Taihu lake 35
Lei Fu, Junmin Wang, Changhuai Wu, Aiju You, Shang Gao

Study on emission characteristics and health effects of volatile organic compounds in child seats.................. 41
Chen Cui, Lei Wang, Wei Liu, Shujie Xu, Xuefeng Liu

Research and application for key technology of anti-cracking of acrylic emulsion mortar............................... 46
Xingyan Liu, Yukun Zhao

Analysis of volatile components in host plants of Xylotrechus quadripes 54
Zijun Luo, Yingzhi Chen, Liping Liu, Genyun Pan, Ruifang Wang

Experimental study on the influence of injection advance angle to diesel engine equipped with electronic unit pump 59
Yupo Ma, Yu Feng, Zengbin Wu, Changcheng Wang, Tongzhou Liao, Zhanyu Zhan, Zhenyu Zuo, Guoning Shi

Graphene-modified CPE as a disposable electrochemical sensor for rapid determination of fluoranthene 67
Wenhan Zhong, Xiaoying Yan, Tao Han, Mengfan Wu, Zhihao Xu, Ying Wang, Xiaohong Jin

New technology of magnetotelluric polaron exploration-taking Huashan piedmont basement fault as an example 74
Lin Zhang, Laixin Li, Dongmin Ma, Pei Zhang, Kai Shao, Xiaolin Ma

Research on reasonable development strategy of offshore buried hill fractured reservoir 81
Lina Yang, Xinran Wang, Yanjun Yin, Li Wei, Shengli Xu, Linlin Dan

The design of a portable seismograph used in aftershock seismic array observation... 89
Congqing Li, Yuntian Teng, Xiaomei Wang, Xiaoyong Fan, Yiyang Tang

Study on gravity anomaly changes and crustal structure characteristics in Weihe Basin.................................. 95
Yongqi Zhang, Jianping Cao, Enhui Zhang, Meitao Han, Yuxiang Zhang

Automatic stem mapping using single-scan terrestrial laser scanning data and Mean Shift clustering............ 108
Xiangjiang Liu, Maolin Chen, Chunsen Tan, Xinyi Zhang, Wenguang Yang

Magnetostratigraphy and AMS study of a late Cretaceous-Paleogene succession in the eastern Xining basin: constraint on K-Pg boundary tectonic event in NE Tibet Plateau... 115
Chicheng He, Zhonghang Wang, Yanguo Fang, Maohua Li, Shaokai Li, Yueqiao Zhang

Direct hydrocarbon detection (DHI) techniques for prospective exploration targets in Chaoshan Depression.. 122
Guangjian Zhong, Renqi Jiang, Hai Yi, Jincai Wu, John Castagna, Changmao Feng, Gang Zhou, Kun Wang, Lina Liu, Ming Sun

Does environmental regulation slow down marine environmental pollution?--Empirical study based on panel threshold model.. 136
Xiaohao Liu, Shujuan Wu, Jianhua Xiao

A Systematic approach to world regional marine resources development models through case studies... 143
Yawen Kong, Chunyu Liu, Shuguang Liu, Lei Song

Analysis on the spatial accessibility of parks and green spaces and optimal site selection—Taking Weifang prefecture-level city as an example.. 155
Wenqi Yan, Shijie Wang

Stage characteristics and causes of a mixed severe convective weather process... 162
Hongwu Liu, Jingyu Xu, Jie Tang, Rong Yao, Yan Hu

The effects of meteorological factors on ozone during summer in Shanghai, China 172
Shengyang Duan, Chenhang Zhao, Yonghang Chen, Qiong Liu

Accuracy analysis of 3D model of open-pit mine based on tilt photogrammetry.. 182
Jinjin Lu, Yizhan Zhao, Lv Zhou, Kejia Wei, Jun Ma, Xin Li

Quick determination of phenanthrene in coloured latexes using graphene-modified glassy carbon................ 189
Furong Liu, Yu Wang, Qiuxiang Liu, Fangping Yan, Xiangyu Ye, Qiwang Wu, Wei Shi

Case study on the landscape of Yuhang Entrance of Shanghai-Hangzhou expressway under the background of the park city 196
Yang Wang, Lingzhi Shen, Xiaomin Ye, Yun Xie

Analysis of the stability of Zhangjiapo landslide 207
Qian Hou, Lixin Liu, Jinxin Chong, Xueping Li

A novel reduced-order method for analysis of hyperspectral images... 215
Yanming Zhang, Lijun Jiang

Classification of terrestrial laser scanning data based on relative projection density 221
Xinyi Zhang, Maolin Chen, Xiyu Wang, Xiangjiang Liu

Evaluation of marine green economy efficiency and its influencing factors in China 228
Jianhua Tang, Shujuan Wu, Jianhua Xiao

The analysis of impact factor on temporal and spatial distribution characteristics of petroleum pollution in surface sediments of Laizhou Bay 235
Fan Gao, Qinbang Sun, Pengfei Zhao, Yumiao Zhang, Wenbo Jiang, Chong Zhang

Assessment of volatile organic pollution control technologies in automotive industry 243
Chonglei Ren, Xuefeng Liu, Jiabao Ren, Lei Wang

Comparative analysis of two hazard rainstorm processes in Hunan affected by the Southwest China vortex 251
Yan Hu, Hongwu Liu, Huanqian Liu, Minghui Tang, Yuting Zeng

Effects of hydraulic uncertainty on slope stability in an embankment dam 262
Xiaoying Chen, Haoqing Yang, Xiaohui Lu, Anling Lu

The effect of sedimentary facies on coalbed methane in Binchang mining area 268
Kai Shao, Dongmin Ma, Laixin Li, Dongdong Wang, Lin Zhang, Gang Yang

Analysis on the geophysical characters of Qiyueshan fault.. 275
Yunsheng Liu, Jun Chen, Ziguang Zhang, Meiheng Zhu, Aiqiong Chen

Research on glacier elevation changes in Southeast Tibet based ICESat and CryoSat-2 Data 282
Lanyu Li

Geochemical behavior of Mo and W in the CO_2-rich mineral waters occurring in the East Elbrus area 289
G A Chelnokov, V Y Lavrushin, A S Aidarkozhina, Q Guo

Characteristics of heavy metal pollution in soils of a typical copper smelting site in China........................... 294
Jieqiong Zheng, Qianling Ma, Li Li, Shangyong Jin

Land suitability and economic value of ecotourism development in Masjid Lake of Kuantan Singingi District, Indonesia 300
Zulkarnaini, Nawari

Efficient removal of lead ions from aqueous solution using C-TiO_2 adsorbent 315
Qi Luo, Guiqin Liu, Li Wang, Wukang Wang, Ying Wang, Hao Wang, Wen Xu

Environmental and economic sustainability in pricing strategy of carbon tax: application of environmental Kuznets curve and real options approach 321
Chuan-Chuan Ko, Chien-Yu Liu, Chunmei Liu, Xiangyong Yuan, Yuanshan Li

Engineering performance's evaluation of integrated backfill materials 328
Xiang Fan, Jinyang Du, Facheng Yi, Yangjie Feng

Research on the influencing factors of reclaimed wastewater reuse and countermeasure in China 335
Leixiang Wu, Panwei Zhang, Xingchen Liu, Ang Li, Zhuowei Wang, Ronghao Guan

Mesoporous TiO_2 sphere with high adsorbability: Adsorption behavior study of Pb(II) from water.............. 342
Hao Ding, Xiangyu Ye, Xiaodong Wu, Quan Liang, Tianyi Hu, Difan Wang

Object-based multi-features Wetland classification method of GF-2 PMS imagery....................................... 348
Leishi Chen, Qinzhe Han, Yilin Lin, Dongsheng Du, Jianbo Deng

Simulation on plateau power recovery of turbocharged diesel engine matched with electronic unit pump.. 355
Yupo Ma, Yu Feng, Zengbin Wu, Changcheng Wang, Tongzhou Liao, Zhanyu Zhan, Zhenyu Zuo, Guoning Shi

Research on ecological design of Junhe Wetland Park in Pingyi County, Linyi, Shandong Province............ 363
Laichen Zhang, Jianwu Ma, Yalin Wang

Analysis of soil nematode community of alpine meadow in the middle of the Tibetan Plateau 370
Lei Hou, Huiying Xue

Influence of underground engineering on surrounding water environment in sedimentary rock distribution area and its treatment principle 380
Maozhong Tian, Lihua Zhang, Zheng Hu, Guo Guo

Extraction of open-pit mining area based on optimal scales selection and hierarchical classification: A case of Nonoc lateritic nickel 400
Xian Zhang, Wei Li, Li Chen, Yaqin Sun

Artificial intelligence prediction of air traffic flow rate at the Hong Kong International Airport 407
Kai-Kwong Hon

GPS/BDS displacement monitoring of railway slopes based on coloured noise analysis 412
Jun Ma

Author Index

IOP Conf. Series: Earth and Environmental Science **865** (2021) 011001 doi:10.1088/1755-1315/865/1/011001

Preface

Christophe Claramunt

Editor, Proceedings of the Fourth International Workshop on Environment and Geoscience (IWEG2021)

and

Conference Chair

Professor, Naval Academy Research Institute, France

Email: christophe.claramunt@ecole-navale.fr

The Fourth International Workshop on Environment and Geoscience (IWEG 2021) was held successfully on 17-19 July 2021 in Hangzhou, China. This year's conference was held as a hybrid conference to avoid international travelling during the COVID-19 situation, delegates could attend the conference in person or online via VooV Meeting.

Conference program was divided into 3 sessions: keynote speeches, oral presentations and poster presentations. We greatly appreciate Prof. Youmin Chen from Henan University, who presided over the conference, and all participants for attending and sharing!

We are honored to invite 5 experts to give the impressive keynote speeches: Prof. Xiaxin Tao from Harbin Institute of Technology; Prof. Bin Zou from Central South University; Prof. Long Ding from University of Science and Technology of China; Prof. Youmin Chen from Henan University; Prof. Chien-Yuan Chen from National Chiayi University.

Following the keynote speakers, 9 scholars made oral presentations respectively in the afternoon. And 8 poster presentations came after it. Reporters shared their latest research findings and audiences were actively involved in discussions.

The proceedings of this conference comprise 52 accepted contributions from 91 submissions. We would like to express our gratitude to the reviewers of these manuscripts, who provided rational, constructive comments & suggestions to the authors. We were extremely grateful as the organizers and technical program committee who make this conference to an unforgettable experience for all participants. And extend our sincere thanks to the authors for their valuable contributions. Finally, our sincere gratitude goes also to the IOP Publishing editors and managers for their helpful cooperation during the preparation of the conference proceedings.

On behalf of the Organizing Committee of IWEG 2021.

IOP Conf. Series: Earth and Environmental Science **865** (2021) 011001 doi:10.1088/1755-1315/865/1/011001

Technical Program Committee

Conference Chair
Christophe Claramunt, Naval Academy Research Institute, France

Conference Co-chairs
Zhenhao Xu, Shandong University, China

TPC Chair
R. A. Stephenson, King's College University of Aberdeen, UK

TPC member
Ramesh Agarwal, Washington University, USA
Mikals Scholz, Lund University, Sweden
Tao Zhu, China University of Mining & Technology (Beijing), China
Sebastián E. Bellú, National University of Rosario, Argentina
Vojin Grković, Consultant in fields of energy and energy technology, Serbia
Yi Wang, China University of Geosciences, China
Vijayan Gurumurthy Iyer, Engineering Council of India, Chennai, India
Poi Sim Khiew, University of Nottingham Malaysia, Malaysia
Yang Zhi, Institute of Huai River Water Resources Protection, Huaihe River Commission, China
António M. C. Oliveira e Sousa, National University of Tucuman, Argentina
Jin Su Jeong, Technical University of Madrid, Spain
Jarosław Krzywański, Jan Dlugosz University in Czestochowa, Poland
Changzhi Li, Institute of Water Resources and Hydropower Research, China
Aleksandr Arakelyan, IGS of NAS RA / Geocom, Ltd, Armenia
Marios C. Phocas, University of Cyprus, Cyprus
H.Y. Gao, Shanxi University, China
Ronaldo Goncalves dos Santos, Centro Universitario da FEI, Brazil
José Carlos Magalhães Pires, University of Porto, Portugal
RUSCĂ Florin Valentin, University POLITEHNICA of Bucharest, Romania

IOP Conf. Series: Earth and Environmental Science **865** (2021) 011001 doi:10.1088/1755-1315/865/1/011001

Conference Overview

The 4th International Workshop on Environment and Geoscience (IWEG2021) was successfully held as a hybrid conference to avoid international travelling during the COVID-19 situation, delegates attended the conference online via VooV Meeting and in person on July 17-19, 2021 in Hangzhou, China. Conference program was divided into 3 sessions: **Keynote Speeches, Oral Presentations** and **Poster Presentations**.

Conference Schedule

<table>
<tr><th>Date</th><th>Time</th><th>Program</th><th>Place</th></tr>
<tr><td>July 17</td><td>10:00-18:00</td><td>Author Registration</td><td>The hotel lobby
VooV Meeting
(腾讯会议)
ID: 408 941 279</td></tr>
<tr><td rowspan="4">July 18</td><td>9:00-9:10</td><td>Opening Ceremony</td><td rowspan="4">Conference hall
VooV Meeting
(腾讯会议)
ID: 408 941 279</td></tr>
<tr><td>9:10-12:00</td><td>Keynote Speech</td></tr>
<tr><td>14:00-16:00</td><td>Oral Presentation</td></tr>
<tr><td>16:10-17:10</td><td>Poster Presentation</td></tr>
<tr><td>July 19</td><td>08:30-17:00</td><td>Social& Cultural Activities</td><td>Hangzhou</td></tr>
</table>

IOP Conf. Series: Earth and Environmental Science **865** (2021) 011001 doi:10.1088/1755-1315/865/1/011001

Keynote Speeches

It was our great honor to have the following impressive keynote speeches during the conference.

Prof. Xiaxin Tao from Harbin Institute of Technology
Speech Title: Test of PSHA map of China in case study of Sichuan region

Prof. Bin Zou from Central South University
Speech Title: Detection of soil heavy metal contamination using remote sensing technology

IOP Conf. Series: Earth and Environmental Science **865** (2021) 011001 doi:10.1088/1755-1315/865/1/011001

Prof. Long Ding from University of Science and Technology of China
Speech Title: Risk assessment and management of domino effects in chemical industry parks

Prof. Youmin Chen from Henan University
Speech Title: Modelling the tendency of climate change and short-term weather forecast with WRF mode

Prof. Chien-Yuan Chen from National Chiayi University
Speech Title: Climate change and initiation of slope-land disasters in Taiwan

Oral Presentations

Poster Presentations

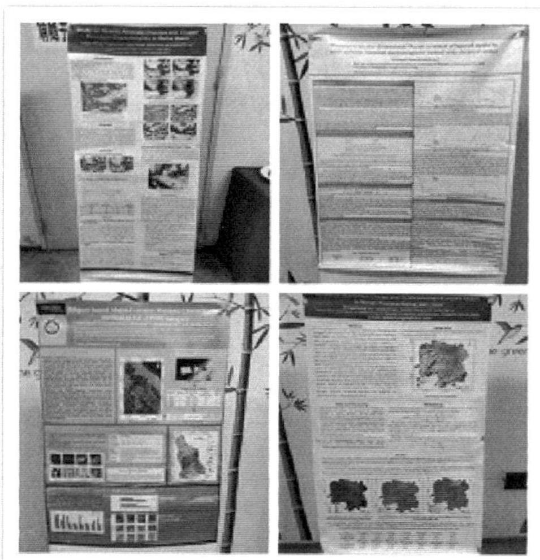

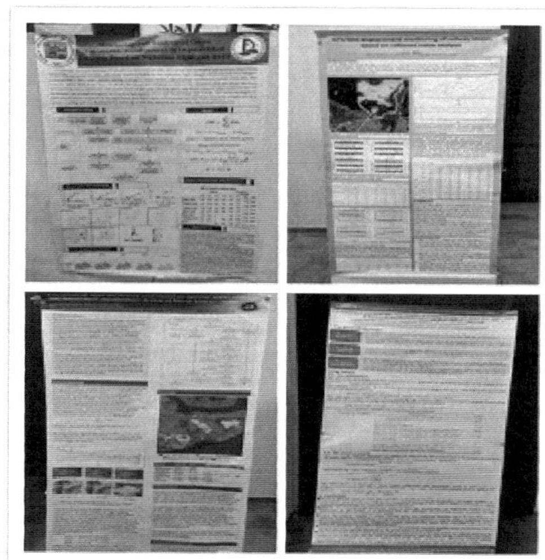

We would like to express our sincere gratitude to the distinguished keynote speakers, oral speakers, poster presenters as well as all the audiences. This conference has made a big success because of their supports and assistance. We are expecting more and more experts and scholars from around the world to join this international conference next year.

IOP Conf. Series: Earth and Environmental Science 865 (2021) 011002 doi:10.1088/1755-1315/865/1/011002

Peer review declaration

All papers published in this volume of IOP Conference Series: Earth and Environmental Science have been peer reviewed through processes administered by the Editors. Reviews were conducted by expert referees to the professional and scientific standards expected of a proceedings journal published by IOP Publishing.

•Type of peer review:
Double-blind

•Conference submission management system:
http://papersub.pmsconf.org/

• Number of submissions received:
91

• Number of submissions sent for review:
85

• Number of submissions accepted:
52

• Acceptance Rate (Number of Submissions Accepted / Number of Submissions Received X 100):
Approximately 57.1%

• Average number of reviews per paper:
3

• Total number of reviewers involved:
256

• Any additional info on review process:
After preview by editorial assistant, 85 papers were sent for peer reviews, 256 reviewers were involved in the process, and review process for each paper took around 1 month. The decision to accept or reject was based on the reviewers' comments. If differences of opinion occurred between reviewers, the editors weighed all comments and arrived at a balanced decision based on all comments, or a second round of peer review was initiated.

• Contact person for queries:
Ms. Lily Gong
Association for Science and Engineering, China
Secretary@iwegconf.org

IOP Conf. Series: Earth and Environmental Science **865** (2021) 012001 doi:10.1088/1755-1315/865/1/012001

Spatial-temporal variation patterns of groundwater tables in the middle section of the Hexi Corridor

Lingqi Li[1,3], Kai Wu[1], Huijuan Yin[1], Bo Qu[1], Liyong Jiang[2] and Fei Yang[1]

[1]Key Laboratory of Soil and Water Conservation of the Loess Plateau of MWR, Yellow River Institute of Hydraulic Research, Zhengzhou, 450003, China;
[2]College of Hydrology and Water Resources, Hohai University, Nanjing, 210098, China

[3]Corresponding author's email: lqli@whu.edu.cn

Abstract. It is of great significance to research spatial-temporal variations of groundwater tables from the perspective of the whole groundwater variable field in a basin. In this study, the original observation matrix was constructed using 38 monitoring wells data from 1984 to 2019 in the Heihe River Basin in the middle section of the Hexi Corridor and processed to be centered and standardized, respectively. The empirical orthogonal function (EOF) method was used to decompose the transformed matrices to obtain spatial and temporal variation patterns of groundwater tables which were represented by the spatial weights of typical uncorrelated modes with the corresponding temporal functions (TF). The resulting main variation pattern showed general drops in regional groundwater tables for both centered and standardized cases, especially in the southeast regions when using centered EOF modes. The second to fifth modes presented reverse distribution characteristics from the northwest to southeast directions. The downward trend in the first TF series indicated that the groundwater depth of the whole basin varied to be deeper. The local rise occurring at some sites demonstrated the positive effects of continuous ecological water delivery since 2000, but was still not recovered to the previous level due to long-term serious groundwater overdrawing.

1. Introduction

Under the background of a severe global water crisis, water resource has become a major issue concerning the human living environment, society sustainable development, and national poverty and wealth. In Northwest China, especially the inland river basin of the Hexi Corridor, the degree of ecological water shortage has been reported to reach more than 80% due to the high water consumption by industry, agriculture, etc [1]. The poor water resources endowment (e.g., the small total amount of available water resources, the uneven spatial and temporal distribution of water resources) and anthropogenic impact on the excessive expansion of irrigated areas both undoubtedly have exacerbated the contradiction between water resources development and the downstream ecological deterioration [2]. Water shortage has become the key factor in continental river basins in the northwest ecological economic development and ecological protection, and meanwhile, led to a high dependence of this regional development on groundwater exploitation. However, the problems of excessive exploitation and abuse of groundwater resources have gradually become prominent and brought many adverse effects, such as the drought of land, ground subsidence, degradation of natural vegetation and oasis, deterioration of ecological environments, and reduced agricultural production [3-4].

Several studies have shown sharp declines of groundwater tables in inland river basins in the Hexi Corridor of Northwest China, and even different levels of local groundwater depression cones [5]. This phenomenon of groundwater over-extraction has therefore seriously hampered the development of the Hexi Corridor region. The Heihe River Basin is a typical example where the terminal Juyan Lake has dried up intermittently since the 1960s, and dried up completely in the 1990s, accompanied by surrounding desertification [6]. To restore the ecological environment and recover river health and life, water-resources scheduling for ecological water conveyance in the Heihe River Basin has been implemented under unified government management annually since 2000, and so far it has lasted 20 years. According to relevant studies, this unified water resources management enables raising the groundwater table and moderating the downstream water stress [7-8]. In this context, research on the assessment of the spatial-temporal variations in groundwater tables and the comparison of effects of ecological water conveyance before and after 2000 is of great practical significance to improve effective utilization and rational development of groundwater resources.

The empirical orthogonal function (EOF) method, as a kind of multivariate statistical analysis method, is earliest applied to extract the main signal characteristics of meteorological variable fields. It is powerful for dimensionality reduction in the field of pattern recognition and data mining, given the multidimensional variables [9]. In addition, the orthogonal eigenvectors (decomposed by the EOF method) corresponding to the first several eigenvalues with large variance contributions can identify main variation patterns in the complex variable field through the separate description of spatial and temporal information without a loss of physical significance [10-11]. Therefore, on account of these advantages, the EOF method has been gradually promoted and employed in the field of hydrological and water resources to carry out studies on the temporal and spatial distribution of hydro-meteorological events.

The objectives of this paper are as follows: (1) analyze spatial and temporal variation characteristics using the EOF method in the Heihe River Basin in the middle section of the Hexi Corridor; (2) compare the differences in spatial-temporal variation patterns before and after the implementation of unified water resources management since 2000.

2. Empirical orthogonal function (EOF) method

Supposing a matrix $\mathbf{Q}_{m\times n}$ of original series $Q(t,\mathrm{u})$ (where t=1, 2,..., n and u=1, 2,..., m denote the number of yearly observations and gauging stations, respectively), $\mathbf{Q}_{m\times n}$ is first processed to be a centered matrix $\mathbf{X}_{m\times n}$ in which each vector $X(t,\mathrm{u})$ is obtained by decomposing $Q(t,\mathrm{u})$ into the long-term mean value of the discharge $\mu_Q(\mathrm{u})$ for a certain site u and the corresponding centered observation $X(t,\mathrm{u})$ as follows:

$$X(t,\mathrm{u})=Q(t,\mathrm{u})-\mu_Q(\mathrm{u}) \tag{1}$$

Similarly, the matrix $\mathbf{X}_{m\times n}$ can also be standardized $\mathbf{Q}_{m\times n}$ [10], thus $X(t,\mathrm{u})$ is calculated by

$$X(t,\mathrm{u})=[Q(t,\mathrm{u})-\mu_Q(\mathrm{u})]/\sigma_Q(\mathrm{u}) \tag{2}$$

where $\sigma_Q(\mathrm{u})$ is the standard deviation of the long-term series $Q(t,\mathrm{u})$.

In the EOF method, this centered or standardized matrix $\mathbf{X}_{m\times n}$ can be expanded into double orthogonal series of empirical orthogonal functions $\boldsymbol{V}_{m\times m}=\{V_1,V_2,...,V_m\}$ and principal components $\boldsymbol{P}_{m\times n}=\{P_1,P_2,...,P_n\}$, which include the information for spatial weights and temporal functions, respectively [10]. The spatial weights $\boldsymbol{V}_{m\times m}$ are determined via the following formula:

$$\mathbf{C}_{m\times m}\times\boldsymbol{V}_{m\times m}=\boldsymbol{V}_{m\times m}\times\boldsymbol{E}_{m\times m} \tag{3}$$

IOP Conf. Series: Earth and Environmental Science **865** (2021) 012001 doi:10.1088/1755-1315/865/1/012001

where $\mathbf{C}_{m\times m}$ means the variance-covariance matrix equal to $\mathbf{C}_{m\times m} = \mathbf{X}_{m\times n}\mathbf{X}^T_{m\times n}/n$ for the case of Eq.(1) or the correlation matrix when using Eq.(2). $\mathbf{X}_{m\times n}{}^T$ is the transposed matrix of $\mathbf{X}_{m\times n}$.

Each vector V_m of the spatial weights $\boldsymbol{V}_{m\times m}$ exactly equals to the eigenvector of the matrix $\mathbf{C}_{m\times m}$. $\boldsymbol{E}_{m\times m}$ is a $m\times m$ diagonal matrix constituted by the eigenvalues $\lambda_i; i = 1,2,...,m$ (arranged from the largest to the smallest) having the following form as:

$$\boldsymbol{E}_{m\times m} = \begin{bmatrix} \lambda_1 & 0 & \cdots & 0 \\ 0 & \lambda_2 & 0 & \vdots \\ \vdots & 0 & \ddots & 0 \\ 0 & \cdots & 0 & \lambda_m \end{bmatrix} \tag{4}$$

that each non-zero eigenvalue corresponds to an eigenvector (i.e., spatial weight from the EOF method).

The bigger the kth eigenvalue λ_k, the more important the corresponding eigenvector, yielding a larger contribution rate (R_k) to the total variance (i.e., relative variance explained):

$$R_k = \lambda_k \Big/ \sum_{i=1}^{m} \lambda_i \tag{5}$$

The error band of λ_i in Eq. (6) is tested to confirm the significance of the decomposition results of the EOF method that the adjacent eigenvalues should at least differ by e_i:

$$\begin{gathered} e_i = \lambda_i (2/mn)^{0.5} \\ \lambda_i - \lambda_{i+1} \ge e_i \end{gathered} \tag{6}$$

The temporal functions $\boldsymbol{P}_{m\times n}$ are then derived by a weighted average of the centered or standardized observed series:

$$\boldsymbol{P}_{m\times n} = \boldsymbol{V}^T_{m\times m} \times \mathbf{X}_{m\times n} \tag{7}$$

3. Study area and data

The Heihe River Basin is the second-largest inland arid basin located in Northwest China [4, 9]. The study area is upstream of the Zhengyixia hydrological station in the Heihe River Basin in the middle section of the Hexi Corridor and consists of the Gaotai County, Linze County, and Ganzhou District, with a total land area of 11316km^2 (Figure 1).

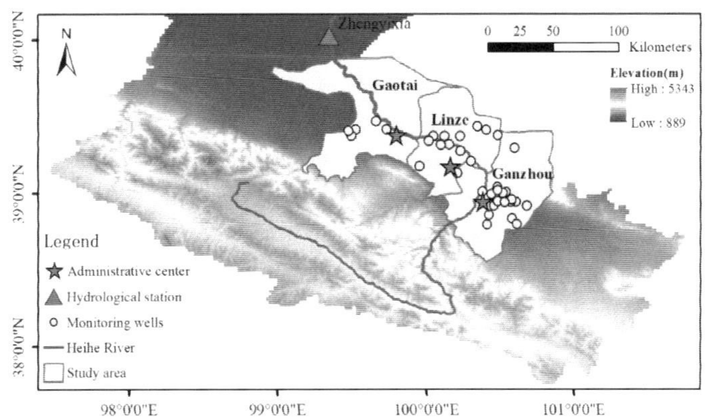

Figure 1. The study area in the Heihe River Basin of the Hexi Corridor, China.

This region belongs to the warm temperate continental drought climatic zone, with strong radiation, less precipitation, and dry air [1-2]. The annual precipitation is 300-700 mm while the potential evaporation reaches 700-1800 mm. With the rapid growth of population and socioeconomic development, there has been a high level of water-resources utilization and serious water contradiction in the Heihe River Basin. Due to overexploitation of water resources, surface water draining into the lower reaches has gradually decreased over recent decades, and the heavy dependence on groundwater use resulted in a sharp decrease in the groundwater table, which greatly degraded local ecosystems and caused the lower reaches and terminal lake to completely dry up [8-9]. Given the fact that the government has implemented the unified management of ecological water conveyance in the Heihe River Basin for about 20 years, it is necessary to assess the spatial-temporal variations in the groundwater table before and after the year 2000 for comparison of the effect of ecological water conveyance. The observed groundwater table data from 1984 to 2019 collected from 38 monitoring wells were provided by the Hydrology and Water Resources Survey Bureau of Gansu Province.

4. Results

4.1. Decomposition of groundwater table based on EOF analysis

The average annual groundwater table data of the Heihe River Basin were centered and standardized using Eq.(1) and Eq.(2), respectively. Each of the two processed data constituted a 38×36 matrix according to the wells' quantities and length of the annual series. The resulting matrix was decomposed by the EOF method, which gave 38 eigenvalues and the corresponding 38 eigenvectors. The spatial distribution of the orthogonal eigenvectors was named as the modes represented by spatial weights and the modes were mutually independent in groundwater table variable fields in the Heihe River Basin. It was numbered to be mode 1 for the eigenvector corresponding to the first (largest) eigenvalue, mode 2 for the eigenvector corresponding to the second eigenvalue, and so on.

The relative variance explained and cumulative variance explained for different modes were calculated using Eq. (5). The eigenvalues were arranged with the values in descending order, as shown in Table 1. In general, to avoid information redundancy and pursue dimensionality reduction, only the first few modes that can approach an above-85% cumulative variance explained would be adopted in the EOF analysis, rather than the use of the whole 38 modes (certainly a 100% cumulative variance explained).

Table 1. The EOF decomposition results.

Data processing	Centered data					Standardized data				
Number of mode	1	2	3	4	5	1	2	3	4	5
Eigenvalue	161.08	9.54	4.86	1.95	0.82	17.62	6.25	3.96	1.90	1.51
Relative variance explained (%)	88.96	5.27	2.68	1.07	0.45	47.69	16.92	10.72	5.14	4.08
Cumulative variance explained (%)	88.96	94.23	96.91	97.98	98.43	47.69	64.61	75.33	80.47	84.55

In Table 1, for the centered data, the cumulative variance contribution rate reached 88.96% merely using the first mode (mode 1), but it can be close to 85% for standardized data which needed more modes, i.e., the first five modes. This result illustrated that the most changeful features were captured faster by centered processing than by data standardization. For ease of comparison, the first five modes were adopted to describe the spatial-temporal variation patterns of groundwater table in the Heihe River Basin from 1984 to 2019.

4.2. Spatial distribution characteristics of groundwater table

The coincidence of sign (all positive or negative) for each weight value contained in a mode indicates that the eigenvector reflects the consistency of the changes in the groundwater table (rise everywhere or all fall). If the weights are spatially distributed between positive and negative, it shows an opposite trend

IOP Conf. Series: Earth and Environmental Science **865** (2021) 012001 doi:10.1088/1755-1315/865/1/012001

of increase or decrease in groundwater table in certain regions. The extreme value (absolute value) of each mode discerns the sensitive location of variations under this mode.

Figure 2 shows the results for different modes when using the centered and standardized data for the EOF analysis. In Figure 2a, the main spatial distribution characteristics of groundwater table variable fields in the Heihe River Basin were shown in mode 1. This result of centered EOF decomposition mode displays that almost all the weight values were negative, which represented a general decline in the groundwater table (i.e., the increase in groundwater depth) in the whole region. The small absolute values often located in the northwest region, whereas the high extreme value center (absolute value) appeared in the southeastern direction of the study area which was located in the sites of Ganzhou District and always presented descending tendencies in the groundwater table with frequent annual fluctuations and large variation ranges. In the case of standardization, the EOF results (Figure 2b) also showed general drops in regional groundwater tables but were different from those with the centered processing. Because of the removal of the large influence of local standard deviation via the standardization, high absolute values of weights did not demonstrate any spatial trends but were widely distributed in Gaotai County, Linze County, and Ganzhou District. It is thus seen that the variations in groundwater depth in different local regions can be described better through standardization, although the EOF modes with centered processing captured more variabilities on account of the local sites with larger data variance (see the relative variance explained in Table 1).

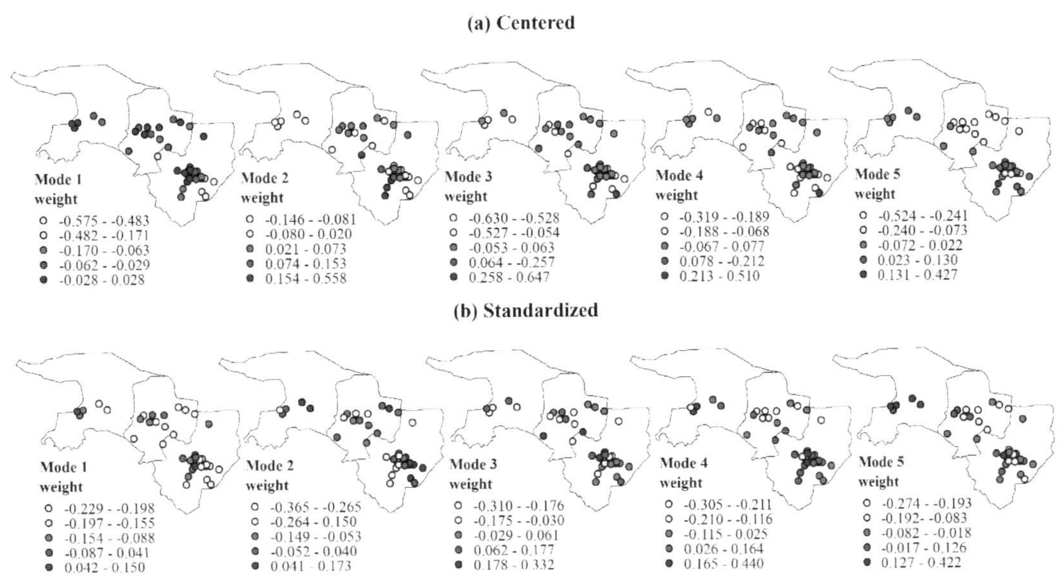

Figure 2. The spatial weights for mode 1 to mode 5 obtained from the EOF decomposition using the (a) centered and (b) standardized observations.

The weights in mode 2 to mode 5 were characterized by both positive and negative values, reflecting different degrees of spatial variations in the groundwater table. Taking Linze County as the transition zone, the spatial distribution of these modes was approximately negative in the northwest areas and positive in the southeast areas, which presented reverse distribution types for northwest-southeast directions. In other words, if the groundwater table falls in the southeast of the region, there could exist a rise in a certain northwest area and vice versa.

4.3. Temporal variation patterns of groundwater table

The temporal functions (TF) corresponding to eigenvectors in Figure 3 reflect the serial variation characteristics of the modes represented by the spatial weights in the study area. The coefficient symbol reflects the direction of the mode, i.e., the positive value indicates the same direction as the mode, while the negative value is the opposite. The greater the absolute value of the TF is, the more typical such mode is at the annual timescale.

In Figure 3, the 1st TF series showed an obvious downward trend, indicating that the groundwater table of mode 1 changed from the shallower groundwater depth of the whole basin to the general descending variation pattern, and the 1st TF from the centered EOF result showed a greater decline than that obtained by standardized EOF analysis. Combined with the spatial distribution result of mode 1 (Figure 2), it was found that in the past 36 years, the most prominent fluctuation and decline in groundwater table occurred in Ganzhou District. The 2nd to the 5th TFs can well capture the changes in the groundwater table with both rising and dropping variation patterns in certain local sites in the study area. The phenomenon of groundwater table rising appeared in some years for the 2nd to the 5th TFs. For example, in the case of standardization, the 4th and the 5th TFs showed upward trends after 2000 and reached their maximum values in 2008 and 2010, respectively. All these results demonstrated that ecological water transport played a positive role in the recovery of the groundwater table.

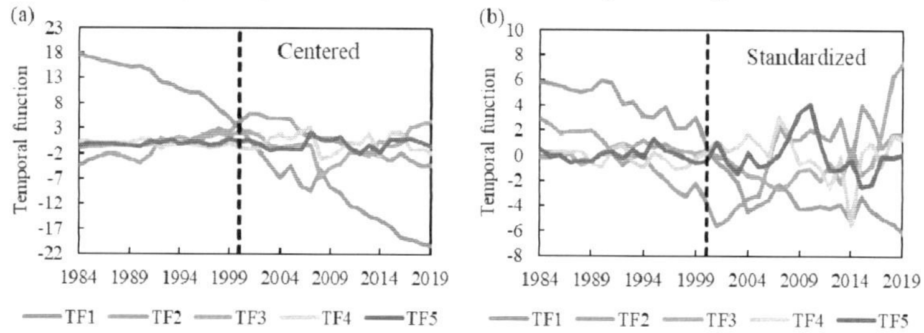

Figure 3. The temporal functions (TF) obtained from the EOF decomposition using the (a) centered and (b) standardized observations (The black dotted lines show the starting year 2000 for the unified management of ecological water conveyance).

5. Discussion

Some studies have shown that the runoff in the inland river basins of the Hexi Corridor, affected by climate warming and melting icebergs, presented an overall increasing trend, and the recharge of groundwater increased [2]. Meanwhile, with the implementation of the Groundwater Over-exploitation Management Program, the amount of groundwater use has been drastically reduced [8]. Under the combined effect of natural and man-made factors, the downward trend of groundwater table in most groundwater over-exploitation areas has slowed down, the groundwater table in some areas has risen slightly, and the area of over-exploitation areas has significantly decreased [7]. However, due to the rapid economic and social development in the 20th century, there has been serious groundwater overdrawing that caused historical long-term unbalanced replenishment and deficits of groundwater, the groundwater table was still lower than that before the water transfer, such as for the Ganzhou District in the middle reaches of the Heihe River [2, 4]. Our study was consistent with these findings. It can be seen from the result in Figure 3 that the 1st TF values before 2000 were all positive, while the 1st TF values after 2000 were almost negative, indicating a tendency of groundwater depth from shallow to deep. Despite the continuous water conveyance since 2000 that has alleviated a certain amount of pressure on the ecological environment, the groundwater table has still dropped significantly in comparison with the periods of the 1980s and 1990s. Therefore, over-exploitation issues should not be underestimated and the relevant supervision should be strengthened. Otherwise, the negative effects of ecological and environmental degradation represented by water quality deterioration, soil salinization, vegetation degradation, and land desertification could spread rapidly in the Hexi Corridor [3].

6. Conclusions

The spatial-temporal variation patterns of the groundwater table were analyzed by applying the empirical orthogonal function (EOF) method to the Heihe River Basin in the Hexi Corridor, China. In the EOF decomposition, both centered and standardized annual groundwater table data matrices were considered to derive the spatial weights of mutually independent EOF modes and the corresponding temporal functions. Because of information redundancy and dimensionality reduction, the typical first five modes with relatively high cumulative variance explained (above 95% for the centered case and approximately 85% for the standardized case) were adopted to represent the main spatial-temporal variation patterns of the groundwater variable field. The results showed that in the main variation pattern, general drops in regional groundwater tables existed for both centered and standardized cases, especially in the southeast regions when using centered EOF modes. The second to fifth modes presented reverse distribution characteristics from the northwest to southeast directions. The downward trend in the first temporal function series indicated that the groundwater depth of the whole basin varied to be deeper. However, there also appeared local rise in some sites and thereby demonstrated the positive effects of continuous ecological water delivery since 2000, but the facts that the recovered groundwater tables were still lower than those before 2000 declared the severity of long-term groundwater overdrawing.

Acknowledgments
This research is financially supported by the Basic Scientific Research Special Fund of Central Nonprofit Research Institutes (HKY-JBYW-2020-04).

References
[1] Bao C and Fang C 2007 Water resources constraint force on urbanization in water deficient regions: A case study of the Hexi Corridor arid area of NW China *Ecol Econ* **62(3-4)** 508-517
[2] Huang S, Feng Q, Lu Z, Wen X and Deo R C 2017 Trend analysis of water poverty index for assessment of water stress and water management polices: A case study in the Hexi Corridor China *Sustainability* **9(5)** 756
[3] Zhu Y, Chen Y, Ren L, Lü H, Zhao W, Yuan F and Xu M 2016 Ecosystem restoration and conservation in the arid inland river basins of Northwest China: Problems and strategies *Ecol. Eng.* **94** 629-637
[4] Han L, Zhang Z, Zhang Q and Wan X 2015 Desertification assessments in the Hexi corridor of northern China's Gansu Province by remote sensing *Nat Hazards* **75(3)** 2715-2731
[5] Guo X, Feng Q, Liu W, Li Z, Wen X, Si J, Xi H, Guo R and Jia B 2015 Stable isotopic and geochemical identification of groundwater evolution and recharge sources in the arid Shule River Basin of Northwestern China *Hydrol. Process* **29(22)** 4703-4718
[6] Zhu Q and Li Y 2014 Environmental restoration in the Shiyang River Basin China: conservation reallocation and more efficient use of water *Aquatic Procedia* **2** 24-34
[7] Zhou W, Sun Z, Li J, Gang C and Zhang C 2013 Desertification dynamic and the relative roles of climate change and human activities in desertification in the Heihe River Basin based on NPP *J Arid Land* **5(4)** 465-479
[8] Guo Q, Feng Q and Li J 2009 Environmental changes after ecological water conveyance in the lower reaches of Heihe River, *northwest China Environ Geol* **58(7)** 1387
[9] Xi H, Feng Q, Si J, Chang Z and Cao S 2010 Impacts of river recharge on groundwater level and hydrochemistry in the lower reaches of Heihe River Watershed, *Northwestern China Hydrogeol J.* **18(3)** 791-801
[10] Li L, Gottschalk L, Krasovskaia I and Xiong L 2018 Conditioned empirical orthogonal functions for interpolation of runoff time series along rivers: Application to reconstruction of missing monthly records *J Hydrol* **556** 262-278
[11] Li L, Krasovskaia I, Xiong L and Yan L 2017 Analysis and projection of runoff variation in three Chinese rivers *Hydrol Res* **48(5)** 1296-1310

Attention-based DenseNet network for multi-source remote sensing classification

Lishuo Zhang[1], Hong Lin[1] and Fanyang Zeng[1, 2]

[1]Guangzhou Urban Planning & Design Survey Research Institute, No.10, Jianshe Damalu, Guangzhou 510060, China

[2]Corresponding author's e-mail: fanyangz@126.com

Abstract. Remote sensing data are abundant in spatial, image information and widely applied in earth observation covering military and civil fields. As the hot spot and critical point, scene classification has attracted increasing attention. Convolutional neural networks (CNNs) behave as representative technologies in scene classification. Nevertheless, many CNNs are lack of the ability to distinguish key information from redundant information. To overcome the problem and achieve efficient feature extractions from multi-source data, we integrate DenseNet and attention mechanism into a CNN-based network. Multi-scale convolution and three-dimensional convolution are applied to extract features; and then, to explore and obtain feature information in a deeper level, a Dense structure are utilized; Afterwards, the features extracted from previous procedures are weighed due to importance and merged through the convolutional attention model. The experiment validates that the presented method poses advantages of efficiently merging various feature by Dense structure and the convolutional attention model.

1. Introduction

In recent years, with the continuous development of remote sensing technology and the continuous improvement of sensor hardware and software performance, remote sensing can obtain Earth observation data with higher resolution and more sources, which provides a better way to classify remote sensing features. Rich data resources. However, as the research on the classification of ground objects becomes more and more in-depth, it is difficult to meet the requirements of refined classification using a single remote sensing data. How to fuse multiple remote sensing data to extract complementary information and remove redundant information to better.

The proposal of AlexNet opened a precedent in the application of convolutional neural networks [1]. Subsequent networks such as GoogleNet and VGG used smaller convolution kernels and increased depth, proving that convolutional neural networks have better performance in processing image problems [2]. In 2015, the residual network was proposed to solve the problem of gradient dispersion caused by too deep network layers, and provided a new idea for network training [3]. Unlike the traditional method of directly learning network parameters, ResNet connects input and output, so as to learn only the changing disturbance, which greatly improves the training speed and training accuracy Its basic idea is the same as ResNet, but it establishes a dense connection between all the front layers and the back layers. Realize feature reuse (feature reuse) through the connection of features on the channel. These features allow DenseNet to achieve better performance than ResNet with fewer parameters and computational costs [4].

Inspired by the significant progress made in natural image processing, remote sensing image scene classification model using CNNs [5, 6], DenseNet, [7] ResNet and attention mechanism [8-10].

The proposed attention model contains multiple layers of Dense structure and an attention module. Considering that the classification needs to consider not only the weight of the feature dimension but also the mutual influence between different pixels, this paper introduces the CBAM module [11]. First, the extracted spectral features and spatial features are analysed through the attention mechanism on the channel. The importance of features and height information is learned, and then the influence of other pixels on the target pixel is considered through the spatial attention mechanism, so as to complete the classification of the features according to the weighted result.

The rest of this paper is structured as follows. Section 2 shows our methodology. Experiment and analysis are conducted in Section 3. The conclusion is obtained in Section 4. References are given at the end of the paper.

2. Methodology

The main architecture of our model is shown in Figure 1.

The proposed method contains two parts: Feature Extraction module and Attention Module.

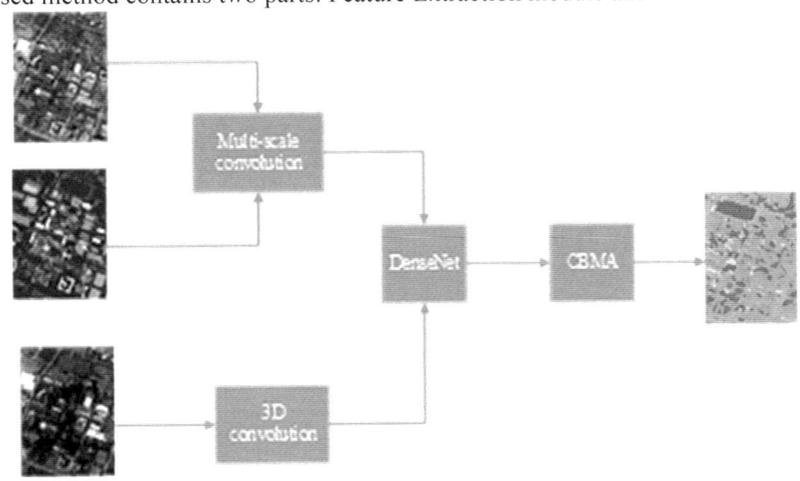

Figure 1. The main architecture of our model.

2.1. Feature Extraction

The spectral information of high-resolution data and LiDAR data is weak, but they have strong spatial characteristics. The commonly used 3x3 convolution kernel can only perceive the information of the surrounding 8 pixels, which limits the extracted features. Therefore, in order to fully extract the spatial information of the data, this paper introduces the multi-scale convolution method, and superimposes the results of the convolution extraction under the multi-scale to obtain richer spatial information.

Aiming at the relatively continuous characteristics of the spectral response of hyperspectral data, the spatial and spectral features of hyperspectral data are extracted simultaneously by means of three-dimensional convolution in this article, and the features between adjacent bands can be learned by sliding on the channel to extract Rich band information.

2.2. DenseNet

DenseNet is a convolutional neural network with dense connections. In this network, there is a direct connection between any two layers, that is, the input of each layer of the network is the union of the outputs of all the previous layers, and the feature map learned by this layer will also be directly passed to All subsequent layers are used as input.

2.3. Convolutional Block Attention Module (CBAM)

At present, the commonly used fusion methods for the extracted features in the neural network are addition and concatenation, both of which are direct processing of various features, but in the actual remote sensing image classification, different types of ground objects have different features. The response is different. For example, vegetation and other ground objects with obvious spectral characteristics can get a better classification effect according to the spectral characteristics. Therefore, spectral characteristics should be considered more when classifying, but materials like cement roads and roofs It is often more accurate to use spatial information to classify similar ground objects with a large difference in shape and texture. At this time, the spatial characteristics need to be enhanced. At the same time, the same convolution kernel of the convolutional neural network shares spatial weights, so it is often necessary to increase the number of convolution kernels to extract different features, which leads to the generation of redundant information. This is achieved through simple feature superposition fusion. The classification of remote sensing images has caused some impact. Therefore, an attention mechanism is introduced in this article to weight the fusion features, so that different features have different importance in the classification of different features. Considering that the classification needs to consider not only the weight of the feature dimension but also the mutual influence between different pixels, this paper introduces the CBAM module. First, the extracted spectral features and spatial features are analysed through the attention mechanism on the channel. The importance of features and height information is learned, and then the influence of other pixels on the target pixel is considered through the spatial attention mechanism, so as to complete the classification of the features according to the weighted result.

3. Experiment and analysis

3.1. Datasets

The data used in this experiment is the public data set grss_dfc_2018 [12] provided by the 2018 IEEE Fusion Competition, including high-resolution aerial images (0.05 m), multispectral LiDAR point cloud data (0.45 m), and hyperspectral data(1m). The reference sample data contains 20 urban LULC classes with a spatial resolution of 0.5m.

3.2. Experimental settings

80% of all samples are selected for model training and 20% for testing. The learning rate is set to 0.0001, including a total of 800 epochs. Due to the large image size, it cannot be directly imported into the network for training. Therefore, the training sample is first divided. The size of the original training image is 4172×1202, which is divided during training. 20 sub-blocks are trained, and the calculated loss function value is the average value of all sub-blocks, optimized by Adam algorithm. All implementations are based on TensorFlow and NVIDIA RTX2060.

3.3. Experimental results

We evaluated our model on grss_dfc_2018 Datasets. Figure 2 show the accuracy and loss curves of our attention model.

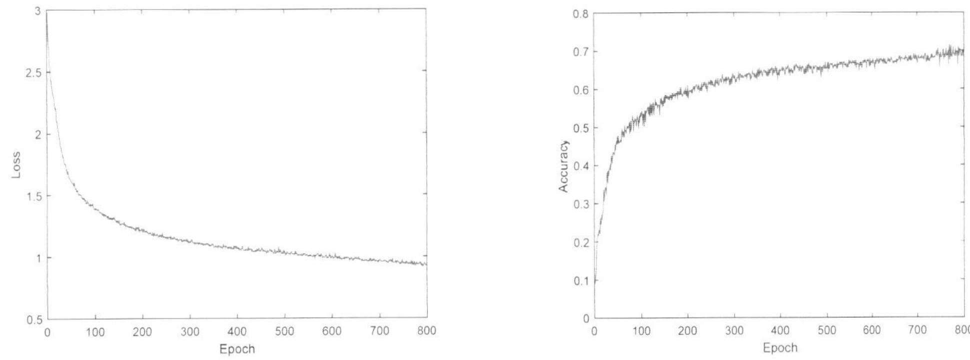

(a) Loss curve (b) Accuracy curve

Figure 2. Accuracy and loss curves.

The feature extraction of high-resolution images, hyperspectral images and LiDAR are shown in Figure 3. High-resolution data and hyperspectral data have spatial and spectral information, so the feature extraction results are relatively similar, the spatial characteristics of the high-resolution data are more obvious. The hyperspectral images can better express highly reflective objects. The features extracted from LiDAR data represent the height information of the object. By fusing different features using the attention mechanism, the fused features are shown in Figure 3(g). The final classification result is shown in Figure 4.

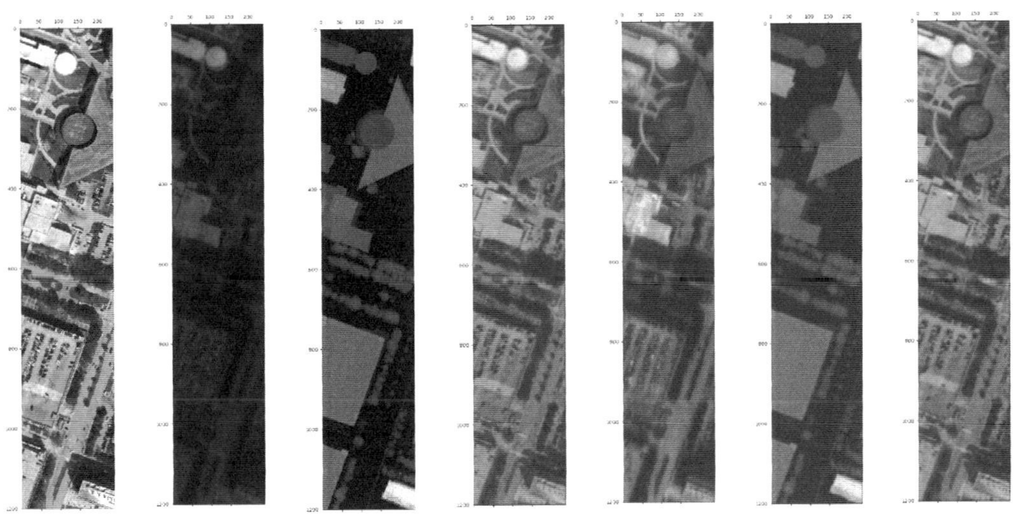

(a) (b) (c) (d) (e) (f) (g)

Figure 3. Feature Extraction (a: High resolution image; b: Hyperspectral image; c: LiDAR image (Canopy height model); d-f: Feature Extraction of (a)-(c); g: CBAM).

Figure 4. Classification result.

As we can see from Figure 2, the training accuracy continues to rise, and the loss value steadily declines. 800 epochs have converged, indicating the feasibility of the model. The overall classification accuracy is 78.62%, This just is the rough classification result. More fine classification result can be obtained by using Markov random field and so on.

4. Conclusion

This paper uses deep learning as the basic framework and uses feature-level fusion to integrate multi-source remote sensing data to complete the classification. Using different feature extraction methods for different data can get richer feature information. By introducing the attention model for feature fusion, a better fusion effect can be achieved, which provides a new idea for remote sensing image fusion. The sample difference between diffident classes is large, and the problem of sample balance needs to be dealt with. In the future, we will set different parameters to test this model.

References

[1] A. Krizhevsky, I. Sutskever and G. E. Hinton 2012 Imagenet classification with deep convolutional neural networks

[2] C. Szegedy, S. Ioffe and V. Vanhoucke 2016 Inception-v4, Inception-ResNet and the Impact of Residual Connections on Learning *CoRR* vol. abs/1602. 07261

[3] K. He, X. Zhang, S. Ren and J. Sun 2016 Deep Residual Learning for Image Recognition

[4] G. Huang, Z. Liu, V. D. M. Laurens and K. Q. Weinberger 2016 Densely Connected Convolutional Networks

[5] W. Han, R. Feng, L. Wang and L. Gao 2018 Adaptive Spatial-Scale-Aware Deep Convolutional Neural Network for High-Resolution Remote Sensing Imagery Scene Classification in *IGARSS 2018 - 2018 IEEE International Geoscience and Remote Sensing Symposium* pp. 4736-4739

[6] X. Xu, W. Li, Q. Ran, Q. Du, L. Gao and B. Zhang 2018 Multisource Remote Sensing Data Classification Based on Convolutional Neural Network vol. 56 pp. 937-949

[7] W. Tong, W. Chen, W. Han, X. Li and L. Wang 2020 Channel-Attention-Based DenseNet Network for Remote Sensing Image Scene Classification vol. 13 pp. 4121-4132

[8] F. Li, R. Feng, W. Han and L. Wang 2020 An Augmentation Attention Mechanism for High-Spatial-Resolution Remote Sensing Image Scene Classification vol. 13 pp. 3862-3878

[9] R. Fan, L. Wang, R. Feng and Y. Zhu 2019 Attention based Residual Network for High-

Resolution Remote Sensing Imagery Scene Classification in *IGARSS 2019 - 2019 IEEE International Geoscience and Remote Sensing Symposium* pp. 1346-1349

[10] X. Tang, Q. Ma, X. Zhang, F. Liu, J. Ma and L. Jiao 2021 Attention Consistent Network for Remote Sensing Scene Classification p. 1-1

[11] S. Woo, J. Park, J. Lee and I. S. Kweon 2018 CBAM: Convolutional Block Attention Module *Cham* pp. 3-19

[12] 2018 IEEE GRSS Data Fusion Contest. Online: http://www.grss-ieee.org/community/technical-committees/data-fusion (accessed on 12 August 2020)

IOP Conf. Series: Earth and Environmental Science **865** (2021) 012003 doi:10.1088/1755-1315/865/1/012003

Design and application of sewage treatment technology in fieldbus control system

Kai Guo[1, 2] and Fuxiao Li[1]

[1]Humanities college, Dongying technician college, Dongying 257446, Shandong, China

[2]Corresponding author's e-mail: 375540163@qq.com

Abstract. At present, most sewage treatment plants in China generally have low sewage treatment automation and low operating efficiency. In order to solve this problem, this article takes a sewage treatment plant in Hangzhou as an example to analyze the functions and control requirements of each process in detail. Adopting three-level sewage treatment process and multiple disinfection technology at the same time, a set of sewage treatment field bus automatic monitoring system for SBR comprehensive process is designed, and the overall structure of the control system, control system hardware, control program and host computer monitoring system are given. The system adopts siemens S7-300 as the main control station, and uses industrial ethernet to complete the data exchange between the bus control and the control station on the network configuration. With various testing equipment, the inverter and MCC serve as the slave stations form a profibus control network, and complete functions such as automatic control of sewage treatment equipment and dynamic process monitoring through Wincc7.4 and Step7 V5.5. At the same time, automatic online monitoring and sampling are adopted, and video monitoring replaces manual inspection. Practice shows that the system The operation is in good condition, the effluent indicators, reliability, safety, etc. have reached the design requirements, and good economic and social benefits have been achieved. It is an integrated control system with strong practicability and high degree of automation, which is useful for further improving the automation of sewage treatment. The degree has important reference value. At the same time, the technical achievements of automatic sewage control and comprehensive treatment can fully support the sewage treatment needs during the new coronavirus epidemic, and have a wide range of application prospects.

1. Introduction

In the 21st century with rapid economic development, severe water pollution has become a major factor restricting the sustainable development of cities. At the same time, my country is a country with relatively lack of water resources. Wastewater reuse has great potential. Wastewater treatment has become an indispensable part of the economic development process. An indispensable part [1], water treatment technology can treat sewage and wastewater to the level of drinking water, and can also be processed to the level of non-drinking water, for toilet flushing, green land watering, road cleaning, etc. [2]. Developed countries abroad have accumulated relatively mature experience in dealing with wastewater treatment problems brought about by industrialization and economic development. At present, the wastewater treatment capacity of the Netherlands, Germany, and Japan is at the world's leading level. These countries attach great importance to the research of sewage treatment automatic control systems [3]. After more than 30 years of development, China's sewage treatment technology has formed its own technology while absorbing and learning from the sewage treatment technology of developed countries such as Europe and the United States. However, compared with foreign advanced technology, our research on automatic control systems is still There is a long way to catch up. The country is stepping up its efforts in sewage treatment, and the research of sewage treatment automatic control system is imminent [4].

The project covers an area of 1 million square meters, with a daily treatment capacity of 200,000 cubic meters. The total scale of the sewage treatment plant is 400,000. Based on the actual research on the sewage treatment process, the SBR three-level integrated sewage treatment technology suitable for urban sewage treatment is designed. On the basis of the existing equipment, a three-level field bus control system composed of industrial computers, plc and intelligent equipment is introduced. The field bus uses industrial ethernet and profibus, the controller uses siemens S7-300, and the central control level is divided into engineer stations. And the operator station, using wincc as the man-machine interface to display the real-time process of sewage treatment, practice has proved that the system can accurately automatically control the various processes of the SBR integrated process, and the sewage treatment field bus control system is a huge society The research direction of benefits, environmental benefits and economic benefits.

2. Design of sewage treatment process

Sewage treatment is a combination of system engineering, the design of the process plan should be determined according to comprehensive factors such as water quality and treatment scale. The process plan of this project is designed for the treatment of urban sewage. A three-stage sewage treatment process is adopted, and physical methods are first adopted. Chemical methods and biological methods are used again, combined with activated sludge method and biological contact oxidation method, as the advanced treatment method of refractory organic sewage, the use of strong oxidants to disinfect, decolor and deodorize the sewage, while introducing ultraviolet light In the biochemical treatment system, the process flow of the sewage treatment control system is shown in Figure 1.

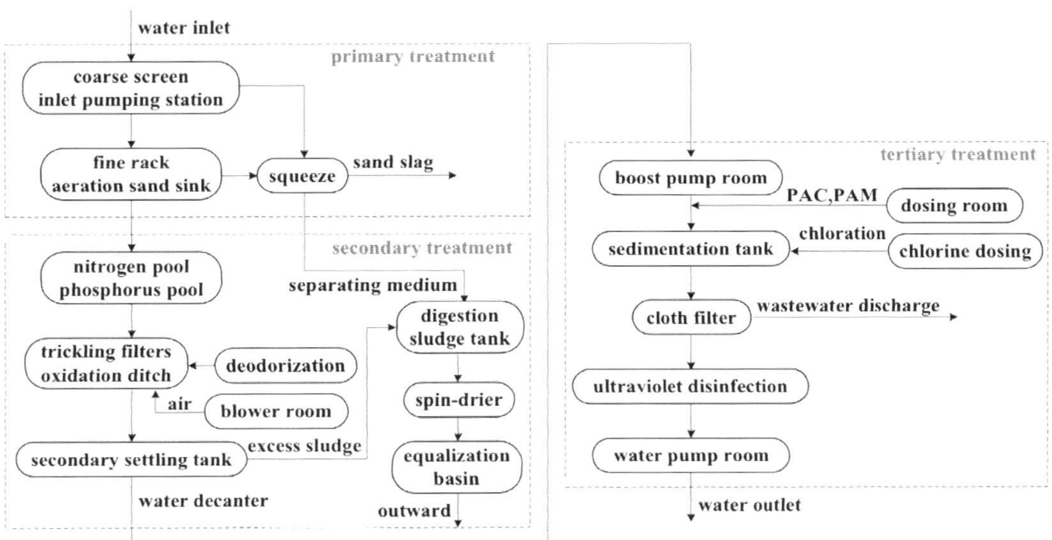

Figure 1. Wastewater treatment process.

3. Scheme design of fieldbus control system

3.1. System design ideas
This control system is a sequential control for the SBR comprehensive process. In order to improve the control accuracy of the system, more sensors are added, such as DO detector, ORP detector, Mlss measuring instrument and other related detection equipment. At the same time, industrial Ethernet is adopted. The fieldbus technology combined with profibus can not only reduce costs, but also improve the reliability and real-time performance of the system, making the system easy to expand and maintain.

3.2. Control system structure
The system adopts a three-tier structure and a two-tier network. The three-tier structure includes a process equipment layer, a field control layer, and an operation monitoring layer. The process equipment layer includes smart sensors, actuators, and analytical instruments that conform to the fieldbus communication protocol. Each device is partitioned MCC, frequency converter, field detection instruments, etc. are connected by profibus; the field control layer completes basic control functions and executes various control algorithms; the operation monitoring layer includes operation stations and engineer stations, which are mainly used for monitoring and control systems of the production process configuration and maintenance [5]. The secondary network includes management information network and real-time control network. The management information network adopts the form of optical fiber ring network to realize the communication between the field control layer and the operation monitoring layer. The communication between the control station and the control unit adopts the profibus DP mode, and the structure of the control system is shown in Figure 2.

3.3. Network topology
The sewage treatment control system adopts the master-slave structure of the upper and lower computers, and uses the industrial ethernet based on TCP/IP to realize the communication between the upper and lower computers, thereby realizing the integrated control of the entire sewage treatment plant. The central control room consists of an engineer station and two operator stations as the upper computer. It forms a wired data communication network with the four field control stations through industrial ethernet, and is connected to the lower computer through an ethernet switch. The lower computer is selected with an open for the S7-300 series of network structure, the topological structure of the control network is shown in Figure 3.

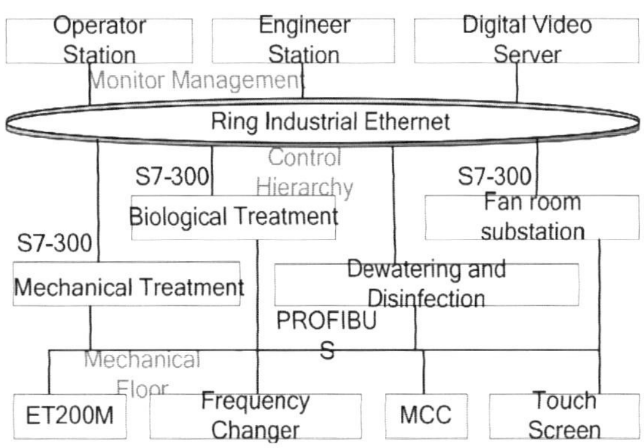

Figure 2. Structure of sewage treatment field bus control system.

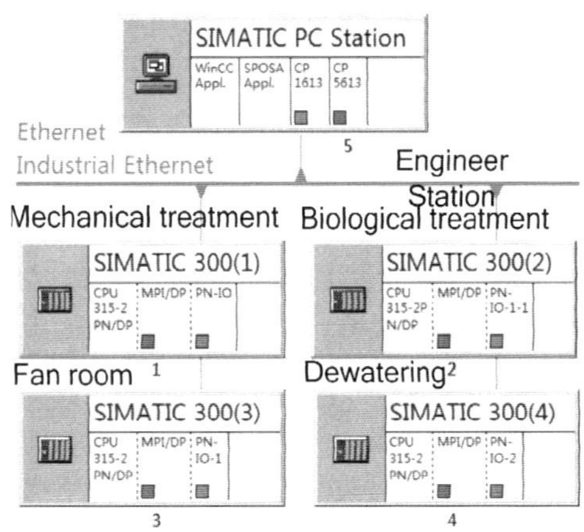

Figure 3. Ethernet topology.

IOP Conf. Series: Earth and Environmental Science **865** (2021) 012003 doi:10.1088/1755-1315/865/1/012003

4. Step7 software system design

4.1. Hardware configuration and address allocation

The control system adopts simatic step7 V5.5 software for hardware configuration, completes the address and station address allocation and program design and development, and realizes the processing of process data by plc. Siemens S7-300 is used as the master station and profibus-DP is the field bus. ET200M series and various inverters, actuators, etc. are used as slave stations, and the IM365 module provides its expansion interface. According to the needs of the process and control requirements, the system is equipped with 4 plc stations, 12 analog input SM331 modules, 2 analog output SM332 modules, 28 digital input SM321 modules, and 15 digital output SM332 modules, as shown in Figure 4. Shown is the hardware configuration of the biochemical processing sub-station.

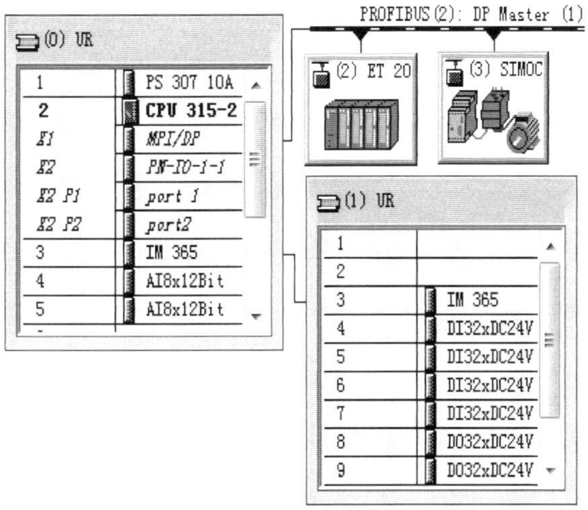

Figure 4. The hardware configuration of the biochemical processing sub-station.

4.2. Communication program design

4.2.1. Ethernet communication between plc. The ethernet communication between S7-300 is realized by calling FB14/FB15 function block [6], the communication addresses of the 4 S7-300 are w#16#1, w#16#2, w#16# 3, w#16#4, timed to read once in 1 second, read two bytes to the start address of DB4, part of the procedure of ethernet communication is shown in Figure 5.

4.2.2. Communication design between frequency converter and plc. The siemens inverter is connected to the Pprofibus network through the DP interface as a slave station. The master station CPU315-2DP and the inverter slave station perform cyclic data exchange. It is necessary to call the function blocks SFC14 and SFC15 [7], the S7-300 and inverter the DP communication program is shown in Figure 6, laddr is the starting address of the process data W#16#100 (that is, 256), the length is 4B, and record represents the address of the slave data area in the data block, when M0.1 is 1 state At the time, call SFC15, package the control word and speed setting value in MW30 and MW32 and send it, and call SFC14 at the same time, unpack the received data and save it in MW4 and MW6.

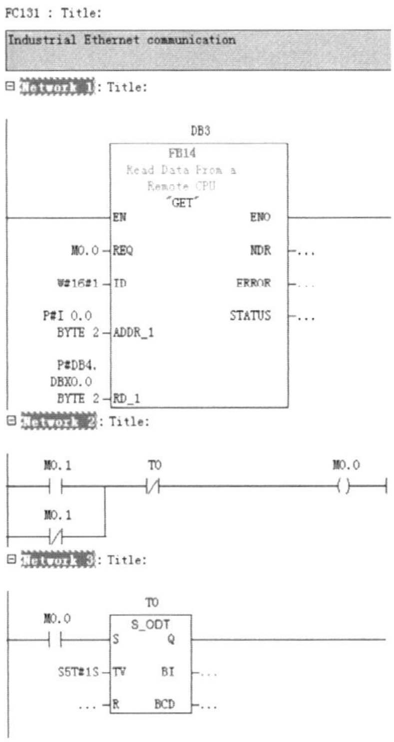

Figure 5. PLC ethernet communication program.

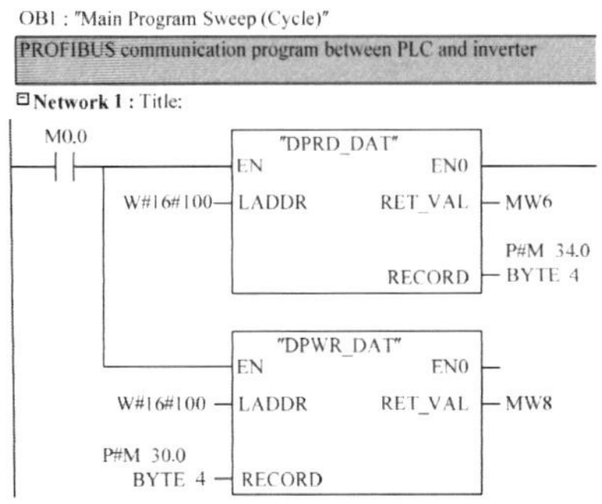

Figure 6. PROFIBUS communication program between PLC and inverter.

4.3. Control program design

Step7 software programming adopts a modular structure. According to the actual situation of the controlled equipment, the entire sewage treatment is divided into several sub-processes. Each sub-process corresponds to a function or functional block, which combines various control functions and communication data between each site. Written in different subroutines respectively, the main control program calls each function through the organization block.

The design of the control program adopts the form of modular structure and subprogram calling. The functions of each main function block are as follows: FC1 is an analog quantity processing program block, FC11～FC15 are respectively pH value control and adjustment program blocks, and the sewage pH value is controlled by the function block FB41 Realize continuous pid control. FC181～FC193 function blocks respectively complete the monitoring of circulating pumps, flow meters, pressure sensors, and temperature sensors. FC201～FC223 function blocks realize other functions, such as agitator control, high and low limits of liquid level gauges, For pH value control and fault alarms, FC1 calls the function block FC32 to convert the input analog quantity. The sensor converts the actual value of the analog quantity into a 4-20mA signal, which becomes a 0～27648 signal after A/D conversion. The number of internal codes, the actual value of the internal code is sent to the host computer for display through programming, and the field instrument reading is processed as follows: field instrument reading = (reading value/27648) × (actual maximum range value-actual minimum range value) +The actual minimum range value [8], the analog quantity processing program is shown in Figure 7.

Take the temperature sensor as an example. The signal transmitted on site is an analog quantity of 4-20mA. The PIW address of this signal is 256. The analog quantity processing program block FC1 is called through the main program OB1 to process the temperature signal from the scene and convert it to the engineering quantity readable by plc, Figure 8 is the ladder diagram program called by analog quantity.

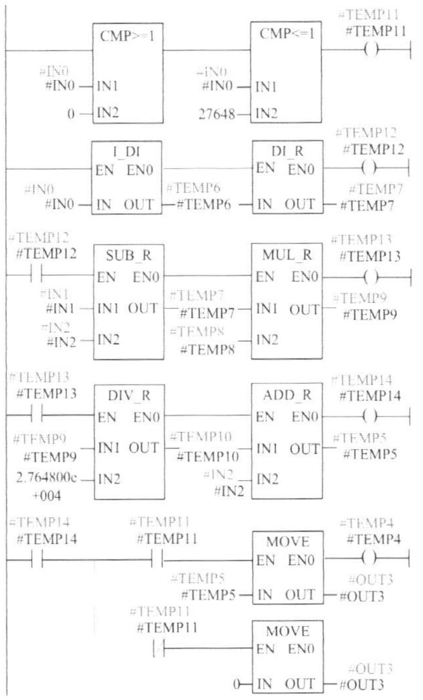

Figure 7. Analog quantity conversion subroutine.

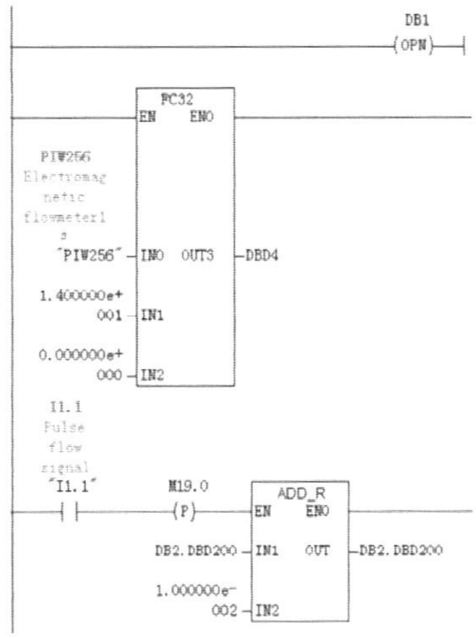

Figure 8. Analog quantity calling program.

5. Wincc monitoring system configuration

5.1. Selection and function of configuration software

According to the actual needs of the sewage treatment control system, the upper computer selects the dedicated Advantech industrial computer for the industrial site, and the graphical interface selects the wincc V7.4 of Simense to centrally monitor the operating conditions of the entire plant's process equipment. Wincc has the powerful process automation control, it also has a very cost-effective SCADA-level operation and monitoring system, and has complete monitoring and management functions [9]. The engineer station and the operator station use the full user version and the running version respectively.

5.2. Configuration realization of monitoring system

The sewage treatment monitoring system is equipped with multiple real-time monitoring screens, including process screens, alarm screens, report screens, event screens and operation record screens, among which the process screen can display the process flow of the entire system, as shown in Figure 9, through the button below different functional areas can be entered. The instantaneous values of equipment parameters are marked on the general process drawing according to the actual installation position, and the real-time data and historical data are made into corresponding sub-screens.

The process flow picture truly reflects the operating conditions of the field equipment, and is presented in the monitoring system in the form of animation. The left part is the three-way water inlet pipe, the distribution tank and the primary sink, and the data frame is set above the distribution tank. To display the liquid level from the site, a data frame is set in the inlet pipe and the dilution pipe of the sedimentation tank to display the real-time flow. The middle part is the H+DN tank and the contact

IOP Conf. Series: Earth and Environmental Science **865** (2021) 012003 doi:10.1088/1755-1315/865/1/012003

oxidation tank. The data frame above each reaction tank is used to display Temperature, acid valve opening degree, alkali valve opening degree, oxygen content, flow rate and other information. The acid-base valve is controlled and adjusted by the PID loop. The output value of the acid-base valve is a percentage (range 0-100%), PID parameters need to be determined by on-site debugging. The treated sewage is discharged in the rightmost discharge channel, and the sludge enters the lowermost sludge treatment channel. PV is the pH feedback value of the sedimentation tank, and SP is the pH setting value.

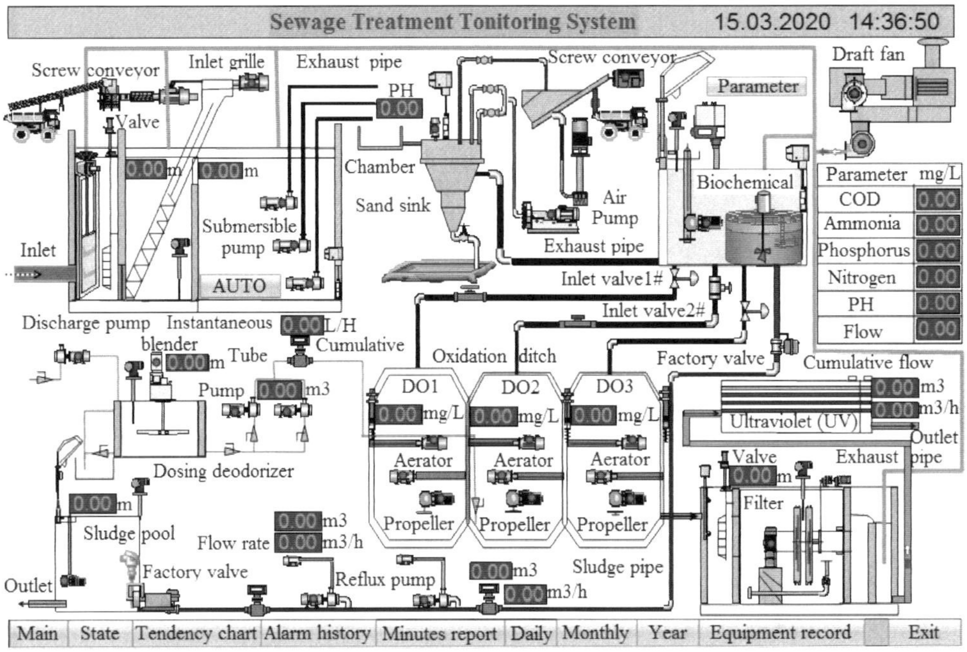

Figure 9. Wincc monitoring system process flow overall screen.

6. Debugging and maintenance

System debugging is generally divided into two steps: static debugging, dynamic debugging and running. Static debugging mainly uses running programs to check the logical relationship of various hardware interfaces. Program debugging mainly uses s7-plcsim and step7 through signal generators. Simulate field signals to debug the hardware in the plc control cabinet. First, debug the subprograms and function module programs, then debug the initialization program, and finally debug the main program. The dynamic debugging adopts manual and then automatic debugging methods until the process control requirements are met. After the system has been debugged separately and on-line, the effluent quality meets the testing requirements of the "Urban wastewater treatment plant pollutant discharge standard" GB18918-2002 class a standard.

7. Conclusion and discussion

Aiming at the problems in sewage treatment, this paper analyzes and studies the entire process flow of sewage treatment according to the characteristics and requirements of urban sewage treatment [10], designs the SBR three-level integrated sewage treatment process [11], and designs the industrial ethernet and profibus phase according to the process requirements. Combined with the overall architecture and control plan, the system selects four s7-300 as the field control station, which realizes the industrial Ethernet communication between the lower computer plc and the upper industrial computer, the lower computer and the field equipment profibus real-time communication [12], and the use of modular I wrote the step7 control program and designed the host computer monitoring interface using wincc configuration software [13], which realized the remote monitoring and fully automatic operation of the system, and also realized the data transmission with the environmental protection department. The entire control system can be adjusted according to the process requirements. The parameters are especially suitable for the situation of insufficient water in the Longmen sewage treatment plant and unstable influent water quality, thereby effectively ensuring the discharge of sewage treatment standards.

Since the system was actually put into operation in October 2019, the control system has been operating stably and reliably, and all indicators have met the design requirements, solving the problem of scattered, complex, and difficult to control sewage treatment plant equipment, and greatly improving the efficiency of sewage treatment The effluent index has reached the national first-level discharge standard, which is of great significance in solving the increasingly serious water pollution problem and saving investment [14]. More importantly, as the automation of the entire system increases, it will lay the foundation for future system transformation and expansion. A solid foundation is of great practical significance under the background of relatively backward domestic sewage treatment control technology and has a wide range of application prospects.

References

[1] Li Qianzhen 2018 Design of PLC control system for sewage treatment based on PROFIBUS *Dalian Jiaotong University* 14
[2] Wang Jinren 2019 *J.* Automatic control system based on SBR sewage treatment process. *Mechanical Design* S1 26-28
[3] Li Ming and Kang Kai 2019 *J.* Application of PROFIBUS and PROFINET field bus in sewage monitoring system *China Instrumentation* **6** 26
[4] Sun Jing 2019 Experimental study and feasibility analysis of Angang's sewage treatment system *Harbin: Harbin Institute of Technology* 28
[5] K J Hidalgo, T. Saito, R S Silva, Tiago P. Delforno, Iolanda C S Duarte, VM de Oliveira and Dagoberto Y. Okada 2020 *J.* Microbiome taxonomic and functional profiles of two domestic sewage treatment systems. *Biodegradation* **3** 35
[6] Rui Wang, Li Jiaqi and Gao Xianbin 2020 *J.* Design and Simulation of an Ozone Catalytic Oxidation System Based on Programmable Logic Controller. *Journal Européen des Systèmes Automatises* **53** 4
[7] Meng Xiaoxi 2019 Research on technical transformation of ultraviolet sterilization technology in sewage treatment plant. *Harbin: Harbin Institute of Technology* 50
[8] Chuan Zhang 2020 *J.* Oilfield sewage treatment technology and improvement countermeasures. *International Journal of Education and Economics* **3** 4
[9] Abdel Shafy Hussein I.,Mansour Mona S. M. 2020 *J.* Rehabilitation and upgrading wastewater treatment plant for safe irrigation reuse in remote area. *Water Practice and Technology* **15** 4
[10] Li Zhanying 2018 Distributed Control System and Field Bus Control System and their engineering design. *Electronic Industry Press* 237
[11] Wang Shanbin 2018 Configuration software application guide. *Chemical Industry Press,* 315
[12] Tang Haifeng 2019 Siemens PLC Mastery Case Course. *Publishing House of Electronics Industry* 277
[13] Sunit Kumar Sen 2020 Fieldbus and Networking in Process Automation. *CRC Press* 12
[14] Zhang Wei, Ma Tao 2019 Proceedings of the 31st China Control and Decision Conference Northeastern University *Chinese Society of Automation* 566

IOP Conf. Series: Earth and Environmental Science **865** (2021) 012004 doi:10.1088/1755-1315/865/1/012004

Monitoring and analysis of green sustainable development in impoverished counties based on nighttime light and RSEI

Zelin Jiang[1], Jian Deng[1, 2, 3], Haijun Luan[1, 2] and Zhangjin Qiu[1]

[1]College of Computer and Information Engineering, Xiamen University of Technology, Xiamen 361024, China;
[2]Big Data Institute of Digital Natural Disaster Monitoring in Fujian, Xiamen 361024, China

[3]Corresponding author's email: dengjian163@126.com

Abstract. Sustainable assessment of the quality of green sustainable development in poor counties is essential for urban environmental management, planning as well as achieving sustainable development goals. Currently, there are shortage of related researches on poverty alleviation and ecological environment monitoring in poverty-stricken counties, this study which is based on the nighttime light images and Landsat8 OLI images, constructing a light index, remote sensing ecological index (RSEI) and coupling coordination model of the two. Analyze the situation from 2013 to 2019 poverty alleviation in poor counties in Fujian Province and the ecological environment and explore its coupling and coordination relationship. The results show that from 2013 to 2019, the economy of poor counties has doubled and the most obvious accomplishments of poverty alleviation were in the year 2015 to 2017; the RSEI value of poor counties has all increased, the overall ecological level is at a high-quality level, and the coverage rate is high. The coupling degree C and coordination degree D of the light index and RSEI show a fluctuating upward trend, and both of them are at a good coupling and coordination level, achieving a win-win situation for economic and ecological benefits.

1. Introduction

Eliminating poverty is the dream of humanity, and the history of human development is also the history of constant struggle against poverty. 17 Sustainable Development Goals (SDGs) have been put forward in the agenda of the UN 2030 Sustainable Development Goals, which aims to solve the development problems of the three dimensions of society, economy and environment in an integrated manner from 2015 to 2030 and shift to the path of Sustainable Development [1]. Monitoring and evaluating the achievements of poverty alleviation and changes in the ecological environment is of great significance for the poverty-stricken counties from getting rid of poverty to realizing rural revitalization [2].

Nighttime Light (NTL) is the use of remote sensing technology to capture the Earth's night light from outer space, which can quickly, accurately and objectively obtain information on the surface and human activities [3,4]. Using NTL data, researchers from various countries have explored changes in the external urban spatial structure and changes in internal social and economic development [5,6], especially in the analysis of human activity characteristics [7,8], urbanization process [9,10], war damage assessment [11] and other related research applications. Pan et al. took the districts and counties of Chongqing as the research area and introduced the night light data to construct a poverty measurement index system[12]; Yu et al. used NPP-VIIRS data to verify that there is a greater correlation between the light index and the comprehensive poverty index of poverty-stricken counties [13]. RSEI is an index based on greenness, humidity, heat and dryness, which is completely dependent on remote sensing information and natural factors [14,15]. It can objectively couple each index to monitor regional ecological quality quickly and easily. Based on the RSEI model, Sun Congjian studied the temporal and spatial changes of the ecological environment in the poverty-stricken areas of Lvliang Mountain with the vegetation coverage index, and found that the overall RSEI value is low and the regional differences are significant, and economic development has a negative effect on the ecological environment [16]; Xu et al. found that changes in the proportion of impervious surfaces in the new area of Xiongan New District will cause main changes in RSEI, which will seriously affect the ecological environment [17]; Wei et al. have realized the remote sensing dynamic monitoring of the ecological environment quality of the Lijiang River Basin in the past 30 years, and pointed out that RSEI and vegetation coverage are positively correlated [18]; Liu et al. analyzed the ecological changes and their spatial heterogeneity during the development of Pingtan Island, and showed that the RSEI value of the area increased from 2013 to 2017 [19]; Mohammad et al. assessed and compared the spatial differences of land use between cities in combination with the normalized differential soil index [20]; Li et al. found that the urbanization process of the Bohai Rim urban agglomeration has an inhibitory effect on the seasonal changes of vegetation primary productivity [21].

The above studies mainly focus on the monitoring of urban urbanization which is based on luminous light or the monitoring of ecological environment which is based on RSEI. No studies have evaluated the economic development and ecological environment quality of poverty-stricken counties in the process of poverty alleviation by combining light index and RSEI. Therefore, this paper adopts NPP-VIIRS DNB night-light remote sensing image and Landsat8 OLI image as the basic data source to study economic development and ecological changes in the process of poverty alleviation in the poverty-poor counties of Fujian Province. This also monitors and evaluates the quality of green and sustainable development in the region so as to provide auxiliary reference and decision support for the evaluation of rural revitalization and sustainable green development goals.

2. Data and methods

2.1. Overview of the study area and data sources

Fujian Province is located in southeastern China on the coast of the East China Sea. The land area is between 23°33′ N to 28°20′ N and 115°50′ E to 120°40′ E. There are towering peaks and ridges, continuous hills, river valleys and basins interspersed between them. Mountains and hills account for more than 80% of the total area of the province. The climate conditions are superior, and it is popularly known as "eight mountains, one river and one field". In 2012, Fujian Province for the first time identified 23 counties in the 6 urban districts as key counties for poverty alleviation and development at the provincial level, strengthened the improvement of conditions in poverty-stricken areas, accelerated support for economic development, and eliminated all poverty-stricken counties by 2019. This paper takes 23 poverty-stricken counties in Fujian Province as the research area, and selects one county in each of the six urban districts as the object: Yongtai, Shouning, Guangze, Qingliu, Pinghe and Changting, to analyze the sustainability of ecological and economic development among them.

As shown in Table 1, the luminous remote sensing year data used in this study were synthesized by the Suomi-NPP satellite using its Visible Infrared Imaging Radiometer (VIIRS), from the Penn Institute of Public Research, USA; Landsat8 OLI remote sensing images were downloaded from China's geospatial data cloud website; Social and economic data such as GDP come from Fujian Statistical Yearbook and monthly statistical data from 2012 to 2020; The vector map of the administrative boundary of the study area comes from the 1:4 million county-level administrative boundary data released by the National Basic Geographic Information Center in 2019.

Table 1. Data sources and information.

Data source	Description	Time	Source
Night light data	Luminous year data is synthesized from the monthly average of cloudless radiation	2012-2020	https://eogdata.mines.edu/nighttime_light/annual/v20/
Landsat8 OLI image	Spatial resolution 30m, cycle 16 days	2013-2020	Geo spatial Data Cloud http://www.gscloud.cn/home
Socioeconomic data	Population and GDP data	2012-2020	Fujian Statistical Yearbook, Fujian Municipal Statistical Yearbook
Administrative boundary	Fujian county-level vector data	2019	National Basic Geographic Information Center

2.2. *Research route*

In this study, data preprocessing was carried out to eliminate image noise for the annual data images of luminous and Landsat8 OLI remote sensing images; Secondly, the nocturnal index model and the RSEI model obtained from the principal component analysis of greenness, humidity, dryness and heat were constructed, and the RSEI equally-spacing classification and difference value were tested and analyzed. Meanwhile, the coupling coordination model of the two was constructed; Finally, the three models are used to monitor and analyze the poverty alleviation and ecological environment of the poverty-stricken counties and evaluate their green and sustainable development. As shown in Figure 1.

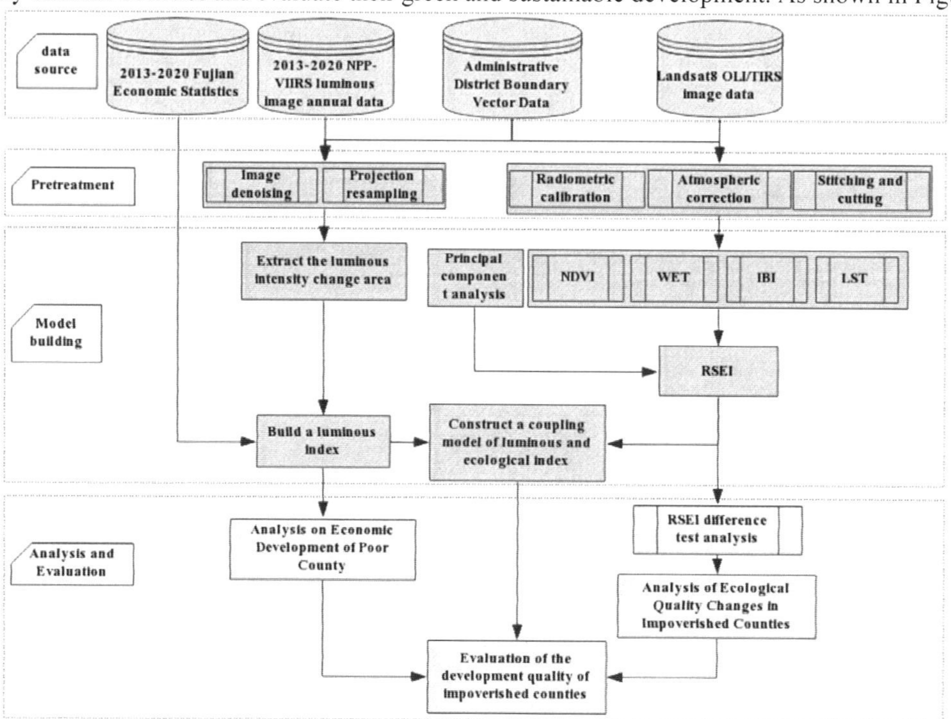

Figure 1. Research route.

2.3. Data preprocessing

2.3.1. Luminous image preprocessing. Firstly, Fujian county administrative zoning map is used to cut out the corresponding NPP/VIIRS night light annual data. To make the data more accurate, avoid the influence of image grid deformation and facilitate the calculation of the pixel area of the bright value in the image, the projection coordinate system of the light raster image data is usually converted into an Albers equal-area cut conic projection. The sampling method adopts the NEAREST proximity method suitable for discrete data processing, and the spatial resolution is 700m [22,23]. The luminous data has negative and extreme values due to the influence of stray light, lightning, moon illumination and clouds. The threshold is selected as the image radiation value equal to 0, that is, the luminous pixel is the luminous pixel if the radiation value is greater than 0, and the rest is assigned as 0 as the background value. The extremely bright value is removed by removing the 0.1% pixels with the highest luminous intensity [24].

2.3.2. Landsat8 OLI image preprocessing. The original image was preprocessed by ENVI software, such as atmospheric correction and radiation calibration [25]. Since there is a large area of water in the research area, the humidity index will have errors, so it is necessary to use the MNDWI index to identify and mask the water part in the research area before extracting the RSEI, and then cropping the corresponding research administrative area with ArcGIS software [26].

2.4. Model construction

2.4.1. Construction of light index. The construction of the luminous index should make full use of the GDP distribution and human activity information carried by image pixels. The sum of the radiance values of the night light pixels in each area is the total nighttime light intensity value (TNL), and the ratio of the total night light intensity to the area is the average nighttime light intensity (ANL), and the two attributes can reflect GDP and population spatial information, the calculation formula is as (1) (2). Development Intensity (DI) refers to the difference in the total light Intensity in poverty alleviation. The greater the difference, the faster the Development speed of the city. In this paper, DI values are divided into 7 levels as shown in Table 2:

$$TNL = \sum_{i=1}^{n} DN_i \tag{1}$$

$$DI = TNL_{past} - TNL_{pre} \tag{2}$$

Table 2. DI classification.

DI	-20 - -5	-5 - -2	-2 - -0.5	-0.5 - 0.5	0.5 - 2	2 - 5	5 - 20
Level	Severe lag	Moderate lag	Mild lag	Unchanged	Mild development	Moderate development	Heavy development

2.4.2. Construction of remote sensing ecological index. Remote Sensing Ecological Index is an ecological index that can integrate multiple ecological factors to monitor and evaluate the regional ecological environment based on remote sensing information. It is composed of the vegetation index (NDVI), humidity index (Wet) and dryness index (IBI), Heat Index (LST). The calculation formula for each index is as (3~7).

$$RSEI = f\{NDVI, Wet, NDSI, LST\} \tag{3}$$

$$NDVI = (\rho_5 - \rho_4)/(\rho_5 + \rho_4) \tag{4}$$

$$Wet = 0.1511 * \rho_2 + 0.1973 * \rho_3 + 0.3283 * \rho_4 + 0.3407 * \rho_5 - 0.7117 * \rho_6 - 0.4559 * \rho_7 \tag{5}$$

$$IBI = \frac{2\rho_5/(\rho_5 + \rho_4) - \rho_4/(\rho_4 + \rho_3) + \rho_2/(\rho_2 + \rho_5)}{2\rho_5/(\rho_5 + \rho_4) + \rho_4/(\rho_4 + \rho_3) + \rho_2/(\rho_2 + \rho_5)} \tag{6}$$

$$LST = \frac{T_b}{1 + (\lambda T_b/\rho) * \varepsilon} - 273.5 \tag{7}$$

Principal component analysis (PCA) can use less calculation to select the quantity and obtain the effect of selecting the best subset of variables. The PCA of the four indexes corresponding to the RSEI index is carried out, and the first component is the initial remote sensing ecological index $RSEI_{PFIRST}$, as shown in formula (8). Use equation (9) to normalize $RSEI_{PFIRST}$ to get RSEI, and its value is between 0 and 1. The closer to 1, the better the ecological environment quality, and vice versa, the worse the ecological quality. The RSEI change value is divided into 7 levels as shown in Table 3.

$$RSEI_{PFIRST} = PFIRSTf\{NDVI, Wet, NDSI, LST\} \tag{8}$$

$$RSEI = (RSEI_{PFIRST} - RSEI_{PFIRST\_min})/(RSEI_{PFIRST\_max} - RSEI_{PFIRST\_min}) \tag{9}$$

Table 3. RSEI classification.

RSEI	-1 - -0.5	-0.5 - -0.2	-0.2 - -0.05	-0.05 - 0.05	0.05 -0. 2	0.2 - 0.5	0.5 - 1
Level	Severely degraded	Moderate degradation	Mild degeneration	Stable zone	Mild improvement	Moderate improvement	Severely improvement

2.4.3. Construction of coupling coordination model. the coupling coordination degree model is used to analyze the coordinated development level of things[27,28]. In this paper, the coupling coordination degree function of the interaction between ecological remote sensing index and luminous index can be obtained by quoting the coupling model of economics, as shown in formula (10,11):

$$C(U_1, U_2) = 2 * [\frac{U_1 U_2}{(U_1 + U_2)^2}]^{\frac{1}{2}} \tag{10}$$

$$D = \sqrt{C * T} \tag{11}$$

In the formula, C represents the interaction coupling degree of each function, reflects the strength of the interaction coupling of each system, and U_1, U_2 is the evaluation function of each index. The closer the value of C is to 1, the better the coupling degree. D represents the degree of coordination, $T=mU_1+nU_2$, m=n=1/2, m+n=1, $T\in[0,1]$. Discuss the coupling degree C and the coordination degree D, as shown in Table 4.

Table 4. Coupling and coordination level.

Coupling		Coordination	
Value	Level	Value	Level
[0,0.3]	Low	[0,0.4]	Maladjustment
(0.3,0.5]	Antagonistic	(0.4,0.5]	Mild coordination
(0.5,0.8]	Running-in	(0.5,0.8]	Well coordination
(0.8,1]	High	(0.8,1]	Quality coordination

3. Results and analysis

3.1. Analysis of economic development of impoverished counties

As shown in Figure 2, the total luminous intensity TNL of the 23 poor counties in Fujian Province and the GDP statistics of the corresponding counties in the four years are logarithmically fitted, and it is found that they have a strong correlation from 2013 to 2019, and the R2 value is above 0.75. The higher the TNL of a poor county, the higher its corresponding GDP and the better the level of economic development. Therefore, luminous data can better reflect the economic development level of poor counties.

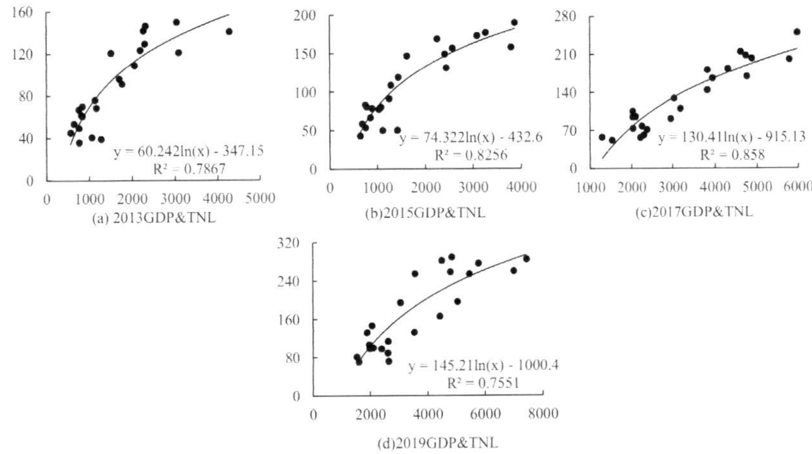

Figure 2. 2013-2019 GDP and TNL fitting.

In order to facilitate the coupling and coordination, the total luminous intensity is normalized to obtain TNL_NORM as shown in Table 5. It is found that the total value of noctilucent light in all the six poverty-stricken counties has more than doubled from 2013 to 2019, and the lower the original total value of the light, the greater the increase. Among them, the biggest increase occurred from 2015 to 2017, indicating the fastest economic development during this period. This is mainly because the 13th Five-Year Plan in 2015 emphasized targeted poverty alleviation and implemented poverty alleviation projects. It also held a poverty alleviation and development work conference to study the poverty alleviation of the poor, develop the urban economy, help urban areas to get rid of poverty, and achieve poverty alleviation in poverty-stricken counties. At the same time, it can be found that there are regional differences in poverty alleviation, and the development of poor counties is unbalanced. In other words, the overall economic development of Shouning, Guangze and Qingliu is lower than that of Yongtai, Pinghe and Changting, but after poverty alleviation, the gap shows a trend of narrowing. This is closely related to the implementation of the mountain-sea cooperation and linkage development model initiated by President Xi Jinping. That is, the coastal economic development counties will provide one-to-one assistance to the 23 key counties of provincial-level poverty alleviation and development work, and promote the mountain-sea cooperation in industrial docking, labor assistance, talent exchange and technical assistance.

Table 5. Normalized total light intensity.

TNL_NORM	2013	2015	2017	2019
Yongtai county	0.22	0.27	0.49	0.57
Shouning county	0.04	0.07	0.25	0.23
Guangze county	0.04	0.05	0.35	0.30
Qingliu county	0.09	0.03	0.21	0.22
Pinghe county	0.54	0.37	0.63	0.76
Changting county	0.25	0.25	0.64	0.62

The development intensity DI from 2013 to 2019 is shown in Figure 3. In addition to the mountainous areas, the overall light intensity and range of residential areas in the six poverty-stricken counties are in an increasing trend, and the increase is most obvious in Yongtai and Changting, especially in the light range and brightness of the county center. Comparing with the national highway network map, it is found that the areas with increased lights mainly revolve around highways and town centers, indicating that urbanization has a significant impact on poverty alleviation. At the same time, there has been a lagging development in some areas, especially the Pinghe County area, showing a trend of lagging development in some areas, which is related to the relocation of urban centers.

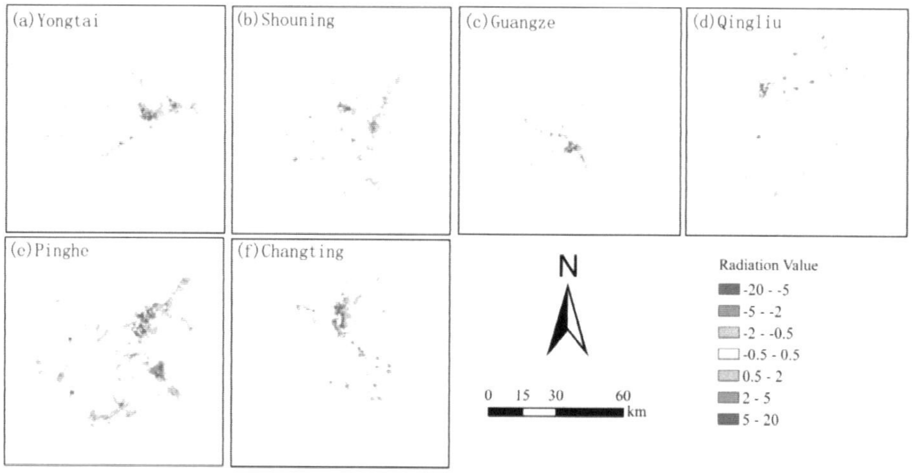

Figure 3. DI of impoverished counties.

3.2. Analysis of ecological conditions in typical poor counties

The statistical results of the RSEI value, the mean value of each index and the contribution to the first principal component of the 6 poverty-stricken counties are shown in Table 6.

From an overall point of view, the RSEI of the 6 poor counties was at a relatively high level in 2013-2019, and the RSEI value was all higher than 0.6, and all had a slight increase. From 2013 to 2015, Yongtai, Shouning and Guangze focused on ecological environment construction and increased the average RSEI to above 0.7, while the ecological environment of Pingping, Qingliu and Changting was damaged to varying degrees. From 2015 to 2017, except for Qingliu County, the RSEI value of other counties decreased to varying degrees, while TNL increased to a large extent, indicating that economic development had a significant impact on the ecological environment.

Judging from the average values of the ecological indicators in the 6 poor counties, the average values of NDVI and WET from 2013 to 2019 are basically above 0.7 with little fluctuation, indicating that the vegetation coverage of poverty-stricken counties is stable and at a high level after poverty alleviation and cap removal. However, the mean value of LST and IBI is about 0.4, without significant increase, indicating that human development activities and urban expansion do not have obvious damage to the ecological environment of the poor counties on the whole.

From the perspective of the first principal component of the ecological indicators in the six poverty-stricken counties, both NDVI and WET indicators are positive, indicating that vegetation coverage and humidity have a promoting effect on the ecological environment. Among them, NDVI has a higher contribution rate, indicating that vegetation had the most obvious promoting effect on the ecological environment. The LST and IBI indicators are both negative, indicating that dryness and heat inhibit the ecological environment. These two indicators are mainly due to the effects of impervious surfaces such as urban buildings, while urban expansion and land development have a destructive effect on the ecological environment.

As shown in Figure 4, the ecological environment of the six impoverished counties is generally good, and there is no situation that the entire area is ecologically poor. From 2013 to 2019, the remote sensing ecological index RSEI has all increased after the poverty-stricken counties were lifted out of poverty. This is because Fujian Province adheres to the new development concept and continues to take a high-quality development path with ecology as the priority and green development as the orientation. Among the six poverty-stricken counties, the RSEI in Shouning County improved most

IOP Conf. Series: Earth and Environmental Science **865** (2021) 012004 doi:10.1088/1755-1315/865/1/012004

obviously, and the ecological environment in most areas of the county improved obviously, and the RSEI increased significantly. The overall ecological environment of Yongtai, Guangze and Changting has been improved, and there is no obvious decline in any area. Although the overall RSEI of Pinghe County in Qingliu County is good, the ecological environment has not improved. Among them, the environment in the southwest of Pinghe County has been significantly improved, but the ecological environment in the southeast has declined significantly.

Table 6. Statistics of RSEI indicators.

Impoverished counties	INDEX	2013 Mean	2013 PFIRST	2015 Mean	2015 PFIRST	2017 Mean	2017 PFIRST	2019 Mean	2019 PFIRST
Yongtai county	NDVI	0.899	0.495	0.874	0.554	0.828	0.477	0.869	0.402
	WET	0.825	0.315	0.811	0.383	0.788	0.303	0.864	0.333
	LST	0.382	-0.463	0.376	-0.428	0.408	-0.545	0.342	-0.566
	IBI	0.294	-0.664	0.394	-0.602	0.448	-0.619	0.299	-0.637
	RSEI	0.69		0.72		0.64		0.72	
Shouning county	NDVI	0.855	0.346	0.877	0.812	0.901	0.366	0.867	0.412
	WET	0.889	0.267	0.559	0.146	0.826	0.418	0.784	0.353
	LST	0.414	-0.702	0.329	-0.432	0.417	-0.647	0.327	-0.476
	IBI	0.387	-0.562	0.453	-0.364	0.273	-0.521	0357	-0.691
	RSEI	0.61		0.74		0.68		0.72	
Guangze county	NDVI	0.823	0.724	0.835	0.655	0.835	0.435	0.808	0.557
	WET	0.727	0.128	0.838	0.115	0.849	0.312	0.854	0.198
	LST	0.399	-0.427	0.368	-0.506	0.423	-0.552	0.342	-0.388
	IBI	0.334	-0.526	0.394	-0.549	0.427	-0.637	0.423	-0.705
	RSEI	0.69		0.71		0.67		0.72	
Qingliu county	NDVI	0.854	0.545	0.805	0.684	0.816	0.563	0.822	0.504
	WET	0.804	0.320	0.543	0.085	0.823	0.277	0.708	0.338
	LST	0.435	-0.426	0.371	-0.513	0.361	-0.389	0.399	-0.425
	IBI	0.402	-0.647	0.452	-0.509	0.409	-0.673	0.389	-0.671
	RSEI	0.68		0.60		0.71		0.69	
Pinghe county	NDVI	0.811	0.725	0.846	0.591	0.825	0.616	0.881	0.599
	WET	0.723	0.111	0.667	0.152	0.821	0.234	0.773	0.175
	LST	0.394	-0.409	0.408	-0.559	0.382	-0.448	0.387	-0.576
	IBI	0.409	-0.542	0.298	-0.561	0.476	-0.603	0.323	-0.527
	RSEI	0.67		0.64		0.64		0.67	
Changting county	NDVI	0.882	0.597	0.771	0.633	0.849	0.487	0.900	0.527
	WET	0.863	0.206	0.817	0.218	0.817	0.298	0.861	0.210
	LST	0.363	-0.445	0.374	-0.521	0.401	-0.553	0.359	-0.478
	IBI	0.341	-0.635	0.469	-0.529	0.402	-0.606	0.308	-0.669
	RSEI	0.76		0.64		0.64		0.77	

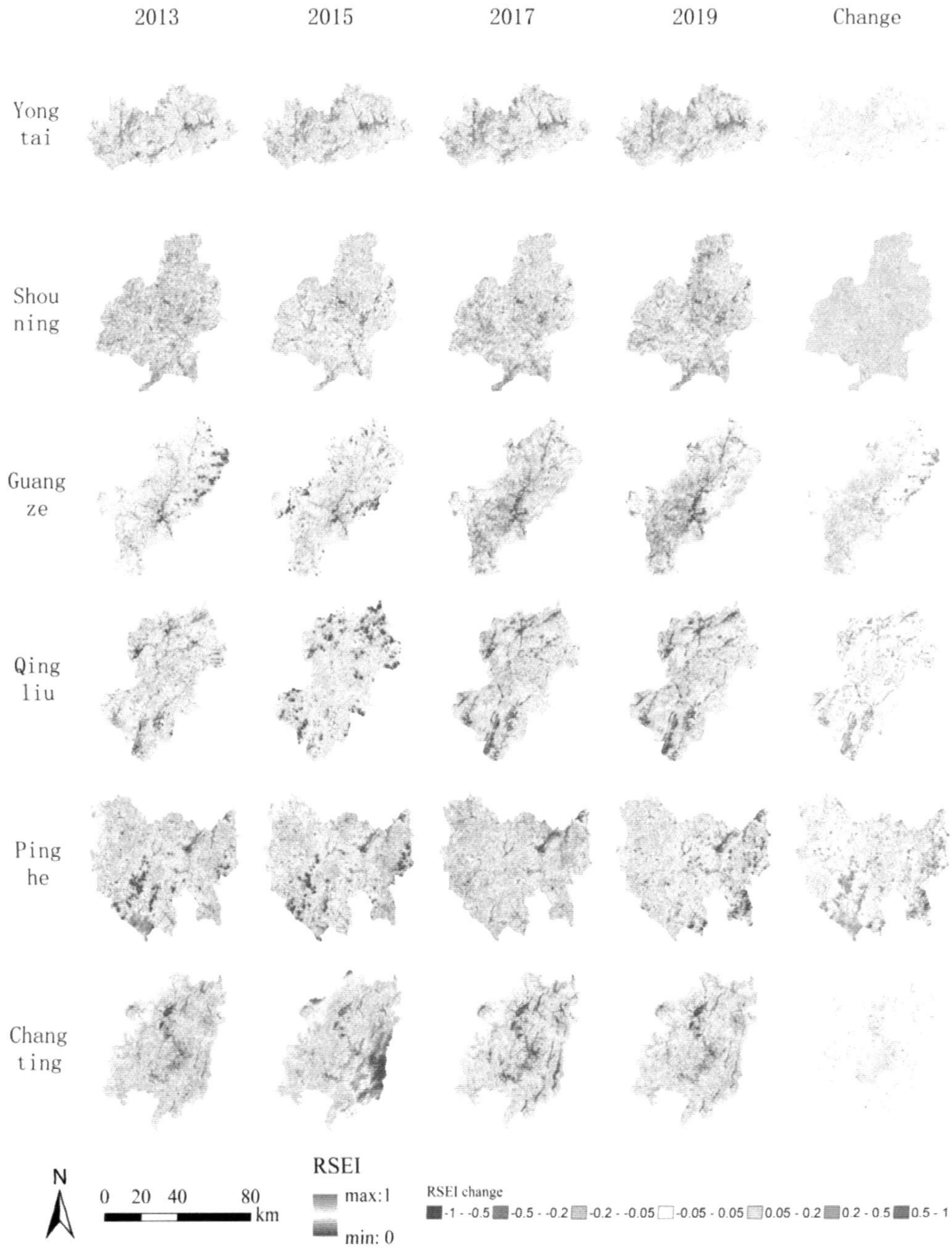

Figure 4. RSEI of 6 poor counties.

3.3. Analysis of Green Sustainable Development

After 2013, Fujian Province gradually integrated the concept of ecological civilization into the development process of new-type urbanization. According to the carrying capacity of resources and environment, Fujian Province established the concept of ecological civilization and green and low-carbon development, focused on promoting green development, and built resource-saving and environment-friendly new towns. As can be seen from Figure 5, TNL_NORM of the six impoverished counties increased greatly during 2013-2019, while RSEI remained unchanged at 0.6-0.8. In Table 7, the coupling degree C and coordination degree D of the study area has increased by more than 30% from the average value. From the specific values of each county, the coupling degree C increased from 2013 to 2019, among which the coupling degree increased fastest from 2015 to 2017 and approached 1 in 2019, indicating that the two were in the high coupling stage, signifying a close correlation between them. The coordination level D increased by 10%-20% from 2013 to 2019, showed a fluctuating upward trend, indicating that the coupling coordination level between poverty alleviation and ecological environment has been continuously improving, and reached above the good coordination level in all cases in 2019. This is because we have adhered to the concept that clear waters and green mountains are gold and silver mountains in the course of development. We have explored the establishment of a diversified compensation mechanism for ecological protection, adopted measures such as closing mountains for forest cultivation, developing green industries, and implementing ecological resettlement and relocation. We have promoted soil erosion control and achieved win-win ecological and economic benefits.

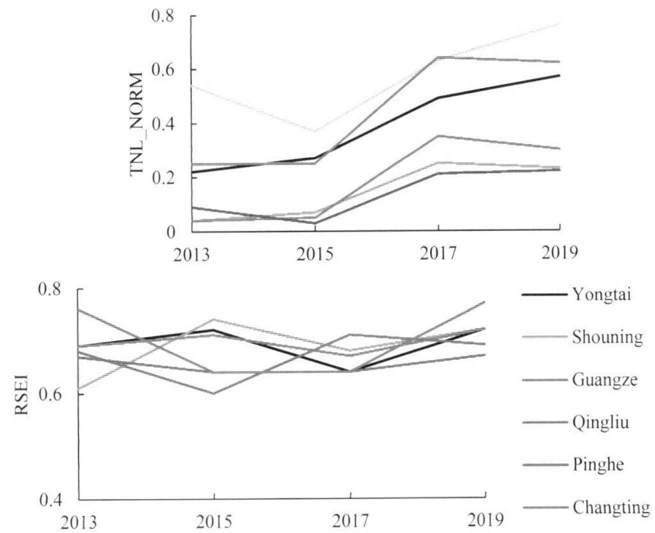

Figure 5. TNL_NORM and RSEI trend.

Table 7. Coupling degree.

	Coupling				Coordination			
	2013	2015	2017	2019	2013	2015	2017	2019
Yongtai county	0.85	0.89	0.99	0.99	0.62	0.66	0.75	0.80
Shouning county	0.49	0.56	0.88	0.85	0.40	0.48	0.64	0.63
Guangze county	0.44	0.49	0.95	0.91	0.40	0.43	0.70	0.68
Qingliu county	0.64	0.43	0.84	0.85	0.49	0.37	0.62	0.62
Pinghe county	0.99	0.96	1	1	0.78	0.70	0.80	0.84
Changting county	0.86	0.90	1	0.99	0.66	0.63	0.79	0.83
MEAN	0.71	0.71	0.94	0.93	0.56	0.55	0.72	0.73

4. Conclusion and discussion

4.1. Conclusion

In this paper, nighttime light data and Landsat 8 OLI remote sensing images were used as data sources. The light index and RSEI method were used to analyze the ecological environment in the study area, and the two methods were coupled and coordinated to realize the monitoring and evaluation of poverty alleviation and sustainability in poverty-stricken counties. Studies have shown that:

1) The luminous index can be used to monitor the economic development of poor counties on a county-level scale, and the correlation R2 value of its logarithmic fitting effect is about 0.8. DI chart can be used to better monitor the economic development degree of poverty-stricken counties. It can be found that poverty-stricken counties doubled their economic growth from 2013 to 2019, their infrastructure was greatly improved, and rural areas were successfully lifted out of poverty, with the most obvious achievements from 2015 to 2017.

2) Monitoring the ecological environment of impoverished counties based on the RSEI shows that the RSEI value of impoverished counties in Fujian Province has stabilized and increased slightly, the overall ecological level is at a high-quality level, and the vegetation coverage rate is high.

3) The coupling and coordination study of the lighting index and RSEI found that the coupling degree C and the coordination degree D present a fluctuating upward trend, and the two are at a good coupling and coordination level, indicating that adhering to the concept of green water and green mountains is a golden mountain and silver mountain can achieve ecological benefits and economics Benefit and win-win.

4.2. Discussion

The image data selected in this article can basically represent the situation in poor counties, but the low resolution of the sensor to obtain luminous data makes it impossible to effectively extract the lights in some rural areas; Although most of the aurora and fire noises are filtered out from the synthesized nocturnal annual data, there is still some noise inevitably. At the same time, there will also be thick clouds in the Landsat8 OLI remote sensing images, resulting in noise in the data; In future work, we can improve the denoising model algorithm and increase the sensor resolution, to provide more accurate data for the assessment of economic development and ecological assessment of poor counties.

Secondly, in terms of model construction. This article uses the total radiation value normalization method to normalize the lighting index normalization, which can further dig into the new poor county lighting index evaluation method to better monitor the economic development of the region. The RSEI model can be combined with multi-source data to improve the representativeness of the ecological index to the ecological environment. In the construction of the coupling model, a new model of the coupling and coordination relationship between the two systems can be explored.

Last but not least, this article selects one poverty-stricken county in each of the six urban areas in Fujian Province. Although it can basically represent the economic and ecological environment development of the 23 poverty-stricken counties in Fujian Province, the specific conditions of the 23 counties are not the same, so they cannot fully reflect the situation of all counties. Therefore, it is possible to research the economic and ecological monitoring of all poor counties to compare the differences in the development of different counties.

References

[1] Fu B 2020 UN sustainable development goals and historical mission of geography *Science & Technology Review* **38 (13)** 19-24(In Chinese).

[2] Tong Y and Long H 2003 Study on sustainable development in the poor areas coupled with vulnerable eco-environment *China Population Resources and Environment* **13 (2)** 47-51(In Chinese).

[3] Hillger D, Kopp T and Lee T 2013 FIRST-LIGHT IMAGERY FROM SUOMI NPP VIIRS. *Bulletin of the American Meteorological Society* **94 (7)** 1019-1029.

[4] Elvidge C, Baugh K and Zhizhin M 2013 Why VIIRS data are superior to DMSP for mapping nighttime lights *Proceedings of the Asia-Pacific Advanced Network* **35** 62-69.

[5] Li X, Xu H and Chen X 2013 Potential of NPP-VIIRS Nighttime Light Imagery for Modeling the Regional Economy of China *Remote Sensing* **5 (6)** 3057-3081.

[6] Zheng Y, He Y and Wang X et al. 2020 Application Review and Prospect of Nighttime Light Remote Sensing Data *Remote Sensing Information* **35 (3)** 1-14(In Chinese).

[7] Wang C, Chen Z and Yang C 2020 Analyzing parcel-level relationships between Luojia 1-01 nighttime light intensity and artificial surface features across Shanghai, China: A comparison with NPP-VIIRS data *International Journal of Applied Earth Observation and Geoinformation* **85 (C)** 101989.

[8] Shi K, Yu B and Huang Y 2014 Evaluating the Ability of NPP-VIIRS Nighttime Light Data to Estimate the Gross Domestic Product and the Electric Power Consumption of China at Multiple Scales: A Comparison with DMSP-OLS Data *Remote Sensing* **6 (2)** 1705-1724.

[9] Mohammed A and Peter M A 2019 Three-Fold Urban Expansion in Saudi Arabia from 1992 to 2013 Observed Using Calibrated DMSP-OLS Night-Time Lights Imagery *Remote Sensing* **11 (19)** 2266.

[10] Yu B, Tang M and Wu Q 2018 Urban Built-Up Area Extraction From Log-Transformed NPP-VIIRS Nighttime Light Composite Data *IEEE Geoscience and Remote Sensing Letters* **PP** 1-5.

[11] Li X and Li D 2014 Can night-time light images play a role in evaluating the Syrian Crisis? *International Journal of Remote Sensing* **35 (18)** 6648-6661.

[12] Pan J and Hu Y 2016 Spatial identification of multidimensional poverty in China based on nighttime light remote sensing data *Economic Geography* **36 (11)** 124-131(In Chinese).

[13] Yu B, Shi K and Hu Y 2015 Poverty Evaluation Using NPP-VIIRS Nighttime Light Composite Data at the County Level in China *IEEE Journal of Selected Topics in Applied Earth Observations and Remote Sensing* **8** 1-13.

[14] Xu H 2013 A remote sensing urban ecological index and its application *Acta Ecologica Sinica* **33 (24)** 7853-7862(In Chinese).

[15] Xu H 2013 A remote sensing index for assessment of regional ecological changes *China Environmental Science* **33 (5)** 889-897(In Chinese).

[16] Congjian S, Xiaoming L and Wenqiang Z et al. 2019 Evaluation of ecological security in poverty-stricken region of Lüliang Mountain based on the remote sensing image *China Environmental Science* **39 (12)** 5352-5360(In Chinese).

[17] Xu H, Wang M and Shi T 2019 Prediction of ecological effects of potential population and impervious surface increases using a remote sensing based ecological index (RSEI) *ECOLOGICAL INDICATORS* **93** 730-740.

[18] Wei Y, Qian J and Fan W et al. 2021 Dynamic monitoring of ecological environment quality in Lijiang River Basin based on RSEI *Science of Soil and Water Conservation* **19 (01)** 122-131(In Chinese).

[19] Liu X, Zhang X and He Y 2020 MONITORING AND ASSESSMENT OF ECOLOGICAL CHANGE IN COASTAL CITIES BASED ON RSEI *ISPRS - International Archives of the Photogrammetry, Remote Sensing and Spatial Information Sciences* **XLII-3 (W10)** 461-470.

[20] Mohammad K F, Solmaz F and Majid K 2020 Remotely Sensed Urban Surface Ecological Index (RSUSEI): An Analytical Framework for Assessing the Surface Ecological Status in Urban Environments *Remote Sensing* **12** 2029.

[21] Li J, Zhang B and He C et al. 2007 The Use of Multisource Satellite and Geospatial Data to Study the Ecological Effects of Urbanization:A Case of the Urban Agglomerations in Bohai Rim *Journal of Remote Sensing* **11 (01)** 115-126(In Chinese).

[22] Chen Z, Yu B and Yang C 2021 An extended time series (2000–2018) of global NPP-VIIRS-like nighttime light data from a cross-sensor calibration *Earth System Science Data* **13 (3)** 889-906.

[23] Elvidge C D, Zhizhin M and Ghosh T 2021 Annual Time Series of Global VIIRS Nighttime Lights Derived from Monthly Averages: 2012 to 2019 *Remote Sensing* **13 (5)** 922.

[24] Zheng Y, Shao G and Tang L 2019 Rapid Assessment of a Typhoon Disaster Based on NPP-VIIRS DNB Daily Data: The Case of an Urban Agglomeration along Western Taiwan Straits,

China *Remote Sensing* **11 (14)** 1709.
[25] Xu H, Wang Y and Guan H 2019 Detecting Ecological Changes with a Remote Sensing Based Ecological Index (RSEI) Produced Time Series and Change Vector Analysis. *Remote Sensing* **11 (20)** 2345.
[26] Wang S, Zhang X and Zhu T et al. 2016 Progress in Geography in the Changbai Mountain Nature Reserve based on remote sensing technology *Progress in Geography* **35 (10)** 1269-1278(In Chinese).
[27] Xiaonan C 2019 Expression and Mathematical Property of Coupling Model, and Its Misuse in Geographical Science *Economic Geography* **39 (4)** 18-25(In Chinese).
[28] Chun-Lin L 2017 Common Mistakes in Coupling Degeee Calculation *Journal of Huaiyin Teachers College(Natural Science Edition)* **16 (1)** 18-22(In Chinese).

Analysis of cyanobacteria outbreak effects in southern offshore area of Taihu lake

Lei Fu[1, 2], Junmin Wang[1], Changhuai Wu[1], Aiju You[1] and Shang Gao[1]

[1]Zhejiang Institute of Hydraulics & Estuary, Hangzhou, China, 310020

[2]Corresponding author's email: F51@qq.com

Abstract. The eutrophication of lake water bodies has been emphasized for decades in our country due to its severe environmental effects. Indeed, 85.4% of lakes have met the eutrophication standard, and 40.1% of the lakes have met the severe eutrophication standard in China nowadays. Taihu lake, as the most important freshwater lake in eastern China, has been embarrassed because of the serious cyanobacteria outbreak problem every year. In this study, a statistic analysis of cyanobacteria outbreak in the offshore area of Taihu lake is conducted. Both the satellite image and field monitoring data are cited and analyzed. As a result, it is confirmed that the cyanobacteria outbreak effects are serious in spring and summer due to the high temperature and the high ratio of TN/TP. Additionally, the variations of TN/TP and cyanobacteria density are both computed and compared to show the environmental issues close to the Taihu lake.

1. Introduction

Eutrophic lakes usually have higher TN (Total Nitrogen) and TP(Total Phosphorus) concentration, previous studies show that the ratio of TN/TP will affect cyanobacteria in the eutrophic lakes[1-4]. Taihu lake has been embarrassed by the cyanobacteria outbreak problem for decades although its water area reaches 2338.1 km^2 [3,5,6]. The offshore line of Taihu lake is almost 393.2km, numerous sources of pollutant lead to serious water environmental issues in Taihu lake. In this study, Taihu lake and its southern offshore area is focused, the TN/TP is computed and the cyanobacteria density is analyzed in order to clarify the environmental issues due to the outbreak of cyanobacteria. A detailed comparison between satellite image and field monitoring data is also represented to show the local environmental issues close to the southern offshore area of Taihu lake as shown in Figure 1 [7-9].

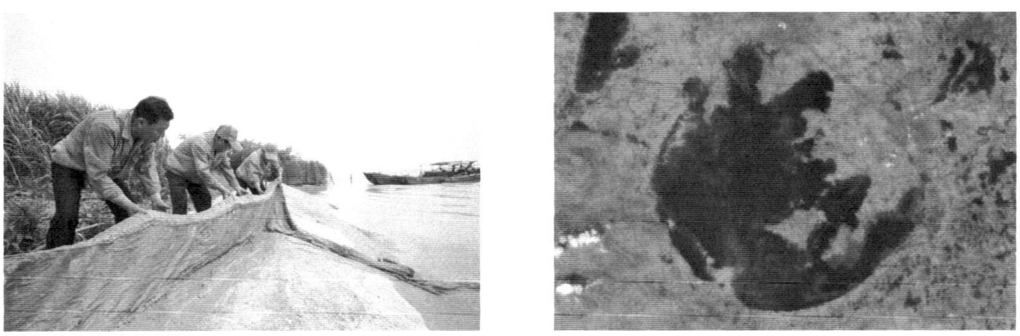

Figure 1. Cyanobacteria disaster in southern offshore area of Taihu lake.

2. Distribution of Cyanobacteria in Taihu lake

According to the Taihu Lake Health Status Report (2018) issued by the Taihu Lake Basin Administration [5], the average density of cyanobacteria in Taihu Lake in 2018 was 86.24 million/L, which decreases of 26.7% from 2017. Among the lake areas which are shown in Figure 2, the number of cyanobacteria in the western coastal area and Zhushan Lake is on a relatively high level. Satellite remote sensing images show that the frequency and intensity of water blooms in Taihu Lake in 2018 have decreased compared with 2017. The largest water bloom area in 2019 was 775.37km^2, which appeared on May 23rd. [10-12]

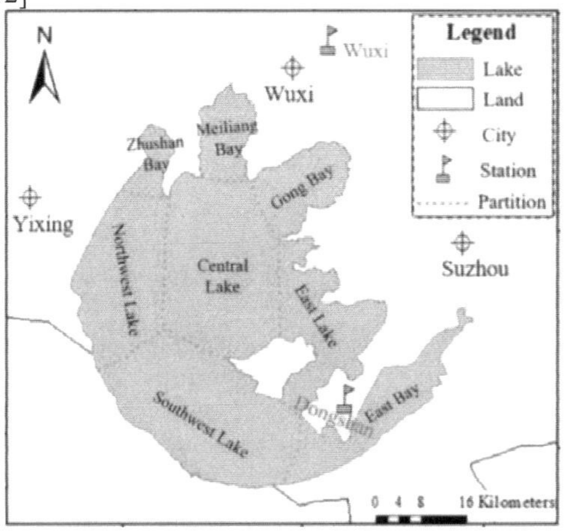

Figure 2. Taihu lake and its offshore areas.

The problem of cyanobacteria in Taihu Lake in 2020 has an descend trend compared with previous years. The density of cyanobacteria and the concentration of chlorophyll a decreased compared with 2019. According to the statistical data of cyanobacteria in each lake area of Taihu Lake, the average density of cyanobacteria in the eastern and southern parts of Taihu Lake is classified as medium and low-severe (the density in most areas is less than 100 million per liter), while the cyanobacteria in the western and Zhushan Lake areas are classified as moderate and severe, which means that the cyanobacteria density is greater than 100 million cells/liter in most areas, and is greater than 150 million cells/liter especially in west and southern areas [4, 6, 13-15].

As shown in Figure 3, combined with the time of the cyanobacteria outbreak period in Lake Taihu over the years, it can be seen that the cyanobacteria outbreaks of Taihu Lake are mostly concentrated in May to July (Summer season) and October to November (Autumn season) each year, and the cyanobacteria outbreaks are mostly close to western and southern regions of Taihu Lake, which occupies almost 3/4 parts of the lake. There are relatively few cyanobacteria outbreaks in the western and northern regions. It is also confirmed that most of the lake areas except the east bank of Taihu Lake are affected by the outbreak of cyanobacteria over years. When considering the southern offshore area of Taihu lake, the river network in Changxing city is affected by Taihu Lake water inversion, and cyanobacteria blooms have a greater impact on the water quality of the river network. According to statistics data, it is found that the river network in Changxing city affected by the cyanobacteria intrusion of Lake Taihu has an impact range up to 15km above the inflow sections, where the cyanobacteria density is usually greater than 20 million cells/liter. Additionally, during the cyanobacteria outbreak period, the rivers in Changxing city were particularly severely affected by the influx of cyanobacteria.

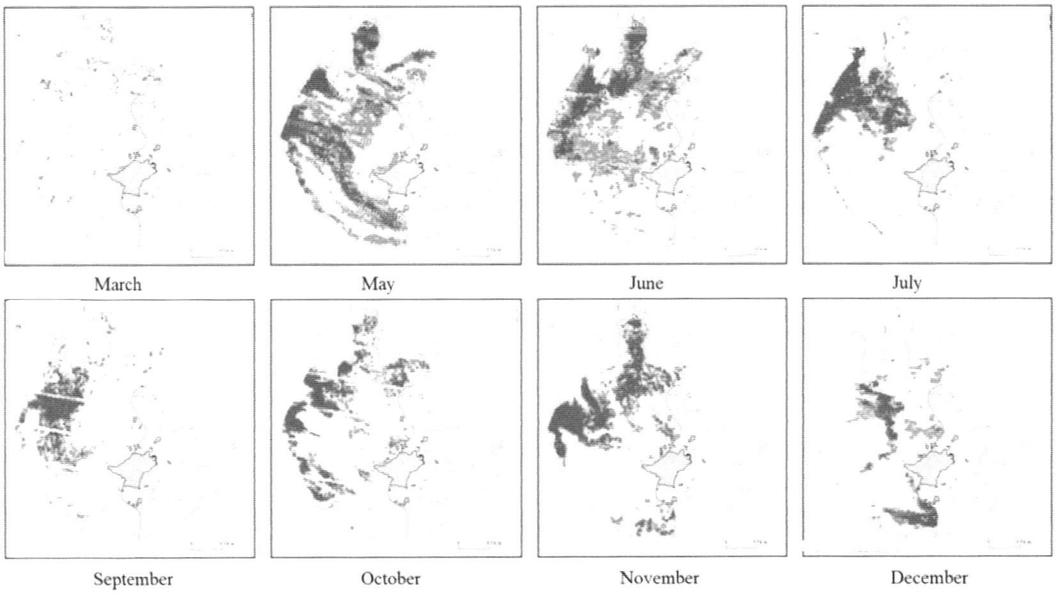

Figure 3. Satellite remote sensing images of cyanobacteria distribution in Taihu lake[1].

3. Analysis of TN/TP and eutrophic index

According to the 2019 annual water quality monitoring data (25 indicators, excluding chlorophyll A) of Xintang Section in Changxing city. A variation curve of TN/TP(Total Nitrogen and Total Phosphorus ratio) of Xintang Section from January to December 2019 is drawn. While the eutrophic index are also calculated on the basis of the water quality monitoring data. The comparison between TN/TP and eutrophic index are shown in Figure 4.

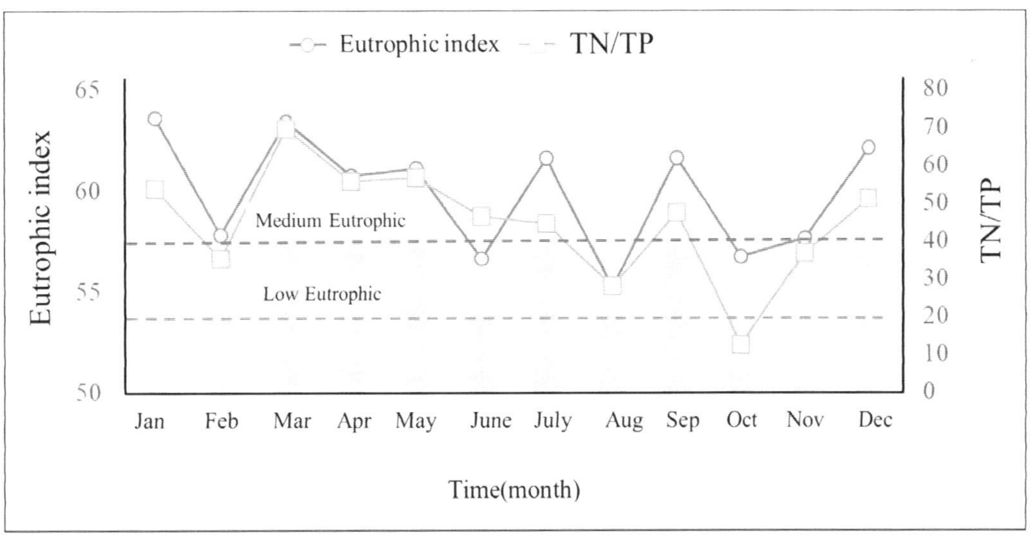

Figure 4. Comparison of TN/TP and eutrophic index.

[1] Cited by Taihu Lake Health Status Report, issued by the Taihu Lake Basin Administration

From the above figure, in 2019, the monthly monitoring data of the TN/TP and the nutritional status index of the Xintang section showed a clear synchronization trend, and the peak and valley value of the change were basically the same, both of them appears in July. The relative peak value occurs in January, March, and September, and the relative valley value occurs in February, August, and October.

When considering the comparison data in Figure 4, firstly, when TN/TP is between 7 and 10, the phytoplankton in the water body is in a nitrogen-restricted state, and biological nitrogen fixation may occur in order to adjust TN/TP to absorb relatively more phosphorus in the water body. Secondly, when the TN/TP is between 23 and 30, the phytoplankton in the water body is limited by phosphorus, and the lower phosphorus concentration will inhibit the organic synthesis of nitrogen. Thirdly, when TN/TP is greater than 29, cyanobacteria outbreaks are widely existed, the release of phosphorus in sediments is easily activated during the cyanobacteria outbreak, while leads to an environmental disaster.

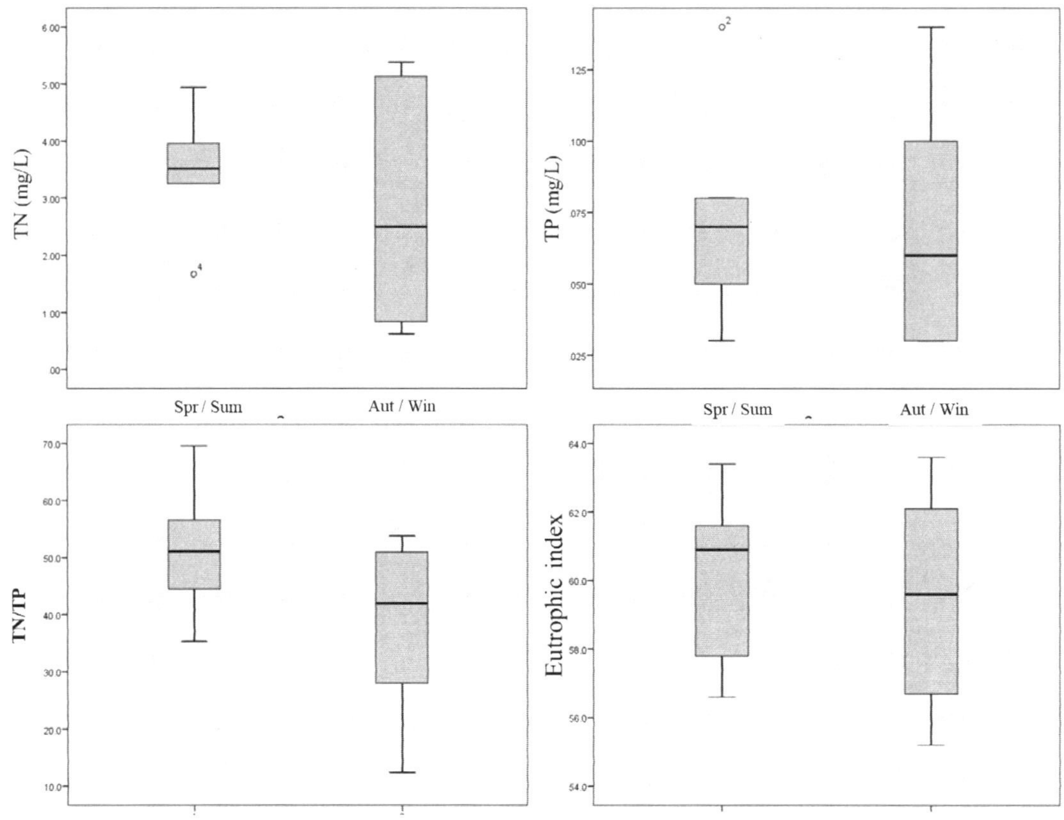

Figure 5. Analysis of TN concentration, TP concentration, TN/TP ratio and eutrophic index.

From Figure 5, according to the quarterly statistical analysis of TN concentration, TP concentration, TN/TP ratio, and eutrophic index at Xintang Section in 2019, it is confirmed that,

①In spring and summer (February to July), the average concentration of TN and TP in Xintang section is higher than autumn and winter (January, August to December). The concentration changes in spring and summer are relatively small, while the concentration varies greatly in autumn and winter.

②The TN/TP ration and the eutrophic index of Xintang section in spring and summer are maintained at high values and the fluctuation is small, the median nutritional status index is 60.90, which belongs to the moderate eutrophication status.

③When considering autumn and winter seasons, lower concentration values are obtained, and the ratio of TN/TP and eutrophic index are also lower than spring and summer season. However, the fluctuation of the above indices are greater. The median eutrophic index is 59.60, which belongs to a mild eutrophication state.

It can be seen that the nutrient status of water bodies in spring and summer is more conducive to the growth of cyanobacteria than in autumn and winter. The outbreak period of cyanobacteria from 2016 to 2020 is usually from May to June each year. Therefore, the southern offshore area of Taihu lake should focus on spring and summer season to reduce the deteriorating effect of the water environment caused by the outbreak of cyanobacteria.

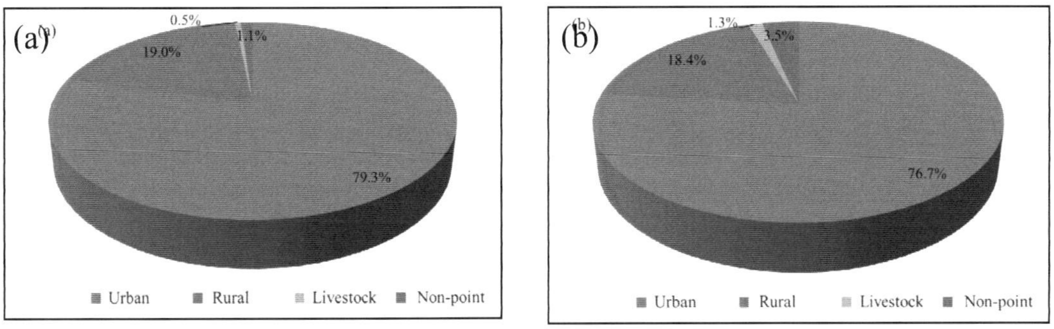

Figure 6. Pollution load in the southern offshore area of Taihu lake (a) TN, (b) COD.

As can be seen from the above Figure 6, the main source of TN pollution load into the river in southern offshore area is the urban citizen pollution which accounts for 79.3% of the total pollution load. While the second main source of pollution load is rural citizen pollution, which accounts for 19.0%. When considering COD(Chemical Oxygen Demand), similar component is obtained, the uraban the rural pollution sources attribute almost 95% of the COD pollution, while the livestock and non-point sources are less than 5%. Indeed, livestock and poultry breeding becomes the third pollution source, and the agricultural non-point sources ranked as fourth. Hence, in the southern offshore area of Taihu lake has embarrassed by the urban and rural citizen pollution, which also lead to a serious cyanobacteria outbreak problem. A systematic control plan of pollution sources in these areas is preferred in the future.

4. Conclusion

Eutrophication is a severe problem in lots of lakes in China. In this study, the relationship between TN/TP and cyanobacteria density has been focused and studied. In addition, the seasonal analysis of Eutrophication in southern offshore area of Tiahu lake is conducted. The comparisons between satellite image and monitoring data show that a medium TN/TP with appropriate TP concentration will lead to a serious cyanobacteria outbreak problem.

As the most developed zone in China, the Taihu basin is embarrassed wastewater pollution problems for decades. For the southern offshore area of Taihu lake, the main pollution source lead to the cyanobacteria outbreak problem is urban and rural pollution. It is also found that the cyanobacteria outbreak problem usually occurs in spring and summer season. Therefore, the pollution control plan as well as the water diversion project in the southern offshore area becomes important in the future to minimize the eutrophication effects in the local zones.

Acknowledgments

This research was supported in part by the Technology Research and Development Program of Zhejiang Province Institute (ZH A20003), and the Research Program of Department of Water Resources of Zhejiang Province (RA2007).

References

[1] Deng X W, Liang G D, Chen J, Qi M and Xie P 2011 Simultaneous determination of eight common odors in natural water body using automatic purge and trap coupled togas chromatography with mass spectrometry *J Chromatogr A* **1218** 3791-3798

[2] Ding Z, Peng S, Jin Y, Xuan Z, Chen X and Yin L 2014 Geographical and seasonal patterns of geosmin and 2-methylisoborneol in environmental water in jiangsu province of China *J Anal Methods Chem* 743924

[3] Silva A F, Lehmann M and Dihl R R 2015 Geosmin induces genomic instability in the mammalian cell microplate-based comet assay *Environ Sci Pollut Res* **22** 17244-17248

[4] Smith J L, Boyer G L and Zimba P V 2008 A review of cyanobacterial odorous and bioactive metabolites: impacts and management alternatives in aquaculture *Aquaculture* **280** 5-20

[5] Taihu Lake Health Status Report (2017-2019) Taihu Lake Basin Administration

[6] Li L, Wan N, Gan N, Xia B D and Song L R 2007a Annual dynamics and origins of the odorous compounds in the pilot experimental area of lake *Dianchi China Water Sci Technol* **55** 43-50

[7] Li L, Wan N, Gan N Q, Xia B D and Song L R 2007b Annual dynamics and origins of the odorous compounds in the pilot experimental area of lake *Dianchi China Water Sci Technol* **55** 43-50

[8] Li Q, Wang W J, Jiang X H, L and Lu D L, et al. 2019 Optimizing the reuse of reclaimed water in arid urban regions: a case study in Urumqi *Northwest China Sustainable Cities and Society* **51** 110

[9] Lyu S, Chen W, Zhang W, Fan Y and Jiao W 2016 Wastewater reclamation and reuse in China: opportunities and challenges *J Environ Sci (China)* **39** 86-96

[10] Zomorodian M, Hin S, Homayounfar M, Ibrahim S, Ehsan S and El-sha A 2018 The state-of-the-art system dynamics application in integrated water resources modeling *J Environ Manag* **227** 294-304

[11] Rousso B Z, Bertone E, Stewart R and Hamilton D P 2020 A systematic literature review of forecasting and predictive models for cyanobacteria blooms in freshwater lakes *Water Research* **182** 115959

[12] Mw A, Hz A, Cd B, Wei Z C, Js A, and Sy A, et al. 2020. Spatiotemporal differences in phosphorus release potential of bloom-forming cyanobacteria in lake taihu. *Environmental Pollution* **271** 116294

[13] Moa B, Mt C, Ae C, Sh D, Wo B and RS A, et al. 2021. High nutrient loading and climatic parameters influence the dominance and dissimilarity of toxigenic cyanobacteria in northern bays of lake victoria *Journal of Great Lakes Research* **286** 117555

[14] Hr A, Mafa B, Kr A, Jt A, Mb C and BK D et al. Suppression of bloom-forming colonial cyanobacteria by phosphate precipitation: a 30years case study in lake barleber (germany) *Ecological Engineering* **162** 106171

[15] Nkm A, Sjm B and Cl A 2020 Using bayesian hierarchical modelling to capture cyanobacteria dynamics in northern european lakes *Water Research* **186** 116356

IOP Conf. Series: Earth and Environmental Science **865** (2021) 012006 doi:10.1088/1755-1315/865/1/012006

Study on emission characteristics and health effects of volatile organic compounds in child seats

Chen Cui[1, 2], Lei Wang[1], Wei Liu[1], Shujie Xu[1] and Xuefeng Liu[1]

[1]China Automotive Technology Research Center Co., Ltd

[2]Corresponding author's email: cuichen@catarc.ac.cn

Abstract. This paper takes two mainstream child seats on the market as the research object, based on the test standard of volatile organic compounds (VOCs) and the objective odor test standard of components and parts assembly, and under 25℃ standard condition, the emission concentration of VOCs and the odor objective level of the two seats were detected to understand the VOCs emission and odor performance of the child seats. Gas chromatography-mass spectrometry (GC-MS) was used to analyze the whole spectrum, and the odor threshold database was used to screen out the key odor components contributing to the odor in the car. According to the WHO-IARC to analyze the carcinogenicity of key odor contributing substances. The results show that the odor performance of the two types of child seats is in average level, and the objective odor intensity were 3.3 and 3.7. The emission of Formaldehyde, Acetaldehyde and Benzene of the two seats were significantly lower than the average value of the seat industry, while dimethylbenzene and Styrene were higher than the industry average. In addition, Benzothiazole, 2,6-Ditert-butyl-4-methyl phenol, Isooctanoic Acid, Cyclohexyl Isothiocyanate, etc. The above-mentioned high-risk odorants are harmful to health, which should be paid more attention by manufacturers.

1. Introduction

Affected by the COVID-19, the consumers' attention to the health of vehicles has been increasing.The Volatile Organic Compounds (VOCs) are the main pollutants in the vehicles, which mainly come from seats, carpets, roofs and other interior parts, as well as child seat and other parts that are in direct contact with human's body. Most of these substances are carcinogens, which do great harm to human liver, kidney, brain and nervous system, and seriously affect human health [1], in addition, some substances also have certain irritation, which will lead to car odor pollution. At present, China's vehicle odor level is between level 3-3.5 (level 1-6, level 1 is the best), and the odor performance of children's seats in this test is lower than the industry average level, so the situation is not optimistic.

China is a big country in the production and export of child seats, and its production and export volume are in the forefront of the world. In recent years, with the increasing awareness of the importance of using child seats and the improvement of relevant laws, the consumption of children's seats in China will gradually increase. Therefore, in order to pay attention to the development of children's physical and mental health and promote enterprises to strengthen management and control, it is necessary to carry out in-depth research on the health performance of child seats.

2. Experiment process

This test focuses on the emission characteristics of volatile organic compounds (VOCs) of child seats, mainly including odor performance, eight substances of VOCs and VOCs full spectrum substances [2]. As shown in Table 1.

Table 1. A table with child seat test index.

Test index	Index items	Focus points
Odor performance	Subjective odor intensity, Objective odor intensity	Olfactory sensory stimulation level
Eight substances of VOCs	Benzene, Toluene, Ethylbenzene, Dimethylbenzene, Styrene, Formaldehyde, Acetaldehyde and Acrolein	Emission performance of eight substances of VOCs controlled by vehicle regulation GB/T 27630
Full spectrum substances of VOCs	Other VOCs substances that can be detected by GC-MS	Impact on odor and health

The samples of this test are two best-selling child seats purchased from regular market channels, one for domestic and one for foreign brands, as shown in Table 2.

Table 2. A table with test sample information.

No.	Brand	Country	Type	Price (RMB)	Picture
1	Goodbaby	China	C8723	1499	
2	COSATTO	Britain	CT4080	840	

The test method is mainly based on the two standards of "GB / T 39897-2021 Determination of Volatile Organic Compounds and Carbonyl Compounds in Nonmetallic Parts of Vehicles" and "T / CAS 406-2020 Interior air of road vehicles-Odour determination-Coupling assessment of sensory and photo ionization detector".

3. Results and discussion

3.1. Analysis of odor performance

When there is unpleasant odor in the vehicle, it will stimulate people's olfactory organs, and even cause psychological and physiological harm to the human body [3]. The current odor evaluation methods in automobile industry are mainly based on subjective evaluation of human beings, which have some problems such as great subjective influence, poor statistics and occupational health hazards, and can not meet the requirements of odor evaluation. The odor performance results of the two child seats in this test are shown in Table 3.

Table 3. A table with the comparison of odor performance results of child seats with industry average.

Brand	Subjective odor intensity	Objective odor intensity	Industry average
Goodbaby	3.5	3.3	3.0-3.5
COSATTO	4.0	3.7	

It can be seen from Table 3 that the odor performance of the two child seats is average, compared with the average level of the industry, Goodbaby Brand is at the average level of the industry, while COSTATTO Brand performs poorly, exceeding the odor range of the industry, and the odor performance still needs to be improved. The deviation of subjective and objective odor intensity grade of the two products is less than 0.3, which indicates that the objective odor evaluation has good accuracy. At the same time, the objective odor evaluation method can accurately determine the odor level of samples, which has a great prospect for later application.

3.2. Analysis of eight VOCs

At present, China has established the national control standards for eight VOCs substances in vehicle, and various automobile enterprises have also established the control limits for VOCs in interior parts and materials. In the early stage, China Automotive Technology and Research Center Co., Ltd.(CATARC) accumulated VOCs data of 40 types of driver's fabric seats. Under the same detection conditions, the comparison data between the two types of child seats in this study and the industry average is shown in Table 4.

Table 4. A table with the comparison of the results of eight sustances of VOCs in child seats with the industry average. (Unit: g/m^3)

Brand	Benzene	Toluene	Ethylbenzene	Dimethylbenzene	Styrene	Formaldehyde	Acetaldehyde	Acrolein
COSATTO	N.D.	31.0	34.0	276.0	58.0	27.0	13.0	N.D.
Goodbaby	N.D.	202.0	83.0	234.0	452.0	20.0	15.0	N.D.
Industry average	7.1	114.7	40.1	213.8	38.7	195.5	139.4	N.D.

It can be seen from Table 4 that the emission of Formaldehyde, Acetaldehyde and Benzene of the two child seats are significantly lower than the industry average, while Dimethylbenzene and Styrene are higher than the industry average. The Toluene, Ethylbenzene and Styrene of Goodbaby Brand were significantly higher than the industry average and COSATTO Brand, especially the Styrene concentration was as high as 452 g/m^3, more than 10 times higher than the industry average. In addition, Styrene is a suspected carcinogen (published by WHO-IARC), which can stimulate the nervous system and respiratory tract. Long term exposure may cause lung diseases, and enterprises need to focus on control [4].

3.3. Analysis of VOCs full spectrum substance

VOCs full spectrum analysis refers to the process of analyzing all the VOCs substance in the sample. CATARC has built a threshold database containing more than 2500 kinds of vehicle odor substances, the database was used to calculate the threshold dilution multiple of VOCs (threshold dilution multiple = detection concentration / olfactory threshold), and it is considered that VOCs with threshold dilution multiple greater than 1 have odor contribution, which will cause interference to human body. The specific analysis results are shown in Table 5.

Table 5. A table with the VOCs full spectrum analysis in child seats.

No.	Chinese Name	CAS Number	Threshold Dilution Multiple	
			COSATTO Brand	Goodbaby Brand
1	Benzothiazole	95-16-9	247.4	/
2	2,6-Ditert-butyl-4-methyl Phenol	128-37-0	229.5	185.3
3	2,4-Ditertbutyl Phenol	96-76-4	47.8	13.8
4	Isooctanoic Acid	149-57-5	36.4	40.1
5	Longifolene	475-20-7	28.6	7.5
6	Cyclohexyl Isothiocyanate	1122-82-3	22.1	63.1
7	Propanoic Acid	79-09-4	8.8	/
8	2-Phenyl-2-propanol	617-94-7	4.0	/
9	2-Methylnaphthalene	91-57-6	3.6	/
10	Phenol	108-95-2	2.3	/
11	Hexadecane	544-76-3	1.9	/
12	Nonyl Aldehyde	124-19-6	/	30.3
13	Octyl Aldehyde	124-13-0	/	27.8
14	Hexanal	66-25-1	/	19.4
15	Decyl Aldehyde	112-31-2	/	5.1
16	Styrene	100-42-5	/	4.4

It can be seen from Table 5 that some VOCs in child seats have strong odor, such as Benzothiazole, 2,6-Ditert-butyl-4-methyl phenol, Isooctanoic Acid, Cyclohexyl Isothiocyanate, etc. Benzothiazole has a quinoline like odor and is a highly toxic substance [5], 2,6-Ditert-butyl-4-methyl phenol has a certain odor and belongs to toxic substance, Isooctanoic Acid has bad smell and belongs to toxic substance, Cyclohexyl Isothiocyanate has pungent odor. The above-mentioned high-risk odor and health hazards should be paid great attention by manufacturers, and the product performance should be improved by strengthening technical research.

4. Results and discussion

The content of VOCs in the two child seats analyzed in this paper is relatively high. VOCs have a huge impact on human health, it can damage the liver, kidney, brain and nervous system, and cause memory loss and affect children's intellectual development and other serious consequences [6]. Therefore, it is urgent to formulate relevant standards for child seats.

References

[1] Chongchong Liu, Feijie Liu, Jiangang Shang, et al. 2018 Investigation and Analysis of Volatile Organic Compounds in National Industrial Area[J]. *Guangdong Chemical Industry*, **045(010)**: 53-54.
[2] Yongquan Zhou 2011 *Study on Degrandation of Volatile Organic Compounds with Vacuum Ultraviolet*[D].
[3] Zhenyu Zhu, Peng Zhang, Xuefeng Liu 2018 Study on the Method of Odor Source Analysis in Vehicles[J]. *Environment and Sustainable Development*, **43(006)**: 210-213.
[4] Jinhai Zu 2014 *The Design of Automobile VOC Evaluation Methods and Measures to Control*

IOP Conf. Series: Earth and Environmental Science **865** (2021) 012006 doi:10.1088/1755-1315/865/1/012006

Emissions[D].
[5] Luyan Sang, Bingxin Sun, Xiang Bai, et al. 2017 Research and Development Status of Isothiocyanate[J]. *Journal of Food Safety & Quality*, **2017(11)**: 4111-4114.
[6] Nan Li 2013 Air Pollution and Prevention of Indoor Environment[J]. *China Science & Technology Panorama Magazine*, **000(012)**: 14-15.

IOP Conf. Series: Earth and Environmental Science **865** (2021) 012007 doi:10.1088/1755-1315/865/1/012007

Research and application for key technology of anti-cracking of acrylic emulsion mortar

Xingyan Liu and Yukun Zhao[1]

Engineering Construction Center, Yellow River Shandong Bureau, Jinan, 250011

[1]Corresponding author's e-mail: 63673236@qq.com

Abstract. Acrylic emulsion mortar is an ideal material of repair and reinforcement for concrete anti-carbonation and erosion, and it has been widely used in the reinforcement of culvert and sluice in recent years. When it is used, shrinkage stress occurs inside the mortar. Once this shrinkage stress exceeds the tensile strength of the mortar, cracks will appear in the mortar, which will seriously affect the construction quality and increase the repair rate or warranty workload. In this paper, the key technology of anti-cracking of acrylic emulsion mortar is studied, characteristic is analyzed of acrylic emulsion mortar protection, and various anti-cracking parameters are proposed. This key technology is verified through the application in the Dongchi sluice reinforcement project. When the technology has achieved economic benefits, it also achieved social benefits. In the future, the application prospects of the key technology of acrylic emulsion mortar anti-cracking in the culvert and sluice reinforcement project in the Yellow River Basin are very broad.

1. Project research purpose

As a new type of repair material for concrete buildings, acrylic emulsion mortar is widely used in culverts and sluices in the Yellow River flood control project with its excellent impermeability, tensile performance, and bonding ability. However, some defects in the material, such as cracks and crazing, were gradually discovered during the repair and use. This is because the acrylic emulsion mortar shrinks due to water loss during its condensing and hardening process in use. When this shrinkage is constrained by the matrix, shrinkage stress will be generated inside the mortar. Once this shrinkage stress exceeds the tensile strength of the mortar, the mortar will crack, which will have a serious impact on the quality of construction and increase the repair rate or warranty workload. [1-2]

The purpose of this paper is to conduct system analysis through a professional laboratory combined with on-site implementation operations, analyze various influencing factors, propose various anti-cracking parameters, research reasonable anti-cracking measures, and calculate economic benefits. And in the future, it will be popularized and applied in culvert and sluice reinforcement project in the Yellow River Basin in the "14th Five-Year Plan" period.

2. Project research background

Dongchi Sluice is located at No. 40+522 of Beijindi Stake in Shen County, Liaocheng City, Shandong Province. It was built in 1979. In May 2009, the relevant expert team of the Shandong River Affairs Bureau conducted a safety appraisal of the Dongchi Sluice. The safety appraisal was classified as the third type of sluices. One of the appraisal conclusions was that the pier head and the surface of the culvert concrete were freeze-thaw damage and carbonization. In 2015, the Dongchi Sluice began to

IOP Conf. Series: Earth and Environmental Science **865** (2021) 012007 doi:10.1088/1755-1315/865/1/012007

implement the danger-removing and strengthening project, one of the contents is the use of acrylate mortar to repair the surface defects of the pier and the culvert concrete shedding. Research for the key technology of anti-cracking of acrylic-snnd mortar was developed based on the above project.

3. Project research route

3.1. The main advantages of acrylic mortar

Acrylic emulsion mortar is the abbreviation for acrylic copolymer emulsion cement mortar, which belongs to high molecular polymer emulsion modified cement mortar. Acrylic emulsion mortar is a new type of repair material for concrete buildings. It has a lot of performance such as excellent adhesion, crack resistance, water resistance, chloride ion penetration, wear resistance, and aging resistance. It also has many advantages, low cost, convenient construction, adapting well to the base concrete temperature, anti-air aging and others.

3.2. Analysis on the factors of cracks in acrylic emulsion mortar

The overall displacement of the structure, especially uneven settlement, will cause settlement cracks and structural cracks in the concrete structure. The causes of cracks appearing in the concrete structure include temperature cracks, shrinkage cracks, steel corrosion cracks, alkali aggregate reaction cracks, construction cracks, etc.. Shrinkage is one of the important reasons that cause cracks in the acrylic emulsion mortar, resulting in a decrease in durability. Studying the factors affecting the shrinkage performance of acrylic emulsion mortar has guiding significance on how to inhibit and reduce the shrinkage of acrylic emulsion mortar. The factors affecting the shrinkage of cement mortar are internal and external reasons. Internal factors refer to cement matrix parameters: water-cement ratio, lime-sand ratio, cement fineness, cement type, mineral admixtures, emulsion content and admixtures. External factors mainly refer to environmental conditions: curing time before drying, curing temperature, drying conditions, and relative humidity of the environment. [1-2]

3.3. Research route design [3-4]

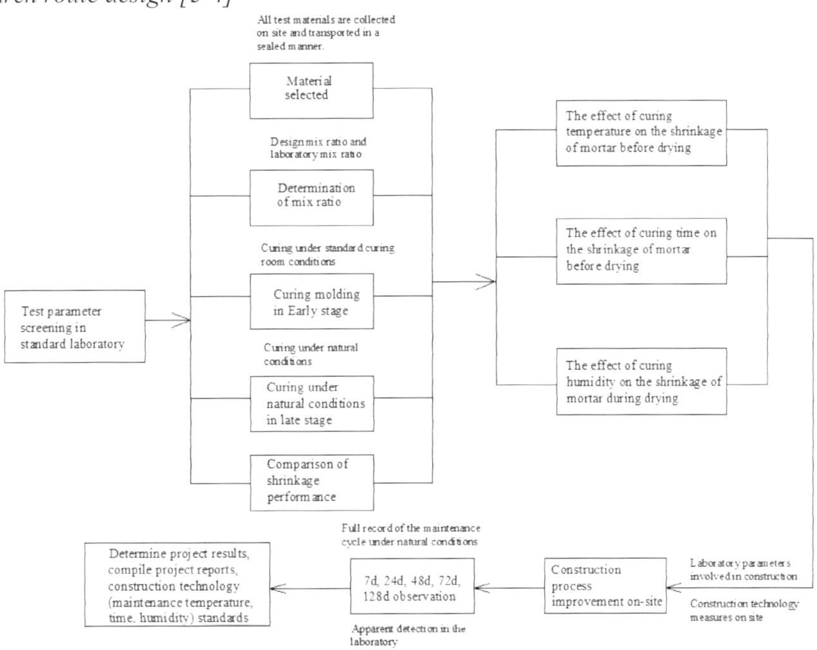

Figure 1. Project research flow chart.

After preliminary understanding and repeated research on the performance of acrylate mortar, the project research route is drawn up as Figure 1.

4. Experimental process

4.1. Laboratory test process

The laboratory test flow chart is drawn up as Figure 2.

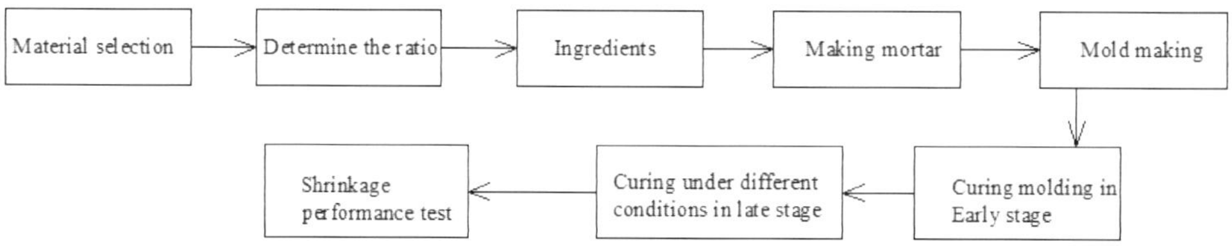

Figure 2. Laboratory test flow chart.

The field test flow chart is drawn up as Figure 3.

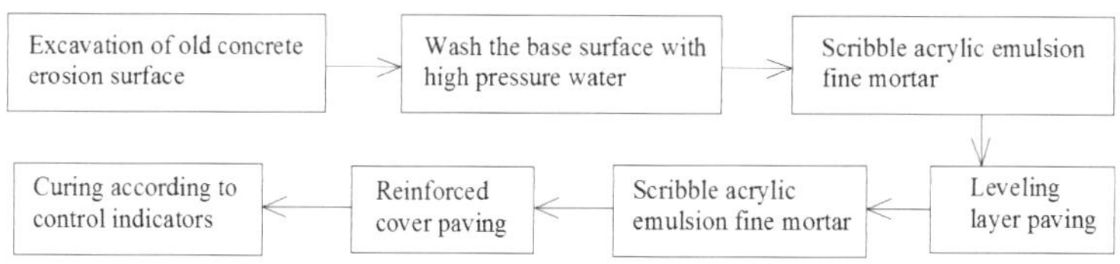

Figure 3. Field test flow chart.

4.1.1. Main test items. Through the experimental analysis, the modification effect of acrylic emulsion mortar under different mix ratios (water-cement ratio), and at the same time examine whether the curing methods match.

The three influencing factors on the shrinkage rate of acrylic emulsion mortar are the curing temperature before drying, the curing time before drying, and the curing humidity during drying. The influence of each factor is grouped as follows:

1)the curing temperature before drying(3 sets of test pieces:10℃, 20℃, 40℃)

2)the curing time before drying (3 sets of test pieces:1d, 2d, 4d)

3)the curing humidity during drying (3 sets of test pieces:45%, 70%, 90%)

4.1.2. Mix ratio. The conclusion of this mix ratio design test is that the ratio of acrylic emulsion mortar is cement: sand: polyacrylate emulsion: water = 100:150:31:14.

4.2. Experimental process [3-4]

4.2.1. Specimen production. For each influencing factor, three sets of test pieces are set up and named in the form of X-X-X. The name of each group of specimens starts with the initials of the Chinese Pinyin of the influencing factors, which are temperature (W), days (T), and humidity (S). The different levels of the same influencing factor are taken as the second number, and the three specimens of each level are named 1, 2, and 3. For example, W-10-1, it means that the main factor affecting of the mortar shrinkage is the curing temperature before drying, the specimen numbered 1 in the group of specimens with a curing temperature of 10° C. A total of 27 sets of specimens were made, corresponding to the curing temperature of 10℃, 20℃, and 40℃, corresponding to the curing days of 1d, 2d, 4d, corresponding to thecuring humidity of 40%, 75%, 90%, the permutation and combination of various factorsand are made and then shrinkage test would be carried out.

4.2.2. The result of the test. a. Test results of acrylic emulsion mortar under different curing temperatures

The correlation formula between the shrinkage rate and age of acrylic emulsion mortar was obtained by numerical fitting from the test data under different curing temperatures before drying.

Using polynomial fitting method, the correlation formulas of age x and shrinkage rate y under curing conditions of 10℃, 20℃ and 40℃ are obtained respectively.

10℃: y = -7E-07x2 + 4E-05x + 2E-05

20℃: y = -6E-07x2 + 3E-05x + 9E-06

40℃: y = -4E-07x2 + 3E-05x - 2E-06

The shrinkage curve and three fitting curves of mortar under different curing temperatures are as Figure 4-7. The dry shrinkage rate is as Table 1.

Figure 4. Shrinkage curve of mortar with different curing temperature.

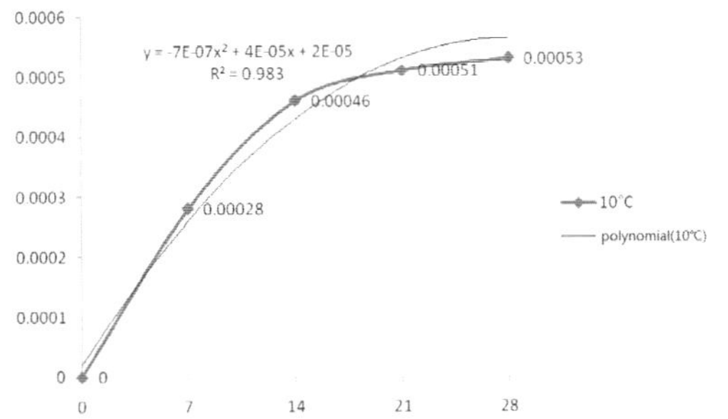

Figure 5. Fitting curve of age x and shrinkage rate y under 10℃ curing condition.

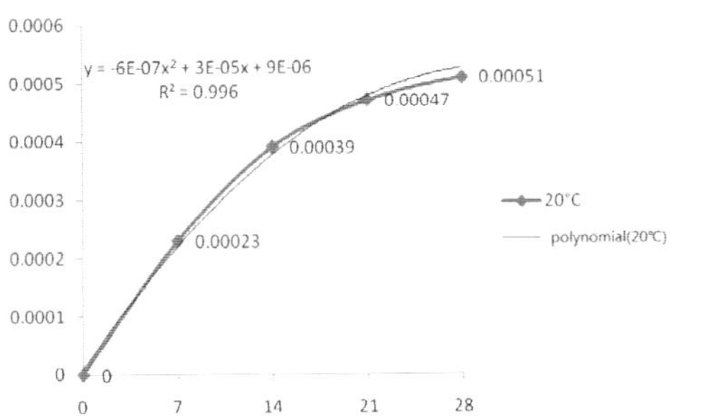

Figure 6. Fitting curve of age x and shrinkage rate y under 20℃ curing condition.

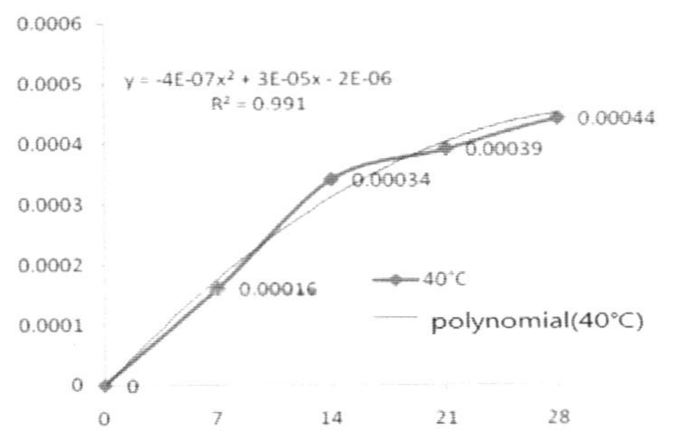

Figure 7. Fitting curve of age x and shrinkage rate y under 40℃ curing condition.

Table 1. Reduction rate of shrinkage.

curing temperature\days	7d	14d	21d	28d	35d
10°C	100%	100%	100%	100%	100%
20°C	82.14%	84.78%	92.16%	96.23%	96.36%
40°C	57.14%	73.91%	76.47%	83.02%	83.64%

b. Test results of acrylic emulsion mortar under different curing days

The correlation formula between the shrinkage rate and age of acrylic emulsion mortar was obtained by numerical fitting from the test data under different curing time before drying.

Using polynomial fitting method, the correlation formulas of age x and shrinkage rate y under curing conditions of 1d, 2d and 4d are obtained respectively.

1d: y = -6E-07x2 + 4E-05x + 1E-06
2d: y = -6E-07x2 + 4E-05x + 3E-06
4d: y = -6E-07x2 + 4E-05x + 1E-05

The shrinkage curve and three fitting curves of mortar under different curing time are as Figure 8-11. The dry shrinkage rate is as Table 2.

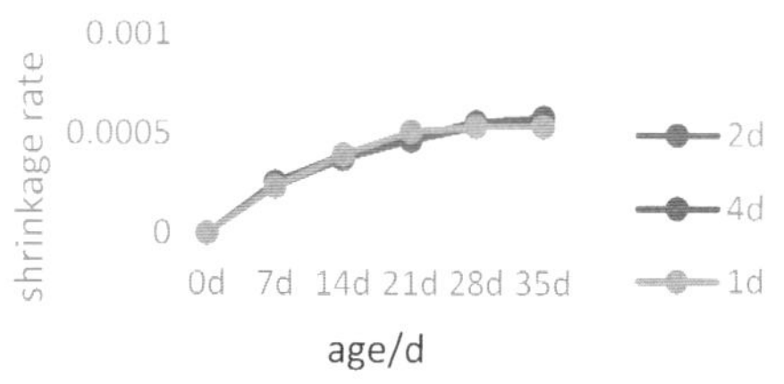

Figure 8. Shrinkage curve of mortar with different curing time.

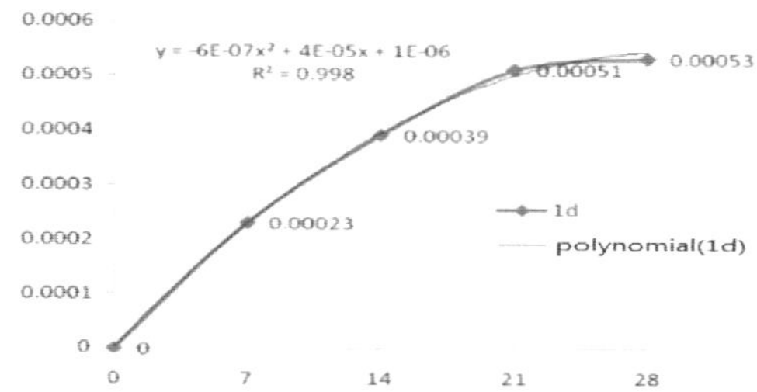

Figure 9. Fitting curve of age x and shrinkage rate y under 1d curing condition.

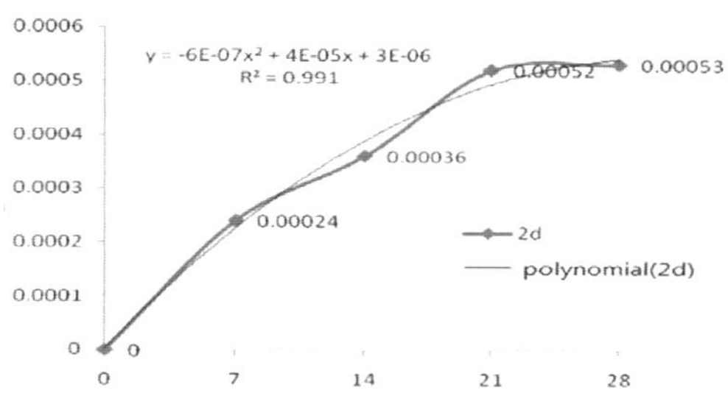

Figure 10. Fitting curve of age x and shrinkage rate y under 2d curing condition.

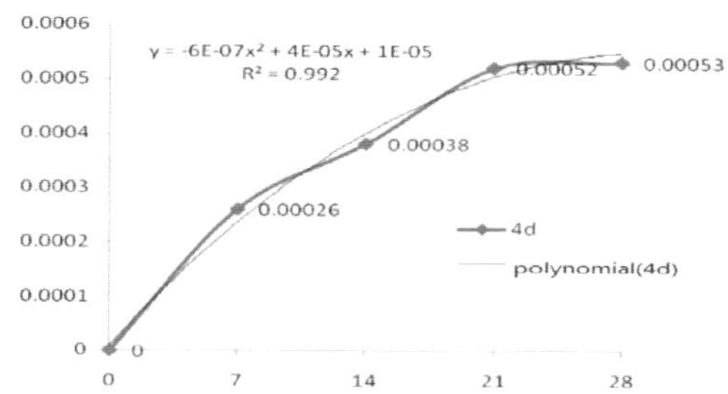

Figure 11. Fitting curve of age x and shrinkage rate y under 4d curing condition.

Table 2. Reduction rate of shrinkage.

curing days\days	7d	14d	21d	28d	35d
1d	100%	100%	100%	100%	100%
2d	104.35%	92.31%	101.96%	100.00%	100.00%
4d	113.04%	97.44%	101.96%	100.00%	101.89%

c. Test results of acrylic emulsion mortar under different curing humidity

The correlation formula between the shrinkage rate and age of acrylic emulsion mortar was obtained by numerical fitting from the test data under different curing humidity during drying.

Using polynomial fitting method, the correlation formulas of age x and shrinkage rate y under curing conditions of 40%, 75% and 90% humidity are obtained respectively.

40%: y = -6E-07x2 + 4E-05x + 2E-05
75%: y = -5E-07x2 +3E-05x +5E-06
90%: y = -3E-07x2 + 2E-05x -1E-06

The shrinkage curve and three fitting curves of mortar under different curing time are as Figure 12-15. The dry shrinkage rate is as Table 3.

IOP Conf. Series: Earth and Environmental Science **865** (2021) 012007 doi:10.1088/1755-1315/865/1/012007

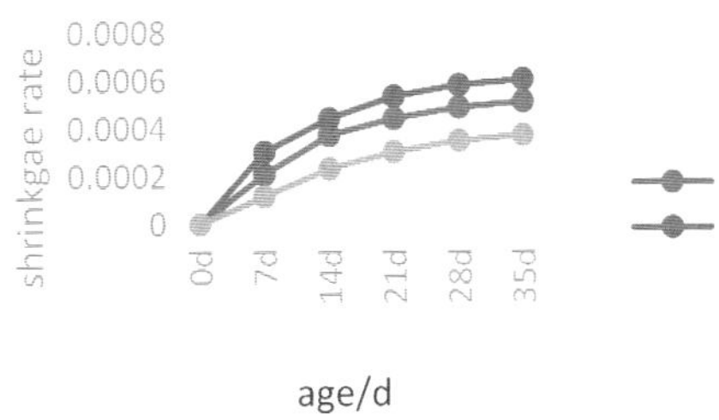

Figure 12. Shrinkage curve of mortar with different curing time.

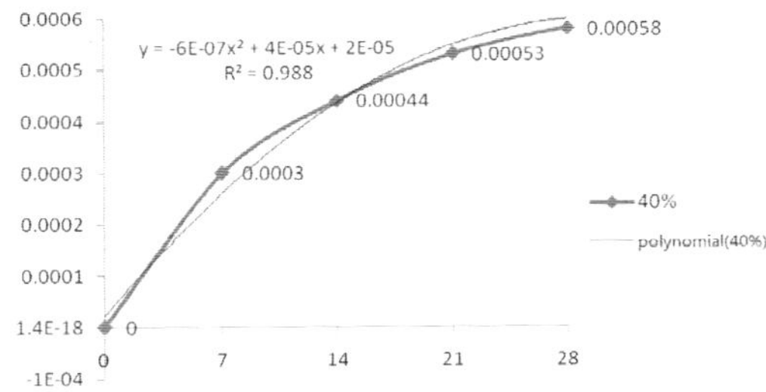

Figure 13. Fitting curve of age x and shrinkage rate y under 40% humidity curing condition.

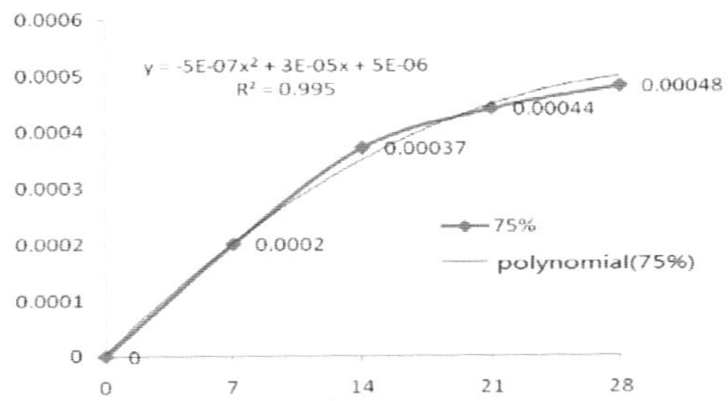

Figure 14. Fitting curve of age x and shrinkage rate y under 75% humidity curing condition.

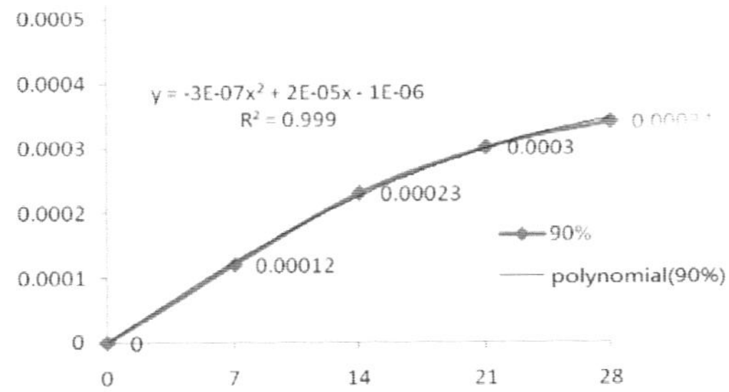

Figure 15. Fitting curve of age x and shrinkage rate y under 90% humidity curing condition.

Table 3. Reduction rate of shrinkage.

humidity\days	7d	14d	21d	28d	35d
40%	100%	100%	100%	100%	100%
75%	66.67%	84.09%	83.02%	82.76%	85.00%
90%	40.00%	52.27%	56.60%	58.62%	61.67%

4.3. Implementation on-site

According to the data provided by the laboratory, the designed repair method is improved in the construction details, time nodes, construction technology, and construction process. Repair the surface defects of concrete shedding of the pier, culvert body and roof of Dongchi Sluice from the following aspects: cutting off the old concrete erosion layer before construction, mixing acrylate mortar, layering paving when mortaring, the culvert divided into construction unit blocks, and the mortar surface immediately covered with plastic film after the wiping off [5]. On May 1, 2016, the construction of the acryl emulsion mortar for reinforcement of Dongchi Sluice began; on June 1, 2016, the construction of the acryl emulsion mortar project was completed; January 11, 2017, it was passed the unit engineering acceptance; May 25, 2017, there was no cracks appeared after one year.

5. Project test conclusion

1. The water-cement ratio is one of the main parameters affecting the shrinkage and deformation of acrylic emulsion cement mortar. Reducing the water-cement ratio can reduce the shrinkage and tensile stress of the acrylic emulsion cement mortar, thereby improving the crack resistance of the mortar. The water-cement ratio is recommended to be 0.35-0.45 during construction.

2. Temperature influence analysis: When the curing temperature of acrylic emulsion mortar is more than 20℃, the shrinkage rate gradually decreases. In view of the permitting degree of construction conditions, the temperature of construction and maintenance should be controlled at 30℃±5℃.

3. Time influence analysis: The curing time of acrylic emulsion mortar is mainly concentrated in the 1 to 3 days of curing before drying. Therefore, early curing measures must be taken for acrylic emulsion mortar to effectively control the early shrinkage cracks of the mortar.

4. Humidity influence analysis: When the acrylic mortar is dried, the curing humidity has the greatest influence on the cracking of the acrylic mortar. As the curing humidity increases during drying, the drying shrinkage rate is significantly reduced. The drying shrinkage rate drops to a minimum on the 7th day, reaching the 40%, and after 35 days it decreases to 62%. The effect of curing humidity on drying shrinkage during drying is greater in the early period of the curing period, and gradually smaller in the later period. When the curing humidity is above 90%, the shrinkage is the smallest. During the construction process, the curing humidity during drying should be kept above 75%.

6. Benefit analysis

6.1. Economic benefit

Through unit price analysis, the unit price of traditional acrylate mortar surface protection is 199.68 yuan/m^2 (double acrylate mortar). According to the research results of the key technology of anti-cracking of acrylic emulsion mortar in this project, by optimizing the proportion of acrylic emulsion mortar and selecting specific conditions for construction and maintenance, and repairing the concrete carbonization and erosion surface, it can reduce the drying shrinkage rate of the acrylic emulsion mortar and improve the cracking situation. The unit price of the improved acrylic emulsion mortar is 154.68 yuan/m^2. The unit price saving rate is 22.54%[(199.68-154.68)×100% /199.68] or more.

6.2. Social benefit

The research results of this project provide a lot of reference data and basis for the construction of acrylic emulsion mortar surface protection in the future. The improved construction technology and quality control measures are guiding for the construction of acrylic mortar in the future. More importantly, the cracking phenomenon of acrylic emulsion mortar surface protection has been effectively solved. While reducing the cost, it has improved the flood control strength of the culvert and sluice and prolonged the overall service life of the culvert and sluice. Therefore, the social benefits of the key technology of anti-cracking of acrylic emulsion mortar are outstanding and obvious in the reinforcement of culverts and sluices.

7. Conclusion

1. This project mainly solves the problem that acrylic emulsion mortar is easy to produce cracks in the culvert and gate reinforcement project.

2. The results of this research have found the key factors affecting the cracking of acrylic mortar during construction. The shrinkage performance under standard curing conditions was compared with that under natural conditions, and the shrinkage performance of acrylic emulsion mortar in the repair and reinforcement of culvert and sluice was tested and analyzed.

3. Through repeated periodic tests, after extracting the data of the standard curing environment and natural environment, the drying shrinkage rate was calculated, and the data was extracted by polynomial fitting method, and the limit of construction (and curing) temperature, humidity and time of acrylic emulsion mortar were obtained.

4. According to professional laboratory tests and on-site application in the Dongchi sluice, the key technology of anti-cracking of acrylic emulsion mortar has achieved economic benefits while also achieving social benefits. In the future, the key technology of anti-cracking of acrylic emulsion mortar will be applied in the culvert sluice reinforcement project in the Yellow River Basin. The promotion and application prospects of key technologies are very broad.

References

[1] Yuexiong Yu 2015 Application of Acrylic Emulsion Mortar in Miaozitou Hydropower Station

IOP Conf. Series: Earth and Environmental Science **865** (2021) 012007 doi:10.1088/1755-1315/865/1/012007

Repair *Water conservancy and hydropower construction* **1** 31, 40

[2] Wei Liang and Aiying Qiao 2016 Application of Acrylic Emulsion Mortar Plastering Technology in the Reinforcement Project of Dongpinghu Sluice on the Yellow River in Shandong *Inner Mongolia Water Resources* **8** 50

[3] Weidong Liu, Zhiguang Zhao and Wendong Yang 2002 Experimental Study on Proerties of Arcylic-sand Mortar *Jurnal of Hydraulic Engineering* **5** 43-46

[4] Xuan Xun and Xin Zhang 2013 Experimental Study on the Thermal Expansion Coefficient of Acrylic-emulsion Cement Mortar at Early and Hardened Ages *Structural Engineer* **4** 172-177

[5] Cheng Li 2016 Application of Acrylic Emulsion Mortar in Water Conservancy Project Rehabilitation with Concrete *Heilongjiang Hydraulic Science and Technology* **10** 85-88

Analysis of volatile components in host plants of *Xylotrechus quadripes*

Zijun Luo[1], Yingzhi Chen[1], Liping Liu[1], Genyun Pan[1] and Ruifang Wang[1, 2]

[1]Pu'er University, Pu'er Yunnan 665000

[2]Corresponding author's email: peurdr_wang@foxmail.com

Abstract. *Xylotrechus quadripes* has a wide range of host plants. As a drillstem pest, it erodes the xylem of the trunk, resulting in the reduction of coffee production, and the control process is also facing difficulties. In this study, solid phase microextraction and gas chromatography were used to analysis the volatile components of host plants of *X. quadripes*. The results showed that 36, 23, 46 and 40 kinds of volatiles were identified from the four plants, mainly terpenes, some esters, alcohols and ketones. The terpenes were mainly monoterpenes and sesquiterpenes. The same volatiles of four plants were (+)-limonene, styrene, β-pinene, (-)-α-muurolene, (-)-α-cubebene and β-caryophyllene, which provides a basis for the prevention and control of *X. quadripes* using plant volatiles as attractants.

1. Introduction

Yunnan coffee has a unique natural environment. After more than 60 years of development, Yunnan Province has become a major coffee producing area in China. There are more than 100 varieties of coffee planted in Yunnan Province. *Acalolepta cervina* and *Xylotrechus quadripes* are the main drillstem pest of coffee in Yunnan Province, the latter is difficult to control. Longicorn is very active in bright sunlight and likes to lay its eggs on the main stem of the exposed coffee tree. Therefore, in the process of planting coffee in Yunnan, intercropping with other crops or planting shade plants are often used to provide shade for coffee trees, reduce the harm of longicorn pests and ensure stable and high yield of coffee. The main shade tree species are *Ricinus communis* L., *Alstonia scholaris* (L.)R. Br., *Artocarpus heterophyllus* Lam., *Adina pilulifera* (Lam.) Franch. ex Drake and *Psidium guajava* Linn. et al.

Plant volatiles, called herbivore induced plant volatiles (HIPVs), play an important role in attracting natural enemies. The main components of HIPVs are monoterpenes or sesquiterpenes. Plant volatile substances, also known as plant aromatic substances, mainly include alkenes, alcohols, aldehydes, terpenes and other organic components containing benzene or aromatic rings. Studies have shown that plants release specific volatile compounds to attract longicorn, terpenes and alkanes are the main components.The control methods of longicorn include intercropping shade trees, scraping egg marks, chemical control, catching adults artificially, cleaning coffee garden and so on.It is urgent to study the volatile compounds of host plants and secondary host plants, which has an important prospect in the development of green agriculture in the future. By studying the types of volatile compounds among plants, we can expand the application of plant source pheromones in pest control. One method is to control the localization of the host plant volatiles by insects, and the other is to apply exogenous plant volatiles to attract pests. In order to achieve the goal of green pest control.

2. Research materials and methods

2.1. Research materials

2.1.1. Materials select C. arabica, A. heterophyllus, A. scholaris and *P. guajava* branches with good growth. Before sunrise, cut the branches with a clean shears and taken it back to the laboratory immediately soak them wounds in deionized water and collect volatile compounds by SPME. The sampling site was Xishuangbanna Tropical Botanical Garden, Chinese Academy of Sciences.

2.1.2. Research instruments. Agilent Techologies GC-MS 5977B gas chromatography mass spectrometry, Manual solid phase microextraction injector, 65μm PDMS/DVB Solid phase microextraction head.

2.2. Research method

2.2.1. Collection of volatile substances. The volatile compounds were collected by solid-phase microextraction. For volatile components, the equilibrium time of headspace extraction is shorter under the same sample mixing conditions [1].The specific operation method is as follows: cover the branch with transparent polyethylene bag, tie up and fix the upper and lower ends tightly, then insert the manual SPME injector with 65 μm PDMS/DVB solid-phase microextraction head and extract for 30 minutes. At the end of sampling, the sample was injected into the chromatograph for 0.5 min, and the injection time was 0.5min.

2.2.2. GC-MS Analysis conditions. The operation time of GC was 50.5 min, the initial temperature of column box was 50 ℃ and the holding time was 2 min. Firstly, the temperature was raised to 230 ℃ every minute at 4 ℃ for 0 min, and then to 280 ℃ every minute at 20 ℃ for 1min.Equilibrium time is 0.25 min, max temperature is 325 ℃. Front SS inlet He non-shunt injection.Heater is 250 ℃, pressure 7.6522 psi, total flow 64 mL/min, spacer purge flow 3 mL/min, split outlet purge flow 60 mL/min maintain 1 min. GSV quantitative tube volume 1 mL, sample loading 0.5 min, injection 0.5 min. Chromatographic column flow rate 1mL/min, Chromatographic outlet pressure 0 psi. MSD initial temperature is 50 ℃ pressure 0.5735 psi flow rate is 1 mL/min Average linear velocity is 36.445 cm/sec, residence time is 1.3719 min.MSD output and MSD transmission line is 250 ℃. MS ion source is 230 ℃, four lever is 150 ℃.

2.3. Volatile matter analysis and screening

The mass spectrum data were obtained by GC-MS analysis and searched in NIST08. L library to determine the volatile components. The relative content was calculated by peak area normalization method with the all peak area as the index. The siloxanes and a small amount of impurity peaks produced by the heating of the solid phase microextraction head, and the components with similarity less than 90 were eliminated.

3. Discussion and conclusion

3.1. Discussion

Ding Jiawen et al. and others used GC-MS technology to analyze 47 kinds of volatile components in the flower of *Elaeagnus angustifolia*. Among them, dimethyl phthalate, isopropyl myristate, ethyl cinnamate, decanoate and linalool had better attractant activity to longicorn beetle under appropriate ratio [2]. Wang Baoxin et al. studied the volatile components of healthy *Pinus massoniana* Lamb. By using gas chromatography-mass spectrometry, seven components were screened out, including (1R) - (+) - α - pinene, β - pinene, myrcene, Terpinene, isohexene, myrcene, (1R) - (+) - α - longleaf pinene and caryophyllene. *Monochamus alternatus* had strong reaction to (1R) - (+) - α - pinene in healthy *P.*

massoniana [3]. Among the attractants prepared by Wei Dan [4] and others, the attractant containing *Elaeagnus angustifolia*, (E)-2-hexenal, (z) - 3-Hexen-1-ol was the best. Ma Yan et al. used dynamic headspace method to investigate the effect of Juglans regia on the behavioral response of *Apriona germari*. The results showed that terpenes and aromatic compounds were the most of the volatiles released from *Juglans mandshurica* under different damage degrees, and there were significant differences between them. *A.germari* had physiological and electrical responses to pinene, Folin acetate, nonanal, α - terpineol, 2-ethylhexyl acrylate and n-hexadecane released from Carya cathayensis under the damage state [5].

3.2. Conclusion

Four kinds of plant volatile compounds were collected under SPME conditions, and the total ion flow chromatogram of volatile compounds was obtained by GC-MS analysis.The volatile compounds were identified by GC-MS and NIST08. L library. Table 1 show the results of volatiles from experimental plants. A total of 79 volatile compounds were identified, including 37 terpenes, 20 esters, 6 alcohols, 5 olefins, 3 ketones and 8 alkanes.There are 8 kinds of the same volatiles in the four species (+)-Limonene, β-Caryophyllene, (-)-α-cubebene, (-)-α-muurolene, (-)α-Copaene, β-pinene, Styrene,and Methyl 2-ethylhexanoate.

Plants attract insects by volatilizing specific volatiles. At the same time, insects also identify and locate plants by volatiles, thus forming a symbiotic relationship. The volatiles of plants are mainly composed of some small molecular weight substances, such as alcohols, esters, terpenes, etc., which constitute their special marker volatiles in a certain proportion. The results showed that the relative content and species of terpenes were very rich in the five species. Terpenes were important products in plant secondary metabolism. They were mainly composed of isoprene, which were divided into chain and ring structures. According to the number of isoprene, they were divided into monoterpenes, diterpenes, triterpenes, sesquiterpenes and so on The terpenes in the species are monoterpenes and sesquiterpenes.Under the coffee cultivation mode, the volatile compounds of plants in the long-term growth process are similar, which has a certain impact on the selective feeding of insects. Further field experiments are needed to identify whether the selected volatile compounds are attractive to insects.

Table 1. Summary of volatile compounds in experimental plants.

retention time(min)	Name	Molecular formula	Type	Relative content (%)			
				C.a	*A.h*	*P.g*	*A.s*
13.69	Limonene	$C_{10}H_{16}$	Terpenes	--	28.34	2.13	37.39
13.83	(+)-Limonene	$C_{10}H_{16}$	Terpenes	29.37	58.36	6.02	58.36
10.65	S-(-)-Limonene	$C_{10}H_{16}$	Terpenes	--	0.61	--	--
28.28	β-Caryophyllene	$C_{15}H_{24}$	Terpenes	21.14	2.40	30.34	31.01
27.06	α-Panasinsanene	$C_{15}H_{24}$	Terpenes	--	--	0.25	0.19
22.35	(+)-Cyclosativene	$C_{15}H_{24}$	Terpenes	--	--	0.10	--
33.05	Caryophyllene	$C_{15}H_{24}$	Terpenes	--	--	0.54	0.41
25.24	(-)-α-Cubebene	$C_{10}H_{16}$	Terpenes	0.62	0.63	0.24	26.24
27.02	(-)-α-Gurjunene	$C_{15}H_{24}$	Terpenes	--	--	0.85	0.36
30.11	(-)-α-Muurolene	$C_{15}H_{24}$	Terpenes	0.34	0.24	0.32	0.40
8.37	(1R)-(+)-α-Pinene	$C_{10}H_{16}$	Terpenes	0.84	0.69	0.06	--
25.56	(-)-α-Copaene	$C_{15}H_{24}$	Terpenes	1.11	0.84	8.41	1.92
15.23	(E)-3,7-dimethylocta-1,3,6-triene	$C_{10}H_{16}$	Terpenes	--	--	0.27	--

26.77	β-Bisabolene	$C_{15}H_{24}$	Terpenes	--	--	0.17	0.09
15.00	(Z)-β-Ocimene	$C_{10}H_{16}$	Terpenes	--	2.01	0.48	--
29.49	(Z)-α-Bisabolene	$C_{15}H_{24}$	Terpenes	0.21	--	1.12	0.73
26.29	(-)-Isocaryophyllene	$C_{15}H_{24}$	Terpenes	--	--	0.60	--
27.19	(−)-allo-Aromadendrene	$C_{15}H_{24}$	Terpenes	--	--	0.71	0.51
26.34	α-Selinene	$C_{15}H_{24}$	Terpenes	3.72	--	8.22	4.01
28.35	α-Caryophyllene	$C_{15}H_{24}$	Terpenes	1.28	--	3.34	1.23
10.77	α-Pinene	$C_{10}H_{16}$	Terpenes	--	1.16	0.11	0.30
29.01	Azulene	$C_{10}H_{8}$	Olefins	1.71	0.75	0.99	--
15.05	γ-Terpinene	$C_{10}H_{16}$	Terpenes	--	--	0.16	0.06
9.77	α-Terpinene	$C_{10}H_{16}$	Terpenes	--	--	--	0.29
12.24	Cyclooctatetraene	$C_{8}H_{8}$	Olefins	1.33	--	--	0.65
30.20	(+)-Aromadendrene	$C_{15}H_{24}$	Terpenes	1.64	--	11.73	--
23.53	α-Longipinene	$C_{15}H_{24}$	Terpenes	--	--	--	0.08
12.22	(+)-Camphene	$C_{10}H_{16}$	Terpenes	0.25	--	--	--
19.45	Styrene	$C_{8}H_{8}$	Olefins	0.63	0.76	0.22	0.38
12.36	(-)-Camphene	$C_{10}H_{16}$	Terpenes	0.31	--	--	--
37.77	(-)-Calamenene	$C_{15}H_{22}$	Olefins	0.35	0.16	0.56	--
35.27	3,6,6-Trimethyl-bicyclo(3.1.1)hept-2-ene	$C_{10}H_{16}$	Terpenes	--	1.03	--	--
23.80	Triene neo-alloocimene	$C_{10}H_{16}$	Terpenes	--	0.24	0.10	--
28.43	(-)-β-Bourbonene	$C_{15}H_{24}$	Terpenes	--	--	0.28	--
10.81	Dodecane	$C_{12}H_{26}$	Alkenes	1.08	--	--	--
28.86	Tridecane	$C_{13}H_{28}$	Alkenes	0.42	--	--	--
23.46	Tetradecane	$C_{14}H_{30}$	Alkenes	1.88	--	--	0.15
26.47	Pentadecane	$C_{15}H_{32}$	Alkenes	1.31	0.35	--	0.30
29.40	Hexadecane	$C_{16}H_{34}$	Alkenes	0.59	--	--	--
32.18	N-heptadecane	$C_{17}H_{36}$	Alkenes	0.34	--	--	0.30
49.06	Icosane	$C_{20}H_{42}$	Alkenes	0.06	--	--	0.05
30.17	β-Cubebene	$C_{15}H_{24}$	Terpenes	--	--	0.21	0.23
16.02	β-Myrcene	$C_{10}H_{16}$	Terpenes	--	--	0.65	0.15
17.11	1,5-dimethyl-1,5-cyclooctadiene	$C_{10}H_{16}$	Terpenes	--	--	6.44	--
19.81	m-cymene	$C_{10}H_{14}$	Olefins	--	0.23	--	--
37.06	Diisobutyl phthalate	$C_{16}H_{22}O_{4}$	Esters	4.45	--	0.60	2.30
13.60	Methyl 2-ethylhexanoate	$C_{9}H_{18}O_{2}$	Esters	2.72	0.21	0.06	0.22
14.45	Methyl 2-ethylheptanoate	$C_{10}H_{20}O_{2}$	Esters	0.45	--	--	0.23
15.53	Ethyl benzoate	$C_{9}H_{10}O_{2}$	Esters	5.16	--	--	--
18.21	Ethyl phenylacetate	$C_{10}H_{10}O_{2}$	Esters	3.15	--	--	--
39.65	Ethyl palmitate	$C_{18}H_{36}O_{2}$	Esters	2.80	--	--	--
34.71	Ethyl tetradecanoate	$C_{16}H_{32}O_{2}$	Esters	2.46	--	--	--
29.28	Ethyl laurate	$C_{14}H_{28}O_{2}$	Esters	0.71	--	--	--

16.38	Methyl salicylate	$C_8H_8O_3$	Esters	0.60	--	--	--
7.45	Methyl 2,4-dimethylpentanoate	$C_8H_{16}O_2$	Esters	--	--	--	0.59
21.72	Ethyl 3-phenylpropanoate	$C_{11}H_{14}O_2$	Esters	0.85	--	--	--
23.25	Ethyl caprate	$C_{12}H_{24}O_2$	Esters	0.31	--	--	--
19.99	Ethyl nonanoate	$C_{11}H_{22}O_2$	Esters	0.28	--	--	--
13.46	Phenylethyl alcohol	$C_8H_{10}O$	Esters	1.03	--	--	--
13.45	Oct-1-en-3-ol	$C_8H_{16}O$	Esters	--	--	--	0.74
9.56	Decane	$C_{10}H_{22}$	Alkenes	--	1.13	0.25	0.73
21.24	3,4-Dimethylacetophenone	$C_{10}H_{12}O$	Ketones	0.62	--	--	--
48.91	2-Hexyl-1-decanol	$C_{16}H_{34}O$	Esters	0.07	--	--	--
28.77	Linalool	$C_{10}H_{18}O$	Esters	0.98	--	--	--
11.56	(-)-β-Pinene	$C_{10}H_{16}$	Terpenes	--	1.62	0.27	1.11
10.99	β-Pinene	$C_{10}H_{16}$	Terpenes	2.10	2.04	0.40	1.10
32.71	(+)-Calarene	$C_{15}H_{24}$	Terpenes	--	--	0.53	--
26.93	2-Ethylhexanol	$C_8H_{18}O$	Esters	--	--	0.54	1.10
11.75	Acetophenone	C_8H_8O	Ketones	0.78	--	--	--
35.12	(+)-Carvone	$C_{10}H_{14}O$	Ketones	0.51	--	--	0.24

Note:
C.a: *Coffea arabica* L.; *A.h*: *Artocarpus heterophyllus* Lam.; *P.g*: *Psidium guajava* Linn.; *A.s*: *Alstonia scholaris* (L.) R. Br.; --: mans the volatile compounds has not been

References

[1] Xiao Dan 2015 Application and Prospect of headspace solid phase microextraction [J]. *China Health Engineering* **14(01)** 88-92

[2] Ding Jiawen, Liu Qiang, Zhu Gengping, et al. 2014, EAG and behavioral responses of Cerambycidae hongyuanensis to eight compounds in Elaeagnus angustifolia [J]. *Journal of Tianjin Normal University (NATURAL SCIENCE EDITION)* **34(02)** 71-75

[3] Wang Baoxin, Xu Xiulan, Xiao Qiangang, et al. 2019 EAG response of Monochamus alternatus to volatiles from healthy Pinus massoniana [J]. *Agricultural staff* **(19)** 103-104

[4] Wei Dan, Qin Qin, Xun Feng, et al. 2012 EAG and behavioral response of longicorn beetle to different attractants and field trapping effect [J]. *Journal of Tianjin Normal University (NATURAL SCIENCE EDITION)* **32(03)** 71-74 + 95

[5] Ma Yan, Shi Liyang, Zhao Yi, et al. 2018 Comparison of volatile components of Carya cathayensis under different damage conditions and GC-EAD and behavioral response of Apriona germari to its components [J]. *Acta Entomologica Sinica* **61(05)** 574-584

Experimental study on the influence of injection advance angle to diesel engine equipped with electronic unit pump

Yupo Ma[1], Yu Feng, Zengbin Wu, Changcheng Wang, Tongzhou Liao, Zhanyu Zhan, Zhenyu Zuo and Guoning Shi

China North Vehicle Research Institute, Beijing 100072, China

[1]Corresponding author's e-mail: mayupo2015@163.com

Abstract. The fuel injection advance angle of electronic unit pump fuel injection system has a great influence on the running performance of diesel engine as well as the energy consumption, environmental and global warming issues. For a supercharged diesel engine equipped with electronic unit pump, the influence of injection advance angle on the injection characteristics, specific fuel consumption, combustion characteristics and reliability have been studied in this paper, which has important reference significance for the optimization of injection strategy of diesel engine.

1. Introduction

The problem of diesel emission and efficiency has always been the focus in tackling the energy consumption, environmental and global warming issues [1-2]. Fuel injection system plays one utmost important role in diesel engine's general performance, high-pressure direct injection associated with electronic control system has been widely regarded as one promising injection manner to modern diesel engines for its advantages in dynamic power and raw emissions [3-5]. In all diesel fuel injection systems, EUP system has variable and flexible injection timing and high injection pressure, it is one of the main technologies to increase power per engine displacement and meet emission requirements [6-9].

The fuel injection advance angle has a great influence on the running status of diesel engine. If the fuel injection advance angle is too large, it will lead to a longer preparation period, which will lead to rough engine operation [10-11]. However, if the injection advance angle is too small, the combustion process will be delayed too much, and the maximum pressure value will drop, which will obviously reduce the thermal efficiency of diesel engine. The best injection advance angle can help obtain the maximum power and the minimum fuel consumption rate under the condition of constant speed and oil delivery [12].

The optimal injection advance angle of diesel engine changes with oil delivery and crankshaft speed, and increases with the increase of diesel engine speed [13-14]. In addition, the structure of diesel engine also has a certain influence on the optimal injection advance angle. For example, the optimal injection advance angle is different with different combustion chambers; Generally, the best injection advance angle of diesel engine with direct injection combustion chamber is larger than that with separated combustion chamber [15].

In this paper, a V-type eight-cylinder supercharged diesel engine matched with electronic unit pump fuel system is taken as the research object. Firstly, the influence of injection advance angle on injection characteristics have been studied. Then, the change trend of specific fuel consumption and combustion characteristics with injection advance angle have been analyzed. Finally, the influence of injection advance angle on the reliability and working safety of diesel engine have been discussed, which has important reference significance for the optimization of injection parameters of diesel engine.

2. Experiment facilities

As a typical time controlled fuel injection system, as shown in Figure 1, the electronic unit pump system has good working characteristics, including flexible and controllable injection timing and quality, more prominently high injection pressure and better working reliability. The normal injection pressure is between 160Mpa and up to 220MPa. Electronically controlled fuel injection works: the components of EUP fuel system are driven by a cam, and the movement of plunger is determined by cam contour. When the plunger moves upward, the solenoid valve is energized and closed, the passage between the high-pressure and low-pressure oil chambers will be cut off, and the oil will be pressed down by the plunger. The high-pressure fuel opens the outlet valve, enters the fuel injector through the high-pressure oil pipe, and is injected into the cylinder through the atomizing nozzle; When the solenoid valve is disconnected, the oil chamber between high pressure and low pressure is connected, the fuel is unloaded quickly, and the nozzle stops injection.

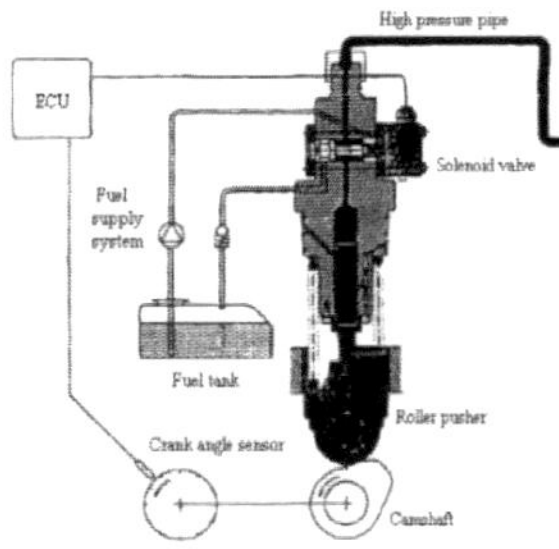

Figure 1. Electronic control unit pump diagram.

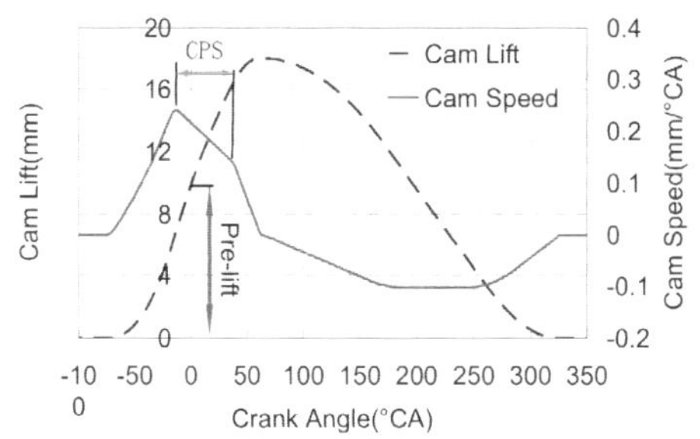

Figure 2. The characteristics of deceleration cam.

A Delphi EUP system was used in this study and its main parameters are shown in Table 1. In order to measure the injection characteristics a Kistler pressure sensor was placed on the high-pressure tube and a needle lift sensor was installed in the injector.

Table 1. Delphi EUP ststem main parameters.

plunger diameter	open injection pressure	injector hole number	Injector hole diameter
11mm	28MPa	8	0.27mm

As analyzing the EUP injection characteristics with the injection advance angel, combining the cam contour to the fuel supply system is very necessary. One kind of deceleration cam has been designed by Qiu Tao, Liu Xing-hua and Liu Fu-shui from Beijing Institute of Technology based on the research of the EUP oil supply system and cam profile matching characteristics. It realized the balance between fuel injection capacity and fuel supply ability [16-18]. Consequently, the fuel delivery per cycle and maximum injection pressure will not change dramatically with the injection advance angle. In this situation, More flexible adjusting methods can be provided for the match work between EUP system and diesel engines. The profile lift and speed curves of the deceleration cam used in this paper are shown in Figure 2.

The cam lift at this moment when the piston arrives at the TDC, is defined as Pre-lift and the segment of deceleration is regarded as working segment or constant pressure segment. At the designed speed (2200r/min crankshaft speed), the fuel delivery per cycle and maximum injection pressure will not vary with different injection advance angles. Theoretically, if the crankshaft speed is smaller or larger than the designed value, the fuel supplying ability will be less or mote than the fuel injection

IOP Conf. Series: Earth and Environmental Science **865** (2021) 012009 doi:10.1088/1755-1315/865/1/012009

capacity. Therefore, the fuel delivery per cycle and the maximum injection pressure will change as the injection advance angle moves. Moreover, due to some uncertain factors, such as velocity fluctuation and pressure oscillation in high-pressure pipe, it is necessary to study the actual injection characteristics experimentally

The test prototype is a V-type eight-cylinder supercharged diesel engine, and its main parameters are shown in Table 2.

Table 2. The diesel engine main parameters.

Bore diameter (mm)	132
Stroke (mm)	145
Compression ratio	14:1
Combustion chamber	double swirl
Ignition sequence	1-8-4-5-7-3-6-2
Start injection pressure (MPa)	20
Nozzle number	8
Nozzle diameter (mm)	0.27
Max torque (N.m)	2415@(1800r/min)
Max power (kW)	500@(2100r/min)

A large number of sensors are installed on the test bench to monitor the state of the engine, Kistler pressure sensors measure the in-cylinder pressure, oil supply pressure in the high-pressure oil pipe, oil pressure and intake and exhaust pressure; The needle valve lift sensor monitors and adjusts the injection position; K-type temperature sensor measures inlet and outlet temperature and inlet and outlet water temperature, etc. The fuel consumption rate of engine is measured by using the Tongyuan transient fuel consumption meter. DEWECA combustion analyzer is used to calculate in-cylinder heat release rate and display combustion pressure curve, oil supply pressure curve and needle valve lift. All measurement results have been calibrated and checked, and the errors are within the reasonable range.

3. Experiment results and analysis

In this part, Firstly, the influence of injection advance angle on injection characteristics have been studied. Then, the change trend of specific fuel consumption and combustion characteristics with injection advance angle have been analyzed. Finally, the influence of injection advance angle on the reliability and working safety of diesel engine have been discussed.

3.1. Influence of injection advance angle on fuel injection characteristics

The experimental results of the EUP with the crankshaft speed 1800r/min and 2500r/min are shown in Figures 3 to 6. where Figure 3 and Figure 4 are the experimental results of injection pressure, and Figure 5 and Figure 6 are the experimental results of the maximum injection pressure and the fuel delivery per cycle, respectively.

IOP Conf. Series: Earth and Environmental Science **865** (2021) 012009 doi:10.1088/1755-1315/865/1/012009

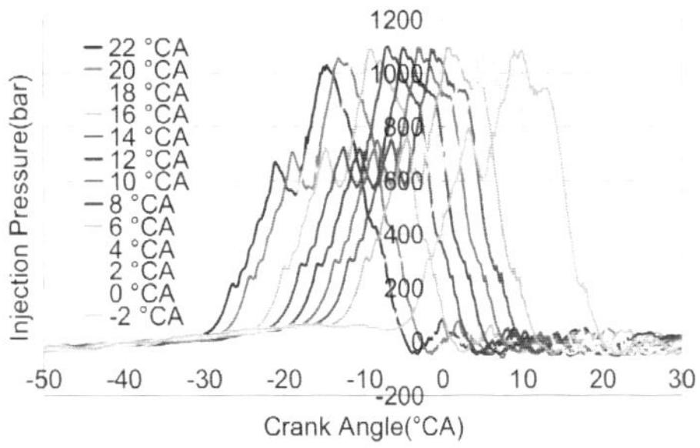

Figure 3. Injection pressure (1800r/min).

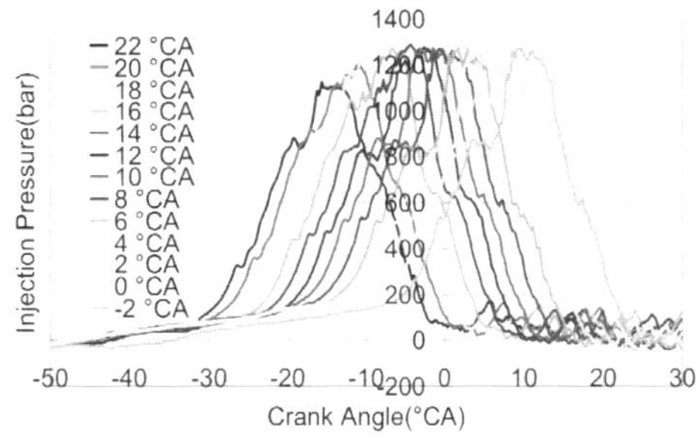

Figure 4. Injection pressure (2500r/min).

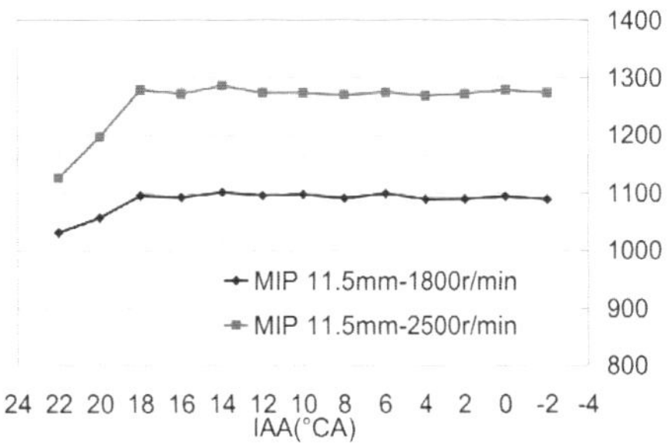

Figure 5. The maximum injection pressure.

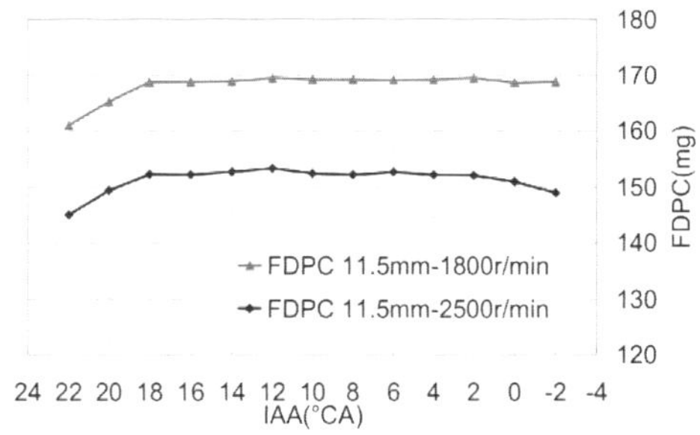

Figure 6. The fuel delivery per cycle.

As shown in Figure 5 and Figure 6, with both the crankshaft speed of 1800r/min and 2500r/min, the injections start too early and the whole process of injection locate in front of the constant pressure segment when the injection advance angle is before 18° CA BTDC. Consequently, both of the fuel delivery per cycle and the maximum injection pressure show a clear downward trend. This is an abnormal working status and can cause more difficulties for the diesel engine mapping process. It is needed to note that for the smaller injection advance angle in the case of 2500r/min, the fuel delivery per cycle begins to drop, although the maximum injection pressure keep the same which did not appear in the case of 1800r/min. It is because that the injection pressure decrease in the late stage of injection process. In other words, although the early injection stage happens in the constant pressure segment, the later stage goes beyond that. Therefore, with the pre-lift increases, the maximum critical value of injection advance angle will increase accordingly. Moreover, a minimum critical value of injection advance angle can be found, if the injection begins after the minimum critical one, the fuel delivery per cycle will decrease, and that happens a little more early than that of the maximum injection pressure. Additionally, the injection duration has a direct impact on the minimum critical value: it gets smaller with a shorter injection duration, which means the fuel delivery per cycle can keep the same in a wider range of the injection advance angle.

3.2. Influence of injection advance angle on fuel consumption and combustion characteristics

Figure 7 to Figure 9 show the test results of fuel consumption, combustion duration and E50 position with injection advance angle under different loads.

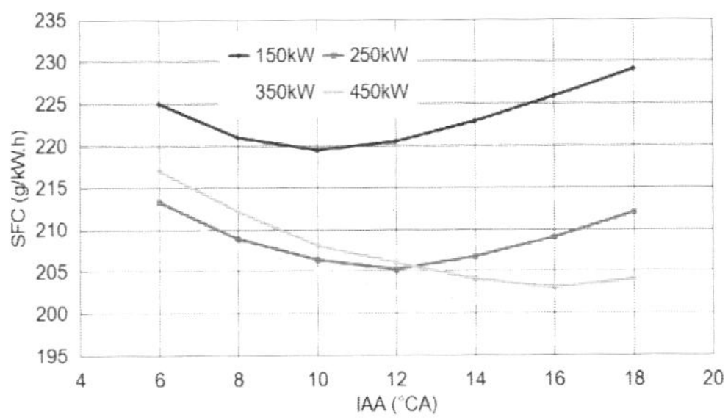

Figure 7. The test result of specific fuel consumption.

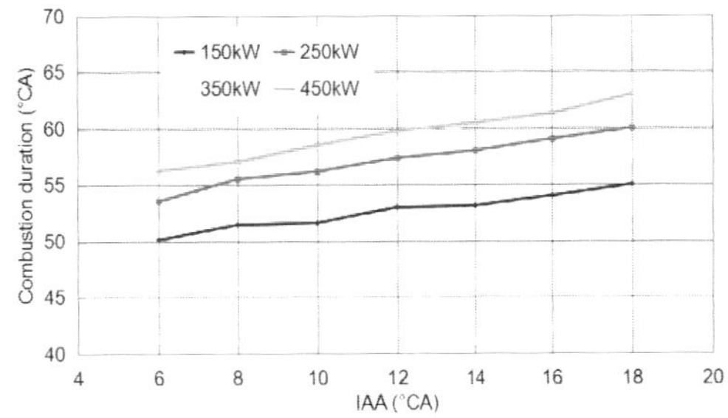

Figure 8. The test result of combustion duration.

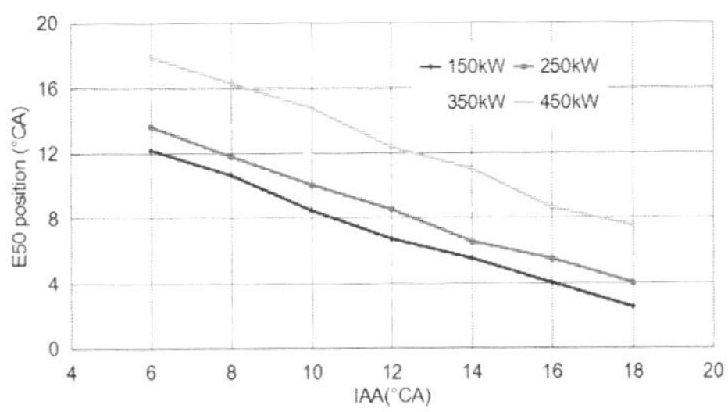

Figure 9. The test result of E50 position.

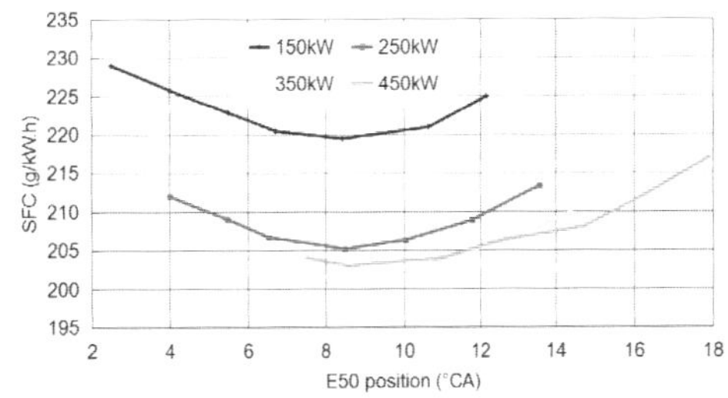

Figure 10. The relationship between specific fuel consumption and E50 position.

As shown in Figure 7, for different loads, with the increase of injection advance angle, the specific fuel consumption decreases first and then increases, that is, for each load, there is a most economical injection advance angle. With the increase of load, the most economical injection advance angle increases correspondingly. And when the load is 350kW, the best specific fuel consumption is the smallest, which shows that when the engine speed is constant, the specific fuel consumption decreases first and then increases with the increase of load. As shown in Figure 8, with the increase of the injection advance angle, the combustion duration increases linearly, and the combustion duration increases by about 0.4° CA for every 1° CA increase of the injection advance angle. And when the load is small, the relationship between combustion duration and load is obvious, that is, the greater the load, the more fuel injection and the longer the combustion duration. However, for medium and large loads, the combustion duration remains basically unchanged, because when the load increases, the combustion temperature in the cylinder increases, so the combustion speed increases and the time required for combustion remains basically unchanged. As shown in Figure 9, with the increase of injection advance angle, the position of E50 decreases, and for every 1° CA increase of injection advance angle, E50 decreases by about 0.8° CA. This is mainly because the fuel injection position is advanced and the combustion duration is slightly increased, so the change of E50 position is smaller than that of fuel injection advance angle.

The above three test results show that when the engine speed is constant, for different loads, the influence laws of fuel injection advance angle on fuel consumption, combustion duration and E50 position are basically the same, but the most economical fuel injection advance angle increases with the increase of load. Figure 10 is a graph showing the relationship between specific fuel consumption and the position of E50 under different loads.

As shown in Figure 10, when E50 appears at about 8.5° CA after TDC, the minimum specific fuel consumption appears for four different loads at 1800rpm. This shows that the position of E50 determines the combustion efficiency of the engine, and when the engine speed is constant, the optimal position of E50 is basically unchanged, which is of great significance for improving and optimizing the combustion efficiency and specific fuel consumption of the electronically controlled unit pump diesel engine.

3.3. Influence of injection advance angle on combustion pressure and exhaust temperature

The above research shows that the combustion duration and E50 position can be changed by adjusting the injection advance angle, and the specific fuel consumption can be optimized. However, the injection advance angle also affects the combustion pressure and exhaust temperature of the engine, and these two parameters have an important influence on the reliability of the engine. Excessive maximum combustion pressure will increase the mechanical load of the cylinder head, while excessive exhaust temperature will increase the heat load of the exhaust system. During the operation of the engine, they need to be strictly controlled and cannot exceed the specified limit for a long time. The maximum combustion pressure of the prototype can not exceed 170 bar, and the exhaust temperature before turbine can not exceed 700℃.

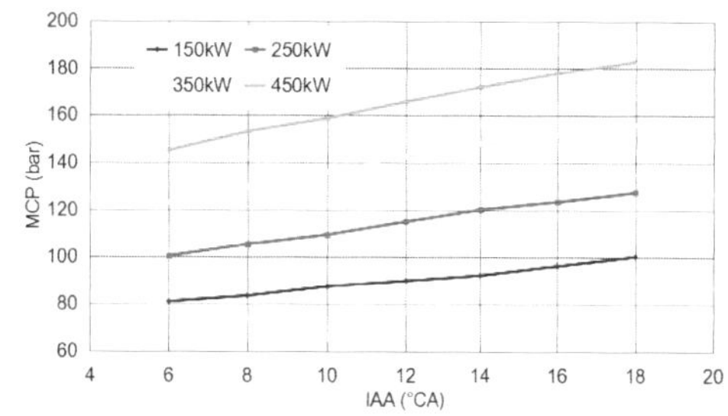

Figure 11. The test result of combustion pressure.

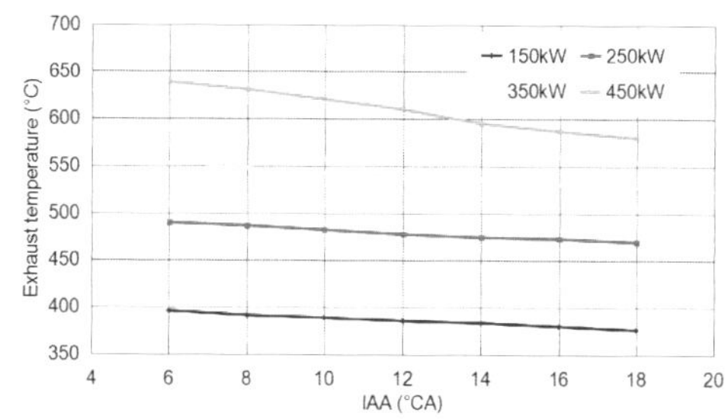

Figure 12. The test result of exhaust temperature.

Figure 11 and Figure 12 show the test results of maximum combustion pressure and exhaust temperature before turbine changing with injection advance angle. It can be found that under different loads, with the increase of injection advance angle,

the maximum combustion pressure and exhaust gas temperature before turbine basically change linearly. When the injection advance angle increases by 1° CA, the maximum combustion pressure increases by 2-3bar, and the exhaust gas temperature before turbine decreases by 2-5℃. The higher the load, the faster they change with injection advance angle. This is because the larger the injection advance angle is, the longer the post combustion will be, and the more fuel will be burned during the lag period, which will cause the increase of combustion pressure. At the same time, the smaller the proportion of post combustion, the lower the exhaust energy, so the exhaust temperature decreases gradually.

In addition, With the increase of injection advance angle, the exhaust temperature decreases slowly, and the maximum combustion pressure increases significantly. Therefore, it is difficult to reduce the exhaust temperature by adjusting the injection advance angle. If the exhaust temperature is too high, we should mainly consider increasing the intake pressure and reducing the intake temperature in order to increase the intake air volume, increase the excess air coefficient, shorten the combustion duration and reduce the exhaust energy. In addition, measures such as optimizing valve timing and increasing injection pressure can be taken.

4. Conclusion

For supercharged diesel engine equipped with electronic unit pump, Firstly the influence of injection advance angle on the injection pressure and fuel delivery per cycle have been studied. Then the change trend of specific fuel consumption, combustion duration and E50 position with injection advance angle have been analyzed. Finally the influence of injection advance angle on the maximum combustion pressure and exhaust temperature have been discussed. The main conclusions drawn from this research are:

(1) When the EUP works in the constant pressure segment, the fuel delivery per cycle and the

maximum injection pressure nearly keep constant, otherwise, both of them decrease dramatically with the injection advance angle.

(2) A minimum critical value of injection advance angle can be found, if the injection begins after the minimum critical one, the fuel delivery per cycle will decrease, and that happens a little more early than that of the maximum injection pressure. Additionally, the injection duration has a direct impact on the minimum critical value: it gets smaller with a shorter injection duration, which means the fuel delivery per cycle can keep the same in a wider range of the injection advance angle.

(3) Combustion duration, position of E50, maximum combustion pressure and exhaust temperature injection advance angle all nearly show linear trend with injection advance angle. The position of E50 determines the combustion efficiency of the engine, when E50 appears at about 8.5° CA after TDC, the minimum specific fuel consumption will be obtained.

(4) The injection advance angle has an important influence on the combustion efficiency, economy and safety of diesel engine. When the specific fuel consumption is optimized by adjusting the injection advance angle, the best injection advance angle can be determined according to the economy for medium and small loads, but for large loads, it is necessary to monitor the highest combustion pressure at the same time to prevent the reliability from dropping significantly.

References

[1] Yan J, Chou S K, Desideri U and Xia X 2014 Innovative and sustainable solutions of clean energy technologies and policies (Part I). *Appl Energy* **130** 447-79

[2] Chou S K, Chua K J, Ho J C and Ooi C L 2004 On the study of an energy-efficient greenhouse for heating cooling and dehumidification applications *Appl Energy* **77** 355-73

[3] Gao Q, Jiang W, Guo H, Ren G and Wu Y 2008 Design of Constant Velocity Fuel Cam for Electronically Controlled Unit Pump System *Vehicle Engine* **179(6)** 15-18

[4] Sun Z, Li G and Chen C et al 2015 Numerical investigation on effects of nozzle's geometric parameters on the flow and cavitation characteristics within injector's nozzle for a high-pressure common-rail DI diesel engine *Energ Convers Manage* **89** 843-61

[5] Pedrozo V B, May I, Nora M D, Cairns A and Zhao H 2016 Experimental analysis of ethanol dual-fuel combustion in a heavy-duty diesel engine: an optimisation at low load *Appl Energy* **165** 166 - 82

[6] Wang P, Liu F and Li X 2012 Driving characteristics for electronic control unit pump fuel injection system *In 2012 International conference on computer distributed control and intelligent environmental monitoring* IEEE 264-7

[7] Fan L, Wen L, Yao C and et al 2012 Study on injection pressure characteristics for electro-controlled unit pump injection system *Chin Intern Combust Engine Eng* **(12)** 58-65

[8] Xu QK, Zhu KQ and Zhuo B 2008 Simulation in thermal design for electronic control unit of electronic unit pump *Chin J Mech Eng Engl Ed* **21(5)** 1-7

[9] Tian B, Fan L, Hayat Q, Ma X, Song E and Wang H 2014 Computation of pressure fluctuation frequency in electronic unit pump for diesel engine *J Mech Sci Technol* **28(4)** 1529-37

[10] Tan P, Lu J and Deng K et al 2004, The Influence of Fuel Injection Timing on Exhaust Emission From a Diesel Engine *Chinese Internal Combustion* **25 No. 2** 9-11

[11] Yuan T, Kai Q, Wang H and Liu S 2017 Study on Fuel Injection Advance Angle of a Turbocharged Diesel Engine under Plateau Condition. *4th International Conference on Manufacturing Engineering and Technology for Manufacturing Growth* 42-44

[12] Ma C, Wang Y and Wang H et al 2015, Effects of Advance Angle of Injection on Common Rail Diesel Engine Performance *Internal Combustion Engine & Parts* **(11)** 36-38

[13] Tan Z, Shen L and Li D et al 2014 Effect of Adjusting Fuel Injection Advance Angle on Characteristics of High Pressure Common Rail Diesel Engine Fueled with Biodiesels *Journal of Kunming University of Science and Technology* (Natural Science Edition) **39(2)** 58-60

[14] Li K, Li B and Sun P 2010 Influence of fuel injection advance angle on nitrogen oxide emission

from marine diesel engine. *Journal of Dalian Maritime University* **(3)** 87-89
[15] Shang H, Dong C and He J et al 2017 Research on Compression Ratio and Fuel Injection Advance Angle Optimization of Two-stage Turbocharged Diesel Engine *Acta Armamentarii* **(36)** 20-25
[16] Qiu T, Liu X, Liu F and Yu L 2008 The Study for Cam Profile Design of EUP. *Transactions of CSICE* **26(5)** 476-478
[17] Qiu T, Dai H, Lei Y and Cao C 2015 Optimising the cam profile of an electronic unit pump for a heavy-duty diesel engine *Energy* **83** 276-277
[18] Zhang Z 2017 Methodology to parametric design of cam profile for electronic unit pump *Energy* **139** 170-172

Graphene-modified CPE as a disposable electrochemical sensor for rapid determination of fluoranthene

Wenhan Zhong[1], Xiaoying Yan[2], Tao Han[3], Mengfan Wu[1, 4], Zhihao Xu[1] and Ying Wang[1] and Xiaohong Jin[1]

[1]Zhejiang Light Industrial Products Inspection and Research Institute, Hangzhou 310018, China;
[2]Zhonghang Monitoring & Testing Research Institute Co., Ltd, Hangzhou 310022, China;
[3]JUSHI Group Co., Ltd, Jiaxing 314500, China

[4]Corresponding author's email: wumengfan95@163.com

Abstract. A disposable electrochemical sensor (ES) was developed by embellishing graphene on carbon paste electrode (CPE), and the electrocatalytic activity of fluoranthene (Frt) was investigated. An anodic peak of Frt-acetonitrile-$NaClO_4$ appeared at 1.6 V based on the cyclic voltammetry. The peak current of Frt at graphene-modified CPE was increased compared with that of unmodified electrode. The differential pulse voltammetry revealed a linear relationship between peak current and Frt concentration within 2–140 ng mL^{-1}, and the LOD was calculated to be 0.95 ng mL^{-1}. Also, the ES method was effective for rapid determination of Frt in black rubbers on modified CPE.

1. Introduction

Polycyclic aromatic hydrocarbons (PAHs) are known as persistent organic contaminations with carcinogenicity and mutagenicity [1-3]. The main sources of fluoranthene (Frt) were biomass combustion and coal burning [4]. Frt is commonly found in black rubber products which include preformed and antiskid rubbers. Electro-oxidation route was shown to be one of the effective manners to the removal of PAHs [5,6].

Current assay methods for Frt include real-time polymerase chain reaction [7], high-performance liquid chromatography (HPLC) [8], gas chromatography-mass spectrometry [9] and atmospheric pressure chemical-ionization mass spectrometry [10]. However, these techniques require large equipment and expensive instrumentation, which are unsuitable for field related studies [11]. Based on facile pretreatment steps of routine quality control, carbon paste electrode (CPE) is mainly used in electroanalysis owing to its slight voltammetric background current and low-cost device [12,13]. For carbon electrodes, the design of chemically functionalized electrodes specific to Frt determination still remains challenges.

This work showed the differential pulse voltammetry (DPV) and cyclic voltammetry (CV) of Frt-acetonitrile-$NaClO_4$, which were characterised using CPE filled with graphite paste and embellished with graphene. The modified CPE was impregnated with Frt at open-circuit potential and then transferred to a three-electrode system. The influence of different mass concentrations of Frt (C_{Frt}) on the sensor performance was investigated, and the electrochemical sensor (ES) data were compared with those of HPLC methodology. Finally, the electroanalysis of Frt in rubber products at modified CPE was expected to establish field test of PAHs.

2. Experimental section

2.1. Instruments and reagents

A CorrTest CS350M electrochemical workstation (China), an Agilent 1290 Infinity II high performance liquid chromatograph (USA), a Hitachi TM4000 scanning electron microscope (Japan), a CEM Mars6 microwave extractor (USA), an Eppendorf 5810R multi-purpose desktop centrifuge (Germany) and a JingChong 3-mm diameter paste electrode shell (China) were utilized.

Graphene, N,N-dimethylformamide, graphite, acetonitrile, HCl and $NaClO_4$ were purchased from Macklin Biochemical, China. 1000 μg mL^{-1} Frt in methanol was purchased from TMRM, China. Water (electrical resistivity of 18.25 MΩ cm) was obtained using a ULUP-I water purifier.

2.2. Preparation of graphene-modified CPE

The CPE was obtained by mixing 70% (w/w) graphite powder with 30% (w/w) paraffin oil in an agate mortar [14]. The well-stirred paste was added to the cavity of electrode body, and the electrode surface was smoothed on a filter paper. The CPE was immersed in 0.2 mol L^{-1} HCl, and the potential value (E) was increased from open-circuit potential to 0.3 V for 1 min.

Next, graphene was added to 50 mL N,N-dimethylformamide (via sonication for 5 min) to prepare a 0.4% (w/w) graphene suspension. The modified liquid (20 μL) was then applied to the electrode surface with infrared drying to obtain the modified CPE.

2.3. DPV measurement

The as-prepared electrodes need to be polished before each measurement. For the electroanalysis of Frt, a three-electrode system is employed, and the DPV parameters are shown in Table 1.

Table 1. Optimized parameters of DPV.

Variables	Conditions
Working electrode	Graphene-modified CPE
Reference electrode	Ag/AgCl electrode
Auxiliary electrode	Pt electrode
Potential range	1.3–1.8 V
Pulse amplitude	70 mV
Pulse width	50 ms
Step time	0.1 s
Potential scan rate	50 mV s^{-1}

The C_{Frt} at 2, 10, 40, 80, 120 and 140 ng mL^{-1} were prepared with 0.1 mol L^{-1} electrolyte ($NaClO_4$ was dissolved in 85% acetonitrile as the electrolyte; the pH value of electrolyte was adjusted at 4.0). The modified CPE was immersed in Frt standards for 10 min and conducted for DPV analysis. The mean value of current response was recorded as the anodic peak current (I). The ES tests were measured in triplicate.

2.4. Sample's pretreatment and HPLC analysis

0.3 g black rubbers and 15 mL acetonitrile was incubated by microwave extraction, and extracted at 80 °C for 20 min. The mixture was centrifuged at 2500 rpm for 3 min, and then filtered by 0.45 μm organic membrane. The extract was next transferred into acetonitrile-$NaClO_4$ for ES tests.

Through silica gel solid-phase extraction, 1.2 mL extract in chromatic bottle was prepared for HPLC analysis. Chromatographic grade acetonitrile and water in certain proportions were used as mobile phases on C_{18} reverse-phase column. The qualitative experiment was determined derived from retention time at 1.5 mL min^{-1} of flow rate. The integral areas of LC peaks were used to acquire the C_{Frt} of black rubbers, comparing with those of ES.

3. Results and discussion

3.1. Microstructure of surface-modified electrode and redox behavior of Frt

The CPE was modified with 20 μL graphene suspension under infrared drying. The microstructure of CPE was observed using scanning electron microscopy (SEM) at an accelerating voltage of 15.0 kV. The bare CPE surface was relatively flat before deposition (Figure 1*A*), but variations occurred on its microstructure after modification. The graphene was dispersed in N,N-dimethylformamide, and the SEM revealed an uneven deposition on CPE surface with a cauliflower-like appearance (Figure 1*B*).

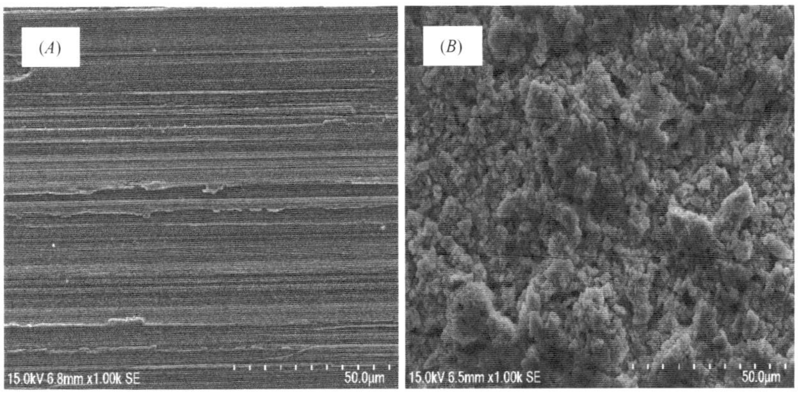

Figure 1. SEM images of the bare CPE (*A*) and modified CPE (*B*) surfaces.

After 10 min open-circuit state, the CV curves of bare CPE and modified CPE were performed in Frt-acetonitrile-$NaClO_4$. No distinct peaks were surveyed in 0.1 mol L^{-1} electrolyte (Figure 2*A*) and 50 ng mL^{-1} Frt-acetonitrile-$NaClO_4$ (Figure 2*B*) on bare CPE. A current signal at 1.6 V appeared that was owing to the oxidation of adsorbed Frt on modified CPE, while a cathodic peak at 0.33 V appeared in the negative potential scan (Figure 2*C*). After modification, graphene provided enhanced the activity of electron-transfer reaction. The oxidation potential of Frt was normally required to ensure high selectivity.

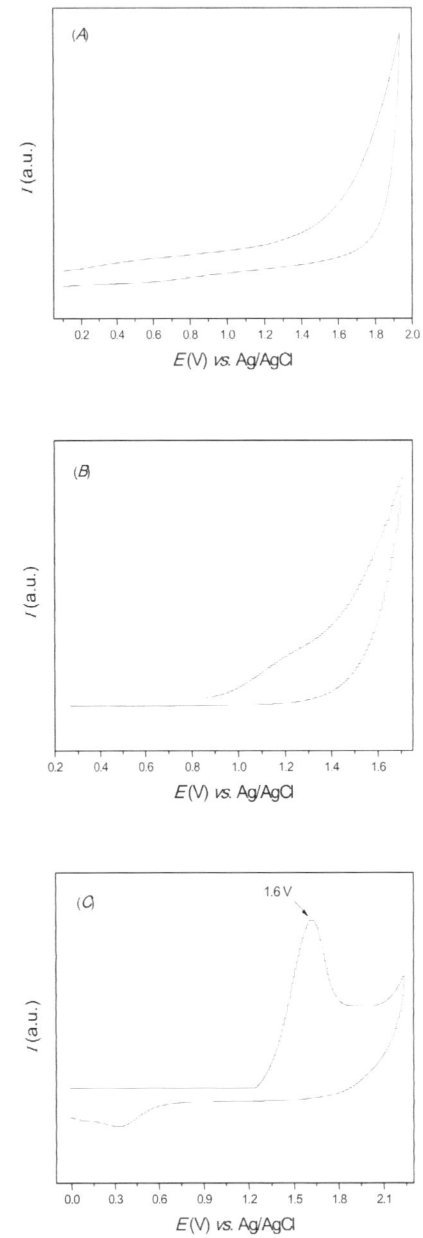

Figure 2. CV curves of 0.1 mol L^{-1} electrolyte (*A*) and 50 ng mL^{-1} Frt-acetonitrile-$NaClO_4$ (*B*) on bare CPE, and Frt-acetonitrile-$NaClO_4$ on graphene-modified CPE (*C*).

For enrichments on CPE surface, the graphene suspensions were embellished and the anodic current was significantly affected by modified liquid volume (Figure 3). The *I* of 50 ng mL^{-1} Frt-acetonitrile-$NaClO_4$ increased within volume of 15–19 μL. The *I* reached a maximum as the volume increased to 20 μL at high redox potential. When the volume was >20 μL, the *I* decreased in a nonlinear manner. This might be the modified film was comparatively thick, which hindered the electron-transfer reactions in electrolyte.

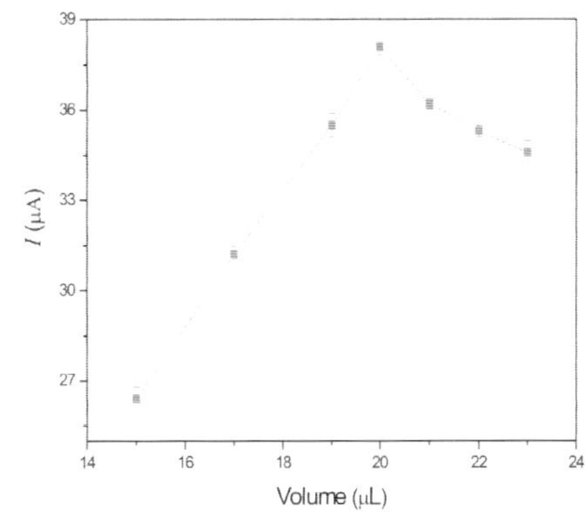

Figure 3. Effect of modified liquid volume on the *I* of graphene-modified CPE.

3.2. Electrochemical determination of Frt in rubbers

The working electrodes were immersed in Frt-acetonitrile-$NaClO_4$ at 2–140 ng mL^{-1}, and the Frt standards were determined by DPV (Figure 4*A*). The relation between I and C_{Frt} was as follows: $I = 0.336C_{Frt} + 15.71$, squared correlation coefficient $R^2 = 0.9982$ (Figure 4*B*). The limit of detection (LOD) was calculated to be 0.95 ng mL^{-1} using formula: LOD = $3\sigma/k$; where σ is the standard deviation of the peak currents for blank sample (n = 9), and k is the slope of the calibration curve. Hence, the ES constructed has high sensitivity for Frt.

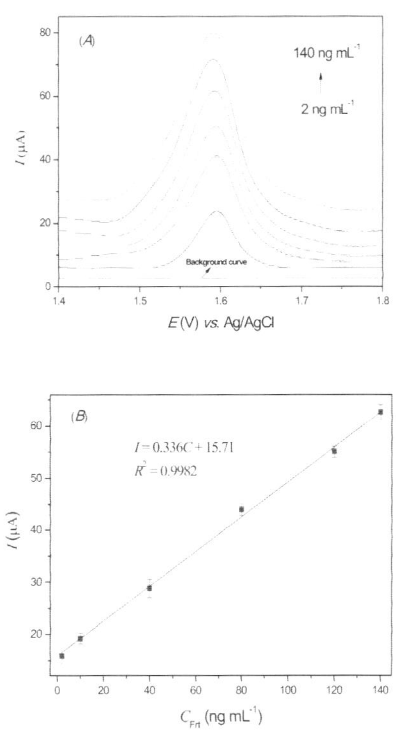

Figure 4. DPV responses of adsorbed Frt-acetonitrile-$NaClO_4$ on modified CPE (*A*), and calibration curve of Frt standards (*B*).

C_{Frt} of the black rubbers were detected by ES and HPLC method. The specific concentrations of spike-in sample at 10, 40 and 80 ng mL^{-1} were sequentially prepared. The mean C_{Frt} of the positive samples was 12.1 ng mL^{-1} by ES and 12.5 ng mL^{-1} by HPLC, of which the deviation was ≤3.4%. The mean recoveries for ES were 93.7%–104.2%, while the mean recoveries for HPLC were 97.0%–108.1% (Table 2). Their deviation was considered acceptable, and the proposed electrochemical method was favourable in terms of inexpensive in price. The relative standard deviations (RSD) of modified CPE and HPLC were ≤5.8% and ≤5.2%. The reproducibility of modified CPE was good in terms of ES detection, indicating that the ES system was applicable for PAHs quantification.

Table 2. Recoveries of ES and HPLC for rapid assay of Frt (n = 3), and n stands for repetitions.

Sample	Added (ng mL^{-1})	Found (ng mL^{-1})		Recovery (%)		RSD (%)	
		ES	HPLC	ES	HPLC	ES	HPLC
Positive rubbers	80	96.0	95.2	104.2	102.9	5.5	4.7
	40	51.4	50.9	98.6	97.0	5.0	5.2
	10	20.7	24.3	93.7	108.1	5.8	4.9

4. Conclusion

A disposable CPE modified with graphene showed good electrocatalytic properties for the oxidation of Frt-acetonitrile-$NaClO_4$ due to its high surface area and charge-transfer reaction. The redox behavior of Frt and the I at modified CPE were enhanced compared with that of unmodified electrode. The mean recoveries of Frt in black rubbers using ES method were from 93.7% to 104.2%. The results confirmed a quantitative approach of Frt and indicated the characteristics of current responses. The oxidizing conditions of Frt were optimized, which was opportune to develop a controlled electro-oxidation system in the domain of PAHs assay.

Acknowledgement

This work was supported by the Zhejiang Technical Project of Market Regulation (No. 20190117).

References

[1] Zhang Q, Liu P, Li S, Zhang X and Chen M 2020 Progress in the analytical research methods of polycyclic aromatic hydrocarbons (PAHs) *Journal of Liquid Chromatography & Related Technologies* **43** 425

[2] Sibiya P N 2013 Modification, development and application of extraction methods for PAHs in sediments and water *International Journal of Remote Sensing* **34** 4812

[3] Wei X, Ding C, Chen C, Zhu L and Sun Y 2021 Environment impact and probabilistic health risks of PAHs in dusts surrounding an iron and steel enterprise *Scientific Reports* **11** 6749

[4] Hu G, Guo J, Luo X, Chen S, Mai B and Li F 2009 Distribution, sources, and risk assessment of polycyclic aromatic hydrocarbons (PAHs) in surface sediments from Baiyangdian Lake *Research of Environmental Sciences* **22** 321

[5] Andreottola G and Ferrarese E 2008 Application of advanced oxidation processes and electrooxidation for the remediation of river sediments contaminated by PAHs *Environmental Letters* **43** 1361

[6] Acimovic D D, Karic S D, Nikolic Z M, Brdaric T P, Kaninski M P and Nikolić V M 2017 Electrochemical oxidation of the polycyclic aromatic hydrocarbons in polluted concrete of the residential buildings *Environmental Pollution* **220** 393

[7] Zhou C and Zhuang H 2009 Determination of fluoranthene by antigen-coated indirect competitive real-time immuno-PCR assay *Journal of Environmental Monitoring* **11** 400

[8] Zima J, Housova A and Barek J 2003 HPLC monitoring of the efficiency of chemical

decomposition of genotoxic fluoranthene derivatives *Chemia Analityczna* **48** 509

[9] Liu S J, Xu H N, Feng K and Wang Y 2013 Consolidated GC-MS/MS analysis of PAHs in gasoline combustion dust *Applied Mechanics and Materials* **448** 337

[10] Pérez S and Barcelo D 2001 Determination of polycyclic aromatic hydrocarbons in sewage reference sludge by liquid chromatography-atmospheric-pressure chemical-ionization mass spectrometry *Chromatographia* **53** 475

[11] Suresh S, Gupta A K, Rao V K, Kumar O and Vijayaraghavan R 2010 Amperometric immunosensor for ricin by using on graphite and carbon nanotube paste electrodes *Talanta* **81** 703

[12] Mardani L, Vardini M T, Es'haghi M and Kalhor E G 2020 Design and construction of a carbon paste electrode modified with molecularly imprinted polymer-grafted nanocomposites for the determination of thyroxin in biological samples *Analytical Methods* **12** 333

[13] Azhari S, Wahab R A and Ahamad R 2020 Determination of cadmium ions using Schiff-base modified carbon paste electrode: A box-behnken design approach *Solid State Phenomena* **307** 231

[14] Gete T G and Kassaw M M 2016 Cyclic voltammetric study of paracetamol at nickel hexacyanoferrate modified carbon paste electrode *Journal of Natural Sciences Research* **6** 28

New technology of magnetotelluric polaron exploration-taking Huashan piedmont basement fault as an example

Lin Zhang[1, 2, 4], Laixin Li[2], Dongmin Ma[1], Pei Zhang[3], Kai Shao[2] and Xiaolin Ma[3]

[1]Xi'an University of Science and Technology, Xi'an, Shaanxi 710054, China
[2]Shaanxi Coalbed Methane Development and Utilization Co., Ltd., Xi'an, Shaanxi 710065, China
[3]Xi'an Petroleum University, Xi'an, Shaanxi 710065, China

[4]Corresponding author's e-mail: 38590673@qq.com

Abstract. Detecting the Huayin Piedmont faults and basement conditions in the Weihe Basin through magnetotelluric polaron exploration technology found that the piedmont faults in the research area are quite developed, with two stepped normal faults with larger angles, along with many small faults; From the terrestrial polarization sub-bands, it is reflected that there are many fractures in this area, and the formation energy fluctuates greatly, indicating that the energy accumulation and formation energy of the formation are discontinuous, and there is a complete anomaly of the upper and lower formation energy at a certain depth. The comparative analysis of drilling and core data believes that it is the interface between the sedimentary rock layer and the underlying Qinling piedmont basement: the overlying stratum is the Tertiary sedimentary rock, and the lower basement is the Qinling Archaean granite layer; the fault structure of the Qinling piedmont is further understood and the basement rock layer and depth also proved the accuracy and reliability of the magnetotelluric polaron exploration technology.

1. Introduction

The Weihe Basin is located in the central part of Shaanxi Province (Guanzhong Plain). It is a Cenozoic faulted basin at the junction of the Qinling orogenic belt and and the Ordos Basin for two geotectonic units [1], south of the Ordos Basin, north of the Qinling Mountains, east to the Yellow River and West to Baoji Canyon; 400km from east to west, the north-south direction is about 20km wide in the west, the east is about 70km wide, and the basin area is about $2\times104km^2$ [1,2]; its stratum belong to the southern edge of the North China stratigraphic area. The research area Huayin City is located in the eastern part of the Weihe Basin, which belongs to the eastern wdge of the Weihe Basin. This area is rich in a variety of resources, including Water-soluble natural gas, helium-rich water-soluble gas, and geothermal water resources, etc. [3,4]. In the process of geothermal exploration in the basin, the stratum depth, fault development and basement depth of the area are unclear [5], resulting in poor geothermal development effects, etc., Cause greater economic losses [6], through advanced magnetotelluric polaron exploration technology to carriy out exploration work, to find out the development of the fault structure and the basement in the piedmont, and to guide the next step of geothermal exploration and development work.

2. Experimental area overview

The exploration area is located in Huayin County, in the eastern part of Guanzhong Basin; it from Tongguan in the east, west to Huaxian, south to Qinling Mountains and Luonan, and north to Weihe

River and Dali. The county is between 109°54'-110°12'east longitude and 34°19'-34°40' north latitude; the exploration area belongs to the Huashan piedmont structure, as shown in Figure 1.

Figure 1. Location map of the research area.

Huashan is the eastern section of the Qinling Mountains. The forward faults in the Huashan piedmont are very developed, forming fault triangles. The progressively decreasing fault structure and combination form are the records of the activities of different stages of the fault. The triangle surface of the entire fault shows a gentle upward and downward steep form, and the fault system is very developed, which is also in line with the wave energy display of the original section. Use magnetotelluric polaron exploration technology to conduct fault zone and base exploration in the above areas to verify the accuracy of this technology for basement exploration.

3. Experimental tasks and data collection

3.1. Principles of magnetotelluric polaron exploration

The principle of magnetotelluric polaron exploration is different from the passive source electromagnetic prospecting method based on traditional physical electromagnetic radiation. In the method of magnetotelluric polaron exploration[Bagdanov, Bavelvitch, 2008], the main body of research is natural electromagnetic Radiation, it is formed during the change of stress and strain state of the lithosphere.This kind of electromagnetic wave is different from the electromagnetic wave of fixed frequency and fixed amplitude, which is formed by the mutagenesis of magnetic field (magnetotelluric field) and atmospheric activity (thunderstorm activity field) [Baghdanov, Pavelvitch, Schumann, 2009].Therefore,it is believed that the medium is also the most active in the formation stress and rock stress.

The rock layer is a favorable environment for electromagnetic disturbances. The nearest electromagnetic disturbance that can reach the surface of the earth forms a complex electromagnetic field structure on the earth, which opens a brand new channel for us to obtain information on the physical properties, structure and performance of the earth.

3.2. Implementation of exploration deployment and data collection

According to the site conditions and work purposes, deployed 3 experimental survey lines, as shown in Figure 2.

According to the detection principle, the detection depth is one-half the distance of the arrangement length, so the longer the arrangement length, the deeper the detection depth will be to receive underground electromagnetic signals. Based on this principle, it is arranged as a double cross survey network to include the fault zone in the area and the depth range of the base.

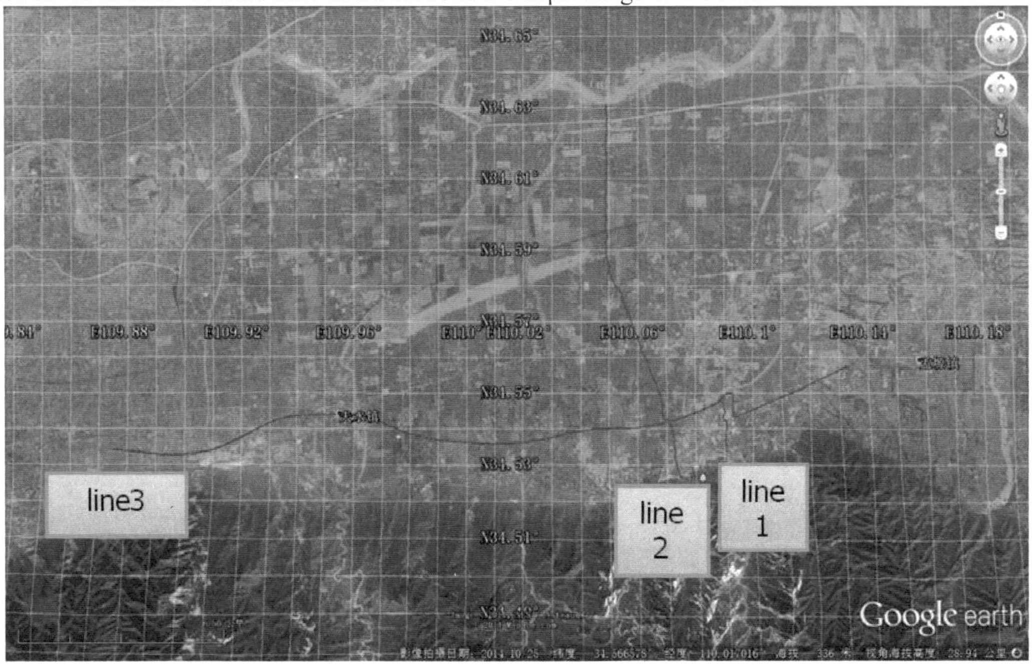

Figure 2. Control map of exploration survey line in experimental area.

Because the experimental site is located in the suburbs of Huayin City, highways and railways are vertical and horizontal, and there are many high-voltage transmission lines in the area. these objective conditions caused stronger electromagnetic interference in many aspects to the signal acquisition of the magnetotelluric polaron exploration technology (Figure 3), has a certain impact on the exploration effect.

The experiment was started from 9:30 am and ended at 18:20 on the same day. The data collection tasks for the above three survey lines were completed. and the survey line length was 47 km in total.

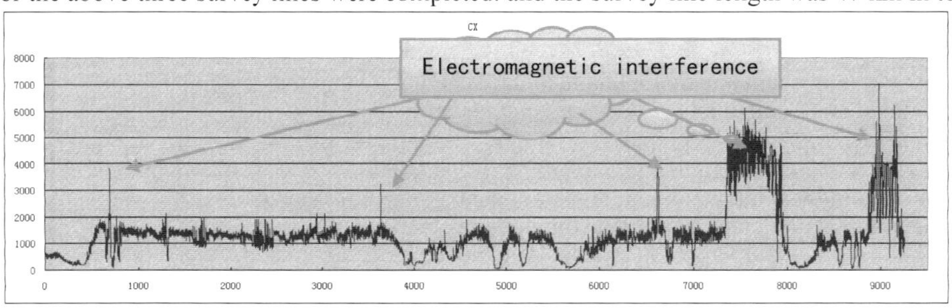

Figure 3. Raw data signal curve.

4. Experimental work and results

4.1. Principles of highlight mining

There are stress and concentrated fields in the geological interface of the underground strata and the fault section, which have also become bright spots, The bright spots have high-energy electromagnetic radiation.

The method of finding bright spots can be simply stated as: two similar peaks in the original data curve, the position center of which is along the lower part of the central axis is the bright spot position, and the relationship between the depth d and the distance L between the two peaks is d= 0.75L (empirical formula, diffraction theorem) (Figure 4).

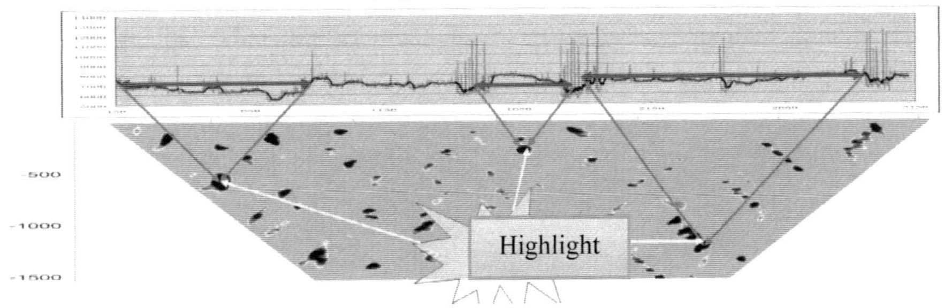

Figure 4. How to find highlights.

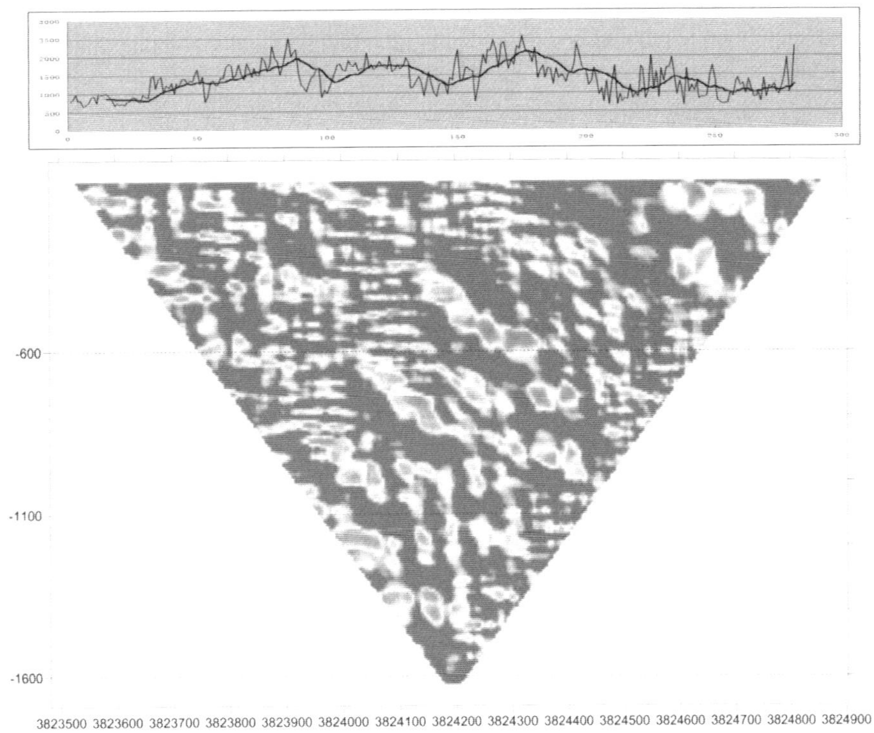

Figure 5. Original profile of survey line 1.

4.2. Data processing

Due to the small fluctuations of the ground surface in the experimental area, the ground leveling method is adopted for data processing (no ground elevation correction), and the ground is set to zero uniformly, and the longitudinal value of the profile is the vertical depth from the ground.

During the de-noising process, many random interferences were found, which affected the overall quality of the data. In contrast, survey line 1 and survey line 2 have better data quality(Figure 5,6), and survey line 3 is because the instrument is mounted in the automobile for exploration, there is more signal interference, less effective data information (Figure 7), and the worst data quality.

It is not difficult to see from the original section that the faults are well developed.

IOP Conf. Series: Earth and Environmental Science **865** (2021) 012011 doi:10.1088/1755-1315/865/1/012011

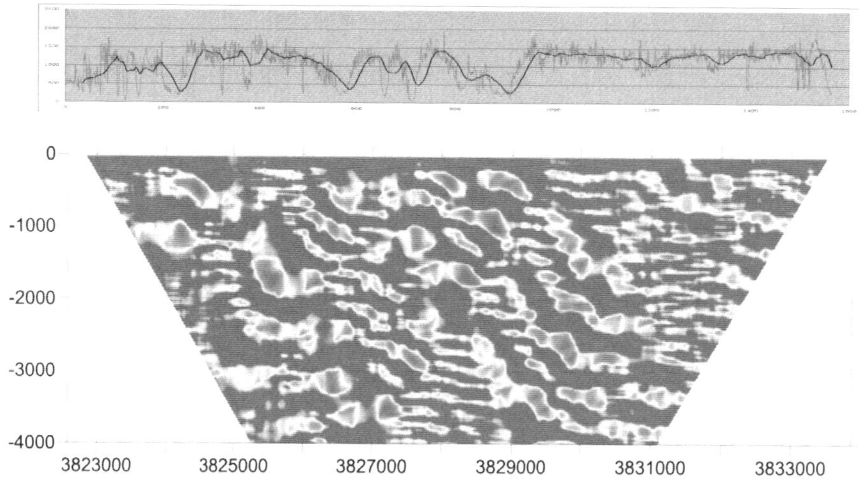

Figure 6. Original profile of survey line 2.

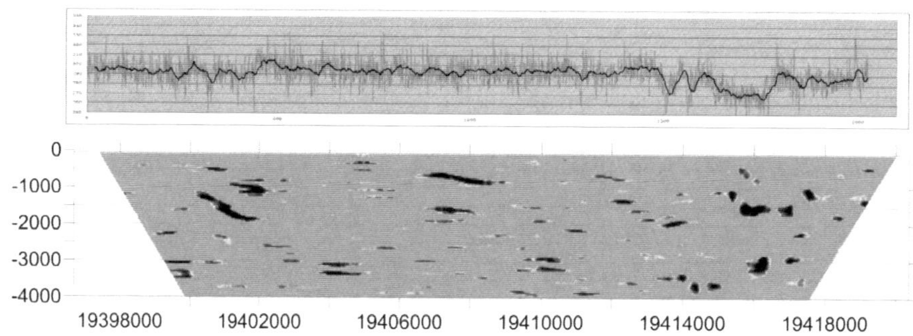

Figure 7. Original profile of survey line 3.

4.3. Geological analysis

From a geophysical point of view, there must be a large difference in physical properties between the basement granite and the upper sedimentary rock. Taking the survey line 2 profile as an example (Figure 8), there are obvious differences in the energy distribution characteristics and combination characteristics of a single point in the longitudinal direction. The interface can be judged as the interface between the basement and the upper sedimentary rock (the black dashed line in Figure 8).

Combining the basic geological data of this area, through energy analysis, a more accurate basement interface depth can be obtained (Figure 9,10,11).

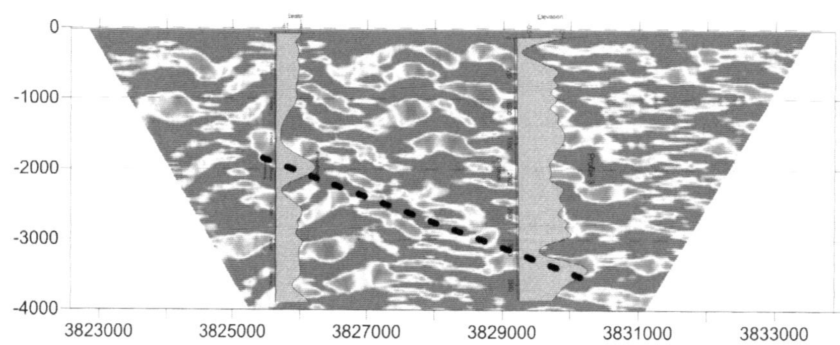

Figure 8. Single-point longitudinal energy analysis of the original profile of survey line 2.

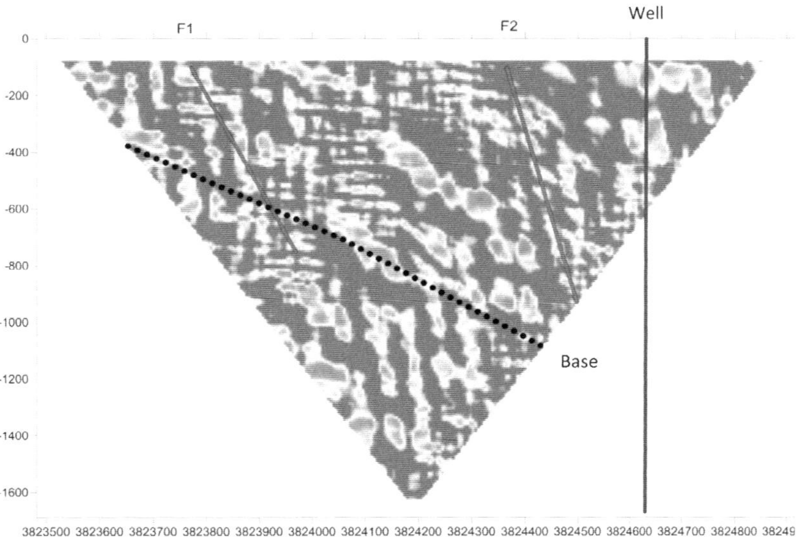

Figure 9. Preliminary interpretation profile of survey line 1.

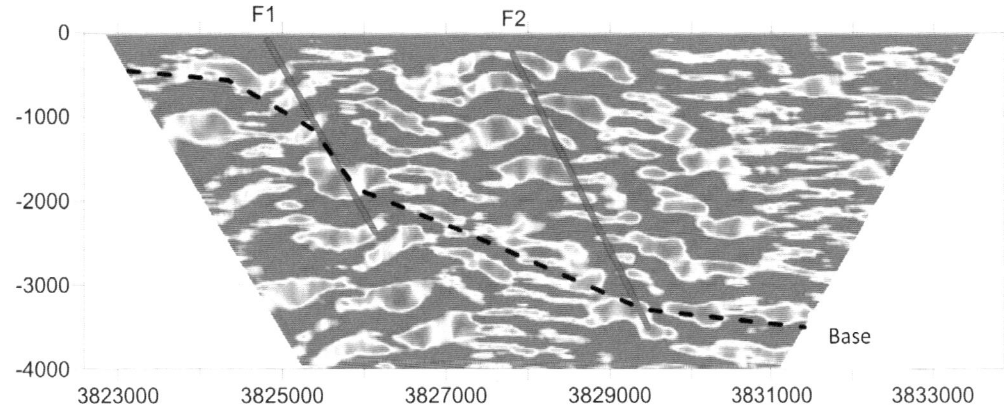

Figure 10. Preliminary interpretation profile of survey line 2.

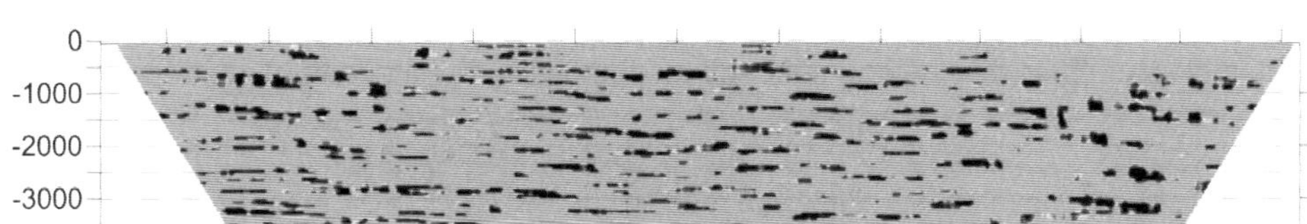

Figure 11. Preliminary interpretation profile of survey line 3.

From the profile, the structure changes drastically in this area, the buried depth of the base gradually increases from south to north, and the developed fault system is basically east-west trending.

There are two geothermal exploration wells in the exploration area, and there is a geothermal exploration well in the front of the Qinling Mountains passing through the 1 and 2 survey lines. According to on-site drilling core data and logging data, granite was drilled to the basement at 1528 meters, and a large fault was shown at about 1730 meters; According to Figure 9, the intersection of the extension line between the fault and the basement and the well 2 basically agrees with the actual drilling data; and there is a geothermal exploration well 1 near the high-speed rail line north of the survey line 2, it encountered basement granite at about 2750 meters, which is also in good agreement with our survey line, proving that the technology has quite high accuracy and reliability (Table 1).

Table 1. Comparison of geothermal well depth and exploration effect of 2 wells in the work area.

Geothermal exploration well	Depth (m)	Fault depth (m)	Substrate depth (m)	Relationship with survey line 2
Well 1	2800	--	2750	parallel
Well 2	1888	1730	1528	intersect

Through the analysis of the actual drilling data and logging data of two wells in the work area, it is proved that the magnetotelluric polaron exploration technology can distinguish the formation and the basement and fault zone accurately, it has a fairly high resolution and Reliability, it also has a clearer understanding of the Qinling piedmont underground stratigraphic structure and basement rock in the Huayin Research Area, guiding the next step of geothermal exploration and development.

5. Conclusion

1. The piedmont faults in the research area are quite developed, with two large-angle step-type forward faults, accompanied by many small faults; From the geopolarization sub-band reflects that there are many faults in this area, and the formation energy fluctuates greatly, indicating that the energy accumulation and formation energy of the formation are not continuous, and there is a complete anomaly in the upper and lower formation energy at a certain depth;

2. Through the comparison between the original well data in the work area and the latest magnetotelluric polaron exploration data, the conditions of the underground strata, structure and basement in the front of the Qinling Mountains in the work area were found out. The analysis suggests that there is an interface between the sedimentary rock layer and the underlying Qinling piedmont basement rock in this area. The overlying strata are tertiary sedimentary rocks, and the lower basement strata are Qinling Archaean granite layers, which will provide further guidance for later geothermal exploration and development;

3. The exploration of this area through the magnetotelluric polaron exploration technology, combined with the drilling data, fault distribution and basic data of two geothermal wells in the area, has been compared and analyzed, which confirms that the advanced technology is more accurate in the identification of strata and faults, and its resolution and reliability are high.

References

[1] Chen W C, Chen J X, Yun Q Y 1977 Fen wei basin oil reconnaissance stage geological results report. Xianyang, The third census brigade 301 brigade

[2] Zhang F L, Sun Q B. The wei he basin water soluble gas drilling stratigraphic division, the layer group correlation and understanding. 2012. The third census brigade

[3] Zhang F L, Sun Q B, Wang X Y 2012-6 The wei he basin water soluble helium gas resources evaluation[J] *Journal of Geomechanics*, **18(2)**

[4] Liu J C, Li R X, Wei G F, Zhao F S 2009-11 Study on the origin and source of water soluble helium in geothermal water in Weihe Basin. *Geological Science and Technology Information*, **28(6)**

[5] Li R X, Liu J C, Wei G F, Zhao F S 2009-100 Study on the origin and source of water soluble hydrocarbon natural gas in Weihe Basin. *Natural gas geochemistry*, **20(5)**

[6] Li Y H, Lu J C, Li J C 2011 The wei he basin gas formation characteristics and significance[J]. *Journal of xi 'an petroleum university*, **26(5)**: 11-16

IOP Conf. Series: Earth and Environmental Science **865** (2021) 012012 doi:10.1088/1755-1315/865/1/012012

Research on reasonable development strategy of offshore buried hill fractured reservoir

Lina Yang[1, 3], Xinran Wang[2], Yanjun Yin[1], Li Wei[1], Shengli Xu[1] and Linlin Dan[1]

[1]CNOOC EnerTech-Drilling & Production Co. Ltd., room 619, Binhai cooperation building, Binhai New Village West, Zhabei Road, Binhai New Area, Tianjin, China;
[2]Tianjin Branch of CNOOC (China) Ltd., No. 2121 Haichuan Road, Binhai New Area, Tianjin, China

[3]Corresponding author's email: yangln5@cnooc.com.cn

Abstract. In order to improve the overall production degree of offshore buried hill fractured reservoir, expand the swept volume of water flooding, and improve the reservoir development effect, according to the actual reservoir characteristic parameters, numerical simulation software and three-dimensional physical model are used to study the reasonable pressure level, injection production mode and water flooding strategy of Oilfield A. The results show that, according to the pressure sensitivity experiment and numerical simulation analysis, the reasonable formation pressure is 70% of the original pressure for the buried hill fractured reservoir, and higher recovery can be achieved at this pressure level. By comparing different injection and production mode schemes, it is determined that under the weak injection and forcible production mode, the swept volume of reservoir water flooding is the largest, the water free recovery period of production wells is longer, and the water cut rising rate is low. The reservoir development is divided into four stages. For different stages, depressurization production, pulse water flooding, cyclic water flooding and asynchronous water flooding strategies are adopted respectively. Compared with the conventional stable water flooding strategy, the balanced development of fractures and matrix can be realized, so as to improve the full cycle development effect of the reservoir. The application of the development strategy results in this study to oilfield production practice has played an obvious role in improving the development effect of fractured reservoir, which has field practical significance and can provide reference for the development of similar fractured reservoir.

1. Introduction

Buried hill fractured reservoir is a special type of reservoir with dual medium. Compared with conventional reservoir, the pore space of this kind of reservoir changes greatly and the permeability heterogeneity is very strong. If conventional development strategy is adopted, the oilfield will face the problems of rapid water cut rise, increased production decline and short water free recovery period [1-4]. For offshore oilfields with high investment cost and short development cycle, it is difficult to bring considerable economic benefits by using conventional development methods. Therefore, it is necessary to research effective development strategies for fractured reservoirs. The existing research results are mainly about the formation mechanism and characterization of reservoir fractures [5-8], and the research objects are mainly onshore oil fields [9-12]. There are few research on the development strategy of offshore buried hill fractured reservoir [13-14]. Oilfield A is the first fractured buried hill reservoir to be developed offshore in China, and there is no mature experience for its development strategy. In order to ensure the oilfield to achieve good development effect in the whole development cycle, the reasonable development strategy is studied. The main research methods are numerical simulation and physical experiment. The research contents include reservoir stress sensitivity, water flooding degree, and development stage division. According to the research results, it is determined that the reasonable pressure maintaining level of Oilfield A is 70% of the original pressure, which can get higher ultimate recovery. Weak injection and forcible production mode is more suitable for Oilfield A, which can obtain higher water flooding degree. Oilfield A is more suitable to adopt reduced pressure production, pulse water flooding, and periodic water flooding and asynchronous water flooding strategy in different stages, which can get better development effect.

2. Brief introduction of Oilfield A

Oilfield A is the largest fractured buried hill reservoir in offshore China. Its lithology is gneiss and cataclastic rock. The main reservoir porosity is fracture and solution pore. Among them, structural fracture is the most developed, followed by cataclastic intergranular pore and solution pore. The porosity of the reservoir changes greatly, the permeability is very heterogeneous, and it has obvious characteristics of dual porosity medium. This study takes Oilfield A as the research object, and studies the reasonable development strategy of buried hill fractured oilfield, so as to determine the main factors to improve the development effect and enhance the oil recovery.

3. Reasonable formation pressure level

3.1. Research on reservoir pressure sensitivity

In the process of production, with the decrease of formation pressure, the pressure on the fracture system and matrix rock block system increases, which makes the fracture system and matrix system produce elastic-plastic deformation, leading to the narrowing or even closing of fracture opening. The deformation of matrix system will reduce the porosity volume, especially the throat, and greatly reduce the permeability of fracture and matrix system.

The development of fracture system in buried hill reservoir is complex. According to the statistics of logging data, the main distribution range of fracture dip angle is 50° to 75°. Therefore, fracture models with different angles are made, as shown in Figure 1. The fracture angles are 50°, 60° and 75° respectively, which are used to study the pressure sensitivity characteristics of fracture systems.

Figure 1. Fractured core model.

According to the experiment results, the pressure sensitivity curves of fracture models with different angles are drawn (Figure 2). It can be seen that the fracture models of different angles show pressure sensitivity, which indicates that maintaining reservoir pressure is very important for the development of fractured reservoirs. When the effective pressure of the model is greater than 25MPa, that is, when the pressure decreases slightly, the permeability decreases slightly, and the average permeability decrease rate is less than 20%. When the pressure decreases until the effective pressure of the model is less than 25MPa, the permeability decrease degree increases greatly, and the average permeability decrease value increases rapidly. According to the results of pressure sensitivity experiment, in order to prevent the reservoir permeability from decreasing greatly, the reservoir pressure should be higher than 25MPa, that is, the pressure maintaining level should be higher than 67%.

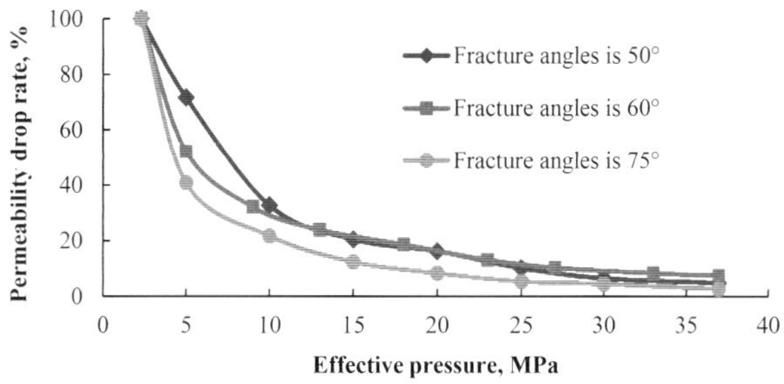

Figure 2. Permeability decline rate curves of fracture models with different angles.

3.2. Formation pressure level and recovery efficiency

In order to quantitatively study the reasonable pressure level of fractured reservoir, the pressure sensitivity experiment results are embedded into the numerical simulation software eclipse, and six comparison schemes are designed. The numerical models of different schemes are set with the same reserves, fracture reservoir ratio, initial pressure, fluid parameters and oil recovery rate, and the pressure levels are retained at 50%, 60%, 70%, 80%, 90% and 100% respectively, recovery index to determine the reasonable pressure maintenance level. The statistical results of each scheme are shown in Figure 3. When the pressure maintaining level is 70%, the cumulative oil production and recovery of the numerical model are the highest. Because the higher the pressure level is, the more sufficient the liquid supply, but the faster the water cut increase in the later stage. The lower the pressure level is, the slower the water cut increase, but it will lead to insufficient liquid supply and affect the final recovery. The results show that the reasonable formation pressure of Oilfield A is about 70% of the original pressure.

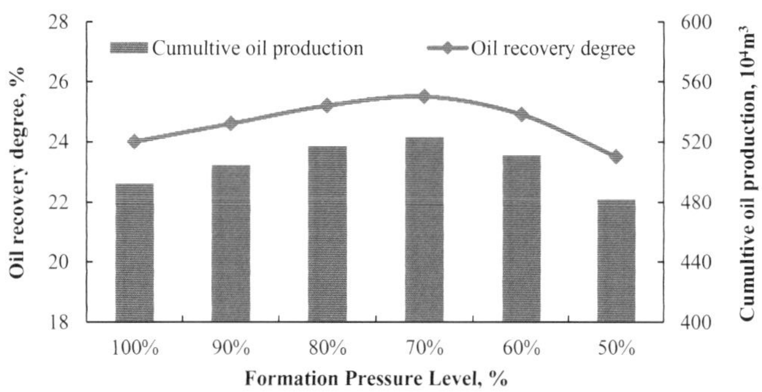

Figure 3. Comparison of development indexes under different pressure levels.

4. Research on injection and production mode

4.1. Numerical simulation study of injection and production mode

Referring to the reservoir and fluid characteristic parameters of the actual oilfield, the geological theoretical model is established by using Petrel software [15]. According to the overall development strategy of the oilfield, combined with the fracture development law of the buried hill, and referring to the development well pattern of water flooding in the bottom horizontal wells and oil production in the top horizontal wells of the reservoir in Oilfield A, the fracture development degree is divided according to the ratio of fracture permeability and matrix permeability, The area where the ratio of fracture to matrix permeability is greater than 80 is defined as the area where fractures are developed, and the area where the ratio of fracture to matrix permeability is less than 30 is defined as the area where fractures are not developed. According to the fracture development degree of the injection wells and production wells, four different injection and production modes are designed, namely, forcible injection and forcible production, forcible injection and weak production, weak injection and forcible production, and weak injection and weak production, as shown in Table 1.

Table 1. Design of numerical simulation schemes for injection and production mode.

No.	Injection and production mode	Injection well location	Production well location
Scheme1	Forcible injection and forcible production	Fractures are developed	Fractures are developed
Scheme2	Forcible injection and weak production	Fractures are developed	Fractures are not developed
Scheme3	Weak injection and forcible production	Fractures are not developed	Fractures are developed
Scheme4	Weak injection and weak production	Fractures are not developed	Fractures are not developed

After different schemes of water flooding, the remaining oil distributions of the numerical model are shown in Figure 4.

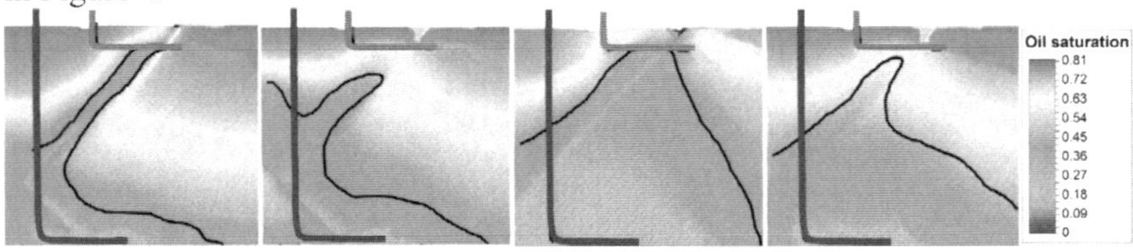

(a) Scheme 1 (b) Scheme 2 (c) Scheme 3 (d) Scheme 4
Production well Injection well Water flooding front

Figure 4. Distribution of remaining oil in different injection and production modes.

The well pattern of Scheme 1 is forcible injection and forcible production. Because of the large fracture communication between injection well and production well, it is easy to form a dominant seepage channel, which makes the production well flooded quickly, resulting in a very low degree of reserves production. The well pattern of Scheme 2 is forcible injection and weak production. Because the production wells are deployed in areas with underdeveloped fractures, the injected water is prevented from rapidly fingering along the large fractures. However, the reservoirs where the production wells are located are poor in physical properties and cannot effectively exploit the potential of the reservoir. The well pattern of Scheme 3 is weak injection and forcible production, and the permeability at the injection well is low, so the fingering is weak during the injection water displacement process, and the permeability at the production well is high, so the reservoir potential is effectively released and the injected water can be affected In a wider area, the degree of water flooding is the highest and the remaining oil is the least distributed. The well pattern of Scheme 4 is weak injection and weak production. Because the physical properties of injection well and production wells are both poor, although the injection water fingering is weak, the overall water flooding spread is not as good as that of Scheme 3.

By judging the change of oil saturation value in the numerical model grid, the number of grids to be affected by water wave is determined, and then the extent of water flooding is determined. The extent of water flooding under forcible injection and weak production mode and forcible injection and weak production mode is 13.6% and 26.2%, respectively, while the extent of water flooding under weak injection and forcible production mode and weak injection and weak production mode is 58.5% and 53.1%, respectively. The weak injection and forcible production mode is more suitable for Oilfield A.

4.2. Research on physics experiment of injection and production mode

The core physical model has more complex permeability characteristics and is closer to the real reservoir conditions. Therefore, simultaneous injection and production mode optimization experiments can effectively verify the accuracy of the numerical simulation research results. According to the Warren-Root model, the complex fracture model is simplified to consist of a mutually perpendicular fracture system and a matrix cut by the fracture system. The physical properties and fluid parameters of the model still refer to the actual situation of Oilfield A. Considering the investment cost and time limit of the three-dimensional physical experiment, water flooding experiments were carried out on the weak injection and forcible production mode and forcible injection and forcible production mode. The relationship between recovery degree and water cut under different scenarios is calculated. As shown in Figure 5, under the forcible injection and forcible production mode, the model has a short waterless recovery period, and the water cut rises quickly after water breakthrough, and the final recovery efficiency is 17.4%. In the weak injection and forcible production mode, the model has a long waterless recovery period, and the water cut rises slowly after water breakthrough. The final recovery efficiency is 21.0%, and the development effect is better. Therefore, the results of physical experiments are consistent with the conclusions of numerical simulations.

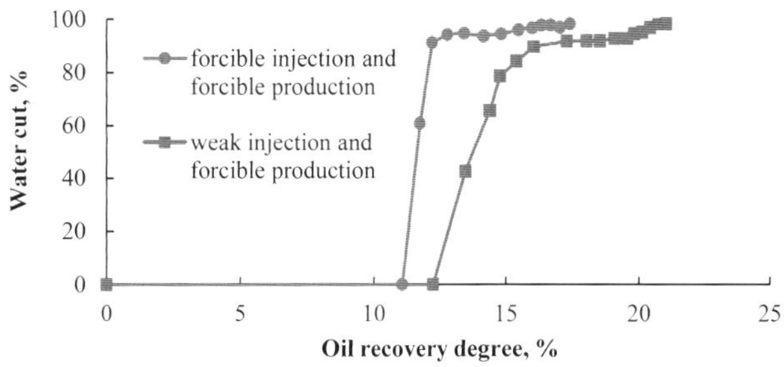

Figure 5. Relationship between recovery degree and water cut under different schemes.

5. Research on water flooding strategy

5.1. Division of development stages

In order to study and optimize the water flooding strategy, the basic scheme is the conventional stable water flooding development mode, that is, the injection production ratio is set at 1.0, and the constant water flooding is used for pressure maintaining production. Tracer technology is applied in Eclipse software to realize the splitting of produced fluid, so as to quantitatively characterize the oil production of fractures and matrix. Under the condition of stable water flooding, the production is finished when the water cut reaches 98%. According to the dimensionless oil production and production change trend in fractures and matrix, the whole production process is divided into four stages (As shown in Figure 6).

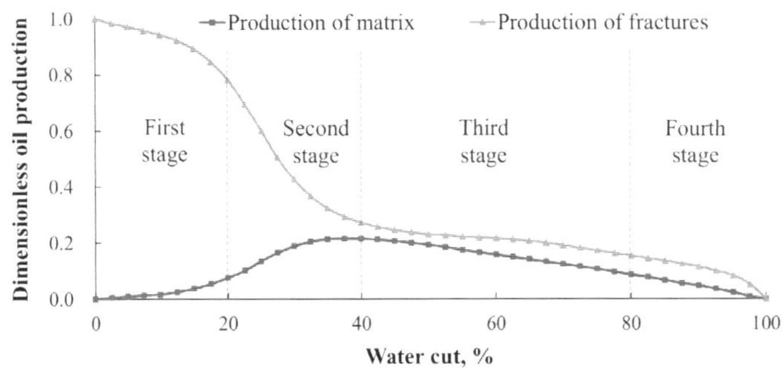

Figure 6. Relationship between recovery degree and water cut under different schemes.

In the first stage (water cut <20%). This stage is the initial stage of development. The oil production rate is relatively high, but the matrix is basically not produced. The fracture oil production is absolutely dominant, and the fracture oil production decline rate is small; in the second stage (water cut between 20% to 40%), The matrix oil production began to rise gradually and reached the peak, while the fracture oil production dropped sharply, the decline rate became larger, and gradually approached the matrix oil production; in the third stage (water cut between 40% to 80%), the fractures The oil production is close to that of matrix. The rate of decline in the oil production of the fractures and the matrix slows down, and the production state is relatively stable; in the fourth stage (water cut> 80%), it enters the low recovery rate stage and gradually moves towards the end of the development. The oil production of the whole reservoir is at a low level.

5.2. Unstable water flooding strategy

According to the main contradictions in the various stages of conventional stable water flooding, construct an unstable water flooding strategy for fractured reservoirs, and determine the strategies for the four stages in order of depressurization exploitation, pulse water flooding, periodic water flooding and asynchronous water flooding, thereby maintaining the formation pressure, restraining the rapid rise of water cut, and increasing the oil production of the matrix in the reservoir.

Depressurization production is applied in the first stage. At the initial stage of development, the reservoir energy is sufficient, and if the water flooding is implemented too early, it will face the risk of breakthrough in advance. Adopting the strategy of depressurization can effectively produce the original oil reserves in the matrix and fractures.

Pulse water flooding is applied in the second stage. The pulse water flooding process can be divided into two stages: pressure rise and pressure reduction: in the pressure rise stage, the oil and water in the fracture are displaced, the pressure in the fracture is higher than the inside of the matrix, and water is injected into the matrix; in the pressure reduction stage, the pressure in the fracture is lower than that of the matrix, and part of the fluid in the matrix flows into the fractures to be produced, thereby prolonging the low water cut oil recovery period.

Cyclic water flooding is applied in the third stage. The process of cyclic water flooding includes two stages: water flooding and stop flooding. In the water flooding stage, the injected water plays the role of oil displacement and pressure increase in the main fractures. In the stop flooding stage, the fluid in the secondary fractures and matrix begins to flow to the main fractures, and the liquid production of production wells decreases, but at the same time, the water cut decreases greatly, which makes the oil production increase.

Asynchronous water flooding production is applied in the fourth stage. At this time, most of the crude oil in the fracture has been produced, and the focus of development begins to shift to improving the displacement degree of oil in the matrix. The asynchronous water flooding strategy injects water into the injection wells in the process of increasing pressure, which the production wells will be shut, and the water channeling can be inhibit. In the process of depressurization, the injection wells will be shut, the production wells will be produced, and the remaining oil in the matrix will be further utilized by using the effect of matrix expansion and capillary imbibition, so as to improve the development effect.

6. Field application practice

The above research results have been applied to the development of Eastern well block of Oilfield A. The original plan of Eastern well block is 2 injection and 5 production well pattern, and the production wells are deployed in the middle of the reservoir. Compared with the original plan, there are many new small faults in the reservoir and the understanding of reservoir structure and fault development degree is clearer. As shown in Figure 7, compared with the original scheme, the injection and production well spacing is increased, and the injection wells are deployed in the fracture undeveloped area at the low part of the structure, while the production wells are deployed in the fracture developed area at the high part of the structure, forming the mode of weak injection and forcible production.

In the initial stage of development, natural energy development is adopted. When the pressure drops to about 70% of initial pressure, water flooding development is started. The strategies of pulse water flooding, periodic water flooding and asynchronous water flooding are implemented in different stages. After the Eastern well area of the Oilfield A was put into development, the average daily oil production of a single well in the first three years increased by 96 m^3 compared with the original plan design, and a longer waterless recovery period was obtained. The actual cumulative oil production is predicted to increase by $20.3\times10^4 m^3$ compared with the original plan. The recovery efficiency of the Eastern well area was increased by 2.5%. It shows that the results of this study can effectively guide the development of buried hill fractured reservoirs.

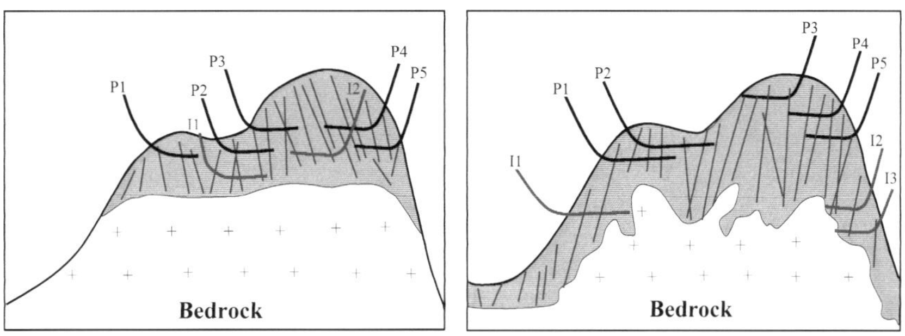

(a) Well pattern designed (b) Actual well pattern

Figure 7. Injection and production mode optimization of Eastern well blocks.

7. Conclusions

Some concluding remarks of the present study are briefly explained as:

(1) According to the actual characteristic parameters of oilfield, the physical experimental study on pressure sensitivity of fractured reservoir in Oilfield A is carried out. It is found that when the pressure drops to 70% of the original pressure, the physical property of reservoir will not be damaged. Combined with numerical simulation, higher oil recovery efficiency can be achieved under the 70% of the original pressure.

(2) Combining numerical simulations and three-dimensional physical experiments to compare the effects of different injection and production modes, it is obtained that the weak injection and forcible production mode can increase the extent of injected water and effectively inhibit the phenomenon of water flooding channeling, thereby improving the water flooding effect of fractured reservoirs.

(3) Based on the permeability characteristics of fractures and matrix in fractured reservoirs, the reservoir development divided into four stages. The different stages adopt reduced pressure production, pulse water flooding, periodic water flooding and asynchronous water flooding, compared with

conventional stable water flooding strategies. It can effectively develop fractures and matrix reserves and improve the overall development effect of fractured reservoirs.

Acknowledgements
This project is supported by CNOOC "13th Five Year Plan" science and technology major project "offshore oil and gas field potential tapping and efficiency increasing technology research".

References

[1] Mejia C, Roehl D, Rueda J, et al 2021 A new approach for modeling three-dimensional fractured reservoirs with embedded complex fracture networks *Computers and Geotechnics* 130 103928
[2] Aghabarari A, Ghaedi M and Riazi M 2020 Prediction of oil recovery in naturally fractured reservoirs subjected to reinfiltration during gravity drainage using a new scaling equation *Petroleum Exploration and Development* 47
[3] Rao X, Cheng L, Cao R, et al 2020 A modified projection-based embedded discrete fracture model (pEDFM) for practical and accurate numerical simulation of fractured reservoir *Journal of Petroleum Science and Engineering 187* 106852
[4] Chang C B, Shu M, Han Y F, et al 2020 Study on the construction and performance of completion fluid system in fractured carbonate reservoir *In Key Engineering Materials* (Vol. 861, pp. 388-394) Trans Tech Publications Ltd
[5] Chen Y, Ma G, Wang H, Li T, Wang Y, et al 2020 Optimizing heat mining strategies in a fractured geothermal reservoir considering fracture deformation effects *Renewable Energy 148* 326-337
[6] Abbasi M, Sharifi M and Kazemi A 2020 Fluid flow in fractured reservoirs: Estimation of fracture intensity distribution, capillary diffusion coefficient and shape factor from saturation data *Journal of Hydrology 582* 124461
[7] Wong D, Doster F, Geiger F, et al 2020 Partitioning thresholds in hybrid implicit-explicit representations of naturally fractured reservoirs *Water Resources Research* 56
[8] Ojha S P, Yoon S and Misra S 2020 Pressure-transient responses of naturally fractured reservoirs modeled using the multistencils fast-marching method *SPE Reservoir Evaluation & Engineering* 23
[9] Cochard J, Léonide P, Borgomano J, et al 2020 Reservoir properties of barremian-aptian urgonian limestones, se france, part 1: influence of structural history on porosity-permeability variations *Journal of Petroleum Geology* 43
[10] Xiaolin Z, Hui C, Xinran W, et al 2019 Research and application of water flooding timing and method for blocky bottom water fractured buried hill reservoir *Journal of Power and Energy Engineering* 7
[11] Gilman J R and Kazemi H 1982 Improvements in simulation of naturally fractured reservoirs *Society of Petroleum Engineers Journal 23* 4
[12] Aguilera, R 1991 An approximate solution of linear flow in naturally fractured reservoirs *Journal of Canadian Petroleum Technology* 30
[13] Hla B, Sga B, Lya B, et al 2021 Change laws of water invasion performance in fractured-porous water-bearing gas reservoirs and key parameter calculation methods - ScienceDirect *Natural Gas Industry B* 8
[14] Farah N and Ghadboun A 2021 The MINC proximity function for fractured reservoirs flow modeling with non-uniform block distribution *Oil & Gas Science and Technology - Revue d'IFP Energies nouvelles* 76
[15] Zhao X , Liu L , Hu J, et al 2014 The tectonic fracture modeling of an ultra-low permeability sandstone reservoir based on an outcrop analogy: A case study in the Wangyao Oilfield of Ordos Basin, China *Petroleum Science* 11

IOP Conf. Series: Earth and Environmental Science **865** (2021) 012013 doi:10.1088/1755-1315/865/1/012013

The design of a portable seismograph used in aftershock seismic array observation

Congqing Li[1, 2, 3], Yuntian Teng[1], Xiaomei Wang[1], Xiaoyong Fan[1] and Yiyang Tang[1]

[1]Institute of Geophysics, CEA, Beijing 100081, China;
[2]Geophysical Exploration Center, CEA, Zhengzhou, 450002, China

[3]Corresponding author's e-mail: congqingli@yeah.net

Abstract. This paper introduces the design of a portable wireless-linked seismograph used in aftershock ultra-dense seismic array observation, which features easy-to-install, low-power consumption, real-time data transmission and remote control capability. Compared to traditional instruments, such as the combination of Reftek 130s data recorder and Guralp CMG-40T seismometer for China Seismic Array Program, it integrates the seismic sensor, seismic data recording unit, GPS timing service and wireless data link into a compact sealed housing. To verify the system's functionality, both laboratory testing and preliminary field experiments were conducted. The exceptional feature of this newly-developed portable wireless-linked seismograph makes it very suitable for aftershock seismic array observation.

1. Introduction

In recent years, upon the occurrence of strong earthquakes in China, Geophysical Exploration Center of CEA rapidly start up the earthquake emergency observation response and researchers were sent to the earthquake region to do aftershock monitoring [1-3]. Substantial valuable raw seismic data are collected for being used in research on focal mechanism and seismogenic fault. The device used in aftershock monitoring is the portable digital PDS seismometer, which features lightweight, compact volume, easy to install and convenient operation. The working mode of using PDS Seismometers is that after deploying the device and conduncting observation for one week, the devices will be taken back, then download the seismic data and do data processing work. Then, next circle begins. This working modes leads to a lag for obtaining the dynamic seismic regime [4-8]. Therefore, in order to realize the real-time monitoring of aftershock seismic activity of earthquake region, and provide the tehcnoligical observation measurements for researchers. A new portable wireless-linked seismograph used in atershock seismic array observation has been developed.

2. Development of the portable seismograph

The instrument system is mainly composed of portable wireless-linked seismograph, remote control server on the PC and back-up power (storage battery, or solar power supply system). The seismograph is designed to do the work of data acquisition, analog-digital conversion, seismic data storage and remote data transmittion and control, etc. On the top cap, there are 2 antennas, one for wireless connection to the 4G mobile communication network and the other for GPS message receiving. The remote control server, running on the persenal computer, is the operating interface between users and seismographs, which has the ability of setting up and checking the device working parameters,

IOP Conf. Series: Earth and Environmental Science **865** (2021) 012013 doi:10.1088/1755-1315/865/1/012013

inquirying the instrument working states etc. The seismograph has a built-in lion battery, but in the case of continuous aftershock seismic array observation and turning on the wireless transmittion function, the built-in battery is not enough. Thus, it is essential to equip with back-up powers, such as large capacity storage battery, or solar power system. The GPS reciever provides time service and geographical position informatinon. Compared to the now available seismographs, such as Reftek 130 and Guralp CMG-40T systems, used in aftershock seismic array observation for megaseisms, this seismograph features low-power, compact volume and esay to operate and deploy. Figure 1 shows the block diagram of the seismograph system.

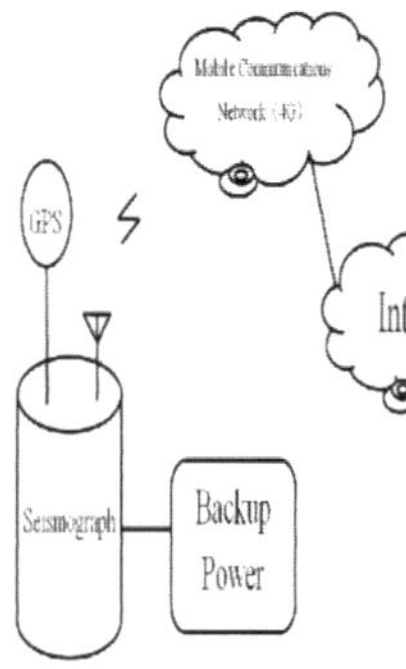

Figure 1. Block diagram of the instrumentation system diagram.

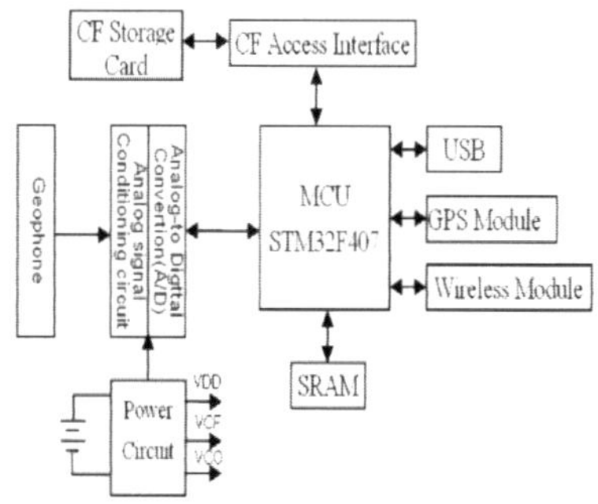

Figure 2. Data acquisition unit block.

2.1. *Hardware design of the seismograph*

The seismograph mainly consists of data acquisition unit, seismic sensor, wireless transmission unit and built-in li-ion baterry. All function blocks are integrated in a compact shell with the wireless antena and GPS interface installed on the top cap.

2.1.1. Data acquisition unit. The data acquisition unit, whose core MCU STM32F407 [9], mainly fulfill the analog to digital convertion of the seismic wave, GPS timing service and geographical location, data storage on the CF card, USB communication interface, RS-232 control interface. As showed in Figure 2.

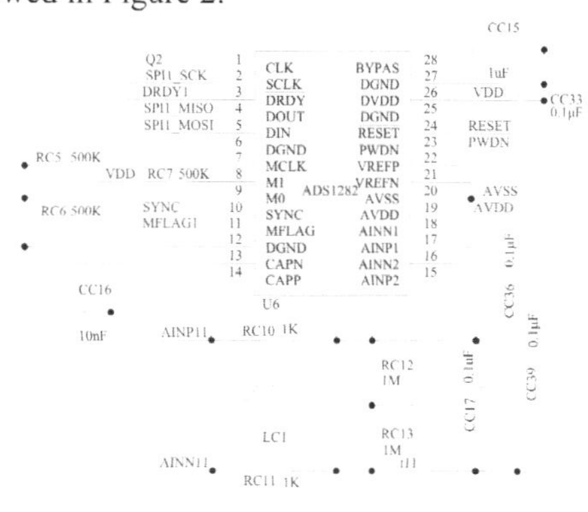

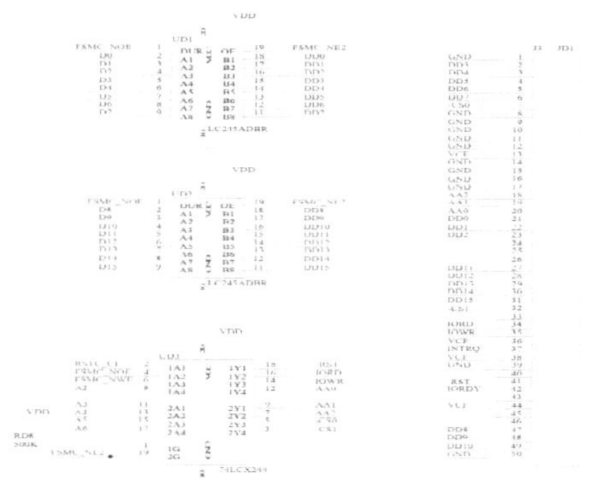

Figure 3. ADS1282 application design circuit. **Figure 4.** CF card interface circuit design.

The seismic sensor converts the velocity of ground motion to electrical signal, which pass through the front-end analog signal conditioning circuit and finally in the form of differential signal feeds into the analog-to-digital convertor. In this design, there are three independent A/D channels used to convert the seismic wave of NS, EW and UV directions to 24bit digital words. The ADC used is the

Analog Devices product ADS1282 [10], very suitable for high-precision, low frequency, feeble seismic signal convertion. Utilizing the SPI interface, ADS1282 makes connection with the core MCU, and on the bootup stage the fireware running on MCU initilizes the gain, sampling rate of the ADC. ADS1282 features excellent low power consuption and is specially fit for uage in portale measuring systems. The application design circuit is as showed in Figure 3.

This seismograph uses CF card as the seismic data storage media, which have the advantages of miniature, high-capacity and low-power [11]. In order to nicely satisfy the requirements for severe environment of field seismic data acquisition, industrial grade CF card is adopted and its working temprature ranges between -40℃-85℃. In this design, by making use of the FSMC of STM32F407 to generate the bus timing for TrueIDE interface, the system realizes the read and write access to CF card, as showed in Figure 4. Besides, the seismograph has a local data-downloading interface through USB device interface. Using the USB cable to connect the seismograph to a lap-top computer,the application software running on the computer can do many operations, such as working parameters setting, running state polling, and access to raw seismic data files, etc. The USB controller is used to realize the function in this design [12], as showed in Figure 5.

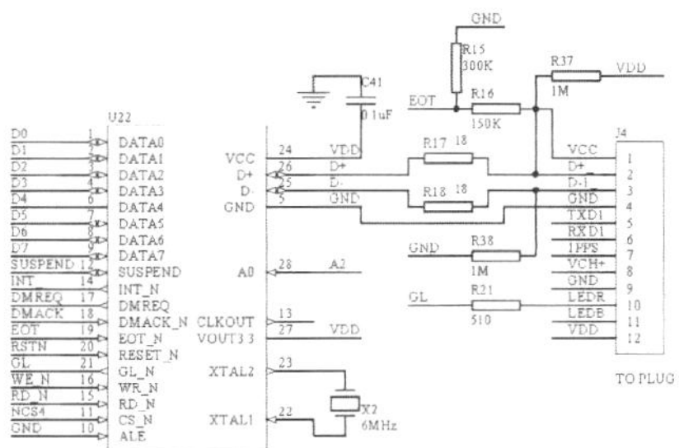

Figure 5. USB interface circuit design.

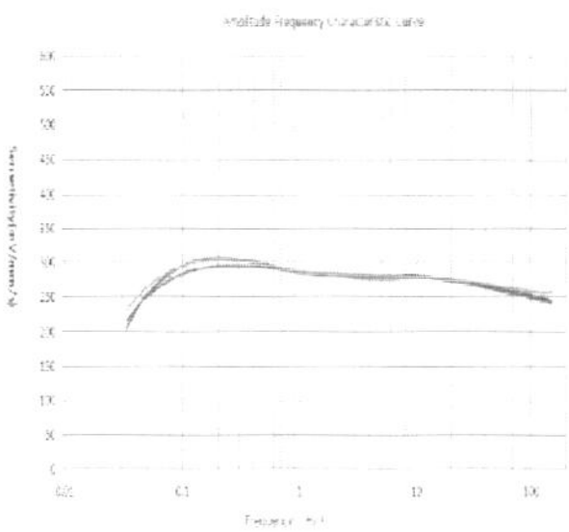

Figure 6. Amplitude frequency characteristic curve.

The seismograph integrates a GPS module to synchronize the system software clock and obtain the geographic information, including the Longitude, latitude and altitude of the observation satation. The power supply of the seismograph is composited of built-in high capacity li-ion battery and backup storage battery, or solar power supply. The power supply voltages of all function modules are generated from LDO circuit, including the voltage of MCU Vcore = 1.8V, ADC reference Vref=2.5V, CF card Vcf=3.3V, and TCXO Vco=3.3V [13].

2.1.2. Seismic senser. The seismic senor is an electromechanical device used to convert the velocity, displacement or acceleration of the ground motion to electrical signal, also known as seismometer. The seismometer used in the design is a velocity output passive transducer CDJ-6B, which can pick up the ground motion velocity of NS, EW and UV directions. The seismic senser has a natural frequency 2.5Hz, internal resistance 5.6kΩ, sensitivity 200V·S/m. In order to collect more low frequency content of aftershock observation, it is necessary to extend the corner requency of CDJ-6B as low as possible, a low frequency integral method, which is accomplished through balanced integration and high precision chips to suppress to suppress the circuit drifting and noise interference, was used to extend the CDJ-6B's frequency response. The low corner frequency is extended from 2.5Hz to under 0.1 Hz, and the sensitivity is enhanced from 200V·S/m to 300V·S/m. The -3dB Bandwidth is tested on the low-frequency shaking table, the amplitude frequency characteristic curve is as showed in Figure 6.

2.1.3. Wireless 4G module and GPS service. The seismograph has the function of remote real-time data transmission and remote control. By means of the antena, the wireless module makes a data connection based on the 4G data service to mobile communication networks, and eventually establishes a virtual communication data link with the monitoring computer on the Internet [14]. In software design, the wireless module fireware interacts with the monitoring application software in the pattern of client/server mode. Upon the power on, the seismograph iniates the connection request to try to setup a data communicaiton link with the monitoring computer. On the other end, the monitoring application server located in the monitoring center listens the connection requset on designated port. Once the wireless connection successfully established, we can do the same operations as what we do using the local USB port. The LEA-6T module is used for GPS service [15-16], and all-direction antennas are used to ensure the harsh observation environment [17].

2.2. Firmware and application design
The software of this seismograph includes two partions. One part is the firmware running on the seismograph MCU, which mainly finishes the data sampling, FIR/IIR filtering, GPS message interpretation, data communication, data strorage controll, etc. The other part is the monitoring application server running on computer, by which the users can do opearations on the seismograph.

2.2.1. Firmware. The firmware is composed of many software function modules, such as control commands processing, communication parameters configuration, sampling parameters control mechanism, seismic data FIR/IIR filtering and CF card drivers. All the firmware codes are developed by ARM C and assembly languages, which leads to several advantages, such as cutting down MCU frequency and furthermore reducing the system power consuption, and saving srtorage space for firmware codes. The firmware structure is as showed in Figure 7.

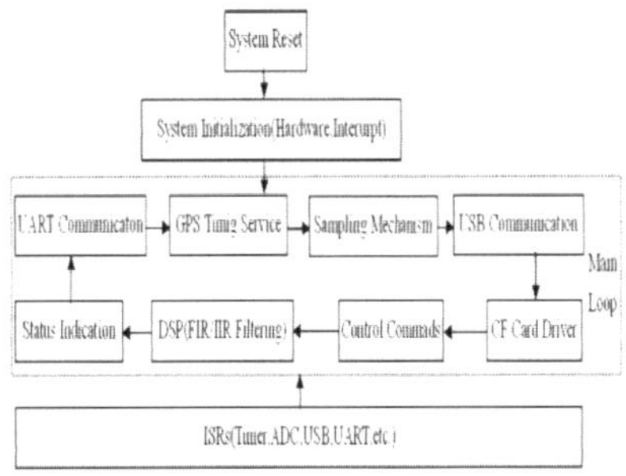

Figure 7. Firmware structure diagram.

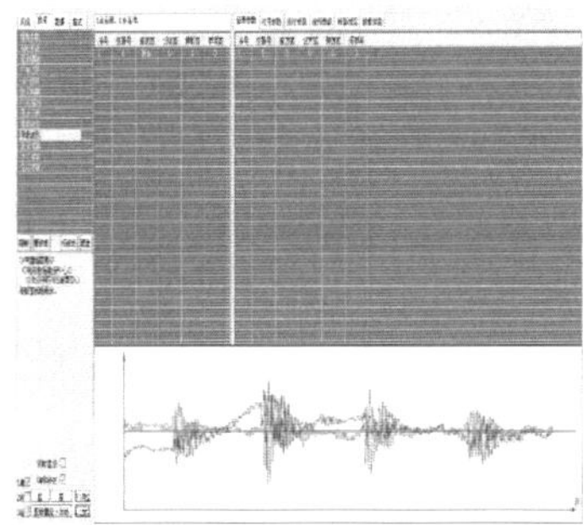

Figure 8. Monitoring application software.

2.2.2. Monitoring application software. The monitoring application software running on PC is used to realize the remote control on the seismograph. When the remote data link is established between the seismograph and the monitoring application, users can conviniently do many operations on the seismograph, such as working parameters configuring, preseting the sampling schedule, filling in station geographical notes, checking working status, and real-time seismic wave display, etc. The Monitoring Application software is as showed in Figure 8.

3. Preliminary testing and field experiment results

Preliminary testing has been conducted on the developed prototype seismographs, which are as showed in Figure 9. The serval key performace indexes of the testing results of the prototype seismographs are as follows: -3dB Bandwidth: 10s -100Hz; Dynamic range :> 132 dB; Self noise: -140 dB (Relative to 1 $[m/s-1]^2$ Hz-1); Digitiser resolution: 24-bit; Sample rates: 50, 100, 200, 250, 500, 1000; Output data Format: MiniSeed; Triggering modes: STA/LTA, level, software; Internal storage capacity:32 GB CF card storage(Non hot-swappable); Power consumption: 450mW; Adjustabel gain: $2^0 \sim 2^7$; Operating temperature: -20 to +65℃. Some preliminary field experiment, which makes use of the artificial seismic sounding prospecting project in October 2020, validates the basic functions and suitability for seismic array observation. The profile result is showed in Figure 10. A thorough comparation experiment with other well-known seismic systems is planned to be done in the second half of the year.

Figure 9. Prototype seismographs.

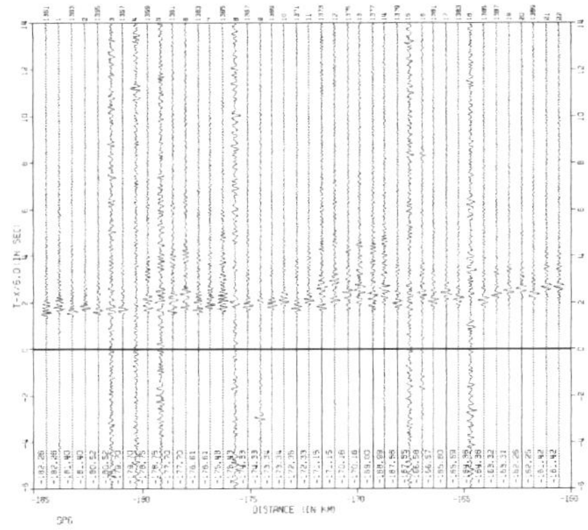

Figure 10. Field experiment Seismic profile.

4. Conclusions and discussion

The development of this portable wireless-Linked seismograph intended for use in atershock seismic array observation has basically finished, which achieves a good balance and tradeoffs between seismograph's performance and portability. This seismograph features miniature, low power consumption, and low-cost compared to the Reftek 130s and Guralp CMG-40T systems. It is very suitable for usage in aftershock dense seismic array observation. More thorough testing and field experiments will be conducted, through improving the hardware and software designs to further enhance the instument stability, reduce power consumption, cut down transmission delay and extend the corner frequency even lower. Continuous efforts will be also devoted to upgrade the system for real-time data trasmission applications, such as large scale dense arrays for abient noise observations.

Acknowledgements

This work was supported by the National Key Research and Development Program of China (2017YFC150 0200) and the Spark Program of Earthquake Technology of CEA(XH20075).

References

[1] Cheng L 2019 Imagming Shallow Crust Structure from Ambient Noise Surface Wave: Methodology and Applications 106.

[2] Tian X B 2018 Short period dense seismic array for prospecting the deep crustal structure, *Annual Meeting of Chinese Geoscience Union (2018)* 2201.

[3] Wei Y H, Tian X B, and Duan Y H 2018 Imaging the Topography of Crust–Mantle Boundary from a High-Density Seismic Array beneath the Middle-Lower Yangtze River, Eastern China, *Seismological Research Letters* **89** 1690-1697.

[4] Yu G P, Xu T, and Zheng T Y 2018 The reseach on Crustal Structure of Jiaodong Area by short period dense seismic array, *Annual Meeting of Chinese Geoscience Union (2018)* 3323-3324.

[5] Sun T W 2019 Study on the sedimentary and crustal structure of the Binchuan area using dense array 62.

[6] Li N 2017, Study on the Crustal and upper Mantle velocity Structure beneath the northern section of the South-North seismic zone using surface wave imaging method based on dense array 97.

[7] Lei H, Pan S Z, and Tian X F 2019 Recent application progress of shortperiod seismic dense array, *The Fourth National Youth Geological Congress* 287-288.

[8] Yuan Q X, Sun H, and Dai P 1999 Portable digital seismic data acquisition system for deep seismic sounding and earthquake survey, *Earthquake Research in China*, **15** 159-166.

[9] STMicroelectronics 2018 STM32F407 ultra-Low power microcontroller datasheet.

[10] Texas Instruments 2020 ADS1282 High-Resolution Analog-To-Digital Converter datasheet.
[11] The CompactFlash Association 2010 CF+ & CompactFlash Specification Revision 6.0.
[12] NXP Semiconductors 2018.TEA19051BT USB-PD 2.0/USB-PD 3.0 controller.
[13] ON Semiconductor 2019 NCP551 150 mA CMOS Low Iq Low-Dropout Voltage Regulator.
[14] u-blox 2018 TOBY-L2 series Multi-mode LTE Cat 4 modules with HSPA+ and/or 2G fallback.
[15] u-blox 2017 LEA-6T Receiver Description Including Protocol Specification.
[16] Sarantel 2009 SL1206(GeoHelix®-S)Active GPS Antenna Product Specification.
[17] National Marine Electronics Association 2008 NMEA Reference Manual Revision 2.2.

IOP Conf. Series: Earth and Environmental Science **865** (2021) 012014 doi:10.1088/1755-1315/865/1/012014

Study on gravity anomaly changes and crustal structure characteristics in Weihe Basin

Yongqi Zhang[1], Jianping Cao[1], Enhui Zhang[1], Meitao Han[1] and Yuxiang Zhang[1, 2]

[1]Shaanxi Earthquake Agency, Xi'an, Shaanxi, 710068, China

[2]Corresponding author's e-mail: zhangyuxiang@sxdzj.com.cn

Abstract. Based on the Earth Gravitational Model (EGM2008) and the global terrain model (ETOPO1), we calculate and obtain the free air gravity anomaly and Bouguer gravity anomaly in the Weihe Basin. The wavelet multi-scale analysis method is used to decompose the Bouguer gravity anomaly with 1st to 5th order wavelet detail and wavelet approximation. At the same time, based on the logarithmic power spectrum method, the average depth of the field source corresponding to the 1st to 5th order wavelet detail and approximation is calculated. The research results show that: (1) the wavelet approximation of Bouguer gravity anomaly of order 1st to 5th in the Weihe Basin corresponds to a field source depth of 12 to 84 km. From the wavelet approximation images, the basic pattern of different tectonic units in the study area can be clearly seen. (2) The field source depth corresponding to the wavelet detail of the 1st to 5th order Bouguer gravity anomalies in the Weihe Basin is 3 to 49 km. From the wavelet detail images, the characteristics of different tectonic divisions and fault distribution in the study area can be clearly seen. (3) Gravity anomaly gradient zones often correspond to relatively large-scale faults, such as the Longxian-Baoji fault, the northern Qinling fault, the Weihe fault, and the Huashan piedmont fault. In addition, from the wavelet detail images, the main structural units such as the Baoji uplift, the Xi'an sag, the Lishan uplift and the Gushi sag can be identified.

1. Introduction

55 to 50 Ma ago, the subduction and collision of the Indian plate and the Eurasian plate caused western China and the entire east Asia to undergo the strongest structural deformation since the Cenozoic [1]. The Himalayan movement not only caused the rapid vertical uplift of the Qinghai-Tibet plateau, but also caused a large amount of material within the Qinghai-Tibet plateau to flow eastward, and the Gansu-Qinghai block produced eastward movement, forming a compression on the western part of the Weihe Basin and making a left-handed strike-slip along the fault zone at the northern margin of the Qinling mountains [2-3]. For activities, the fault zone at the northern margin of the Qinling orogen was generally regarded as a left-handed strike-slip fault. Under the combined action of the high angle upthrust collision of the Philippine sea plate and the significant back arc expansion of the Ryukyu Archipelago arc, the north China and south China blocks in the eastern part of China have a SE trending pushing effect. In addition, the horizontal migration velocity of south China block to SE was obviously stronger than that of north China block, and accompanied by the drag of deep mantle small-scale convection in the Basin, thus forming a NW-SE extensional environment between the Ordos block and south China block [4]. In addition, the differential uplift of the Qinling orogenic belt resulted in the formation of a deeper Weihe Basin between the Qinling orogenic belt and the Ordos block [5-6].The complex crustal tectonic environment and geodynamic evolution process of the Weihe Basin have made this area as a hotspot for scholars at home and abroad (as shown in Figure 1). From the Eocene to the

Oligocene, by tracing a series of preexisting faults near east-west, it was discovered that the Weihe Basin expanded south and north, forming a nearly east-west faulted Basin. In the mid-Miocene, the Weihe Basin strengthened its rifting and accepted sedimentation again. At the same time, the sedimentary range expanded from north to south, and the Basin expanded to the south near the north piedmont fault of the Qinling orogen. During this period, the whole Weihe Basin was in the NEE direction, and at the same time, a normal fault system with NEE syncline series began to appear in the northern margin of the Weihe Basin. The late Miocene was an important stage in the tectonic evolution of the Weihe Basin, with the intensification of the fault-depression structure, the uplift area in the western part of the Weihe Basin began to rift and accept sedimentation. The northern boundary of the Basin also extended to the Liquan-Pucheng fault and moved eastward. The sedimentation of the Weihe Basin's fault depression extended to the Yuncheng Basin. In the early Pleistocene of the quaternary, the western part of the Weihe Basin gradually uplifted and there were sedimentary discontinuities. The Weihe Basin was generally deep in the east and shallow in the west, and the deposition center moved to the east [7-9].

In addition, the Weihe Basin is also an important seismic activity zone in our country. There have been many earthquakes above M6.0 in history. Among them, the Huaxian M8.0 earthquake had occurred at the year of 1556, which killed more than 800,000 people. However, in recent years, the seismic activity of the Weihe Basin has been weak, especially the central part of the Basin has been in a quiet period of earthquakes for a long time. This is extremely inconsistent with the strong background of the surrounding faults since the late quaternary. Therefore, the study of the characteristics of the crustal structure in the Weihe Basin is very important to grasp the deep structure and judge seismic risk of study area. At present, the research on the characteristics of the crustal structure in the Weihe Basin mainly relies on geological exploration, seismic deep reflection profile and other methods, and relatively few studies are carried out using gravity and magnetic data [10-12]. Generally speaking, the Bouguer gravity anomaly reflects the total effect of gravity anomaly of different depths, scales and densities of the crust. It not only reflects the vertical stratification of geological bodies, but also reflects the lateral density heterogeneity of the crustal medium. The gravity field is closely related to geological structure, crustal deformation and deep material migration, which can intuitively reflect the distribution of geological bodies, deep crustal structure and fault structure distribution, and provide an important scientific basis for the study of the earth's internal structure and geodynamics [13-15]. Traditional gravity data processing methods can only simply decompose gravity anomalies into local and regional fields. But gravity anomaly is formed by the superposition of underground anomaly with different depths and scales, so the traditional methods cannot clearly distinguish more detailed parts of the structure in different horizontal and vertical scales. In recent years, the wavelet analysis method has been widely used in the field of geophysical signal processing. Based on the gravity anomaly, geoscientists have done a lot of meaningful exploration and research. Hou et al [16] used wavelet multi-scale analysis method to analyze the Bouguer gravity anomaly in China mainland, and achieved good results. Yang et al [17] applied wavelet multi-scale analysis method to separating and processing of gravity anomaly, and proposed a new way for the separation of gravity anomaly. The basic principle of wavelet multi-scale analysis method is that, it can decompose the signal into multiple scale components that are woven together, and use the corresponding size of the spatial domain sampling step to obtain different details of the research object, so as to achieve the purpose of field separation [18-19].

Therefore, this article first calculates free air gravity anomaly and Bouguer gravity anomaly based on the EGM2008 global gravity field model. On this basis, the wavelet multi-scale analysis method is used to separate the field source of the Bouguer gravity anomaly. Then we study the characteristics of gravity field anomaly at different scales in the Weihe Basin, and analyze the crustal structure characteristics of the Weihe Basin, so as to provide reference information for researching the deep structure of the area's seismic risk.

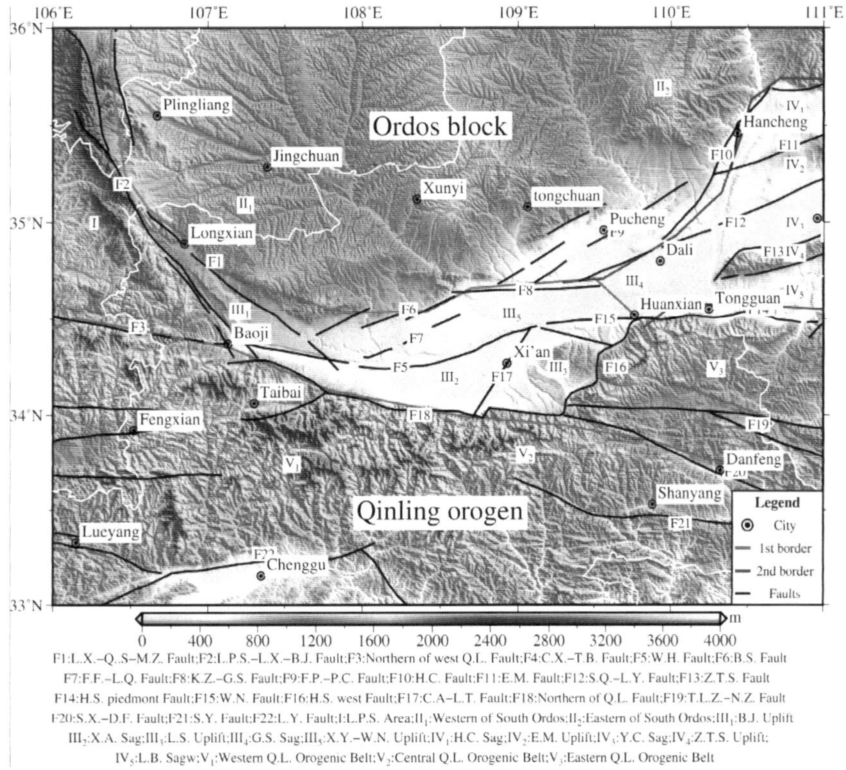

Figure 1. Structural features and faults distribution in Weihe Basin. The topographic figure is drawn based on the topographic data of ETOPO1. The red line represents the boundary of the first-level block in mainland China, and the blue line represents the boundary of the second-level block in mainland China. The block border data comes from *https://gmt-china.org/data/*. The black line indicates the main fault structure in the study area, and the fault data refers to the research results of Deng and others. The structural division of the study area mainly refers to the research results of Li [10] and others.

2. Data and methods

2.1. Research data

This paper mainly uses the EGM2008 (*http://icgem.gfz-potsdam.de/*) gravity field model data. The establishment of the EGM2008 earth gravity model based on multiple sources of gravity data such as GRACE satellite tracking data, satellite altimetry data, and ground regional average gravity data, as well as high-resolution global topographic data. The spherical harmonic coefficient of EGM2008 global gravity field model is up to 2160 orders, and the spatial resolution is about $5' \times 5'$. This model can provide information about gravity anomaly with a wavelength of about 10 km [20-21]. The free air gravity data comes from EGM2008 gravity field model, the highest order is 2160, and the reference system of the free air gravity data is WGS84. The weighted root mean square (WRMS) error of the data is 3.59 mGal. The digital terrain model data used for terrain correction is ETOPO1. The ETOPO1 data model is distributed by the National Geophysical Data Center (NGDC) in August 2008. Its spatial resolution is $1' \times 1'$. ETOPO1 (*https://www.ngdc.noaa.gov/mgg/global/etopo1sources.html*) data covers land and oceans around the world. Based on the free air gravity anomaly (as shown in Figure 2a) and ETOPO1 data, we obtained the Bouguer gravity anomaly through terrain correction and height correction (as shown in Figure 2b). In order to increase the calculation speed on the basis of ensuring the calculation accuracy, the terrain is divided into three areas, distant zone, intermediate zone and inner zone. In each zone, we use the different calculation scheme, the expression presented for the attraction produced by a right rectangular prism. In the distant zone the grid step is 4 km, in the intermediate and inner zone the grid step is 2 km respectively. The limit of the intermediate zone is 20km, the limit of the distant zone is 167 km, and the crustal reduction density is $2670\,kg/m^2$. In this article, the terrain correction procedure mainly refers to the literature of Fullea et al. [22-23]. In this paper we mainly study the Weihe Basin and its adjacent areas, which the range of longitude is from $106^\circ E$ to $111^\circ E$, the range of latitude is from $33^\circ N$ to $36^\circ N$.

2.2. *Wavelet multi-scale analysis method*

The wavelet multi-scale analysis method can decompose the Bouguer gravity anomaly into approximation and detail fields. The approximation fields are the reflection of regional Bouguer gravity anomaly information, which mainly reflect the large-scale low frequency anomaly information caused by deep field sources. The detail fields are the reflection of local Bouguer gravity anomaly information, which mainly reflect the small-scale high frequency anomaly information caused by shallow field sources. Among them, low-level details represent shallow geological body anomaly, and high-level details represent deep geological body anomaly [14-19]. The basic expression of the wavelet multi-scale analysis method is that: the signal $f(x) \in R$ with a resolution of 2^{-j}, using $A_j f(x)$ represents the low frequency approximation part in the space of V_j, $D_j f(x)$ represents the high frequency details in the space of W_j, then we can get [24]:

$$A_j f(x) = A_{j+1} f(x) + D_{j+1} f(x) \tag{1}$$

It can be further written as:

$$P_{V_0} f(x) = A_N f(x) + \sum_{j=1}^{N} D_j f(x) \tag{2}$$

Where, $P_{V_0} f(x)$ indicates that the function of $f(x)$ is mapped to the subspace of V_0, N is the maximum order of wavelet decomposition, for gravity anomaly $\Delta g(x)$, suppose it is decomposed into the space of $N = 4$, formula (2) can be simplified as [24]:

$$\Delta g(x) = A_4 G + D_4 G + D_3 G + D_2 G + D_1 G \tag{3}$$

Where, $A_4 G$ represents the 4th order wavelet approximation part, $D_1 G$ to $D_4 G$ represents 1st to 4th wavelet details. For 2D plane function $f(x, y)$, need using 2D wavelet transform for analysis. For 2D gravity anomaly, wavelet multi-scale decomposition can also be performed [24]:

$$\Delta g(x, y) = A_0 f(x, y) = A_4 f(x, y) + \sum_{j=1}^{4} (D_j^h f(x, y) + D_j^v f(x, y) + D_j^d f(x, y)) \tag{4}$$

Where, $D_j^h f(x, y)$, $D_j^v f(x, y)$, $D_j^d f(x, y)$ represents the high-frequency details in the horizontal, vertical and diagonal directions in the space of W_j^2, respectively.

2.3. *Logarithmic power spectrum method*

The Logarithmic power spectrum method is a gravity and magnetic field analytical processing method first proposed by Spector and Gant [25]. On the basis of wavelet multi-scale analysis method, the power spectrum method is used to calculate the equivalent depth of the field sources. The power spectrum method, as an auxiliary method of the wavelet multi-scale analysis method, adds a quantitative estimate of the average depth of field sources at different depths. The combination of power spectrum analysis and wavelet multi-scale analysis method is beneficial to the deeper and more reasonable crustal structure explanation [26-29]. Consider the spectrum of the gravity anomaly of the ball in the simplest case, the power spectrum expression in polar coordinates is that [25]:

$$E(r) = |S(u, v)|^2 = A^2 e^{-2hr} \tag{5}$$

Take the logarithm of both sides, we can get formula $\ln E(r) = \ln A^2 - 2hr$, $\ln E(r)$ is called the radial logarithmic power spectrum of the anomalous spectrum, A is the size factor, r is the circular frequency or wave number, h is the depth of the field source. The radial logarithmic power spectrum has a linear relationship with the radial frequency, and the slope of the straight line is $-2h$. According to the slope of the straight line, the depth of the field source can be calculated as [25]:

$$h = \frac{\ln E(r_1) - \ln E(r_2)}{2(r_1 - r_2)} \quad (6)$$

3. Characteristic analysis of gravity anomalies in Weihe Basin

3.1. Characteristics of the free air gravity anomaly

The free air gravity anomaly in the Weihe Basin (as shown in Figure 2a), shows obvious topographical-related features. In the whole Weihe Basin the free air gravity anomaly changes mainly in negative values, and in the Zhouzhi area in the middle section of the Weihe Basin, the free air gravity anomaly is at the lowest value, about -150 mGal. The free air gravity anomaly in the Liupanshan area shows a positive change, with a maximum value of 60 mGal, forming a gravity gradient zone with the adjacent north margin of the West Qinling. The free air gravity anomaly changes smoothly in the western area of the southern margin of the Ordos block, basically changing around 0 mGal. Local positive changes occur in the Xunyi-Zhengning area in the central part and the Huanglong area in the east of Weihe Basin, and the maximum values are 20 mGal and 50 mGal, respectively. The free air gravity anomaly in the Qinling orogenic belt changes mainly is positive values, with a maximum value of about 120 mGal. The gravity anomaly changes drastically between the Qinling orogenic belt and the Weihe Basin, forming a very significant gravity gradient zone. However, the change of gravity anomaly between the Shangxian-Danfeng suture zone in the northern margin of the Qinling Mountains and the Mianxian-Lueyang suture zone in the south is relative consistent. In addition, the Qinling orogenic belt shows Basin-ridge-mountain structural characteristics obviously.

3.2. Characteristics of the Bouguer gravity anomaly

The Bouguer gravity anomaly is obtained after the free air gravity anomaly which was corrected by mesosphere and topography [29-30] (as shown in Figure 2b). Compared with Figure 2a, there is a weak correlation between Bouguer gravity anomaly and topography, but there is an obvious mirror relationship between Bouguer gravity anomaly and Moho interface depth (as shown in the last figure). The Bouguer gravity anomaly in the Liupanshan tectonic area shows a wide range of negative change. Especially the triangular area bounded by the Longxian-Baoji fault and the Chengxian-Taibai fault is the lowest Bouguer gravity anomaly zone in the entire study area, with the smallest value which reaches -240 mGal. The southern margin of the Ordos block is bounded by Xunyi-Zhengning. The Bouguer gravity anomalies in the east and west are obviously different, indicating that there are significant differences in the crustal structure and the crustal medium density between these two regions. The Weihe Basin is bounded by the longitude of 109°, with obvious differences of the Bouguer gravity anomalies from east to west. There may be two reasons for the differences of the gravity anomaly: on the one hand, it may be related to the deep crustal structure and medium density, such as upwelling of mantle material, uplift of Moho surface, difference of rock density, and so on. On the other hand, it is related to the long-range comprehensive effect of NE pushing of Indian plate and SW pushing of Pacific plate.

In addition, in the eastern part of the Weihe Basin, there are obvious the Bouguer gravity anomaly change gradient zones near Huashan Piedmont fault, Shuangquan-Linyi fault, Hancheng fault, Zhongtiaoshan fault, and so on. From north to south of these areas, the Bouguer gravity anomaly appears alternately with high value and low value. The Bouguer gravity anomaly in the Qinling orogenic belt has obvious segmentation and zoning features. In particular, the Bouguer gravity anomalies of West, Mid and East Qinling bounded by Chang'an-Lintong fault arc obviously different. In addition, taking the Mianxian-Lueyang suture zone as the boundary, there are obvious differences in the Bouguer gravity anomalies between the north and south sides. From the Bouguer gravity anomaly image, the main structural units in the Weihe Basin and adjacent areas can be identified clearly, such as the first-level structures of the Weibei Uplift (correspond the high gravity anomaly area), the Liupanshan structural area (correspond the low gravity anomaly area), and so on; such as the second-level structures of the Baoji Uplift (correspond the high gravity anomaly areas), the Xi'an Sag (correspond the low gravity anomaly area), the Lishan Uplift (correspond the high gravity anomaly area), the Gushi Sag (correspond the low gravity anomaly area), and so on. The Bouguer gravity anomaly is a superimposed field that reflects the information of different deep field sources. Due to the mutual interference of different signals, the definition of geological structures and faults distribution characteristics are not enough, and the information of different field sources needs to be separated using the wavelet multi-scale analysis method.

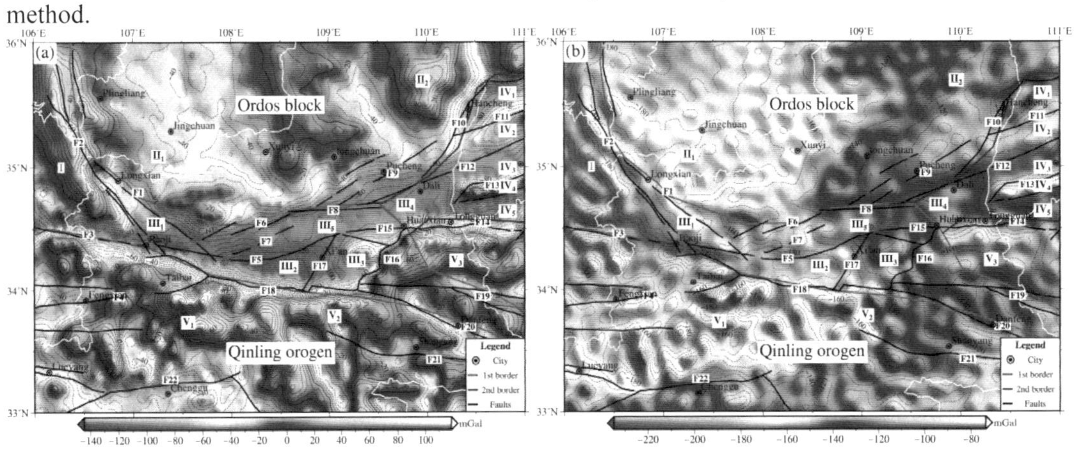

Figure 2. Gravity anomaly in Weihe Basin (a) Free air gravity anomaly; (b) Bouguer gravity anomaly.

4. Characteristics of crustal structure in Weihe Basin

The Bouguer gravity anomaly data is decomposed by wavelet multi-scale of order 1st to 5th, and the wavelet approximation and detail images of order 1st to 5th are obtained. The approximate images of wavelet transform reflect the regional Bouguer gravity anomaly information, which mainly reflect the large-scale low-frequency anomaly information caused by the deep field sources. The gravity anomaly components include the interface depth change of the upper mantle top, the density change of the lower, the middle and the upper crust [16-18, 24, 26-29]. The wavelet transform detail images are the reflection of the local field Bouguer gravity anomaly information, which is mainly reflected the reflection of small-scale high-frequency information caused by shallow field sources. On this basis, the logarithmic power spectrum method is used to calculate the approximate field source depth corresponding to the 1st to 5th order wavelet approximation and detail images. Since the detail images of the 1st order wavelet transform mainly reflect the density changes of the sediments near the surface [28-29], as the same, the depth of the field source corresponding to the approximation images of 1st and 2nd order wavelet is basically the same, so this paper only analyze the features of deep crustal structure revealed by the detail and approximation images of the 2nd to 5th order.

4.1. Characteristics of the Bouguer Gravity Wavelet Approximation

These are 2nd to 5th wavelet approximation images of Bouguer gravity anomaly in the Weihe Basin [22, 30] (as shown in Figure 3). Table 1 shows that the equivalent field source depth corresponding to the 2nd to 5th order wavelet approximation, the range of the equivalent field source depth is from 15 to 84 km. Although the depth of the gravity field reflected in the image of each order of wavelet approximation is different, however the 2nd to 5th order wavelet approximation images can clearly identify the basic pattern of different tectonic units in the study area. With the increase of the wavelet approximation order, the local anomaly information of gravity anomaly in different structural units and large faults has been better suppressed, and the regional anomaly information has become more prominent. The basic pattern of major tectonic units and their distribution characteristics of Bouguer gravity anomaly are becoming more apparent. The characteristics of gravity anomaly belts distributed in different regions are different in their directions, shapes, and gradient changes, which are actually manifestations of geological differences.

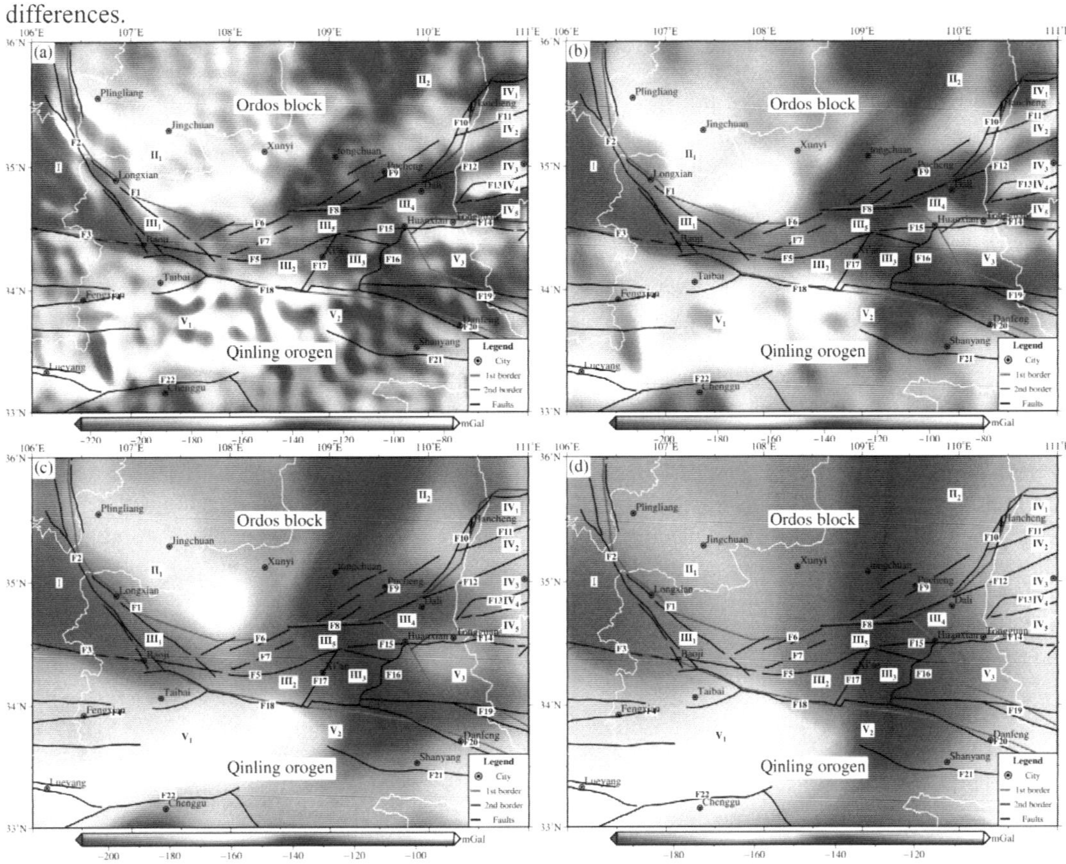

Figure 3. Wavelet approximation of Bouguer gravity anomaly in Weihe Basin.

(a) 2nd order; (b) 3rd order; (c) 4th order; (d) 5th order

Table 1. Equivalent field source depth of Wavelet Multiscale Decomposition.

Order	Equivalent field source depth (km)	Order	Equivalent field source depth (km)
1st order	12	1st order	3~5
2nd order	15	2nd order	5~6
3rd order	31	3rd order	10
4th order	51	4th order	24
5th order	84	5th order	49

This is 2nd order wavelet approximation image of the Bouguer gravity anomaly in the Weihe Basin (as shown in Figure 3a), Figure3a presents obvious zoning characteristics: taking the Xunyi-Xi'an as the boundary, the gravity anomaly in eastern and western of study area has obvious differences. The Bouguer gravity anomaly in the Liupanshan tectonic area appears low value. The Bouguer gravity anomaly gradually increases from west to east of the study area, and the variation range is from -240 to -80 mGal. Taking the Xunyi-Xi'an as the boundary, there are significant differences of the gravity anomalies from south to north in the western part of the study area. Comparison of the southwestern margin of the Ordos block and the Qinling orogenic belt with the Weihe Basin, the Bouguer gravity anomaly in the first two regions is slightly larger, but the internal gravity anomaly has little different, showing obvious horizontal density uniformity. The boundary of gravity anomaly difference has obvious consistency with the geological structure boundary, and the deep and large faults, structural units and the distribution of gravity anomaly have a good corresponding relationship. The Bouguer gravity anomaly in the eastern part of the Weihe Basin is relatively complex, and two relatively low value areas of Bouguer gravity anomaly appear the areas of Dali and Hancheng. The Bouguer gravity anomaly in the central part of the Weihe Basin changes relatively smoothly, indicating that the gradient of the topographical change is relatively slow. The Bouguer gravity anomaly on the southern margin of Ordos gradually increased from west to east, but the magnitude was lower than that of the Weihe Basin. In the east and west of the Qinling orogenic belt, the Bouguer gravity anomaly is significantly different. The south of Huaxian and north of Luonan is also known as the small Qinling tectonic belt [5-6], the Bouguer gravity anomaly value reaches -100 mGal, which is nearly 100 mGal different from the West Qinling orogenic belt. Further observation reveals that the NEE trending Chang'an-Lintong fault seems to be the dividing line between the eastern and western Qinling orogenic belts, which further shows that the EW direction of the Qinling orogenic belt is not continuous, and its orogenic formation mechanism and orogenic period is not the same [31].

This is 3rd order wavelet approximation of the Bouguer gravity anomaly in the Weihe Basin (as shown in Figure 3b), it can be seen that along with the depth of the field source continues to increase, the regional characteristics of the Bouguer gravity anomaly in the Weihe Basin become more obvious, reflecting the basic characteristics and distribution characteristics of deep structures. In particular, some local gravity anomaly in the southern margin of the Ordos block and the Qinling orogenic belt disappeared. The boundary characteristics of the secondary block structure are more obvious, the gravity anomaly in the west of the Weihe Basin has a large-scale consistency. On the contrary, the difference in the Bouguer gravity anomaly in the east of the study area is quite obvious, showing obvious differences in deep density. The structural features of depression and uplift are obvious. From the wavelet approximation of Bouguer gravity anomaly in the Weihe Basin (as shown in Figure 3c, 3d), the gradual increase of Bouguer gravity anomaly from west to east becomes more obvious with the further deepening of the source depth.

4.2. Characteristics of the Bouguer Gravity Wavelet Details

Figure 4 shows the 2nd to 5th order wavelet detail images of the Bouguer gravity anomaly in the Weihe Basin. Table 1 shows the depth of the field source corresponding to the wavelet details is from 5 to 49 km. The 2nd order wavelet detail of the Bouguer gravity anomaly in the study area (as shown in Figure 4a) reflects the diverse and complex field source information, which mainly reflects the lateral anisotropy of the sedimentary layer and upper crust density. The long axis of the abnormal trap has a good corresponding relationship with the distribution of faults and the main structures. The gravity anomaly in the Ordos block is basically concentrated between -4 and 4 mGal, the gravity anomaly range is small and the amplitude is low, showing a single crustal structure with obvious stable cratonic characteristics. The gravity anomaly between the Ordos block and the Qinling orogenic belt presents obvious zoning and block characteristics. The gravity anomaly trend gradually changes from NW to NE in the west and east of the Weihe Basin, and the magnitude of the change is more than that in the Ordos block, the increasing trend basically changes between -10 and 10 mGal. The gravity anomaly changes in the Qinling orogenic belt is the most intense, with the largest scale and amplitude.

According to the equivalent field source depth, the 3rd order wavelet details of gravity anomaly in the Weihe Basin (as shown in Figure 4b) mainly reflects structural information at the depth of about 10 km. Compared with the 2nd order details of gravity anomalies (as shown in Figure 4a), the local small-scale gravity anomalies are significantly reduced. The Ordos block presents a wide range of low-value gravity anomalies, and the cratonic characteristics are further reflected. The gravity anomaly in the Weihe Basin presents a narrow strip with multiple associated gravity cascades. The distribution of faults and structural zoning is becoming more and more obvious. The boundaries of the positive and negative changes of gravity anomalies often correspond to relatively large-scale faults, such as the Longxian-Baoji fault, the Shuangquan-Linyi fault, the Huashan Piedmont fault, the Chang'an-Lintong fault, and so on. In addition, the main structural units such as the Baoji Uplift, the Xi'an Sag, the Lishan Uplift and the Gushi Sag can be identified based on the distribution characteristics of gravity anomalies. The gravity anomaly in the Qinling orogenic belt alternates with positive and negative changes, with obvious Basin-ridge structural features [5-6, 10].

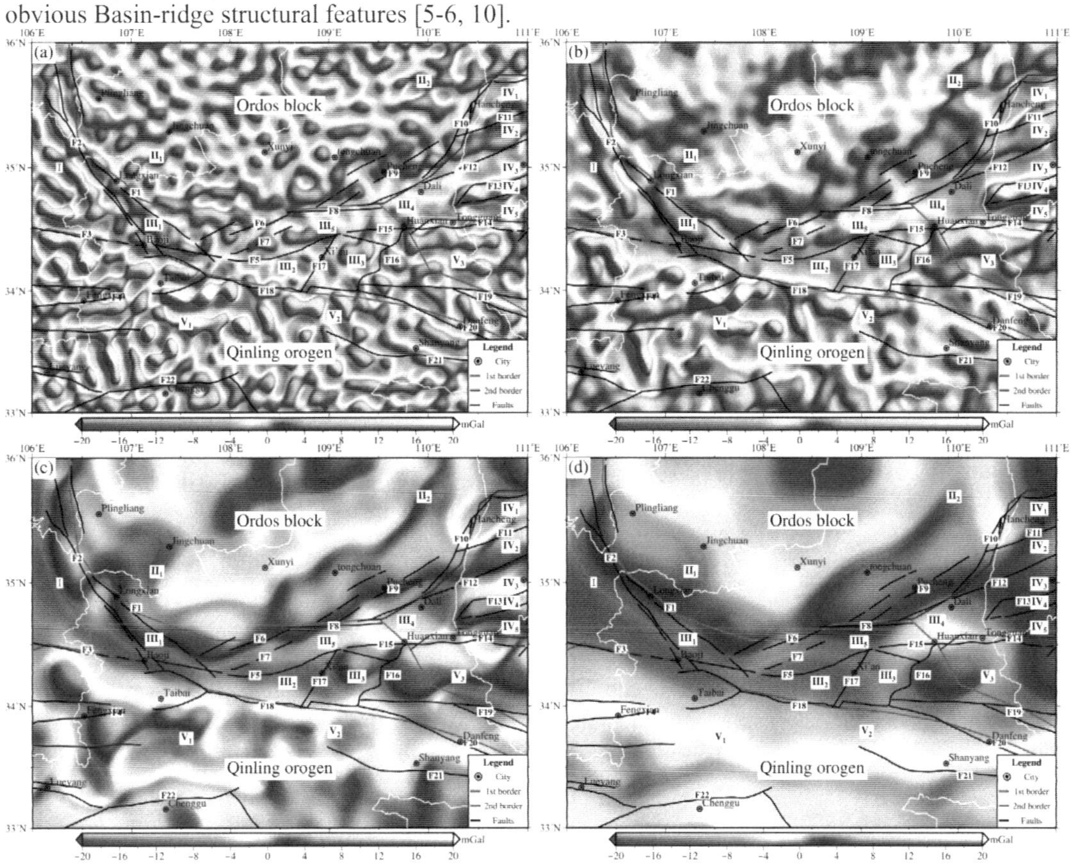

Figure 4. Wavelet detail of Bouguer gravity anomaly in Weihe Basin.

(a) 2nd order; (b) 3rd order; (c) 4th order; (d) 5th order

Figure 4a and 4b mainly reflect the density structure characteristics of the middle and upper crust. The distribution of gravity anomaly indicates that the crustal density in the study area changes drastically laterally, and the geological body is relatively fragmented. The radial logarithmic power spectrum reflects the field source depth is about 10 km, which is similar to the upper and middle crust depths of the study area [11]. The closeness mainly reveals the variation of the density heterogeneity of the upper and middle crust, and shows the difference in the density structure of the crust in the Ordos block, the Weihe Basin and the Qinling orogenic belt. With the increase of the wavelet detail order, the gravity anomaly becomes extensive and smooth, which mainly reflects the large scale low frequency anomaly information caused by deep field sources. According to the equivalent field source depth, the 4th and 5th order wavelet details (as shown in Figure 4c, 4d) mainly reflect the density structure of the lower crust and upper mantle. Compared with the 2nd and 3rd order wavelet detail figures, the gravity anomaly distribution and trend characteristics of the 4th and 5th order are quite different. With the disappearance of small scale stripe-shaped and beads-shaped anomalies, it reflects the relatively poor continuity and penetration of the longitudinal density structure in the study area. In addition, the gravity anomaly in the southern margin of the Ordos block is mainly positive and negative low value anomaly, and the depression-uplift structural units are distributed alternately. There is a clear boundary between the Ordos block and the Weihe Basin, while the boundary between the Weihe Basin and the Qinling orogenic belt is not obvious, it can be seen that the basement density of the Weihe Basin and the Qinling Orogenic Belt tends to be the same at the lower crust depth. In particular, in terms of west of the Chang'an-Lintong fault, the gravity anomaly shows a wide range of low-value anomaly, which is consistent with the Weihe Basin's near NS trending extensional structural environment. The 5th order wavelet detail (as shown in Figure 4d) mainly reflects the density structure of the upper mantle. The density difference of the adjacent areas between the Ordos block and the Weihe Basin is obvious, and the position of the gradient zone corresponds to the block boundary. The gravity anomaly between the Weihe Basin and the Qinling orogenic belt is basically the same, indicating that the two areas belong to the same tectonic system in the deep crust. From north to south of the Qinling orogenic belt, the change of gravity anomaly between Shangxian-Danfeng and Mianxian-Lueyang suture zone is relatively gentle, showing a linear trend from negative to positive.

In recent years, relatively few studies have been conducted on the distribution of faults and structural zoning in the Weihe Basin based on gravity and magnetic data, and most of the studies have been based on geological and seismic methods. In this paper, based on the Bouguer gravity anomaly data, the research results obtained by the wavelet multi-scale analysis method are different from the previous research results. Such as the boundary of the structural zone deviates from the existing results, the spread position and cutting depth of the fault are different from the previous results. The reasons for this difference may be: the resolution and accuracy of the Bouguer gravity data is not enough high, and there are differences between geophysical methods and geological and seismic methods. In the future, we will use higher precision and resolution gravity anomaly data and advanced analysis methods to research the tectonic divisions and faults of Weihe Basin [10-13, 32-34].

4.3. Characteristics of the Moho surface depth

The Moho surface is the interface between the crust and the mantle. Based on the crustal thickness inverted by using the receiver function [35], and on the basis of subtracting the elevation data [36], the real Moho surface depth in the Weihe Basin is obtained (as shown in Figure 5). It can be seen from Figure 5 that the depth of the Moho surface in the study area is from 32 to 50 km, with large fluctuations in the lateral direction. Overall, it shows the characteristics of shallowness and deepness in the east and west of the study area. The Moho undulation and the Bouguer gravity anomaly (Figure 2b) show a "mirror" relationship obviously, which is in good agreement with the research results of Ren et al. [37]. The Liupanshan tectonic area and the Qinling orogenic belt have the deepest depth of the Moho, followed by the western Weihe Basin, and the southern edge of the Ordos block and the middle and eastern part of the Weihe Basin are the shallowest. According to the overall distribution characteristics of the Moho, the whole study area can be divided into four different regions based on the three Moho depth gradient zones, which are the Longxian-Taibai zone, the Tongchuan-Zhashui zone, the Pucheng-Shanyang zone, and the depth of Moho surface in each area presents different characteristics. The area to the west of Longxian-Taibai, which corresponds to the Liupanshan and West Qinling tectonic areas, the depth of Moho surface is from 44 to 50 km. The thickness of the crust gradually becomes thinner from west to east of the study area. The thinning of the thickness is mainly due to the upwelling of asthenosphere material, which causes the stable craton to undergo crustal fold deformation, thinning of the crust, frequent earthquakes, and active magmatic activity. In the middle part of the Weihe Basin and the Mid Qinling Tectonic Zone, the crustal thickness is gradually thinning, the Moho surface is distributed in a long stripe shape, and the depth is from 36 to 44 km. It is the transition zone between the west and east of the Weihe Basin, and the Moho surface changes relatively smoothly. The depth of Moho surface between the eastern part of the Weihe Basin and the southern margin of Ordos are distributed in a "plow shape" along the EW direction. With Lintong-Weinan as the boundary, the Moho is elliptical, with the long axis in the EW direction and the depth is from 30 to 38 km. It is the place where the depth of the Moho is the shallowest in the entire study area, showing obvious local uplift characteristics [11, 32], which is related to the geological structure evolution process and thicker sedimentary layers in this area. In general, the deep crust structure profile of the study area is closely related to the overlying structure, and can clearly show the following characteristics: The undulation of the Moho surface in the study area reflects the thickening of the crustal thickness from east to west. The block features of deep structures reflected by the depth of Moho surface are clear. The long axis direction of the sag area is mainly NE and NW direction, and the long axis direction of the uplift area is mainly NE and NEE direction.

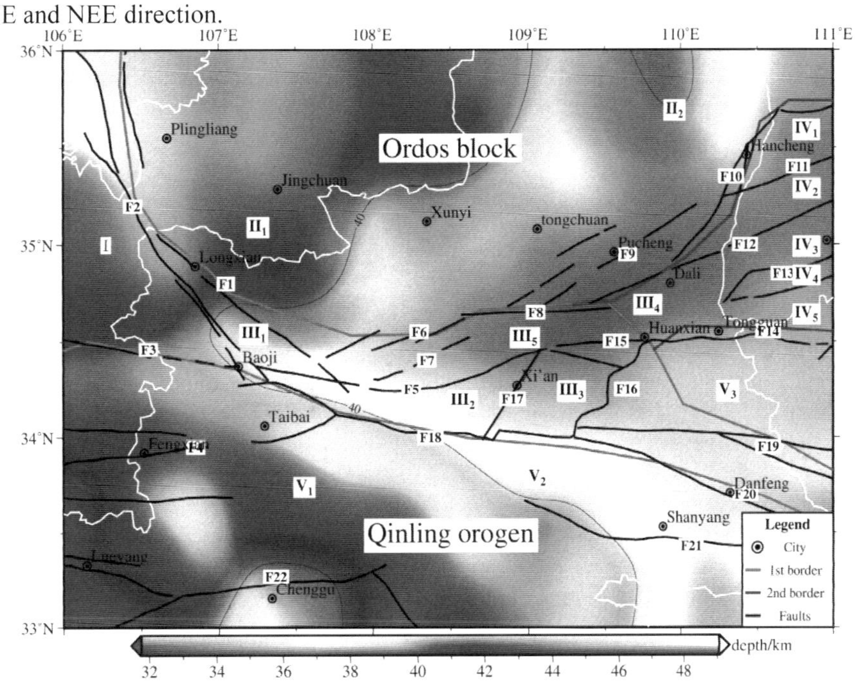

Figure 5. Moho surface depth of Weihe Basin.

5. Conclusions

Based on the data of EGM2008 and ETOPO1, we first calculate and obtain the data of free air gravity anomaly and Bouguer gravity anomaly in the Weihe Basin. On this basis, the wavelet multi-scale analysis method is used to decompose the Bouguer gravity anomaly data, and the 1st to 5th order wavelet detail and wavelet approximation data are obtained. At the same time, the logarithmic power spectrum analysis method is used to quantitatively calculate the average depth of the field source corresponding to the wavelet detail and wavelet approximation of 1st to 5th order. Finally, combined with the geological, seismic and other geophysical data of the Weihe Basin, the characteristics of gravity anomaly and crust structures are analyzed, and the following conclusions are obtained:

(1) The resolution and accuracy of the Bouguer gravity anomaly data obtained based on the EGM2008 and the ETOPO1 data are relatively high. Compared with other field source information separation methods, the wavelet multi-scale analysis method is a very effective technical method. According to the results of this paper, this method has obvious advantages for separating gravity anomalies caused by different depths and different scales of geological bodies.

(2) The free air gravity anomaly in the Weihe Basin has a strong correlation with topography, which can reflect obvious regional topographic features. The Bouguer gravity anomaly in the Weihe Basin has a relatively weak correlation with topography, but there is a mirror image relationship with the Moho depth in the area. And according to the characteristics of the Bouguer gravity anomaly, the main structural boundaries and fault distribution locations in the Weihe Basin can be identified.

(3) The wavelet approximation data of Bouguer gravity anomalies of order 1st to 5th in the Weihe Basin correspond to the average depth of the field source at 12 to 84 km, corresponding to the upper crust, lower crust and upper mantle respectively. The wavelet detail data of Bouguer gravity anomalies of order 1st to 5th in the Weihe Basin correspond to field source depths of 3 to 49 km, corresponding to sedimentary layers, upper crust, middle crust, and lower crust, respectively. The basic pattern of different tectonic units in the study area can be clearly seen from the wavelet detail and the approximation images. The boundary of the positive and negative variation of the Bouguer gravity anomaly often corresponds to a relatively large-scale fault. In addition, according to the distribution characteristics of gravity anomalies, major structural units such as the Baoji Uplift, the Xi'an Sag, the Lishan Uplift, and the Gushi Sag can also be identified. The gravity anomaly of the Qinling orogenic belt changes between positive and negative alternately, and the Basin-ridge structure features obvious.

Acknowledgments
This study was supported by Sail and Innovation Fund of Shaanxi Earthquake Agency (NO.: QC202013) and Innovation Team Project of Shaanxi Earthquake Agency (Project: Observation of Spatial Deformation and Deformation Dynamics).

References
[1] P Molnar and P Tapponnier 1975 *J. Science* **189** 4201
[2] P Tapponnier and P Molnar 1976 *J. Journal of Geophysical Research* **82** 20
[3] P Tapponnier, Z Q Xu, F Roger, et al 2001 *J. Science* **294** 5547
[4] P Z Zhang, Z K Shen, M Wang, et al 2004 *J. Geology* **32** 9
[5] G W Zhang, A L Guo, Y P Dong, et al 2011 *J. Earth Science Frontiers* **18** 3
[6] Q S Wang, J W Teng, Y Q Zhang , et al. 2013 *J. Chinese Journal of Geophysics* **56** 3
[7] Z C Li, W H, Li Y X Li, et al 2015 *J. Journal of Palaeogeography* **17** 4
[8] B Wang, H B Zheng, Z He, et al 2014 *J. Geological Journal of China Universities* **20** 3
[9] J L Mercier, P Vergely, Y Q Zhang, et al 2013 *J. Tectonophysics* **582**
[10] Y S Li 1992 M. Seismological Press Beijing
[11] J Ren, X J Feng, F Y Wang, et al 2013 *J. Chinese Journal of Geophysics* **56** 2
[12] X C Quan 2005 *J. Coal geology of China* **17** 3
[13] B R Xu 1990 J. Journal of Xi'an Petroleum Institute **5** 4
[14] H Lou, C Y Wang 2005 *J. Acta Seismologica Sinica* **27** 5
[15] J W Teng, S L Li, Y Q Zhang, et al 2014 *J. Chinese Journal of Geophysics* **57** 10
[16] Z Z Hou and W C Yang 1997 *J. Acta Geophysica Sinica* **40** 1
[17] W C Yang, Z Q Shi, Z Z Hou, et al 2001 *J. Chinese Journal of Geophysics* **44** 4
[18] Y Xu, T Y Hao, Z W Li, et al 2009 *J. Journal of Geophysics and Engineering* **6**
[19] W L Jiang, J F Zhang, T Tian, et al 2012 *J. Physics of the Earth and Planetary Interiors* **212-213**
[20] N K Pavlis, S A Holmes, S C Kenyon, et al 2008 *C. The 2008 EGU, Vienna, Austria.*
[21] N K Pavils, S A Holmes, S C Kenyon, et al 2012 *J. Journal of Geophysical Research* **117** B4
[22] J Fullea, M Fernàndez, H Zeyen 2008 *J. Computers and Geosciences* **34** 12
[23] W W Zhu, K Qin, T Q Wang, et al 2016 *J. Journal of Geodesy and Geodynamics* **36** 10
[24] T Y Liu 2017 M. Beijing: Science Press
[25] A Spector and F S Grant 1970 *J. Geophysics* **35** 2
[26] Z Z Hou, W C Yang, Y Wang, et al 2015 *J. Chinese Journal of Geophysics* **58** 3
[27] W L Jiang and J F Zhang 2011 *J. Progress in Geophysics* **26** 6
[28] Y Chen, J F Zhang, W L Jiang, et al 2017 *J. Progress in Geophysics* **32** 6

[29] S M Fang, X K Zhang, S X Jia, et al 2002 *J. Journal of Geodesy and Geodynamics* **22** 1
[30] S B Sun, S Chen, L Z Gong, et al 2018 *J. Acta Seismologica Sinica* **40** 3
[31] Y P Dong, G W Zhang, S S Sun, et al 2019 *J. Journal of Geomechanics* **25** 5
[32] L C Lu, P Wang, M X Zhou, et al 2017 *J. Journal of Geodesy and Geodynamics* **39** supp.1
[33] L Z Liu, L C Lu, B K Zhao, et al 2018 *J. Geology of Shaanxi* **36** 1
[34] X L Feng, B Q Yuan, Y H Li, et al 2019 *J. Oil Geophysical Prospecting* **54** 2
[35] Y H Li, M T Gao and Q J Wu 2014 *J. Tectonophysics* **611**
[36] E H Zhang, L Shi, J Luo, et al 2018 *J. Acta Seismologica Sinica* **40** 6
[37] X Ren, Z G Xu, H Yang, et al 2012 *J. Chinese Journal of Geophysics* **55** 12

IOP Conf. Series: Earth and Environmental Science **865** (2021) 012015 doi:10.1088/1755-1315/865/1/012015

Automatic stem mapping using single-scan terrestrial laser scanning data and Mean Shift clustering

Xiangjiang Liu[1], Maolin Chen[1,3], Chunsen Tan[1], Xinyi Zhang[1] and Wenguang Yang[2]

[1]School of Civil Engineering, Chongqing Jiaotong University, No. 66, Xuefu Avenue, Nan'an District, Chongqing, China;
[2]School of Earth Sciences, Cheng Du University of Technology, No. 1, East Third Road, er xian qiao street, Chenghua District, Chengdu, China

[3]Corresponding author's email: maolinchen@cqjtu.edu.cn

Abstract. As an effective tool for evaluation of forest structure parameters, terrestrial laser scanning (TLS) can capture forest data efficiently and automatically. In this paper, a stem mapping method of TLS data is proposed, based on Mean Shift clustering. The ground points first are extracted by cloth simulation filter (CSF) and used to generate ground model. A subsequent classification based on Random Forest (RF) identifies candidate stem points using 21-dimensional geometric features. After obtaining the candidate stem points, we utilize Mean Shift clustering to extract individual stems, instead of the usual Euclidean clustering method, and adopt adaptive density filtering to remove non-stem cluster. Finally, RANSAC cylinder fitting is used from the sliced point cloud, the intersection of the cylinder axis and the ground is taken as the stem location. The results show that Mean Shift clustering is more effective than Euclidean clustering in stem detection. Additionally, the precision and recall of stem mapping based on the proposed method are 88.83% and 93.94% respectively.

1. Introduction

Terrestrial laser scanning (TLS) has the ability to detect stems of below the canopy. It can effectively penetrate the forest interior to obtain high-density and high-precision point cloud data in a fast, nondestructive and automatic way [1, 2], which makes it possible to automatically collect the reference data and quantitative forest inventories [3].

The stem location within forest sample plots can be used as a fundamental parameter for extraction of information regarding the vertical structure [4], registration [5], or calibration [6]. So far, three main method types of stem mapping have been proposed, which are two-dimensional-layer methods, image-based methods and the single-point methods. In the two-dimensional-layer methods, the point cloud is usually sliced horizontally into a series of sections at a certain height, or the 3D points are projected onto a plane for postprocessing. Tanaey et al. [7] used least-square shape-fitting and a circular Hough transform to measure the diameter at breast height (DBH) based on slicing the point cloud. The least square fitting is proved to be the most accurate method, with an RMSE of 0.019 m. Pueschel et al. [8] systematically evaluated the influence of scanning mode and three circle fitting algorithms on the extraction of stem diameter and volume, and proposed the method of extracting stem points based on the vertical slice of point cloud and circle fitting. Though image-based methods use mature image processing technology to stem detection, its applicability is limited. Aschooff et al. [9] used the method of transforming multiple slices extracted from different heights in X/Y projection into regular raster image, mapped the stem contour into Hough-transformation circle rings, and extracted accurately based on the circle and ellipse fitting. In the single-point methods, the feature extraction and classification are usually used to detect the target of interest. Liang et al. [10] proposed a two-scale property to extract stem points, which used cylinder fitting to calculate the stem radius and direction. Due to the greater influence of point density, the overall detection accuracy was 73%. Wang et al. [11] separated stems between 2 m and 4 m above the ground by point projection density and Z-normal vector value, detected complete stems by DBSCAN clustering, and extracted stem parameters based on RANSAC cylinder fitting. Chen et al. [12] used local point features by establishing an adaptive KNN and classified the stem points by support vector machine (SVM) method based on Cuckoo search(CS). Then Euclidean clustering of points was carried out and the least square cylinder fitting was used to merge the stem cluster and build a stem model.

In the study of stem mapping based on TLS data, the variability of scan mode and the point density poses great challenges. It is necessary to further develop the technology of stem mapping with TLS. To avoid multi-scan point cloud matching, a stem mapping method in large areas is proposed to use single-scan TLS data in this paper. Compared with some previous studies, the proposed method need not merge stem points. Instead, Mean Shift clustering is used to separate stems from neighboring stems. Furthermore, the method contains adaptive density filtering algorithm to deal with non-stem cluster. There are four main steps: classification, clustering, filtering and cylinder fitting. The single components and the respective methods are explained in Section 2.

2. Material and methods

2.1. *Study area and data*

The test data was obtained by Riegl VZ-400 scanner with angular density of 0.04° and maximum scanning distance more than 100 m. The main tree species is arbor, a few shrubs in understorey and mainly grass vegetation on the ground. The number and relative position of the reference stems are obtained by interactive measurement on the screen of TLS intensity image.

2.2. *Preprocessing of TLS data*

TLS data are classified by cloth simulation filter (CSF) algorithm [13] to separate the ground points from non-ground points. Then a regular grid is created from ground points. The lowest point in each subgrid is selected to build a triangulation net as the ground model and the non-ground points are used as the input data for classification.

2.3. *Candidate stem points recognition*

In the process of terrestrial laser scanning (TLS), the complexity of forest scenes caused by different point densities, scanning area and the occlusion effect [8, 14-16], which brings many challenges to forest inventory. Based on 3D scene analysis, taking advantage of the geometric features of stem points, the candidate stem points are classified from non-ground points [12, 13]. In this section, we refer to 21-dimensional geometric features proposed by Weinmann et al. [17], and Random Forest (RF) method [18] to classify the non-ground points into stem and crown classes. Then the points labelled as stem are used as candidate stem points for further processing.

2.4. *Mean Shift clustering*

After classification, candidate stem points are obtained, and most of the non-stem points such as foliage, branch and grass are excluded. Due to the adaptability of feature vectors and the uncertainty of noise, there are still a few non-stem points preserved in the candidate stem points. But those points are much sparser than the stem points.

Base on this characteristic, we utilize the Mean Shift clustering [19] to further refine candidate stem points obtained from the classification. As a density-based spatial clustering algorithm, Mean Shift cleverly uses the density of the points to generate a reasonable number of clusters. The result is controlled by one parameter—bandwidth value. The bandwidth value can generally be chosen according to the knowledge of some specific fields. The clustering process is shown in Figure 1.

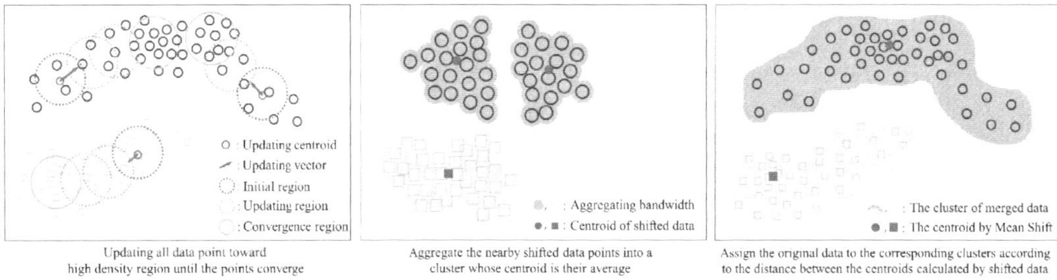

Figure 1. Flowchart of Mean Shift clustering algorithm.

In this section, the candidate stem points within the height of 4 m are projected on the horizontal plane for the Mean Shift clustering.

For a stem point set $X=\{x_i \in \mathbb{R}^2, i=1, 2, \ldots, n\}$ in the horizontal plane, the mean shift vector of any point x_i in the point set is

$$M_h(x) = \frac{1}{t}\sum_{x_i \in S_h}(x_i - x) \tag{1}$$

Where S_h is a high-dimensional sphere with bandwidth h centered on x, which represent the effective region, including t sample points.

The mean shift vector always move toward an area of densest point cloud, the main process is obtained by iteration

- Calculate the mean shift vector $M_h(x)$
- Translate the window $x^{j+1} = x^j + M_h(x^j)$

If the distance between the clustering center at the end of iteration and the previously obtained clustering center is less than h, they will be merged. Otherwise, the result will be treated as a new stem cluster until all the points are classified.

2.5. *Density adaptive filtering*

After Mean Shift clustering, there are still a few non-stem clusters. These small cluster points are relatively sparse and interfere with the stem detection, so it is necessary to set a threshold of point number to eliminate them. Since the point density decreases with an increase of the distance from the scanner and the number of cluster points is significantly different, which could result in poor filtering performance if a fixed threshold is set conventionally. This paper adopts a density-based adaptive method to filter small clusters [12].

2.6. *RANSAC cylinder fitting*

Random Sampling Consensus (RANSAC) [20] is an iterative method that can estimate the parameters of a mathematical model from an observational dataset containing outliers. Additionally, considering that the surface of the stem is approximately a cylinder [21, 22], for each point cluster, slices are generated on different height and used for RANSAC cylinder fitting [23]. Only the cylinders meeting below conditions are preserved:

- Radius is within $[R_{min}, R_{max}]$
- The axis of the cylinder is approximately parallel to the Z axis

For each stem cluster, the intersection point of the cylinder axis that meet the conditions and the ground model is calculated as the stem position.

3. Results and analysis

Seven parameters are defined in the proposed method. In Mean Shift clustering, the bandwidth h is set to 1.2m. Three parameters need to be set in adaptive density filtering. First, $\gamma = 0.35$ is the preseted occlusion scalar, which indicates the degree of invisibility of stems in the plot. Another two are the minimum stem height h_{min} and stem radius r_{min}, which are preseted as 6 m and 0.01m respectively. For the last three parameters, R_{min} and R_{max} depend on the radius of the stem in the scan area which are set to 0.01m and 0.5m respectively, the maximum allowed absolute angular distance in degrees between the direction of fitted cylinder and the Z axis is set to 5.

3.1. Classification results

In the course of the experiment, the number of points is reduced after ground filtering and classification. Table 1 lists the number in two steps. The number of stem points after stem classification for less than 15% of the non-ground points, as shown in Table 1.

Table 1. Points number after different steps.

Original	Ground points	Non-ground points	
		Non-stem points	Stem points
24128206	8270240	13885716	1972250

3.2. Clustering and filtering results

We compare the Mean Shift clustering with the commonly used Euclidean clustering [12, 14], as shown in Figures 2 and 3.

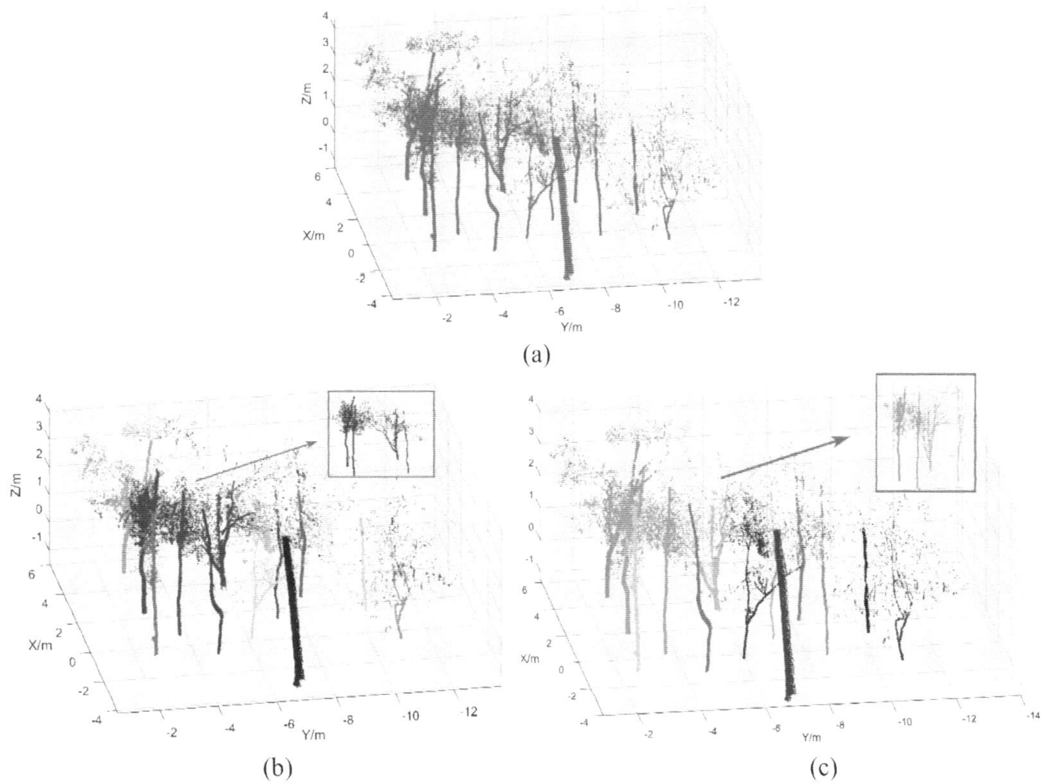

Figure 2. Comparison of two clustering methods (a) candidate stem points (13 stems); (b) Euclidean clustering (875 clusters); (c) Mean Shift clustering (13 clusters).

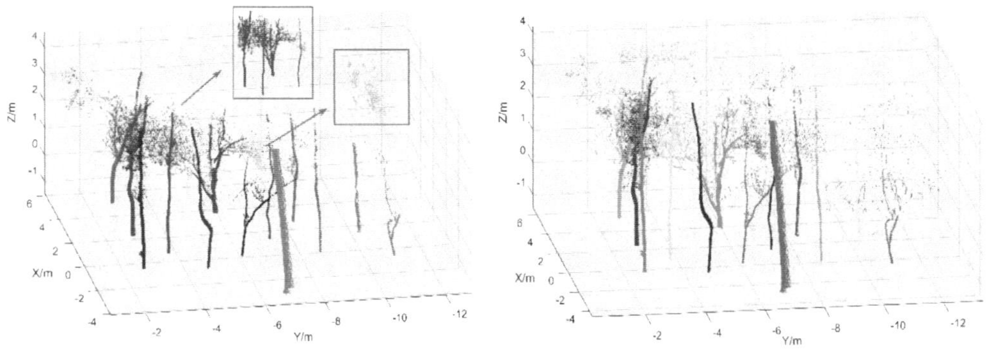

(a) (b)

Figure 3. Comparison of two clustering methods after filtering. (a) Filter results based on Euclidean clustering (11 clusters); (b) Filter results based on Mean Shift clustering (13 clusters).

From the perspective of cluster number, before filtering, candidate stem points formed 875 clusters by Euclidean clustering, while only 13 clusters are generated after Mean Shift clustering, which accurately correspond to the 13 stems in candidate stem points. As shown in the red box in Figure 2 (a), there is an obvious error that four adjacent stems are merged into one cluster in the Euclidean clustering result, but these four stems are accurately divided into four clusters in the Mean Shift clustering (As shown in Figure 2 (b)). After filtering, the cluster number based on Euclidean clustering is reduced to 11 clusters. It means that the adaptive density filtering can reduce a large number of non-stem clusters, but this kind of situation that still exists, as shown in Figure 3 (a). In contrast, Mean Shift clustering method has better clustering effect.

3.3. Stem detection results

The results of stem mapping by our method and the Euclidean clustering based method are shown in Figure 4. To assess the accuracy, precision and recall can be formulated, as follows:

$$Precision = \frac{T}{T + C} \tag{2}$$

$$Recall = \frac{T}{T + O} \tag{3}$$

True stem (*T*) indicates the number of correctly detected stems. Commission error (*C*), or Type I error, means the number of falsely recognized stems. Omission error (*O*), or Type II error, which is actually a stem but not recognized. Figure 4 shows the accuracy evaluation of stem mapping.

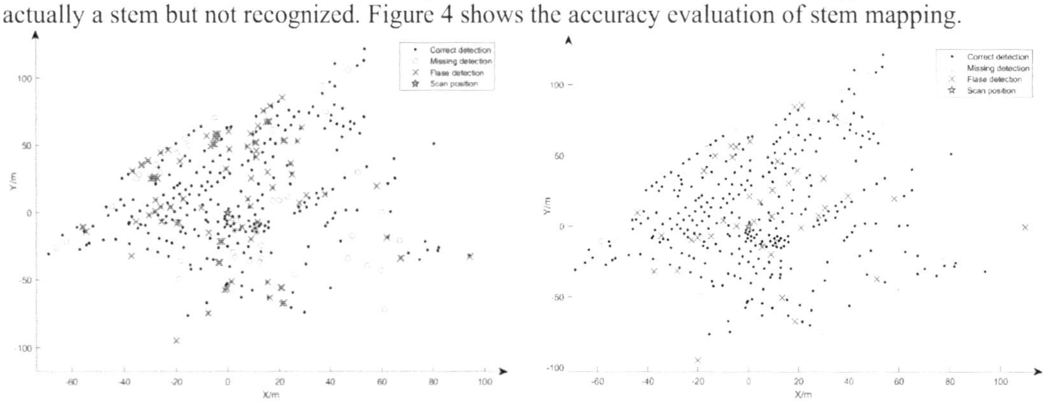

(a) (b)

Figure 4. Stem mapping. (a) Euclidean clustering based method; (b) The proposed method (Mean Shift clustering based method)

Table 2. Results and assessments of stem mapping.

Method	Reference stems	Detected stems	Type I error	Type II error	True stems	Precision	Recall
Euclidean clustering based method	330	382	94	42	288	75.39%	87.27%
The proposed method		349	39	20	310	88.83%	93.94%

Table 2 shows that 288 of 330 reference stems are correctly detected by the Euclidean clustering based method, which is less than the proposed method. Both clustering methods have some error detections. Type I error is mainly caused by the non-stem points formed after classification, but there are also errors caused by the lack of the process of merging stem points, which mainly exists in the Euclidean clustering based method. There are two main reasons for Type II error. First, most of the unrecognized stems are far away from the scan center, especially in Figure 4 (b), and the points on the surface of these stems are usually sparse or separated into several small pieces, which is caused by density change and occlusion effect. Therefore, those points are wrongly identified as a non-stem clusters and deleted at the classification or filtering phases. Second, because of the neighboring stems that are very close to each other or connection of non-stem points, the points from different stems are clustered together. This kind of situation is more significant in the Euclidean clustering based method. Type I error and Type II error of the Euclidean clustering based method more than twice as much as the proposed method as shown in Table 2. The results and assessments show that the proposed method is better than the Euclidean clustering based method.

4. Conclusions

This paper presented an automatic method for stem mapping from the single-scan point cloud. We focus on detection of single stem, which used Mean Shift clustering to extract single stem from the classified candidate stem points. The proposed method is tested in a large plot area in which the distances of stems vary from 1m to 5m. Among 330 reference stems, 310 stems are detected correctly and the precision and recall are 88.83% and 93.94% respectively. The test results indicate that the proposed method has better performance of stem detection and higher accuracy of stem mapping, compared with the Euclidean clustering based method. For future work, we plan to further 1) find suitable feature combinations by feature selection methods to improve the accuracy and efficiency. 2) explore the evolving Mean Shift with adaptive bandwidth to enhance the quality of the density estimate.

Acknowledgments

This work is supported in part by National Natural Science Foundation of China under grant no. 41801394, in part by Chongqing Natural Science Foundation under grant no. cstc2019jcyj-msxmX0370, in part by the Science and Technology Research Program of Chongqing Municipal Education Commission under grant no. KJQN201900729 and in part by Innovative Research Program for Graduates of Chongqing Jiaotong University under grant no. CYS21339.

References

[1] Van Leeuwen M, Nieuwenhuis M 2010 Retrieval of forest structural parameters using LiDAR remote sensing[J]. *European Journal of Forest Research*, **129(4)**: 749-770.

[2] Hopkinson C, Chasmer L, Young-Pow C, et al. 2004 Assessing forest metrics with a ground-based scanning lidar[J]. *Canadian Journal of Forest Research*, **34(3)**: 573-583.

[3] Raumonen P, Kaasalainen M, Åkerblom M, et al. 2013 Fast automatic precision tree models from terrestrial laser scanner data[J]. *Remote Sensing*, **5(2)**: 491-520.

[4] Maas H G, Bienert A, Scheller S, et al. 2008 Automatic forest inventory parameter determination

from terrestrial laser scanner data[J]. *International journal of remote sensing*, **29(5)**: 1579-1593.

[5] Zhang W, Chen Y, Wang H, et al. 2016 Efficient registration of terrestrial LiDAR scans using a coarse-to-fine strategy for forestry applications[J]. *Agricultural and Forest Meteorology*, **225**: 8-23.

[6] Liang X, Hyyppä J. 2013 Automatic stem mapping by merging several terrestrial laser scans at the feature and decision levels[J]. *Sensors*, **13(2)**: 1614-1634.

[7] Tansey K, Selmes N, Anstee A, et al. 2009 Estimating tree and stand variables in a Corsican Pine woodland from terrestrial laser scanner data[J]. *International Journal of Remote Sensing*, **30(19)**: 5195-5209.

[8] Pueschel P, Newnham G, Rock G, et al. 2013 The influence of scan mode and circle fitting on tree stem detection, stem diameter and volume extraction from terrestrial laser scans[J]. *ISPRS journal of photogrammetry and remote sensing*, **77**: 44-56.

[9] Aschoff T, Spiecker H. 2004 Algorithms for the automatic detection of trees in laser scanner data[J]. *International Archives of Photogrammetry, Remote Sensing and Spatial Information Sciences*, **36**(Part 8): W2.

[10] Liang X, Litkey P, Hyyppa J, et al. 2011 Automatic stem mapping using single-scan terrestrial laser scanning[J]. *IEEE Transactions on Geoscience and Remote Sensing*, **50(2)**: 661-670.

[11] Wang D, Hollaus M, Puttonen E, et al. 2016 Automatic and self-adaptive stem reconstruction in landslide-affected forests[J]. *Remote Sensing*, **8(12)**: 974.

[12] Chen M , Wan Y , Wang M , et al. 2018 Automatic Stem Detection in Terrestrial Laser Scanning Data With Distance-Adaptive Search Radius[J]. *IEEE Transactions on Geoscience and Remote Sensing*, **56(5)**: 2968-2979.

[13] Zhang W, Qi J, Wan P, et al. 2016 An easy-to-use airborne LiDAR data filtering method based on cloth simulation[J]. *Remote Sensing*, **8(6)**: 501.

[14] Xia S, Wang C, Pan F, et al. 2015 Detecting stems in dense and homogeneous forest using single-scan TLS[J]. *Forests*, **6(11)**: 3923-3945.

[15] Bienert A, Scheller S, Keane E, et al. 2007 Tree detection and diameter estimations by analysis of forest terrestrial laserscanner point clouds[C]//*ISPRS workshop on laser scanning*, **36**: 50-55.

[16] Brunner A, Gizachew B 2014 Rapid detection of stand density, tree positions, and tree diameter with a 2D terrestrial laser scanner[J]. *European journal of forest research*, **133(5)**: 819-831.

[17] Weinmann M, Jutzi B, Hinz S , et al. 2015 Semantic point cloud interpretation based on optimal neighborhoods, relevant features and efficient classifiers[J]. *ISPRS Journal of Photogrammetry and Remote Sensing*, **105**(jul.): 286-304.

[18] Breiman L 2001 Random forests[J]. *Machine learning*, **45(1)**: 5-32.

[19] Cheng Y 1995 Mean shift, mode seeking, and clustering[J]. *IEEE transactions on pattern analysis and machine intelligence*, **17(8)**: 790-799.

[20] Fischler M A , Bolles R C 1981 Random Sample Consensus: A Paradigm for Model Fitting with Applications To Image Analysis and Automated Cartography[J]. *Communications of the ACM*, **24(6)**: 381-395.

[21] Kelbe D, Van Aardt J, Romanczyk P, et al. 2015 Single-scan stem reconstruction using low-resolution terrestrial laser scanner data[J]. *IEEE Journal of Selected Topics in Applied Earth Observations and Remote Sensing*, **8(7)**: 3414-3427.

[22] Wang D, Hollaus M, Puttonen E, et al. 2016 Automatic and self-adaptive stem reconstruction in landslide-affected forests[J]. *Remote Sensing*, **8(12)**: 974.

[23] Wang D, Hollaus M, Puttonen E, et al. 2016 FAST AND ROBUST STEM RECONSTRUCTION IN COMPLEX ENVIRONMENTS USING TERRESTRIAL LASER SCANNING[J]. *International Archives of the Photogrammetry, Remote Sensing & Spatial Information Sciences*, 41.

IOP Conf. Series: Earth and Environmental Science **865** (2021) 012016 doi:10.1088/1755-1315/865/1/012016

Magnetostratigraphy and AMS study of a late Cretaceous-Paleogene succession in the eastern Xining basin: constraint on K-Pg boundary tectonic event in NE Tibet Plateau

Chicheng He[1, 2, 5], Zhonghang Wang[1, 2], Yanguo Fang[1, 2], Maohua Li[1, 2], Shaokai Li[3] and Yueqiao Zhang[4]

[1]Three Gorges Geotechnical Consultants Co., Ltd, Wuhan, 430074;
[2]Changjiang Design Group Co., Ltd, Wuhan, 430010;
[3]School of Earth and Space Science, Peking University, Beijing, 100871;
[4]School of Earth Sciences and Engineering, Nanjing University, Nanjing, 210023

[5]Corresponding author's e-mail: hcc@pku.edu.cn

Abstract. Cretaceous-Cenozoic basins developed in the NE Tibetan Plateau contain key archives to unravel the growth history of the plateau in response to the India-Eurasia collision. Here we present magnetostratigraphic results of a Late Cretaceous to Paleogene succession of the Zhongba section outcropping at the southern margin of the eastern Xining basin. This succession consists of three lithological units punctuated by two stratigraphic unconformities, which best recorded the deformation history of this foreland basin. Detailed magnetostratigraphic investigation show that the lower terrestrial sedimentary rock unit, the Minhe Group, was deposited in latest Cretaceous in the time span of ~74.5-69.2Ma; the middle unit was deposited in Paleogene in the time span of ~49.3-22Ma; and the upper conglomeratic unit, not dated, possibly was deposited in early Miocene. Accordingly, the K-Pg unconformity, widely observed in the foreland basins of NE Tibet, represents a sedimentary hiatus of ~19.9Ma duration from ~69.2 Ma to ~49.3 Ma, which possibly recorded the far-field response to the tectonic transition from Neo-Tethys oceanic plate subduction to the India-Eurasia collision in southern Tibet. AMS results show a clockwise rotation of the shortening direction from NEN-SWS in Latest-Cretaceous to NE-SW in Paleogene, which support the view that the compressional environment may dominate the NE Tibet in Early-Cenozoic period.

1. Introduction

The northeastern margin of this plateau, bounded to the south by the eastern Kunlun-West Qinling fault zone and to the north by the Altyn Tagh-Qilian Shan-Haiyuan fault zone (Figure 1), has been earlier viewed as the on-going frontier of the outward growth plateau, where the most tectonically vigorous deformation was attributed to a Pliocene-Quaternary plateau-building epoch [1]. In recent two decades, increasing evidences were accumulated in NE Tibet about range exhumation or cooling history, basin formation, and initial ages of fault activities around Paleocene to early-Eocene period, and has revealed a syn-collision tectonism, supporting the view that the plateau margins deformed simultaneously with the initial India-Eurasia collision [2-13].

As a response, many basins in northern Tibet recorded a sedimentary hiatus across the Cretaceous and Paleogene (K-Pg) boundary and revealed a consistent early Eocene age of the initial deposition through magnetostratigraphy studies (Figure 2) [7-13]. This K-Pg unconformity, widely presented in most of the basins in NE Tibet, may represent an important epoch of the tectonic transition from the Neo-Tethys subduction in Late Cretaceous to the India-Eurasia collision in Paleogene [14-15]. However, previous work has mostly focused on the age of Cenozoic sedimentary succession in NE Tibet but ignored the timing of this significant stratigraphic unconformity, and its duration was assigned vaguely between Early Cretaceous and Paleogene. Therefore, the question when did the India-Eurasia collision propagate northward to the region of the NE Tibetan margin, still remains elusive.

Another question concerns "How" the Early Cenozoic deformation generated the basin-and-range morphology in the NE Tibet Plateau. One opinion holds that compressional foreland environment prevailed in the NE Tibet in Paleogene [2, 11]. Another one based on extensional normal faults observed in the Paleogene strata of the Xining and other basins in the NE Tibetan Plateau [16-17], envisaged a transtensional setting possibly attributed to the far-field effect of westward subduction of the Pacific Plate.

To solve the above "When" and "How" questions, it needs a more complete stratigraphic chronology of Cretaceous-Paleogene succession in the NE Tibetan Plateau, which is vital for ascertaining the timing and possible northward propagation of deformation before and during the India-Eurasia collision. In this paper, we present the magnetostratigraphy and anisotropy of magnetic susceptibility (AMS) study of a Late Cretaceous to Paleogene succession of the Zhongba section outcropping in the southern margin of the eastern Xining basin, in an attempt to determine the age of the K-Pg unconformity and to explore its tectonic significance.

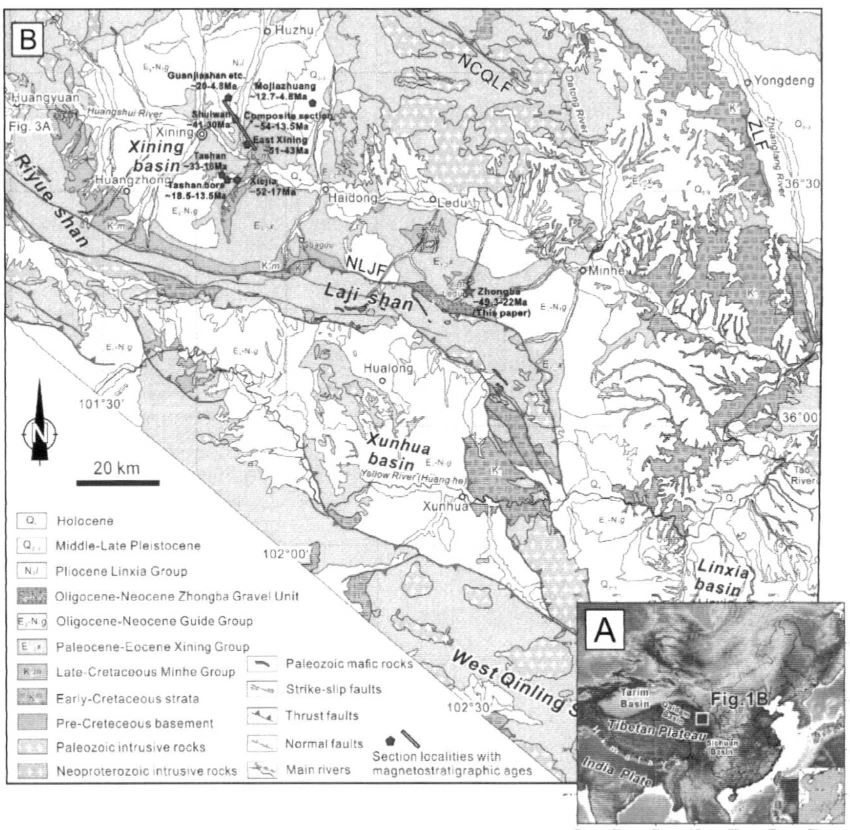

Figure 1. (A) DEM topography of the Tibetan Plateau and the location of Xining basin; (B) Simplified geological map of the Xining basin and its surrounding regions, showing locations of the previously studied magnetostratigraphic sections and our Zhongba section.

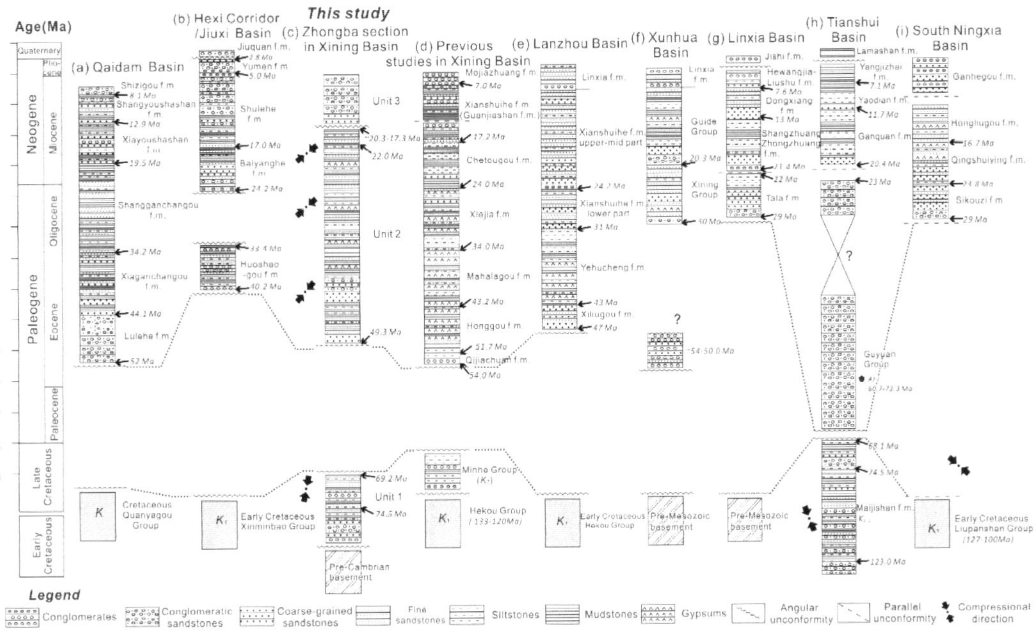

Figure 2. The correlation of different stratigraphic columns of the main Cretaceous-Cenozoic basins in the northeastern margin of the Tibetan Plateau.

2. Geological setting

As Figure 1 shows, the Xining basin was part of the unified Cretaceous-Cenozoic Longzhong Basin in NE Tibet (Figure 1B) [18-19]. It is separated to south from the Xunhua basin by Riyue Shan-Laji Shan and connects to the east with the Lanzhou-Linxia basin. Laji Shan, and its southward bending Jishi Shan, a branch of central Qilian Shan, has been considered as a young morphotectonic range being emerged in late Cenozoic. The Mesozoic-Cenozoic successions in the Xining basin were deformed in Late Cenozoic [8], forming a set of NNE-SSW striking faults and associated folds developed in the central and southeastern basin.

Cenozoic stratigraphy of the Xining basin consists of a thick and continuous succession, characterized by reddish and dominantly fine-grained lacustrine-fluvial to saline-playa deposits, which unconformably overlies the Cretaceous red beds of the Hekou and Minhe Groups or older basement rocks. The Early-Cretaceous Hekou Group had been partly dated, as ~133-120Ma [19], and the Cenozoic succession has been well established based on magnetostratigraphy studies and mammal fossils, as ~54-4.8Ma [8].

Our section lies at the north foothills of Laji Shan in the south of Xining basin. Figure 2 presents three lithological units separated by two distinct unconformities which are exposed in Zhongba section, and it can well be correlated to other sections in the inner Xining basin. The lower unconformity separates the Cretaceous red beds and Cenozoic strata, and the upper one was overlain by a distinct thick conglomeratic succession never seen before in the inner Xining basin, named Zhongba formation by us. This section was discovered earlier by Ye [20], and the bio-fossils found in the middle lithological unit can generally refer to a time span of Eocene to Oligocene period.

3. Magnetostratigraphy results and AMS analyses

Figure 3 show the magnetostratigraphy results of the Zhongba section. Based on the bio-fossils and magnetozones correlation, the Late Cretaceous Minhe Group was deduced to span the period of ~74.5-

69.2Ma. The basal age of the Paleogene succession is assigned to ~49.3 Ma, which suggests a hiatus of about 19.9Ma duration for K-Pg unconformity.

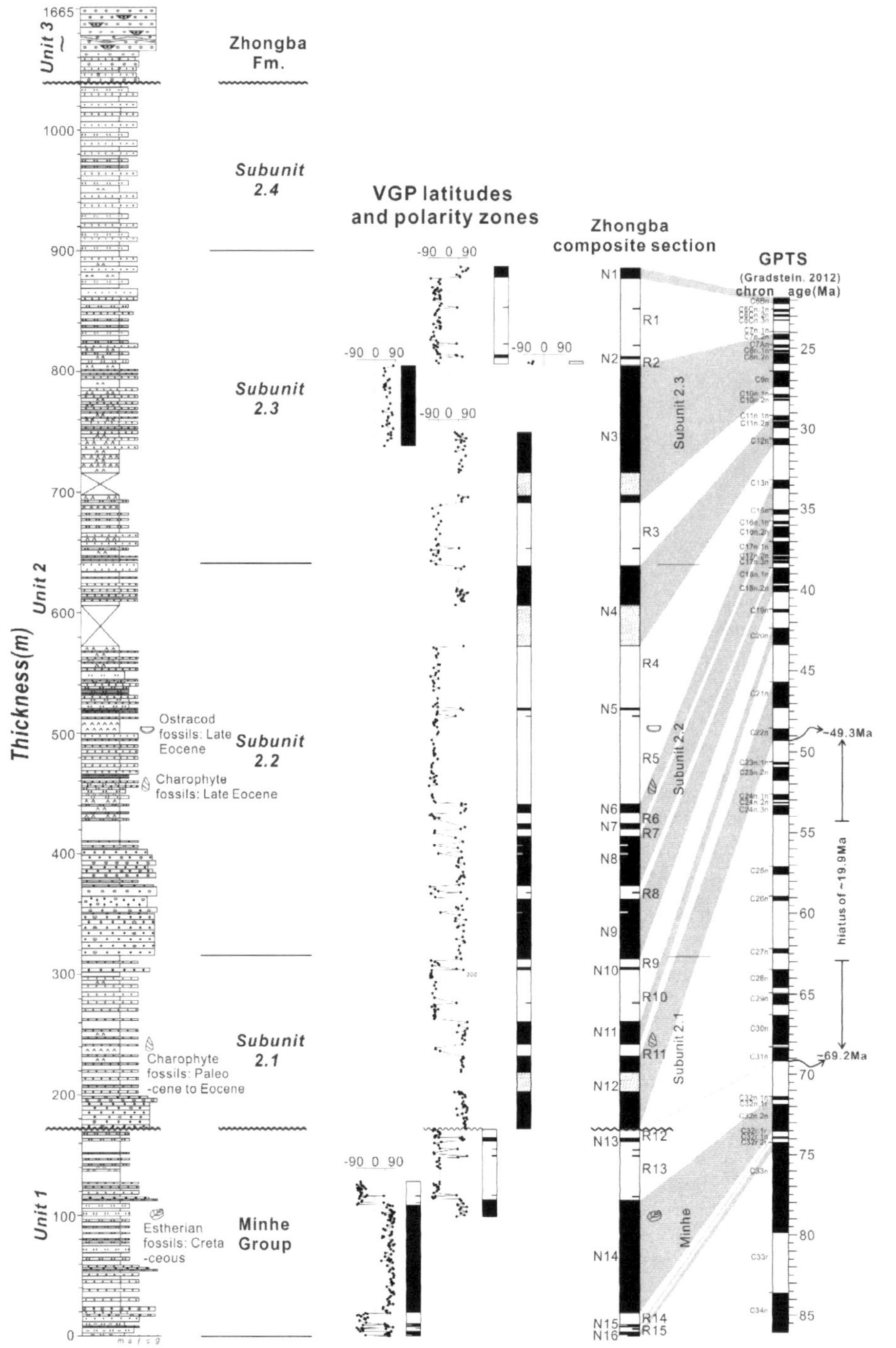

Figure 3. Lithostratigraphy and magnetostratigraphy results of the Zhongba section, and correlations of the observed polarity zones with the geomagnetic polarity timescale (GPTS).

Figure 4 presents the AMS results measured from the 536 specimens of Zhongba section. The Cretaceous AMS fabrics is featured by a cluster of k_1 axe perpendicular to shortening direction and k_3 axe normal to bedding, while the Paleogene AMS fabrics display a weakly-clustered girdle of k_3 axe and show typical "pencil structure", which indicates more intense deformation [21-22]. The mean pitch directions of k_1 calculated for each of the four lithological units by using Fisher statistical method, display a rotational pattern from WNW-ESE (112°) to NW-SE (129.3°) across the K-Pg boundary, and ~10° clockwise rotation at ~32Ma (Figure 4b1, c1).

Unlike the embryonic AMS fabrics which are mainly controlled by the stable hydrodynamic environment and paleocurrent, the AMS results of Unit 1 present that k_1 directions concentrated and parallel to the bedding, and it was probably related to the initial deformation [22]. Because weak strains commonly cause k_1 directions to be perpendicular to the maximum stress (σ_1) [21], we infer that the regional compressional direction recorded in the Upper Cretaceous succession (~74.5-69.2Ma) was in NEN-SWS (in the present coordinates without block rotation) and rotated to the NE-SW direction in the Paleogene. This result supports the view the Xining basin was continuously under the compressional stress environment spanning the period from Late Cretaceous to Early Cenozoic.

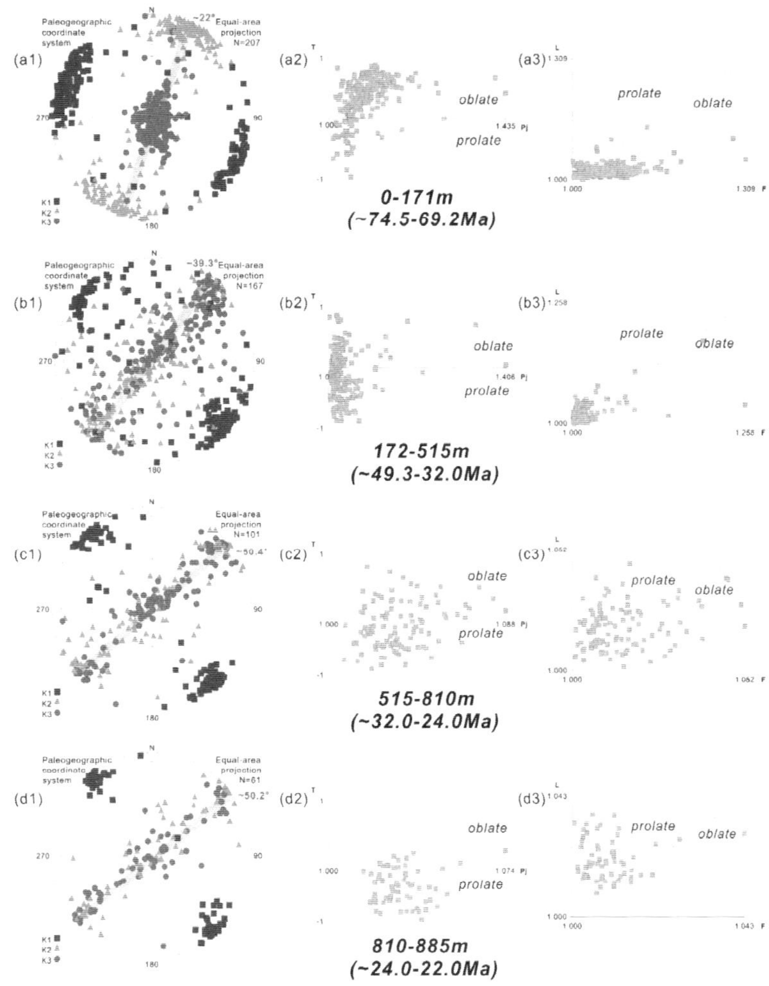

Figure 4. AMS magnetic fabrics of different sedimentary units in Zhongba section after bedding correction, and the corresponding P_j-T and Flinn diagrams.

IOP Conf. Series: Earth and Environmental Science **865** (2021) 012016 doi:10.1088/1755-1315/865/1/012016

4. Discussion: Implication of the K-Pg boundary unconformity in NE Tibet

Stratigraphically, the timing of this unconformity (~69.2-49.3Ma) constrained by the Zhongba magnetostratigraphic section provides the enveloping time-span for the real initial age of this K-Pg tectonic event in northern Tibet. This time interval corresponds to the late Cretaceous to Paleocene regional compressional event documented in central-southern Tibet, with the development of the Gangdese proto-Plateau, significant Late Cretaceous to Paleocene magmatic "flare-up" [23-24], major crustal shortening and various basins infilling [25-28], formation of a flexural foreland basin system along the Bangong-Nujiang suture (BNS) and Tanggula Shan orogen [29-31] (Figure 5A), all of these indicating that the Tibetan crust had been thickened substantially prior to the India-Eurasian collision.

Kapp et al. [32] proposed that northward underthrusting of the Lhasa block beneath the Qiangtang terrane in the latest Cretaceous to earliest Paleogene. This Andean-type retro-arc thrusting system may result from low-angle subduction of the Tethyan oceanic lithosphere beneath the Eurasian continent in late Cretaceous [33]. This process may trigger the basal tractions from mantle flow in a complicated mantle environment, as through "slab suction" of old slab remnants beneath Tibet [34], and then influence the vast areas across the NE Tibet Plateau. In short, the overlapping of the time span across the K-Pg unconformity boundary in NE Tibet must record the tectonic transition from the Neo-Tethys subduction to the India-Eurasian collision.

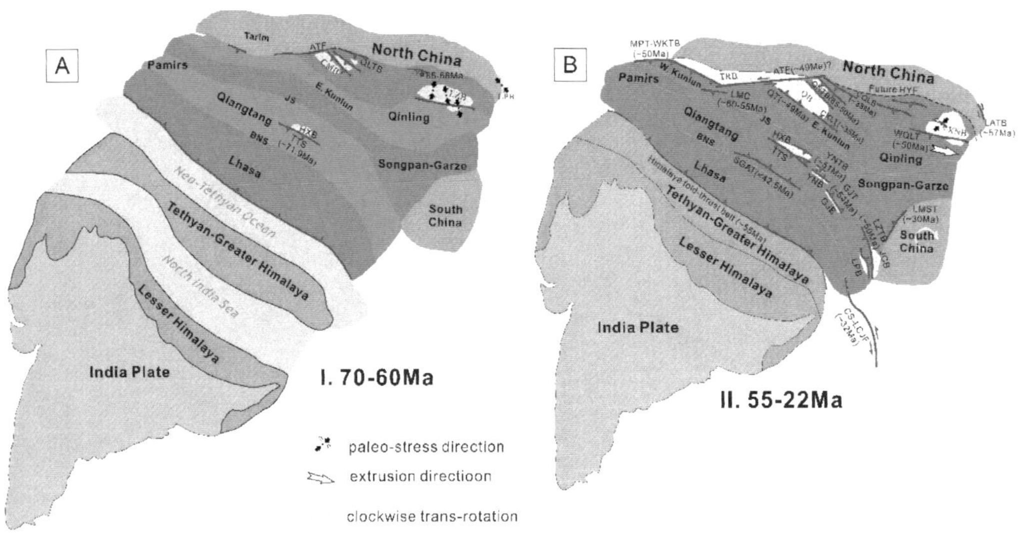

Figure 5. Sketch showing deformation model of the Tibetan Plateau during late Cretaceous 70-60Ma (A) and Paleogene 55-22Ma (B).

As to the Early-Cenozoic period, some authors hold the view that extensional regime may dominated the NE Tibet, as evidenced by the extensional normal faults observed in the Paleogene strata of the Xining and other basins in the NE-margin of the Tibetan Plateau [17, 35]. This NW-SE extensional regime seems consistent with that prevailed in the Weihe graben and in the southern North China basins in late Paleogene [36]. This finding led to an interpretation that the NE margin of the Tibetan Plateau may have also affected by far-field effect produced by the Pacific plate subduction [17]. In fact, these two plate tectonic processes, the India-Eurasia collision in southern Tibet and westward subduction of the Pacific Plate along the eastern margin of the Asian continent, have operated synchronously in Cenozoic and may have been interacted with each other, which exerted significant effects in the region of east and west Qinling. The NE-SW compression in Paleogene, as revealed by AMS results of the Zhongba section, may favour an eastward escape of the crustal fragments in NE Tibet, accommodated by sinistral strike-slip motion along major boundary fault zones, as documented in East Qinling [36] (Figure 5B). We suggest that normal faults may form locally in the foreland basin.

5. Conclusion

Magnetostratigraphy of a Late Cretaceous to Early Cenozoic succession across the Zhongba section in the eastern Xining basin yielded the spanning period of ~74.5-69.2Ma for the Upper Cretaceous unit, of ~49.3-22.0Ma for the overlying Paleogene unit. This sedimentary succession was punctuated by two distinct unconformities. The lower K-Pg unconformity represents a depositional hiatus of ~19.9Ma duration, possibly registered the period of the tectonic transition from the Neo-Tethys oceanic plate subduction to the India-Eurasia collision in southern Tibet. The NE-SW compression in Paleogene, as revealed by AMS results of the Zhongba section, may favour an eastward escape of the crustal fragments in NE Tibet, accommodated by sinistral strike-slip motion along major boundary fault zones, arousing several local transtensional faults in the foreland basin.

References

[1] Tapponnier P, Xu Z Q and Roger F 2001 *Science* **294** 5547
[2] Clark M K, Farley K A and Zheng D W 2010 *Earth. and Plan. Sci. L.* **296** 1-2
[3] Dai S, Fang X M and Dupont-Nivet G 2006 *JGR: Solid Earth.* **111** B11102
[4] Duvall A R, Clark M K and van der Pluijm B A 2011 *E. Plan. Sci. L.* **304** 3-4
[5] Yin A, Dang Y Q and Wang L 2008 *Geol. Soc. of A. Bull.* **120** 7-8
[6] Li W, Dong Y P and Guo A L 2013 *J. of A. Ear. Sci.* **73**
[7] Fang X M, Garzione C and Van der Voo R 2003 *Ear. Plan. Sci. L.* **210** 3-4
[8] Fang X M, Fang Y H and Zan J B 2019 *Earth-Science Reviews* **190**
[9] Dai S, Fang X M and Song C H 2005 *Chinese Science Bulletin* **50** 15
[10] Ji J L, Zhang K X and Clift PD 2017 *Gondwana Research* **46**
[11] Liu S F, Zhang G W and Pan F 2013 *Basin Research* **25** 1
[12] Wang W T, Zhang P Z and Kirby E 2011 *Tectonophysics* **505** 1-4
[13] Wang W T, Zhang P Z and Liu C C 2016 *JGR: Solid Earth* **121** 11
[14] Metcalfe I 2013 *J. of A. Ear. Sci.* **66**
[15] Yuan J, Yang Z Y and Deng C L 2020 *National Science Review* **nwaa173**
[16] Wang W T, Kirby E and Zhang P Z 2013 *Geol Soc of A Bull* **125** 3-4
[17] Fan L G, Meng Q and Wu G L 2019 *Ear. Plan. Sci. L.* **514**
[18] Zhai Y P and Cai T L 1984 *Gansu Geology* **1**
[19] Horton B K, Dupont-Nivet G and Zhou J 2004 *JGR: Solid Earth* **109** B04402
[20] Ye L S 1976 *Qinghai Geology* **3**
[21] Parés J M 2004 Geological Society, London, Special Publications **238** 1
[22] Jia D, Chen Z X and Luo L 2007 *Progress in Natural Science* **17** 5
[23] Zhang K J, Zhang Y X and Tang X C 2012 *Earth-Science Reviews* **114** 3-4
[24] Ma L, Wang Q and Li Z X 2013 *Lithos* **172**
[25] Leier A L, DeCelles P G and Kapp P 2007 *Geol. Soc. of A. Bull.* **119** 1-2
[26] Kapp P, DeCelles P G and Leier A L 2007 *GSA Today* **17** 7
[27] Jin C S, Liu Q S and Liang W T 2018 *Ear. Plan. Sci. L.* **486**
[28] Xu Z Q, Zhao Z B and Ma X X 2019 *Acta Geologica Sinica* **93** 1
[29] Kapp P, DeCelles P G and Gehrels G E 2007 *Geol. Soc. of A. Bull.* **119** 7-8
[30] Li Y L, Wang C S and Dai J G 2015 *Earth-Science Reviews* **143**
[31] Wu Z H, Barosh P J and Ye P S 2015 *J. of A. Ear. Sci.* **114**
[32] Kapp P, Yin A and Harrison T M 2005 *Geol. Soc. of A. Bull.* **117** 7/8
[33] Ding L and Lai Q Z 2003 *Chinese Science Bulletin* **48**
[34] Conrad C P and Lithgow B C 2004 *JGR: Solid Earth* **109** B10407
[35] Zhang J, Wang Y N and Zhang B H 2016 *Basin Research* **28** 2
[36] Zhang Y Q, Dong S W and Li J H 2019 *Tectonophysics* **769**

Direct hydrocarbon detection (DHI) techniques for prospective exploration targets in Chaoshan Depression

Guangjian Zhong[1,2,5], Renqi Jiang[3], Hai Yi[1,2], Jincai Wu[3], John Castagna[4], Changmao Feng[1,2], Gang Zhou[3], Kun Wang[3], Lina Liu[3] and Ming Sun[1,2]

[1]Key Laboratory of Marine Mineral Resources, Ministry of Natural Resources, Guangzhou, China
[2]Guangzhou Marine Geological Survey, Guangzhou, Guangdong, China
[3]New Horizons Ltd., Beijing, China
[4]University of Houston, Houston, United State of American

[5]Corresponding author's e-mail: 2645078906@qq.com

Abstract. Located in northern South China Sea, Pearl River Mouth Basin is a large Meso-Cenozoic basin. And in southeasters part of the basin lies Chaoshan Depression, which is mainly a residual Mesozoic depression, with a construction of Meso-Cenozoic strata over 7000m thick and good hydrocarbon accumulation conditions. The selected DHI technique, amplitude attribute of -90 degree phase component derived by phase decomposition is employed to the zone of interest (ZOI) of B structure, Chaoshan Depression. And it is found that there are evident amplitude anomalies occurring around ZOI of structure B. Phase decomposition is applied to forward modelling results of the ZOI, and high amplitudes occur on the -90 degree phase component more or less when ZOI is charged with hydrocarbon, which shows that the amplitude abnormality in ZOIs of structure B is probably caused by oil and gas accumulation. Meanwhile, another DHI attribute of spectral decomposition peak amplitude is applied also to ZOIs of B structure, and there is also amplitude anomalies coincides with those of -90 degree phase component. The two DHI attributes' coincidence strengthens the confidence of hydrocarbon accumulation prediction results.

1. Introduction

The Chaoshan Depression is located in the southeast of the Pearl River Mouth Basin, northern South China Sea, with an area of more than 3.7×10^4 km^2 and a water depth ranging from 300m to 2000m. China National Offshore Oil Corporation conducted seismic surveys and preliminary studies in the region from 1979 to 1986. With the continuous oil and gas exploration in the Pearl River Mouth Basin [1] and in order to search for new exploration areas, detailed studies of the Chaoshan Depression began in the early 1990s, and it was found that the lower tectonic layers of the Chaoshan Depression were Mesozoic marine sediments, with the largest thickness of about 7000m, which has the potential to generate hydrocarbons and the conditions to form various types of oil and gas traps [2-13].

The Chaoshan Depression is a residual Mesozoic depression, which experienced mid-late Jurassic subsidence, late Jurassic to early Cretaceous uplift, superimposed burial of early to late Cretaceous, and then late Cretaceous Uplift and erosion. The Guangzhou Marine Geological Survey Bureau divided the Chaoshan Depression into five secondary tectonic units, namely Eastern Sag, Western Sag, Central Low Uplift, Northern Slope, Central Slope and Western Slope. Among them, B Structure in Central Low

IOP Conf. Series: Earth and Environmental Science **865** (2021) 012017 doi:10.1088/1755-1315/865/1/012017

Uplift is selected as a favourable exploration target, (Figure 1). The article will elaborate on petroleum potential and DHI analysis of the target.

Due to the shortage of available drilling data, reliable petro physical data and hydrocarbon information in the target area of the Chaoshan Depression (Figure 1), there is a large uncertainty in the direct prediction of oil and gas with conventional techniques. In order to make a breakthrough to the situation of Mesozoic oil and gas exploration in the Chaoshan Depression, this study applies peak frequency amplitude attribute and phase decomposition amplitude attribute analysis techniques for direct oil and gas detection.

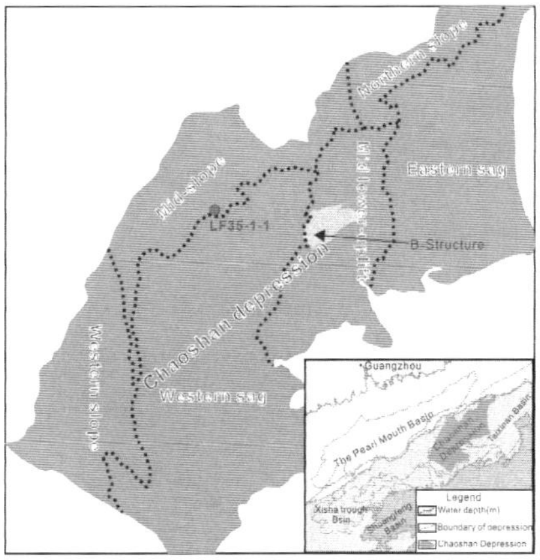

Figure 1. The location map of the favourable target (Structure B) in Chaoshan Depression.

2. Petroleum geology conditions in Chaoshan Depression

Drilled well LF35-1-1 confirmed that Chaoshan Depression is filled with Cretaceous continental sediments and Middle-Late Jurassic marine sediments, among which Middle-Late Jurassic marine sediments have good petroleum geological conditions [14-16].

2.1. Source rocks

The upper part of the Middle-Upper Jurassic strata revealed by well LF35-1-1 is composed of grey-black veined layered mudstone, argillaceous siltstone with siliceous rock, containing a small amount of micrite limestone; the lower part is grey-black veined layered mudstone and shale siltstone with sandstone, limestone. The mudstones are rich in organic debris. A set of river-lake sediments is generally made up of the Cretaceous strata. The upper part is a combination of purple mudstone, siltstone and sandstone with a small amount of marl, and the lower part is a combination of grey grained layered mudstone, siltstone and sandstone, containing some organic debris (Table 1).

Table 1. Dark mud-rock properties at well LF35-1-1.

Interval (m)	Strata	Thickness	Lithology	Paleo-strata thickness	Effective source rock thickness	TOC (%) min−max	TOC (%) Average (Number of samples)	Evaluation
977-1369	K	392	Dark mud-rock	0	0	0.05−0.54	0.10 (25)	Non-source
1369-1698	K	329	Tuff	33	0	0.06−0.30	0.15 (6)	Non-source
			Dark sandy mud-rock	45.5	0	0.06−0.11	0.08 (15)	Non-source
			Tuff	215	0	0.14−0.63	0.27 (28)	Non-source
1698-2412	J_{2+3}	713	Dark sandy mud-rock	529.38 (with tuff and sandstone reduced)	82.87	0.5−1.15	0.70 (36)	Poor-moderate source
					46.16	1−1.48	1.32 (11)	Moderate to good source

2.2. Reservoir conditions

According to the formations revealed by well LF35-1-1 and seismic data interpretation, there may be two sets of reservoirs in the Chaoshan Depression. One set is the continental river-lake sandstone reservoir in Cretaceous, and the other set is coastal-shallow sea sandstone, slope fan and basin-bottom fan sandstone in the Upper-mid Jurassic, and sandstone and limestone reservoirs in transgresses system tracts.

The seismic profile (Figure 2) across structure B shows that a set of limestone layers with stable thickness are distributed on the top of the Middle Jurassic (T_{j2} interface). A set of submarine fan sandstones advancing from SE to NW are developed on the top of the Upper Jurassic. The fan root (inner fan and levee deposits) is located on southeast of the profile. And mid-fan mainly develops in middle and northwest with mid-fan channel deposits, where there are sandstone reservoirs of the best properties. Mid-fan frontal deposits develop in northwest.

Vertically on the seismic profile, the down lap characteristics of the fore set can be clearly seen from SE to NW. The submarine fan can be divided into four stages, among which the second and third are most flourishing, and it is speculated that the sand content is the largest (Figure 3). Laterally, the results of seismic facies analysis show that the submarine fan is distributed in SE-NW direction, and structure B is located in braided channels and non-channel of the mid-fan area, where sandstone reservoirs of good quality develop.

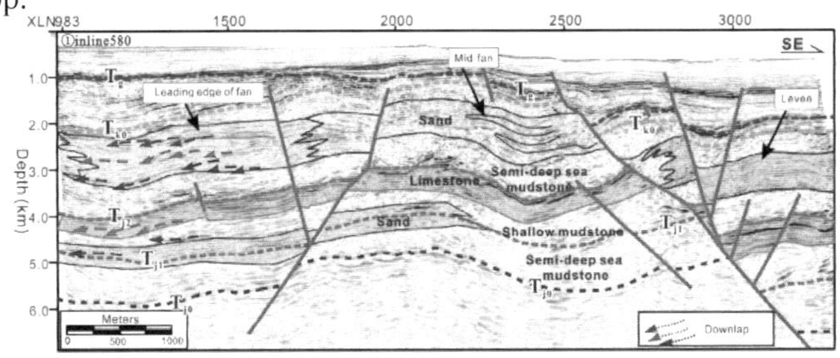

Figure 2. Seismic facies interpretation profile for reservoirs of structure B.

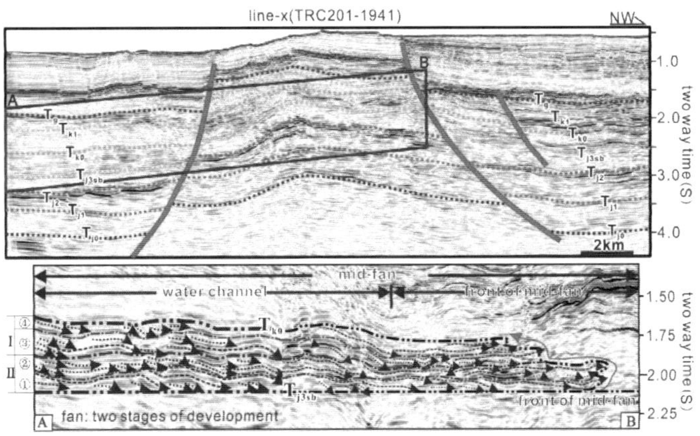

Figure 3. Detailed interpretation for upper-Jurassic fan of structure B.

2.3. Seal conditions

According to well LF35-1-1, multiple layers of mudstones encountered in this well are thick, and the maximum thickness of individual layer is more than 30 m. They should have good sealing capabilities and serve as important regional cap rocks in the area. In addition, there are many sets of mudstones in the Middle-Upper Jurassic, and these mudstones are mostly characterized of continuous or poorly continuous, medium or weak amplitude reflections on the seismic profile. According to the seismic profile, the distribution of these strata is stable, which can also serve as a regional cap rock.

2.4. Reservoir-forming model

According to cores of well LF35-1-1, the Cretaceous and Middle-Upper Jurassic are mostly composed of interbedded sand-mudstone in the Chaoshan Depression, which is very favourable to form good reservoir-seal assemblage. The main accumulation mode is source-reservoir coincidence and reservoir overlaying source (Figure 4).

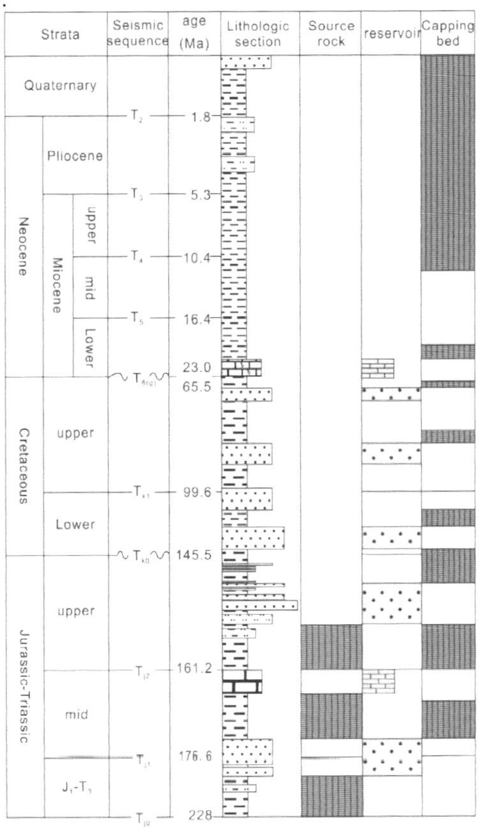

Figure 4. Main source-reservoir-seal assemblages in Chaoshan Depression.

3. Selected DHI techniques

3.1. Amplitude attribute of -90 degree phase component by phase decomposition

Phase decomposition is a unique and creative technology for DHI, originated by Castagna [17]. The acoustic impedance variations caused by pore fluids or lithology changes do not always produce very obvious amplitude anomalies on conventional post-stack profiles. By decomposing seismic traces into different phase components and suppressing unwanted phase components, it is possible to identify some subtle hydrocarbon, lithological changes or anomalous responses caused by sand channels, where significant amplitude anomalies occur at the expected phase components. This technology has become a very effective tool for special geological body identification and hydrocarbon detection [18-19].

Based on spectral decomposition [20], any seismic trace can be decomposed into 2D functions, amplitude spectrum and phase spectrum. A seismic trace S (t) is a 1D function whose amplitude varies with time. The amplitude spectrum A (f,t)is a 2D function whose amplitude varies with both time and frequency, and the phase spectrum θ(f, t) is a 2D function whose phase varies with time and frequency also. If we combine the amplitude spectrum and the phase spectrum into a function A (f, t) * cosθ (f, t), then a two-dimensional function representing reflection amplitude changes with phase and time can be obtained by the following formula:

$$S'(\theta,t)=\int_{f_1}^{f_2} A(f,t)\cos\theta(f,t)df \quad (1)$$

Where f1 and f2 are the starting and ending frequencies of the band. We define this function as phase gather. Any desired phase component in the specified frequency band is extracted as follows:

$$S'(t)=\int_{\theta_1}^{\theta_2} S'(\theta,t)d\theta \quad (2)$$

Where θ1 and θ2 are the starting and ending phases of the phase band. In this way, the seismic reflection events with specific spectral characteristics can be suppressed or enhanced. The process of decomposing a seismic trace into a two-dimensional function of amplitude with time and phase is called phase decomposition. The amplitude change with time at a specific phase is referred to as a phase component. The seismic reflections caused by small impedance changes in the layer below the tuning thickness can be amplified, making specific geological features easier to discover. Any pair of reflection coefficients can be expressed as the sum of an odd and an even component. The reflection coefficients of even component are the same in size and sign, and those of odd component are the same in size and opposite in sign. As to an individual thin layer, the impedance variation changes the reflectivity of odd component, and thus the phase of the local reflectivity, while the even component remains almost unchanged. A phase filter can be designed and applied to seismic data based on the method described above, and the odd component may show an obviously strong amplitude anomaly at the location where weak amplitude appears originally. Especially in the case of low-impedance thin gas or light oil sands (bright spots), the amplitude abnormality corresponding to gas saturation will appear in the -90 degree phase component. Therefore, in practical applications, amplitude attribute analysis of -90 degree phase component is often used as a means for DHI.

Based on this technology, oil layer prediction is successfully carried out in EBS field in southeast Baghdad, Iraq. The primary layer of EBS oilfield for development is the Z5-Z6 sandstone reservoirs, which have obtained high-rate oil flow. However, the reservoir heterogeneity is strong, the sand-bodies are difficult to co-relate laterally, and the reservoir prediction is difficult. The controlling factors and the types of oil accumulation are unclear, and DHI is difficult. For this set of reservoirs, phase decomposition technology is used to for DHI, and a number of drilled wells confirm the reliability of the technology.

The figure on the left is the minimum amplitude map (Figure 5). According to this map, the reservoir distribution is stable with a good continuity. Using phase decomposition technique, the -90 degree phase amplitude attribute map is extracted (Figure 5 on the right). According to the map, the reservoir range is obviously reduced, and at the same time, the local continuity is deteriorated, and the oil accumulation range is significantly reduced. Well EB-B is the first newly drilled based on the prediction map and post-drilling results are consistent with the predicted results. The oil layer thickness is about 12m, equivalent to that of the early well EB02., Other four wells (Z-1, Z-2, Z-3 and Z-4) are drilled one after another afterwards, among which Z-2 and Z-3 are better in production. However Z-1 and Z-4 are located at the edge of the anomaly on the attribute map, and only the 2-3m oil layer is encountered with poor oil-bearing property, although the sand reservoirs are well developed.

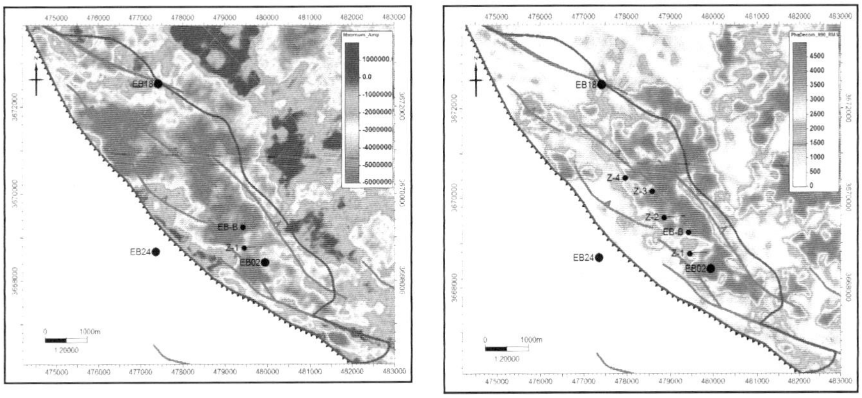

The minimum amplitude map of layer Z5-Z6 in EB02 well area **-90 degree phase amplitude map of layer Z5-Z6 in EB02 well area**

Figure 5. Comparison of the minimum amplitude map and -90 degree phase amplitude map of layer Z5-Z6 in EB02 well area.

3.2. Spectral decomposition peak amplitude analysis

The peak amplitude attribute is an important one by spectral decomposition, and it is the maximum value of the reflection coefficient amplitude spectrum after eliminating the tuning effect [21]. Over the time domain amplitude attribute, it has the advantage to reflect the changes reservoir properties and hydrocarbon accumulations more realistically.

According to Castagana, any pair of reflection coefficients can be decomposed into odd and even parts (Figure 6). The odd component ro is composed of reflection coefficients with opposite amplitudes and polarities, while the even odd component re is composed of reflection coefficients with equal amplitude values and the same polarities.

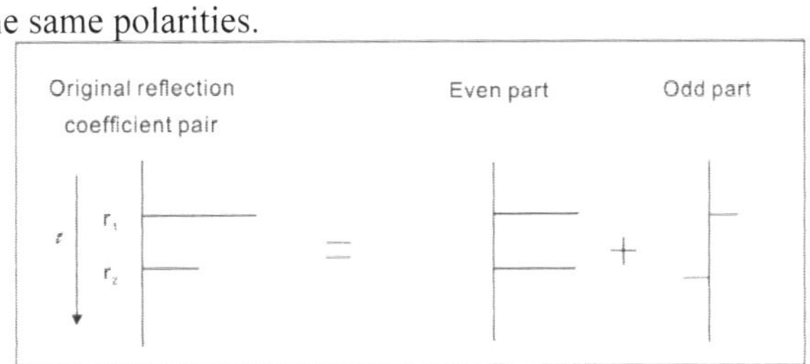

Figure 6. The decomposition of a reflectivity pair into an even and an odd pair.

With Fourier transform, the reflection coefficient expression is represented in frequency domain:

$$G(f) = 2r_e\cos(\pi fT) + i2r_o\sin(\pi fT)$$

Then the amplitude spectrum expression of the reflection coefficient pair is obtained:

$$A[G(f)] = [(r_1 + r_2)^2\cos^2(\pi fT) + (r_1 - r_2)^2\sin^2(\pi fT)]^{\frac{1}{2}}$$

The peak amplitude attribute is the maximum value of the amplitude spectrum expression. From the calculation formula of the amplitude spectrum, it can be seen that for a set of layers, changes in fluid or lithology or reservoir properties will increase absolute values of the top and bottom reflection coefficients, which is reflected by the significantly increased peak amplitude attribute [22].

In practical application in this area, the peak amplitude attribute can be obtained to directly detect reservoir and oil-bearing properties. The original seismic profile and peak amplitude attribute are displayed in Figure 7 and 8 respectively. It is difficult to distinguish the oil and brine layer directly on the conventional amplitude profile since the amplitude of the brine layer and the oil layer is not much different. While on peak amplitude attribute profile, the oil layer is characterized of strong amplitude and the oil layer is characterized by weak reflection. The peak amplitude attribute can be used to effectively distinguish oil layers.

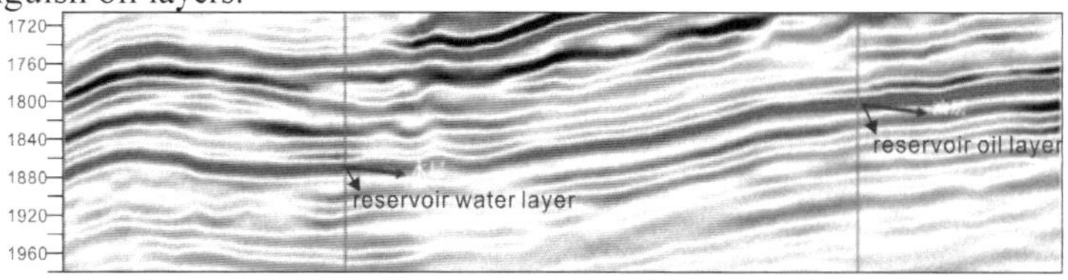

Figure 7. Original seismic section.

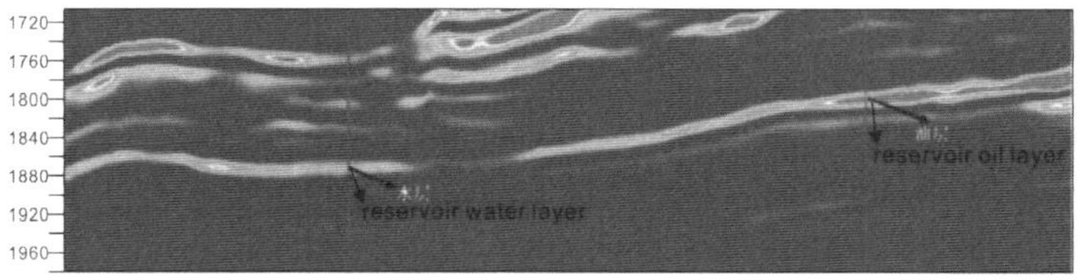

Figure 8. Peak amplitude attribute section.

4. Selected DHI techniques

-90 degree phase component technology for DHI has achieved excellent results in practical applications. In the exploration target area of Chaoshan Depression, there is no well drilled available. It is wondered that the technique of phase decomposition and -90 degree phase component DHI can achieve good results. We conduct forward modeling using the reservoir properties adjacent to the Chaoshan Depression to see what will happen to the amplitude attribute of -90 degree phase component if target reservoir contains oil and gas, and whether it can predict oil and gas.

4.1. Model establishment

The main purpose of establishing the forward model is to investigate the response characteristics of the amplitude of the -90 degree phase component to oil, gas or brine. The forward modeling is implemented using Leonardo technology and then the response characteristics of the two attributes with different pore fluids are summarized.

In practical work, there are three types of impedance changes for reservoir and its surrounding rocks: ① The impedance of surrounding rocks is higher than that of oil, gas or brine reservoirs, and oil-gas reservoirs are characterized of bright spots; ② The impedance of surrounding rock is lower than that of oil, gas or brine reservoirs, and oil-gas reservoirs is characterized of dim spots; ③ The impedance of surrounding rock is lower than that of brine-bearing reservoirs, but higher than that of oil-gas reservoirs, which shows polarity reversal (Figure 9).

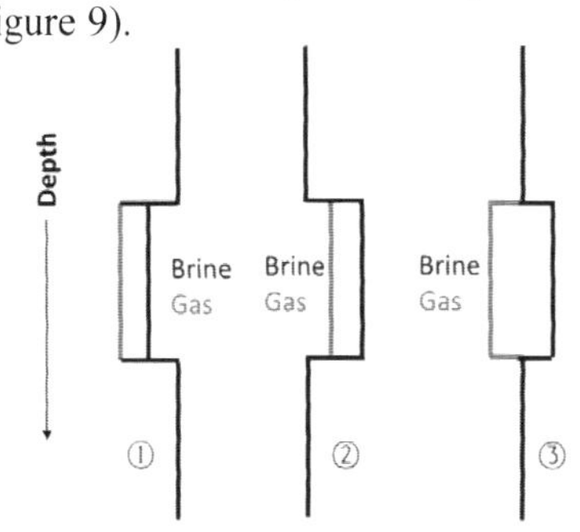

Figure 9. Assemblage patterns of reservoir and surrounding rocks.

IOP Conf. Series: Earth and Environmental Science **865** (2021) 012017 doi:10.1088/1755-1315/865/1/012017

According to the above assemblage patterns for reservoir and the surrounding rocks, and the drilling data of well LF35-1-1 in northern slope, the geological model is established firstly according to the lithology combination and velocity changes in the target area. The upper Jurassic stratum is the target zone, with a depth of about 1700-2300 meters. The lithology of the overlying rocks includes siliceous rock, mudstone, and argillaceous siltstone; and the reservoir lithology includes sandstone, siltstone; and underlying rocks is limestone. Various lithology and its velocity are as follows in Table 2.

Table 2. Rock types and their velocities listed.

Lithology	Depth (m)	Velocity (m/s)
Siliceous rock	1700-1900	5350
Mudrock	1950-2080	4200
Shaly siltstone and siltstone	1900-1950 2080-2200	5250
Limestone	2200-2300	5450

According to the velocity characteristics of siliceous rock and mudstone, the following three models are established (Figure 10-12). For No. 1 model, the overlying rock is siliceous rock with velocity of 5350m/s, which is greater than that of the sandstone reservoir. For No. 2 model, the overlying rock is mudstone, and its velocity of 4200 m/s is less than that of sandstone reservoir. As to No. 3 model, the overlying rock is a combination of siliceous and mudstone with the velocity of 5100m/s, which is the average velocity of siliceous rocks and mudstones, and the caprock velocity are close to that of the sandstone reservoir.

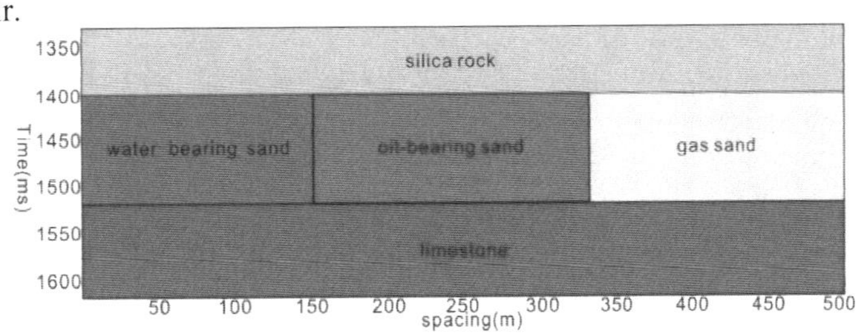

Figure 10. Lithology assemblage No. 1 model.

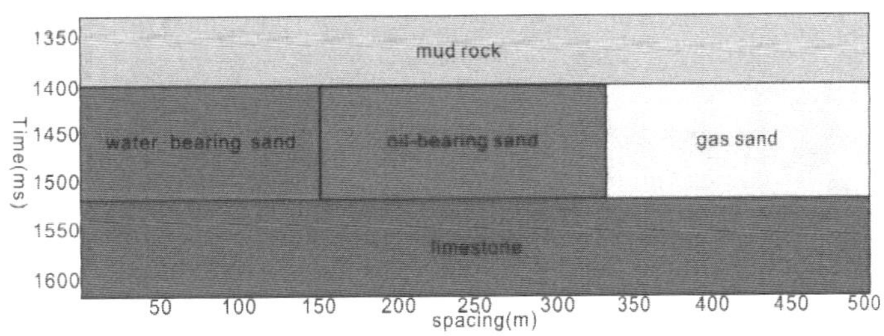

Figure 11. Lithology assemblage No. 2 model.

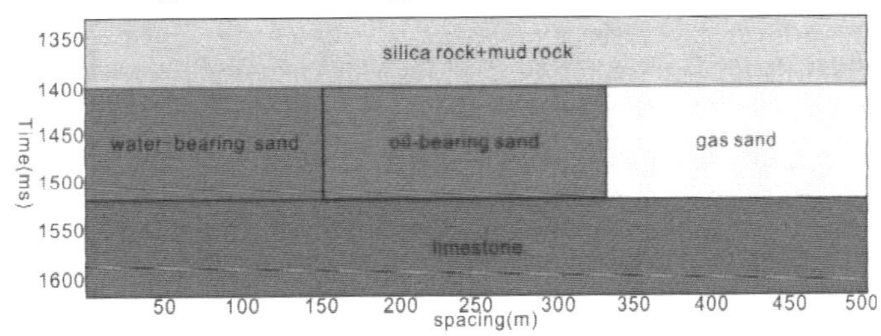

Figure 12. Lithology assemblage No. 3 model l.

4.2. Seismic modeling for DHI in Chaoshan Depression

The forward seismic profiles corresponding to the three models are displayed in Figures 13-15. According to the synthetic profiles by forward modeling, the reservoir top of model 1 is a trough, and the amplitude of oil/gas-bearing case is significantly higher than the brine-bearing reservoir. The reservoir top of model 2 is a peak reflection, and the amplitude when oil/gas-bearing case is significantly lower than brine-bearing case reservoir. The brine layer top of model 3 is a peak reflection, and the gas layer is a trough. The amplitude of the brine layer is slightly different from the gas layer, and the oil layer is a peak of weak amplitude. On seismic sections, it difficult to directly identify the oil and gas layer.

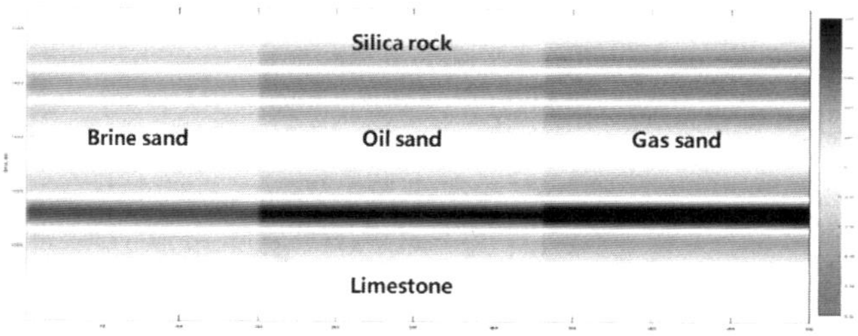

Figure 13. Synthetic section of model 1.

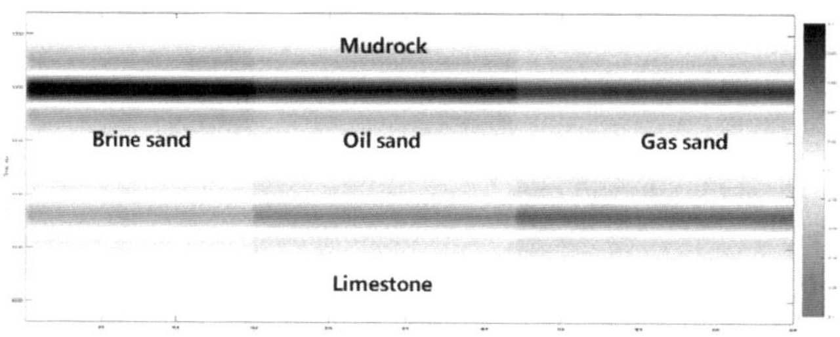

Figure 14. Synthetic section of model 2.

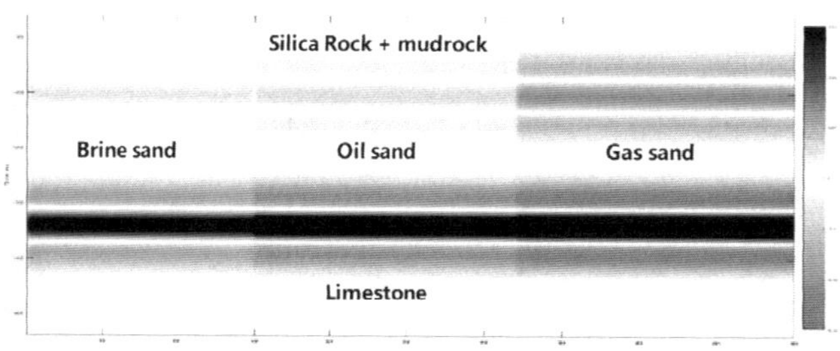

Figure 15. Synthetic section of model 3.

The amplitude attribute of the -90 degree phase component is extracted on forward synthetic seismic data (Figure 16-Figure 18). On -90o phase component profiles of the three models, it is similar that the gas layer has the strongest amplitude and brine layer the weakest amplitude, and the oil layer amplitude between gas and brine.

In Table 3, the absolute amplitude values of synthetic seismic data and -90 degree phase attributes are read from the three models of brine, oil and gas respectively. And amplitude values of -90 degree phase attributes are cross-plotted against those of synthetic seismic data (Figure 19). On the cross-plotting, there is not a uniform threshold to identify oil and gas layers for original synthetic seismic amplitude only, while for -90 degree phase amplitudes, uniform thresholds can be seen between brine and oil and gas. The amplitude value above 0.0005 is gas layer, and the amplitude between 0.0035-0.0005 oil layer, and the amplitude less than 0.00035 brine layer.

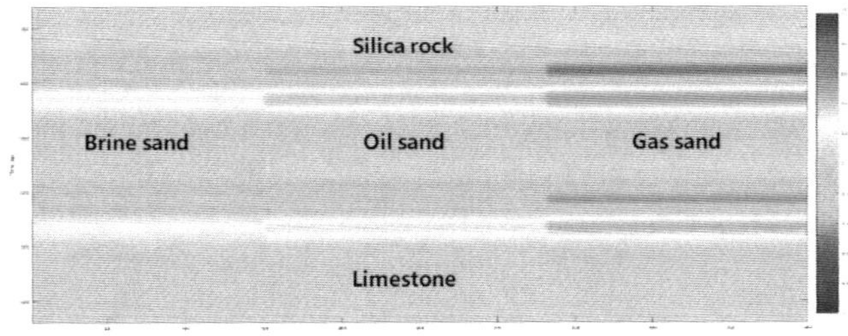

Figure 16. -90° phase component section of model 1 synthetics.

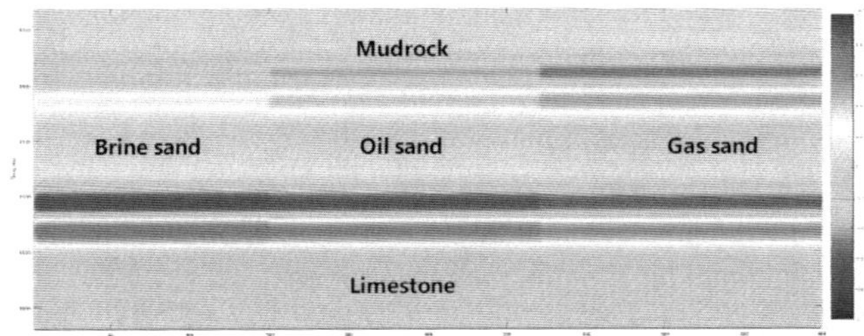

Figure 17. -90° phase component section of model 2 synthetics.

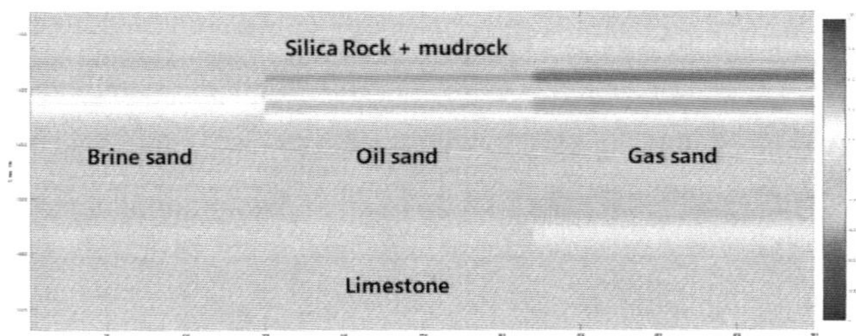

Figure 18. -90° phase component section of model 3 synthetics.

Table 3. Synthetic amplitudes and its phase component for different pore fluids cases.

Model	Fluids	Absolute amplitude of synthetic data	Absolute amplitude of -90 degree phase component
No. 1	Gas	0.053	0.00051
	Oil	0.041	0.00041
	Water	0.028	0.00031
No. 2	Gas	0.078	0.00052
	Oil	0.092	0.00042
	Water	0.11	0.00032
No. 3	Gas	0.021	0.00052
	Oil	0.0078	0.00043
	Water	0.0052	0.00028

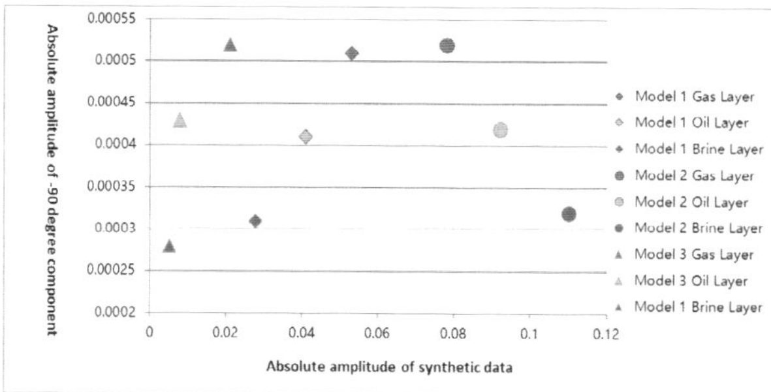

Figure 19. Cross plotting for synthetic amplitudes and its phase component of 3 models.

5. DHI for identification of exploration targets in Chaoshan Depreession

Well LF35-1-1 confirms that there are two sets of reservoir-cap rock assemblages in the Chaoshan Depression. One is interbedded sandstone and mudstone in Cretaceous, the other is the Middle-Upper Jurassic. And the latter is the main target interval for oil and gas exploration. For this main exploration target, the reservoir occurrence is analyzed by extracting conventional seismic amplitude attributes. DHI is performed using -90 degree phase component amplitude attribute analysis and spectrum decomposition peak amplitude attribute analysis respectively, so as to verify the two prediction results mutually and increase DHI confidence.

5.1. Analysis of conventional seismic amplitude attributes

Seismic amplitude attributes can indicate reservoir property variations to a certain extent, so the seismic amplitude attributes are usually applied to qualitatively analyze reservoir development. The seismic section along one favorable exploration target in the Chaoshan Depression is shown in Figure 20. On the section, the ZOI of Upper Jurassic shows strong reflections locally, which is probably the seismic response of the reservoir with better properties. By extracting its amplitude attribute, the lateral distribution range of the strong reflection of ZOI is preliminarily delineated, which probably corresponds to the range of favorable reservoirs of ZOI (Figure 21).

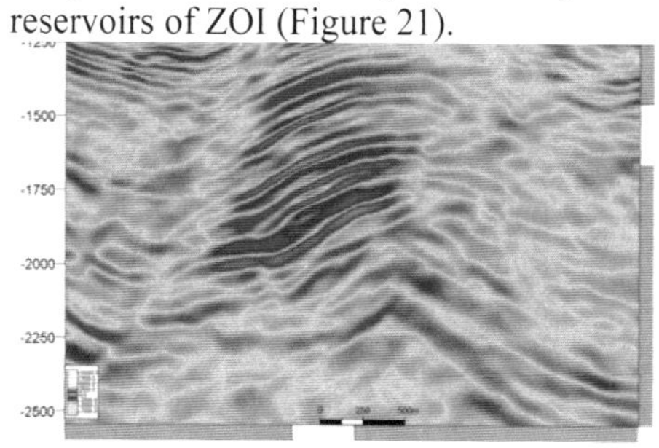

Figure 20. Original seismic section around ZOI.

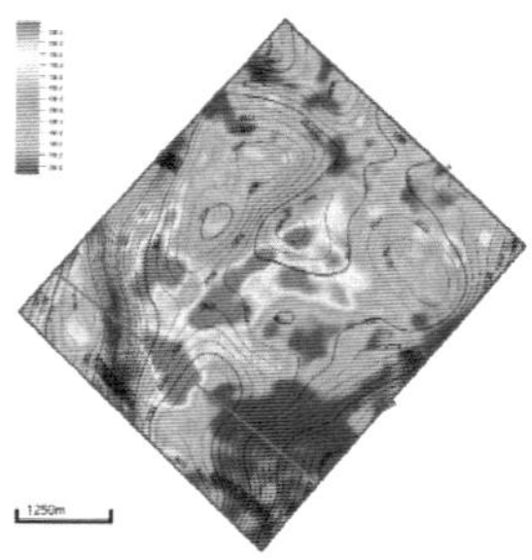

Figure 21. Original seismic amplitude attribute map extracted along ZOI.

5.2. -90 degree phase component amplitude attribute analysis for DHI

The -90 degree phase component amplitude attribute by phase decomposition for DHI is used to study on the seismic section where there are strong reflections around ZOI, to clarify the distribution of abnormal amplitudes of the -90 degree phase component attribute, and then to identify its possible oil and gas distribution scope.

A comparison between the original seismic profile and the -90 degree phase component profile is illustrated in Figure 22. From this comparison, strong amplitude reflections on the -90 degree phase component profile do not coincide with all those on the conventional seismic profile, and some strong amplitudes are no longer strong on -90 degree phase component profile. According to the previous forward modeling results, strong amplitudes on the -90 degree phase component profile is probably caused by presence of oil and gas. By extracting the amplitude attribute of the -90 degree phase component, the possible oil and gas distribution range is identified in Figure 23 on the right.

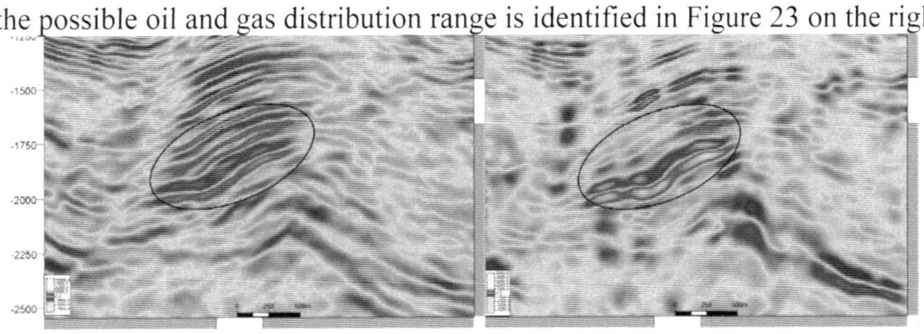

Figure 22. The seismic section (left) and -90 degree phase component section (right) of Up-Jurassic reservoir.

The comparison between the attribute map of conventional amplitude and that of -90 degree phase component along ZOI is illustrated in Figure 23. There are clusters of strong amplitude characteristics on both maps. However, the strong amplitude range on the -90-degree phase component map is reduced, and the boundary of the strong amplitude range is clearer, and it is more consistent structural contours. Obviously stronger amplitudes of -90-degree phase component especially occur on structural highs (Figure 23, on the right). Therefore, it is speculated that the oil and gas accumulation is extremely likely to occur in the areas with high amplitudes.

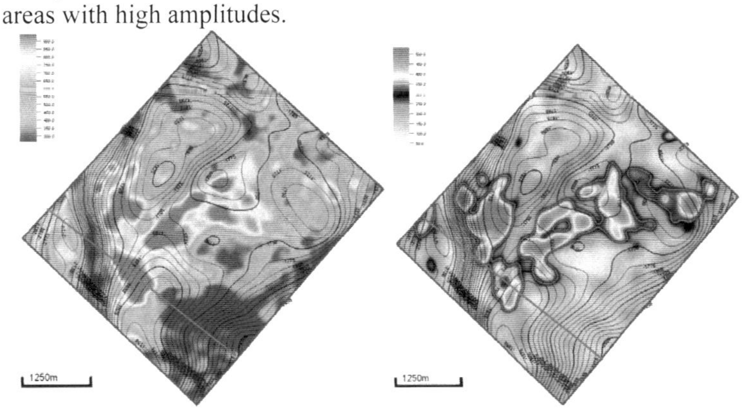

Figure 23. The attribute map (left) and -90 degree phase component map (right) of Up-Jurassic reservoir.

5.3. Spectral decomposition peak amplitude analysis for DHI

The peak amplitude attribute is sensitive to reservoir oil and gas. By comparing the original seismic section with the peak amplitude attribute section, it is found that oil/gas-bearing reservoirs show strong amplitude characteristics on the peak amplitude attribute section. A comparison between original seismic section and the peak amplitude attribute section of favorable exploration targets in Chaoshan Depression is illustrated in Figure 24. The peak frequency amplitude section shows strong reflections locally, but the continuity varies slightly on the top. By comparing peak amplitude attribute map with the original amplitude attribute (Figure 25), the intensity of the peak amplitude attribute anomalies has slight changes laterally, and their continuity in anomalous areas fluctuates slightly. It also indicates that hydrocarbon accumulation scope is smaller than reservoir distribution range.

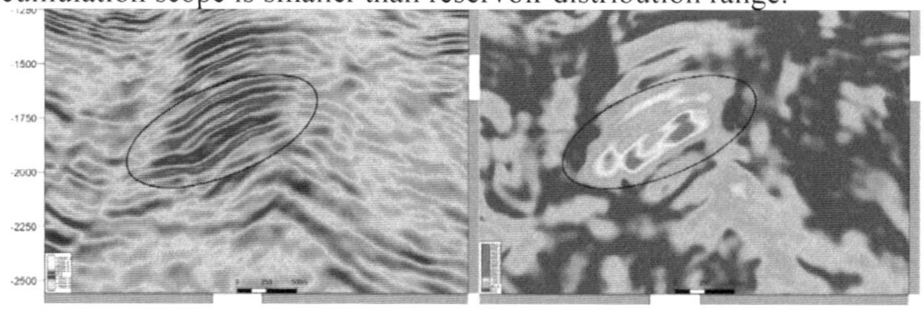

Figure 24. The seismic section (left) and peak amplitude section (right) of Up-Jurassic reservoir.

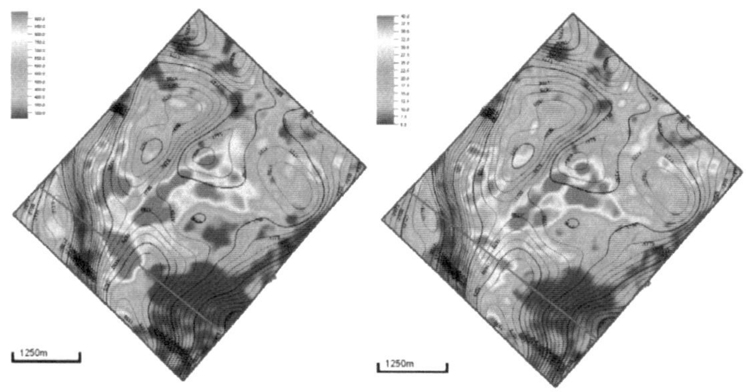

Figure 25. The attribute map (left) and peak amplitude map (right) of Up-Jurassic reservoir.

6. Conclusions

For exploration frontier areas under study, there is much uncertainty to apply conventional seismic attribute analysis for identifying oil and gas prospective targets due to few wells available. Fortunately, the joint application of two methods, -90 degree phase component amplitude attribute analysis by phase decomposition and peak amplitude attribute by spectrum decomposition, is an effective alternative for DHI in these areas. Together with its forward modeling analysis, -90 degree phase component amplitude attribute analysis method is used to predict reservoir oil and gas bearing property of the structure B in Chaoshan Depression, which indicates that there are probably oil-bearing layers in Upper Jurassic reservoirs. At the same time, it is supported by peak amplitude attribute analysis, and thereby the confidence of DHI results is enhanced.

References

[1] Liu G D 2001 The second exploitation of the oil and gas resources in China [J]. *Progress in Geophysics*, **16(4)**: 1-3.

[2] Zhang J Y, SUN Z, Zhang S F 2014 Analysis of Mesozoic tectonic deformation in the Chaoshan Depression of Pearl River Mouth Basin [J]. *Journal of Tropical Oceanography*, **33(5)**: 41-49.

[3] Hao H J, Lin H M, Yang M X, et al. 2001 The Mesozoic in Chaoshan depression: A new domain of petroleum exploration. *China Offshore Oil and Gas (in Chinese)*, **15(3)**: 157-163.

[4] Wang L L, Cheng R H, LI F, et al. 2009 The Mesozoic Sedimentary Sequences, Correlation and Geological Significance for Petroleum of the North Margin of South China Sea [J]. *Journal of*

Jilin University (Earth Science Edition), **39(2)**: 175-182.

[5] Zhang L, Geng A s, Wang LL, et al. 2012 Assessment of Mesozoic source rocks at The Margin of South China Continent [J]. *Marine Geology & Quaternary Geology*, **32(1)**: 99-108.

[6] Ji Z Y, Zhao H Q, et al. 2014 The Mesozoic Petroleum System of Chaoshan depression. *Petroleum Geology and Enginerering*, **28(3)**: 9-15.

[7] Wang P, Xia K Y, Huang C L 2000 The Mesozoic Marine sediment distribution and Geology-Geophysics characteristic at the North-Eastern of South China Continent [J]. *Journal of Tropical Oceanography*, **19(4)**: 28-35.

[8] Duan J C, Mi H F 2012 Seismic Facies and Sedimentary Facies study of Mesozoic in Chaoshan Sag [J]. *Resources & Industries*, **14(1)**: 100-105.

[9] Hao H J, Wang R L, Zhang X T, et al. 2004 Mesozoic marine sediment identification and distribution in the eastern Pearl River Mouth Basin. *China Offshore Oil and Gas*, **16(2)**: 84-88.

[10] Zhang Q L, Zhang H F, Zhang X T, et al. 2018 The upper Cretaceous prototype basin of the Chaoshan depression in the northern South China Sea and its tectonic setting [J]. *Chinese Journal of Geophysics*, vol **61(10)**: p4308-4321.

[11] Zhou D, Chen H Z, Sun Z, et al. 2005 Three Mesozoic sea basins in the eastern and southern South China Sea and their relation to Tethys and Paleo-Pacific domains. *Journal of Tropical Oceanography*, **24(2)**: 16-25.

[12] Yao B C, Zhang L, Wei Z Q, et al. 2011 The Mesozoic Tectonic Characteristics and Sediment Basins in the Eastern Margin of South China Continen [J]. Marine Geology & Quaternary Geology, **31(3)**:47-60.

[13] Zhong G J, Wu S M, Feng C M 2011 Sedimentary model of Mesozoic in the northern South China Sea. *Journal of Tropical Oceanography*, **30(1)**: 43-48.

[14] Hao H J, Shi H S, Zhang X T, et al. 2009 Mesozoic sediments and their petroleum geology conditions in Chaoshan sag: A discussion based on drilling results from the exploratory well LF35-1-1. *China Offshore Oil and Gas (in Chinese)*, **21(3)**: 151-156.

[15] Zhong G J, Yi H, Lin Z, et al. 2007 Characteristic of Source Rocks and Mesozoic in Continental Slope Area of Northeastern the South China Sea and East Guangdong of China. *XINJIANG Petroleum Geology*, **28 (6)**: 676-680.

[16] Yang S C, Tong Z G, He Q, et al. 2009 Mesozoic hydrocarbon generation history and igneous intrusion impacts in Chaoshan depression, South China Sea: A case of LF35-1-1 well. *China Offshore Oil and Gas (in Chinese)*, **20(3)**: 152-156.

[17] Castagna J, Oyem A, Portniaguine O and Aikulola U 2016 Phase decomposition: Interpretation, 4, no. 3, SN1–SN10, doi: 10.1190/INT-2015-0150.1.

[18] Elita Selmara De Abreu, John Patrick Castagna and Gabriel Gil 2018 Case study: Phase-component amplitude variation with angle: Geophysics, 84, B285–B297, doi: 10.1190/GEO2018-0762.1.

[19] Barbato U, Portniaguine O, Winkelman B and Castagna J 2017 Phasedecomposition as a DHI in bright-spot regimes: A Gulf of Mexico casestudy: 87th Annual International Meeting, SEG, Expanded Abstracts, 3976–3980, doi: 10.1190/segam2017-17737608.1.

[20] Partyka G, Gridley J and Lopez J 1999 Interpretational applications ofspectral decomposition in reservoir characterization: The Leading Edge, 18, 353–360.

[21] Barnes A E 1993 Instantaneous spectral bandwidth and dominantfrequency with applications to seismic reflection data: Geophysics, 58, 419–428, doi: 10.1190/1.1443425.

[22] Bracewell, R. N., 1978 The Fourier Transform and its applications: McGraw-Hill.

IOP Conf. Series: Earth and Environmental Science **865** (2021) 012018 doi:10.1088/1755-1315/865/1/012018

Does environmental regulation slow down marine environmental pollution?--Empirical study based on panel threshold model

Xiaohao Liu [1, 2], Shujuan Wu[1]and Jianhua Xiao[1]

[1]School of economics and management, Wuyi University

[2]Corresponding author's e-mail: 572148934@qq.com

Abstract. In order to identify the real impact of environmental regulation on the marine environment pollution in China, the paper constructs a panel threshold model of environmental regulation on the marine environment pollution, by adopting data from year 2006-2016 of 11 coastal provinces and cities in China, perform a linear regression analysis and a threshold regression analysis, taking environmental regulation as a threshold and explanatory variable, population at the end of the year, the GOP and total population, GDP as control variables, the annual amount of wastewater discharged directly into the sea as dependent variable. The results show that environmental regulation cannot slow down marine environmental pollution, China's environmental regulation has a significant threshold effect on marine environmental pollution, GOP and GDP have significant positive impacts on marine environmental pollution, and impact of GOP is higher than that of GDP, and population has insignificant negative impacts on marine environmental pollution. Therefore, regulations should be improved to protect the marine environment.

1. Introduction

Due to the industrialization and urbanization, marine environmental pollution has become an urgent problem to be solved worldwide, and environmental regulation has become a governmental means to interfere with the resources and environment of enterprises. With the diversification of environmental problems, environmental regulation is changing from a single executive order mode to a diversified mode such as executive order mode, market encouragement and moral persuasion, such as emission charge, information publicity, ideological education, etc. The report of the 19th National Congress has clearly emphasized the government should improve the comprehensive governance of watershed environment and coastal waters, but in 2019 there were still 38 red tides, and ending time of green tides was the latest during the past five years. The above problems are still seriously restricting the sustainable development of marine economy. Therefore, academics proposed a problem what environmental regulation impacts on marine environmental pollution.

2. Literature review

The domestic and foreign governments and academic researchers have done extensive research on the issue of environmental regulation. The major controversial focus is whether environmental regulation can relieve marine pollution.

Most of the foreign academics focus on qualitatively researching the effect of environmental regulation on the environment. The Australian government has invested US$3 billion in the

IOP Conf. Series: Earth and Environmental Science **865** (2021) 012018 doi:10.1088/1755-1315/865/1/012018

construction of the "Yanbu" Industrial City, where pollutant emissions are restricted by the laws, manage criteria and guidelines [1]. Brian Garrod [2] argues that corporate environmentalism can never be a perfect substitute for the government intervention in marine environment management, while the government's environmental regulation remains the main way to govern the marine environment. Borsatto Jaluza Maria Lima Silva [3] reviewed the 2011-2019 literature, concluding that environmental regulation is one of the main factors the promote enterprises to make green innovations. Danny Marks [4] discovered in 19 interviews with the actors from the public, private and non-profit sectors that main factor restricting the reduction of ocean plastics is the lack of the government's environmental regulation. Agovino Massimiliano [5] has proven by the generalized method of moments (GMM) that environmental regulation is beneficial to the sorted collection of waste.

The Chinese academic researchers focus on quantitatively researching the effect of environmental regulation on the environment. Fan Qingquan [6] has drawn a conclusion based on the panel threshold model that environmental regulation is a necessary condition for corporate pollution treatment and able to reduce energy consumption. Du Wencui [7] has drawn a conclusion based on the Stackelberg model that strengthened environmental regulation can increase industrial concentration, thereby reducing the intensity of air pollution emissions, but the intensity of water pollution emissions is enhanced. Zhang Zhibin [8] performed regression with the 2011-2017 panel data of 35 key Chinese cities, concluding that the government's environmental regulation helps to reduce local pollutant emissions. Liang Rui [9] has drawn a conclusion based on the threshold regression model that environmental regulation is only beneficial to the reduction of air pollution in a specific economic growth range. Kuang Chang'e [10] conducted a threshold regression analysis with the 2006-2017 data of 14 prefecture-level cities in Hunan Province, concluding that environmental regulation has a significant inhibitory effect on industrial pollution below the threshold level and has an opposite effect above the threshold level.

In summary, there are already abundant studies, and there is a difference in the conclusions caused by factors such as regionalism, climate, and economy. At present, most academic researchers mainly study the effect of environmental regulation on the terrestrial environment, while there are few studies on marine pollution. As the sea-land ties are strengthened, the environmental regulation on land has also exerted a certain effect on marine pollution and on the sustainable development of the marine economy.

Based on that, this paper intends to build a threshold model between environmental regulation and marine pollution by panel threshold regression, and use the statistical data of 11 coastal provinces and cities to empirically test the effect of environmental regulation on marine pollution in the hope of paving the way for the solving of the problem.

The innovations of this paper are as follows: First, the non-linear relationship of environmental regulation to marine pollution is taken into consideration, and a threshold regression model is used for conjecture verification. The second is the research object. Owing to the sea-land connection, GOP, year-end population and GDP, and year-end population are taken as control variables to study the effect of marine and terrestrial activities on marine pollution.

3. Research design

3.1. Threshold regression model

In order to comprehensively explore the effect of environmental regulation on marine pollution, nonlinear regression is used in this paper for a comparative study of the empirical results. On the basis of Hansen's threshold regression model, a panel threshold regression model is built by reference to Qi Shaozhou's model [11]:

$$\mathrm{MP}_{it} = \mu_i + \beta_1 \mathrm{TRI}(\mathrm{TR} \le \gamma) + \beta_2 TRI(TR > \gamma) + \beta_3 x + 3_{it} \quad (1)$$

$$MP_{it} = \mu_i + \beta_1 TRI(TR \le \gamma_1) + \beta_2 TRI(\gamma_1 < TR \le \gamma_2) + \beta_3 TRI(TR > \gamma_2) + \beta_4 x + 3_{it} \quad (2)$$

Equation (1) is for a single-threshold situation, while equation (2) is for a double-threshold situation, where MP_{it} represents the degree of marine pollution in i Province in year t;TR represents the environmental regulation, adopted as an explanatory variable and threshold variable here; x represents control variable, including GOP, GDP, and population; I is an indicative function, divided into different parts according to the threshold value; $\beta_i(i = 1,2,3,4)$, μ_i represents the regression coefficient; $\gamma, \gamma_1, \gamma_2$ represents the threshold value; ε_{it} represents the stochastic disturbance term.

3.2. Data and indicator system description

Marine pollution (MP) is taken as an explanatory variable, and the annual amount of wastewater discharged directly into the sea is used to represent the degree of MP. Environmental regulation (TR) is taken as an explanatory variable and threshold variable. By reference to Gai Mei's method [12], the ratio of 11 coastal provinces and cities' total investment in environment pollution treatment to their fixed investments is used to represent TR. The control variables include: (1) Marine economic status, represented by annual per-capita gross ocean product (GOP). (2) Terrestrial economic status, represented by annual per-capita gross domestic product (GDP). (3) Population (PD): The increase in population leads to an increase in the use of resources, with more and more waste thus generated from consumption, exerting a certain impact on the marine environment. Therefore, the total year-end population of the 11 coastal provinces and cities is used to represent the population size.

11 coastal provinces and cities of China are taken as research objects, including Liaoning Province, Tianjin city, Hebei Province, Shandong Province, Jiangsu Province, Shanghai City, Zhejiang Province, Fujian Province, Guangdong Province, Guangxi Province and Hainan Province. The data come from the 2006-2016 China Marine Statistical Yearbook, China Statistical Yearbook on Environment and Bulletin of China Marine Ecological Environment Status. Logarithm of some variables is taken and is expressed by lnMP, lnGOP, lnGDP, lnPD. The descriptive statistics of the data are shown in Table 1.

Table 1. Descriptive statistics of indicators.

Variable	Observation	Mean	Std.Dev.	Min	Max
lnMP	121	10.1056	1.277	7.080	12.190
lnGOP	121	8.831	0.913	6.457	10.409
lnGDP	121	10.697	0.532	9.233	11.666
TR	121	2.125	1.075	0.419	9.064
lnPD	121	8.328	0.769	6.729	9.306

4. Empirical analysis

4.1. Related variable test

The correlation coefficients of the marine pollution status and its effect variables are analysed in order to test the correlation among the variables. As can be seen from Table 2, there is the greatest correlation between GOP and GDP, 0.8153. Therefore, it is necessary to separate GOP from GDP for threshold regression analysis.

Table 2. Correlation coefficient matrix.

Variable	MP	lnGOP	lnGDP	TR	lnPD
MP	1.0000				
lnGOP	0.2345	1.0000			
lnGDP	0.0464	0.8153	1.0000		
TR	0.1648	-0.3841	0.0179	1.000	
lnPD	-0.0904	0.1700	0.2574	-0.0788	1.000

4.2. Basic regression analysis

To observe whether there is a linear or nonlinear relationship between environmental regulation and marine pollution, first, on the premise of not considering the nonlinear relationship, the static panel model is used to observe the marginal effect of environmental regulation on marine pollution, and select fixed effects for testing. An estimation is made with GOP, population and GDP, and population as control variables and with control variables excluded. Table 3 shows the fixed effect estimation results. The regression coefficient of environmental regulation is positive, and significant at the level of 10% without control variables, indicating that at the present stage, environmental regulation remains unable to reduce marine pollution. The regression coefficient of population is negative but not significant, while the regression coefficients of GOP and GDP are positive, suggesting that economic development contributes to marine pollution.

Table 3. Basic regression analysis results.

Variable	FE1	FE2	FE3
TR	0.066*	0.040	0.044
lnPD	/	-0.859	-0.834
lnGOP	/	0.458***	/
lnGDP	/	/	0.427***
Conf	9.966***	13.131**	12.395***

Note: ***, ** and * indicate high significance at the 1%, 5%, and 10% levels. FE means fixed effect.

4.3. Threshold regression analysis

Stata16 software was used for threshold regression analysis. Environmental regulation is taken as a threshold variable to verify the threshold effect of environmental regulation on marine pollution. The F-statistics and Bootstrap method are adopted for verification. The test results show that the single threshold is significant at the 10% level, while the double threshold is not significant (Table 4). Therefore, the single threshold model is selected, and the threshold value is 1.939.

Table 4. Threshold test results.

variable	environmental regulation	
threshold	Single threshold	Double threshold
Bootstrap	300	300
Trim	0.01	0.01
The threshold value	1.939*	1.874;1.939
95% confidence interval	[1.840,1.942]	[1.848,1.942]
		[1.857,1.949]

Note: ***, ** and * indicate high significance at the 1%, 5%, and 10% levels.

After the threshold value is determined, the parameters of the threshold model are estimated and analyzed. Table 5 shows the threshold regression results. When it is lower than the threshold value (TR$\leqslant$1.939), environmental regulation has an obvious deteriorating effect on the marine environment. When it is greater than the threshold value (TR>1.939), environmental regulation's deteriorating effect on the marine environment is significantly reduced, but it still has a positive promoting effect on marine pollution. This is contrary to the conclusions reached by most scholars, because most scholars think environmental regulation is conducive to slow down environmental pollution [6-10]. According to the some academics' ideas [13-14], the reason is that when China's ER level was still low, local enterprises and governments didn't pay attention to the environmental quality, and enterprises improved their economic income by reducing machinery and equipment and increasing product output. However, the amount of waste increased in this mode of production, which is characterized by emphasis on quantity rather than quality. The government treated environmental pollution by an ineffective means, i.e., "treatment after pollution", causing continuous deterioration of the terrestrial environment. As a result, the waste generated during terrestrial activities was discharged into the ocean through rivers, acid rain, etc., posing a huge threat to the marine environment. With the continuous improvement of the environmental regulation level, the local enterprises and governments have started paying attention to the domestic environmental quality. Most enterprises have discarded obsolete, aged and environmentally harmful machines and equipment and produced high-quality products. The government has also increased investment in environmental governance by getting rubbish out of water and strengthening human supervision, partly improving the quality of the terrestrial and marine environments, but overall, environmental regulation still has a deteriorating effect on China's marine environment.

The regression coefficient of GOP is positive, indicating that GOP has a positive effect on marine pollution. The main reason is that the increase in GOP also cause the generation of certain waste, which is directly discharged into the ocean, and this is not conducive to the control of marine pollution. The regression coefficient of GDP is positive and smaller than that of GOP, indicating that the development of the terrestrial economy also aggravates marine pollution, but the effect is smaller than GOP. This is consistent with most scholars' belief that economic development has a negative correlation with the marine environment within a certain interval [12, 15-16]. The regression coefficient of population is negative, indicating that population increase is conducive to large-scale governance of the marine environment and mitigation of marine pollution.

Table 5. Threshold regression results.

variable	TR is the threshold variable	
	≤1.939 >1.939	≤1.939 >1.939
lnGOP	0.480***	/
lnGDP	/	0.471***
lnPD	-0.655	-0.706
TR1(≤1.939)	0.184***	0.204***
TR2(>1.939)	0.071**	0.078**
Conf.	11.12**	10.703*

Note: ***, ** and * indicate high significance at the 1%, 5%, and 10% levels.

5. Conclusions and policy suggestions

5.1. Conclusions

The nonlinear relationship between the environmental regulation of coastal provinces and cities and marine pollution is taken into consideration in this paper. The 2006-2016 data of 11 coastal provinces and cities are used to a threshold model for marine pollution. GOP, population and GDP, and population are taken as control variables for threshold value grouping according to environmental regulation. The following conclusions are drawn:

(1) Environmental regulation has an obvious marine and terrestrial economic threshold effect on marine pollution. That is, at different levels of environmental regulation, environmental regulation has a single threshold effect on marine pollution. On the left side of the threshold, because the enterprises prefer quantity to quality and the government harnesses environmental pollution based on the principle of "treatment after pollution", environmental regulation's improving effect on the environment falls far short of the deterioration of environmental pollution. Terrestrial pollution is discharged into the ocean through rivers and acid rain, so environmental regulation aggravates marine pollution. On the right side of the threshold, the enterprises carry out advanced production while the government prevents environmental pollution at the early stage and strengthens environmental governance at the later stage. Therefore, environmental regulation has some improving effect on the quality of the terrestrial and marine environments, but on the whole, environmental regulation still has a deteriorating effect on China's marine environment.

(2) The terrestrial economy and marine economy are significantly positively correlated with marine pollution. Economic development exacerbates marine pollution, and the effect of the marine economy on marine pollution is greater than that of the terrestrial economy. There is an insignificantly negative correlation between population increase and marine pollution. The increase in population improves manpower, making it possible to carry out marine environmental governance on a large scale.

5.2. Policy suggestions

Based on the research results, the following policy suggestions are put forward:

(1) The government should strengthen environmental management and earnestly implement the environmental impact assessment system to prevent planning, development and construction from exerting an adverse effect on the environment at the source; strictly implement the "three simultaneous" system to ensure that environmental protection facilities and the main project are designed, constructed and put into production at the same time; fulfill the environmental protection targets set at all levels, and take environmental protection as important content of performance assessment for the Party and government leaders and cadres at all levels. The leaders and cadres who have made a decision mistake, causing a severe pollution accident, seriously disturbing environmental law enforcement, shall be held accountable in accordance with the law.

(2) The government should advocate a circular economy and accelerate high-quality economic development; implement a policy on energy and resource saving in accordance with the principle of "resource conservation, secondary utilization and recycling use" to strengthen the comprehensive utilization of resources and improve the resource recycling system to comprehensively promote clean production, and increase the resource utilization efficiency to reduce the generation and discharge of pollutants. The concept of circular economy development must be practiced in key links such as planning, policy formulation, system design, project decision-making, law enforcement and regulation implementation.

(3) The government should also coordinate the relationship between the economy and the environment. During economic development, environmental pollution shall not exceed its maximum carrying capacity. The government should formulate and implement correct industrial policies, and adjust the industrial structure to reduce environmental pollution and destruction. It is necessary to strictly limit the development of enterprises characterized by high energy consumption, large waste of resources and serious pollution, and implement industrial policies to support enterprises in improving the treatment and comprehensive utilization of the wastewater, exhaust gases and waste residues generated during their production.

References

[1] Ayedh Al 2015 Shehai Royal Commission at Yanbu Environmental Regulations **125** 193-198
[2] Brian Garrod and David Whitmarsh 1995 The economics of marine pollution control[J]. *Marine Pollution Bulletin* **30(6)**
[3] Borsatto Jaluza Maria Lima Silva and Bazani Camila Lima 2020 Green innovation and environmental regulations: a systematic review of international academic works.[J]. *Environmental science and pollution research international*
[4] Danny Marks, Michelle Ann Miller and Sujitra Vassanadumrongdee 2020 The geopolitical economy of Thailand's marine plastic pollution crisis[J]. *Asia Pacific Viewpoint* **61(2)**
[5] Agovino Massimiliano, Ferraro Aniello and Musella Gaetano 2020 Does national environmental regulation promote convergence in separate waste collection? Evidence from Italy 125285-
[6] Qingquan Fan, Chengjun Chu, Jingran Liu and Mingyi Zhang 2020 Environmental regulation, industrial upgrading and smog control [J]. *Acta Ecologica Sinica* **7(04)** 189-213 Impact of economic structure change on Marine environment in China's coastal areas [J]. *Marine Economy* **10(04)** 25-36
[7] Wencui Du and Bo Chen Environmental regulation, industrial concentration and environmental pollution [J/OL]. *Journal of Xi 'an Jiaotong University* (Social Science Edition) 1-12. (in Chinese)
[8] Zhibin Zhang 2020 Government environmental regulation, enterprise transformation and upgrading and urban environmental governance: An empirical study based on panel data of 35 key cities [J]. *Journal of Social Sciences of Hunan Normal University* **49(06)** 146-154
[9] Rui Liang, Ming Gao andf Xueping Wu 2020 The relationship between environmental

IOP Conf. Series: Earth and Environmental Science **865** (2021) 012018 doi:10.1088/1755-1315/865/1/012018

regulation and air pollution reduction: An analysis of threshold effect based on economic growth [J]. *Ecological Economics* **36(09)** 182-187
[10] Chang 'e Kuang, Shuling Liu and Jiayu Bin 2020 Research on the impact of environmental regulation on industrial pollution reduction: empirical evidence from Hunan Province [J]. *Contemporary Economics* **(05)** 76-79
[11] Shaozhou Qi and Yaxue Yan 2017 Study on Kuznets Curve of Chinese Haze (PM_(2.5)) Based on Panel Thresold Model [J]. *Journal of Wuhan University* (Philosophy and Social Sciences Edition) **70(04)** 80-90
[12] Mei Gai and Qing Zhang 2020 Impact of economic structure change on Marine environment in China's coastal areas [J]. *Marine Economy* **10(04)** 25-36
[13] Ajiang Chen, Yajuan Luo and Tao Chen 2011 The path and consequence of "pollution before treatment" -- a case study of environmental protection and the development of Maoji leather industry [J]. *Journal of Guangxi University for Nationalities* (PHILOSOPHY AND SOCIAL SCIENCES EDITION) **33 (03)** 55-60
[14] Zhiwei Xu 2016 Industrial economic development, environmental regulation intensity and pollution reduction effect: theoretical analysis and empirical test based on the development model of "pollution before treatment" [J]. *Financial research* **42 (03)** 134-144
[15] Guangsheng Wang, Peifang Guo, Yingyu Tan and Hongyuan Shi 2014 EKC analysis of economic growth and marine environmental pollution panel data in coastal areas based on unit root test[J]. *Marine Environmental Science* **33(03)** 425-430+445
[16] Shuaibang Fan and Liling Zhao 2015 Research on the Economic Development and Marine Environment of Liaoning Coastal Economic Zone[J]. *Regional Research and Development* **34(02)** 40-44+50

IOP Conf. Series: Earth and Environmental Science **865** (2021) 012019 doi:10.1088/1755-1315/865/1/012019

A Systematic approach to world regional marine resources development models through case studies

Yawen Kong[1], Chunyu Liu[2], Shuguang Liu[1, 3, 4] and Lei Song[1]

[1]School of Economics, Ocean University of China, Qingdao 266100, China;
[2]Graduate School of Arts and Sciences, Columbia University, New York 10027, USA;
[3]KRI Institute of Marine Development, Qingdao 266100, China

[4]Corresponding author's e-mail: 2000046@ouc.edu.cn

Abstract. The marine resources are regarded as the major components of global marine ecosystem within the sphere the earth surface. While the sustainability of regional marine resources is essential for worldwide human-nature co-existence and coordinated development in the long run. Highlighted by principles of modern regional economics, the paper manages to conduct systematic assessment of 12 worldwide cases in 4 categories of coastal zones, islands, offshore waters, as well as high seas & international seabed. Five aspects of indicators are determined in terms of innovation, coordination, adaptation, openness, and sharing during the research. Among the findings of the initial approaches, the coastal zone, high seas & international seabed cases show higher indicator performances than those of near seas and islands, and conservation oriented cases are better in scoring results than those of resource exploitation cases.

1. Introduction

Marine resource and environment system, the main part of surface earth system, is a vital material basis for the development and progress of human society [1]. Sustainable exploitation and utilization of marine resources have become the basic premise for the sustainable development of mankind [2]. The "global commons" nature of marine resources has made them the focus of global economic attention [3]. It is necessary to draw on and evaluate regional marine resource development mode. The development of human civilization is accompanied by the change and transformation of the exploitation of regional marine resources. The Ancient Greek Xenophon discussed the strategic role of the acquisition of marine resources for national economic development [4]. In the Middle Ages, Venice and Genoa broke through "the Braudel Bell Jar" and created the thriving history of a country using Mediterranean Sea resources [5], and "the Hanseatic League" extended the European model of the overall development of land and sea [6]. In modern times, Columbus started the trans-Atlantic Ocean and even the global marine resources development boom [7]. In the 17th century, the East India Company established by Dutch was an early example of the modern capitalism commercially expands the space of marine resources [8]. In the 18th century, Britain achieved the synchronization of the internationalization development of industry and the construction of colonial trade network, and became an important member of the development both in deep and far sea resources [9].With the profound global division of labor and global environmental change in the 21st Century, the exploitation and protection of marine resources become an important topic of sustainable development, and the research on sustainability of marine resources and environmental protection has been integrated into the mainstream research category of economics [10,11]. The research progress of regional marine resource development is in the stage of conceptual discussion and case analysis, resulting in the lack of systematic analysis and evaluation on global classic regional marine resource development model. The purpose of the paper is to categorize the characteristics of world typical regional marine resource cases through applying methodologies of regional sustainable economics, and summarize the world regional marine resources development models through case studies. Hopefully this research could lay foundations for further approaches in regional marine resources development studies.

2. Theory and methods

2.1. Theory discussion

The regional marine resource system is an intrinsic part of the earth surface sphere, and has basic characteristics of open complex system. The space-time integrating attribute of marine resource system provides the possibility for systematically studying the marine resource development mode [12]. The land-based living characteristic of human beings objectively cause the difference between human-sea and man-land spatial relationship, enlarging the complex spatial relationship between marine resource object and marine resource exploitation subject [13]. It is a challenge for traditional analysis paradigm of regional economics, which takes land economic activities as the premise of analysis [14]. Therefore, we integrate analysis paradigm of system science into classic analysis of regional economics, and construct an analytical framework of the coupling between regional marine natural resource system and socio-economic development activity system. We regard issues of regional marine resource development as sustainable development issues of regional marine resources and marine economy complex system, and then apply the analytical paradigm of modern regional sustainable development system to study them [15].

Based on existing research results [16], we combine the five state parameters of "innovation, coordination, adaptation, openness, and sharing" with the sustainable development system evaluation method, integrate basic concepts of marine resources development and protection, and try to construct a corresponding evaluation paradigm and index system that is suitable for regional marine resources development system around the world.

2.2. Research methods

2.2.1. Construction of evaluation model. Based on the above-mentioned general theoretical discussion, we formulate an evaluation and analysis program suitable for regional marine resources development: (1) regard the development and utilization of specific regional marine resources as a sea area resource economic system. According to the spatial relationship between human and sea, the development and utilization of specific regional marine resources can be divided into four types: coastal resources development, islands and surrounding waters resources development, offshore waters resources development, as well as high seas & international seabed area resources development; (2) define the spatial boundary of the sea area resource economic system is the location and scope of the selected cases; (3) determine the system parameters that are suitable for the high-quality development of marine resources as evaluation indicators; (4) construct the equation between the development situation / state parameters of regional marine resources and its main indicators, prepare for experts scoring of through providing them with case descriptions, and (5) evaluate regional marine resource development state and models based on scoring results.

Based on the analysis above, the evaluation model of marine resources development state is:

$$Y = F(X) \tag{1}$$

$$X = \{X_1, X_2, X_3, X_4, X_5\} \tag{2}$$

Here Y is the marine resources development state. X represents characteristic factor indexes that affect the development state of regional marine resources, including innovation index of marine resources development (X_1), coordination index of marine resources development (X_2), adaptation index of marine resources development (X_3), openness index of marine resources development (X_4), and sharing index of marine resources development (X_5). Assuming $X > 0$, the increase in any dimension of characteristic factors will promote the increase of Y , and this positive effect satisfies the marginal diminishing utility.

2.2.2. Evaluation index system. With reference to the construction specifications of the general evaluation model index system for regional economic development, taking into account the complexity of regional marine resources, and taking into account the data availability as well as resolution accuracy of the selected regional sea area, we establish the evaluation index system for regional marine resource development state (Y). The index system structure is divided into three levels. The first-grade indexes include the degree of innovation in marine resource development, coordination, adaptability, openness, and sharing (Table 1).

Table 1. The evaluation index system for regional marine resources development state.

First-Grade Indexes	Second-Grade Indexes	Third-Grade Indexes	Indexes for scoring
Innovation (X_1)	Innovation of development elements	Quality of resource endowment	Area and depth of the sea Sea water quality standard Quality and abundance of marine resources Average straight-line distance between the sea and the world's three major economic centers
		Capabilities of technology research	National Marine Science and Technology Innovation Index
		Capabilities of financial integration	Number of global financial centers Score of global financial centers index
		Strength of human capital	National average human capital index
	Innovation of development organizations	Innovation ability of leading enterprise	Global influence of leading companies (institutions)
		Innovation capability of industrial cluster	Comprehensive strength of the state in the development (protection) of marine resources
Coordination (X_2)	Construction of collaborative organization	Soundness of organizations	Number and influence of organizations, institutions and committees for the conservation of the marine habitats
		Linkage mechanism of inter-organization	Meetings between maritime management and protection authorities
	Level of collaborative activities	Efficiency of collaborative operation	Frequency of meetings
		Degree of Collaborative Stability	Collaboration time between organizations Authority of the conventions and regulations between collaborative organizations
Adaptation (X_3)	Degree of environmental adaptation for development	Awareness of environmental changes	Number of organizations responsible for exploration and inspection of the sea area Number of literatures for exploration and inspection of the sea area Response time of developers to environmental changes

IOP Conf. Series: Earth and Environmental Science **865** (2021) 012019 doi:10.1088/1755-1315/865/1/012019

		Ability to cope with environmental change	Emergency preparedness of developers to environmental changes
	Degree of resource adaptation for development	Renewal capacity of renewable resources	Degree of reasonable utilization for renewable resources Degree of protection for renewable resources
		Protection capability of non-renewable resource	Number of marine protected areas Operational effects of marine protected areas Maturity of resource development technology
Openness (X_4)	Openness of elements flow	Cross-border flow of natural elements	Natural attributes of ocean energy resources flow Natural attributes of marine life migration
		Cross-border flow of economic elements	Degree of funds, technology and talents flow
	Openness of development organization	Openness of industrial division of labor	The degree of internationalization of the enterprise
		Openness of institutional cooperation	The degree of internationalization of organizations Number of countries participating in the organization
Sharing (X_5)	Sharing of direct stakeholders	Participation in the development process	Integrity of the subject's participation in the development or protection process
		Distribution of development results	Fairness of results allocation in the development or protection process
	Sharing of indirect stakeholders	Knowledge of the development process	Number of media reports Degree of knowledge of experts on the cases
		Degree of overflow of development results	Availability of marine resource development reports Number of marine resource development reports

3. Analysis of global classic cases of regional marine resources development

3.1. Selected cases

According to the spatial classification of global regional marine resources development, we select four spatial types: coastal zones, islands and surrounding waters, offshore waters, and high seas & international sea-beds. Based on the principles of significant global impact, balance of spatial classification and development methods, and random distribution of sea areas, 12 regional marine resources development cases were initially selected for analysis and evaluation (Figure 1).

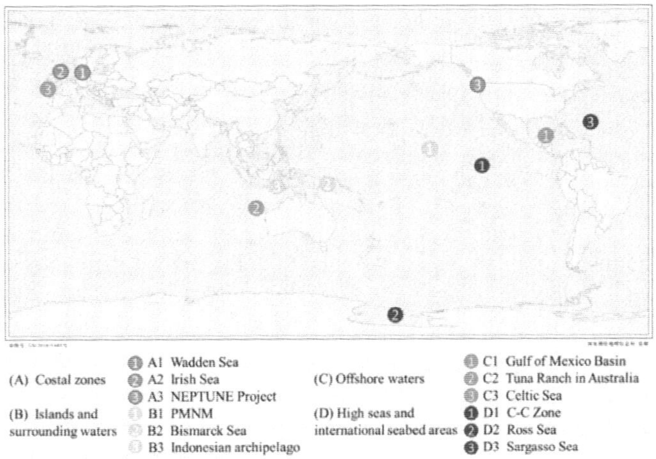

Figure 1. Locations of regional marine resources development cases.

3.2. Costal zones

3.2.1. Marine resources development in the Wadden Sea. The Wadden Sea is a marginal sea in North Sea jointly managed by the Netherlands, Germany and Denmark. Its vast intertidal zone is a habitat for more than 10,000 species. Facing conflicts of ecological and economic interests, such as overfishing led to a large reduction of the blue mussel bed in the intertidal zone, three countries established the Trilateral Wadden Sea Cooperation in 1978 [17], which helped improve ecological and economic sustainability by coordinating local residents' scallop farming, underwater natural gas extraction, and coastal tourism activities [18]. The adopted Wadden Sea Plan carries out protective management, banning wild mussels fishing, limiting mussel farming area and the number of fishing licenses, monitoring the reproduction of migratory bird populations, and controlling natural gas extraction capacity [19]. The practice of the protection plan made the sea area as a World Heritage Site by UNESCO in the early 21st century.

3.2.2. Marine resources development in the Irish Sea. The Irish Sea is located between the island of Ireland and the island of Great Britain on the edge of the Northeast Atlantic Ocean, jointly managed by Ireland and the United Kingdom. The annual economic value of the Irish Sea fishery and aquatic products is about 6 billion euros [20], but overfishing has caused the stock of marine resources facing a crisis of collapse. The European Commission formulated the Irish Sea Cod Recovery Plan in 2000, taking measures such as reducing quotas, banning trawl nets, and closing spawning areas, but the high fishing mortality rate made the plan little effect [19]. Then the Joint Nature Conservation Council (JNCC) formulated a marine spatial planning framework for the Irish Sea with fishermen and other stakeholders. However, there are still problems such as fishing in closed time, difficulty in controlling the scale of fishing, and unfair quota transfers [21].

3.2.3. NEPTUNE project on the Pacific Coast of North America. The North-East Pacific Undersea Networked Experiments (NEPTUNE) project is located in the Juan de Fuca plate on the seabed of the Pacific Ocean off the coast of North America. Canada and the USA jointly invested 300 million US dollars for the project. The NEPTUNE-Canada project operated by Ocean Networks Canada (ONC) was put into operation in 2009. It is the first regional underwater ocean observatory project directly connected to the Internet and provides access for laboratories and universities around the world [22]. The project monitors the marine crustal structure and dynamics, assesses the state of marine biodiversity and fishery activities, etc. [23], and serves the maintenance of national marine sovereignty, marine traffic dispatch, and marine big data services [24]. The project is a milestone in marine science and provides valuable experience for building submarine monitoring systems in the world.

3.3. Islands and surrounding waters

3.3.1. Establishment of Papahānaumokuākea Marine National Monument. Papahānaumokuākea Marine National Monument (PMNM) is located in the northwestern part of the Hawaiian Islands, covering an area of 362,000 square kilometers. Approved as a UNESCO World Heritage of Nature and Culture [25], the Monument is jointly managed by the U.S. Fish and Wildlife Service (FWS), National Oceanic and Atmospheric Administration (NOAA), and the local government [26]. PMNM protects the natural marine resources, local cultural resources and the heritage of maritime activities [27]. Because the rising sea level and temperature has caused coastline erosion, inundation of islands and reefs, degradation of coral reef ecosystems, and surrounding fishing has interfered with bird foraging and migration, the expansion protected area was established in 2016. But it has an adverse impact on longline fishing operations in local traditional fishing grounds [28].

3.3.2. Mineral resources development of Bismarck Sea in Papua New Guinea. The Bismarck Sea is located in the southwestern Pacific Ocean, covering the waters north of Papua New Guinea, and south of the Bismarck Islands and Admiralty Islands, a typical accumulation area of hydrothermal sulfide in the global seafloor. Nautilus Minerals Ltd. of Canada obtained a mining license issued by Papua New Guinea in 2011. The operation area is located in the Manus Basin, and aims to extract 120,000 tons of copper, a small amount of gold, and other minerals [29]. Nautilus emphasizes that mining activities are far away from the public, the mining operations are short and limited in scope, and has passed expert environmental assessment and other permits [30]. However, in view of multiple conflicts with local communities in the actual trial mining process, the company began to abandon this controversial development plan [31].

3.3.3. Marine resources development of the Indonesian Archipelago. The Indonesian archipelago is located between the Indian Ocean and the Pacific Ocean. It has about one-fifth of the coral reef resources on the earth. Indonesia has approximately 82 billion dollars fishery resources per year [32]. As the main producer of tuna in Asia, destructive fishing methods such as explosives and cyanide has caused the depletion of fishery resources and damage to the coral reef system [33]. The Asian Development Bank and the World Bank provided funds to carry out Coral Reef Restoration and Management Plan (COREMAP) issued by Indonesian Ministry of Ocean Affairs and Fisheries [34], but it didn't work well. Subsequently, Indonesia promoted a community-participated marine protected area project. It is helpful for conserving island and reef ecosystems, reducing poverty, as well as raising awareness on ecological and economic sustainability [35].

3.4. Offshore waters

3.4.1. Oil and gas resources development of offshore waters in the Gulf of Mexico Basin. The Gulf of Mexico Basin is the third largest petroliferous basin in the world. The United States, Cuba, and Mexico develop its oil through independent or international cooperation. In 2010, the British operating platform near the US shoreline exploded and sank, resulting in oil spill. This serious accident caused more than 10 billion dollars loss to neighboring coastal commercial and aquaculture industries [36]. BP reached a settlement agreement with US federal and local authorities for 18.7 billion dollars and the cost of cleanup, environmental compensation, economic losses, and fines amounted to 54 billion dollars. The incident exposed the lack of logistical, technical, and financial response capabilities, highlighted the importance of deep-sea environmental baseline awareness [37], promoted the establishment of a deep-sea oil spill regional impact assessment system [38], and reflected on the importance of the environmental safety for both international cooperative offshore oil enterprises and the host country [35].

IOP Conf. Series: Earth and Environmental Science **865** (2021) 012019 doi:10.1088/1755-1315/865/1/012019

3.4.2. Bluefin tuna culture in the offshore waters of Northwest Australia. The southern waters of Australia and New Zealand are the main living place of southern bluefin tuna with highly migratory characteristics [39]. The declining tuna fishing quota prompted Australia, New Zealand, and Japan to sign the " Convention for the Conservation of Southern Bluefin". Then they established the Commission for the Conservation of Southern Bluefin Tuna (CCSBT), a regional marine fishery management organization of the FAO [40]. To improve the scale and quality of migratory tuna catches, Australia carries out pasture farming of bluefin tuna before harvest in Port Lincoln, which has become an international model for the staged industrial aquaculture and research of deep-sea fish [41]. The marine ranching is facing the dual challenges of international supply and demand with increasing tuna breeding bases. The government has maintained its international leading position by improving the management of the marine ranching [42].

3.4.3. Energy development of British Celtic Offshore Waters. The Celtic Sea is located on the southwest side of Ireland and the United Kingdom. It has huge wind energy, ocean current energy and wave energy brought by sea breeze [43]. The government encourages billion-pound private investors, entrepreneurs, academic institutes, and the public to join in resources development. Especially on the coast of Wales near the Celtic Sea, the government focuses on the development of marine energy industrial clusters, promotes the increase and global expansion of the relevant marine energy value chain [44], promotes the land-sea coordination of marine energy development in the UK through the integration of offshore energy development and terrestrial energy grids [45].

3.5. High seas and international seabed areas

3.5.1. Submarine resources development in C-C Zone. The Clarion-Clipperton Fracture Zone is located in the international seabed area of the Northeast Pacific Ocean. Rich in polymetallic nodules dominated by nickel, manganese, and cobalt, the C-C zone is managed by the International Seabed Authority (ISA) in accordance with the United Nations Convention on the Law of the Sea [46]. The ISA has successively promulgated seabed area development regulations and contract terms since 2016, but not yet effective [47]. By the end of 2018, the ISA had signed 16 contracts for the exploration of polymetallic nodules with contractors from multiple countries and international organizations. However, the huge sunk cost of seabed resource development and coupled with opportunity costs such as damage to the deep sea and seabed ecosystems will make the compensation and restoration technologies very expensive [48]. The social and economic feasibility of seabed mining needs to be reassessed [49].

3.5.2. Establishment of the Ross Sea Marine Reserve. The Ross Sea belongs to the deep bay on the edge of Antarctica. Marine plankton promotes the formation of characteristic marine animal ecological communities of Ross Sea. The Ross Sea is managed by Commission for the Conservation of Antarctic Marine (CCAMLR). Due to the special state of the biological diversity and climate regulation function, coupled with the huge pressure of commercial fishing [50], CCAMLR established the world's first large-scale marine protected area on the high seas in 2016, becoming an important supporting case of the United Nations for the negotiation of Biological Diversity of Areas Beyond National Jurisdiction (BBNJ) [51]. The establishment, management, and operation of the protected area embodies the determination and actions of the world's ocean powers to work together for global ocean governance [52].

3.5.3. Conservation and development of Sargasso Sea. The Sargasso Sea is located in the center of the subtropical circulation of the North Atlantic Ocean. The floating "forest" of Sargasso provides food and shelters for the newly hatched red sea turtles in the Atlantic Ocean. It is an endangered European eel and American eel co-laying area [53]. In 2010, the Sargasso Sea Alliance was initiated by the Bermuda government, non-governmental organizations, scientists, and private donors to provide a model for the conservation of high seas [54]. In 2014, the Sargasso Sea Committee was further established in accordance with the Hamilton Declaration, and successively coordinated the management of the sea area with the International Commission for the Conservation of Atlantic Tunas, the Northwest Atlantic Fisheries Organization, the International Maritime Organization, the International Seabed Authority, etc [55]. The Sargasso Sea Alliance became an observer organization of the United Nations in 2016.

4. Empirical results

4.1. *Evaluation process and results*

We make a questionnaire with 12 cases descriptions and relevant quantitative background data, and invite 35 experts in the fields of marine resources and regional economics to score online independently. By summarizing the scoring data and the feedback of experts, we obtain the preliminary results of regional marine resources development state (Table 2).

Table 2. Assessment results of regional marine resources development state.

Marine spatial types	Cases (Exp -Exploration / Con -Conservation)	Score					Total score
		X_1	X_2	X_3	X_4	X_5	
(A) Coastal zones	(A1) Marine resources development in Wadden Sea (Con)	3.44	4.52	4.05	4.05	3.89	**19.95**
	(A2) Marine resources development in Irish Sea (Exp)	3.24	2.58	2.72	3.19	3.33	**15.06**
	(A3) NEPTUNE Project on the Pacific Coast of North America (Con)	4.76	3.98	3.92	4.15	3.95	**20.76**
(B) Islands and surrounding waters	(B1) Establishment of Papahānaumokuākea Marine National Monument (Con)	2.94	3.99	3.64	3.87	2.74	**17.18**
	(B2) Mineral resources development of Bismarck Sea in Papua New Guinea (Exp)	3.00	1.98	2.42	2.99	3.44	**13.83**
	(B3) Marine resources development of the Indonesian archipelago (Exp)	2.62	3.04	2.54	3.01	3.25	**14.46**
(C) Offshore waters	(C1) Oil and Gas Resources Development of Offshore Waters in the Gulf of Mexico Basin (Exp)	2.71	2.70	1.48	3.40	3.11	**13.40**
	(C2) Bluefin tuna culture in the offshore waters of Northwest Australia (Exp)	3.83	3.27	3.03	3.29	3.90	**17.32**
	(C3) Energy development of British Celtic offshore (Exp)	4.21	3.28	3.26	4.25	4.11	**19.11**
(D) High seas and international seabed areas	(D1) Submarine resources development in C-C area (Exp)	3.66	2.95	2.64	3.94	3.46	**16.65**
	(D2) Establishment of the Ross Sea Marine Reserve (Con)	3.11	2.68	3.54	3.97	4.05	**17.35**
	(D3) Conservation and development of Sargasso Sea (Con)	2.39	2.46	2.80	3.47	4.12	**15.24**

4.2. *Results analysis*

As shown in Figure 2a and 2b, scores of the four spatial types cases in (A) coastal zones, (B) islands and surrounding waters, (C) offshore waters, (D) high seas and international seabed areas are respectively summed and averaged, and scores of the two utilization types cases in exploitation (Exp) and conservation (Conservation, Con) are summed and averaged.

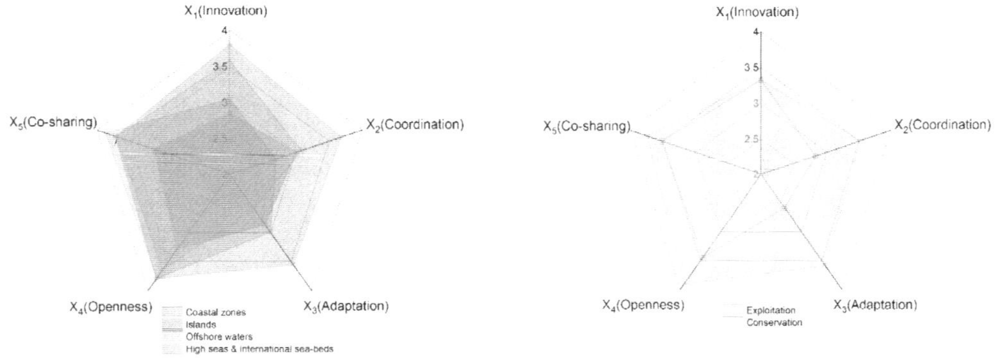

(a) Scores of factors in different spatial types (b) Scores of factors in different utilization types

Figure 2. Correlation coefficients and scores in different spatial and utilization types of factors.

The structural characteristics of the spatial type factors. The evaluation results of the marine resources development state in the four marine spatial types have certain differences (Figure 2b). The score of it from high to low are coastal zones (A), offshore waters (C), high seas and international seabed areas (D), and islands and surrounding waters (B). First, the score of coastal zone cases is overall balanced and outstanding, which means that humans have a deep foundation, sufficient cognition, and strong ability for the development and governance of the coastal zones. Secondly, the high scores of marine resources development state in the high seas and the international seabed area are due to the high degree of openness and continuous sharing. It shows the high openness of the high seas and the international seabed area resource endowment, and reflects people's "collective action" efforts to avoid the "tragedy of the commons". Thirdly, the resource-exhausted mining in offshore waters and the environmental pollution tragedies reflected in case studies objectively show the excess production capacity in offshore development process and the lack of environmental response capabilities. Fourthly, the performance of islands and surrounding waters is limited and fragile toward economic development and technical support, unequal resource development benefit sharing mechanism, etc., making it difficult to extract or maintain island resources.

The structural characteristics of the utilization type factor. The types of development and utilization activities also have differences both in overall resource development state and structural factor performance (Figure 2c). First, the overall index scores of conservation-type cases are higher than those of resource mining cases, indicating that resource development is relatively easy, but it is more likely to cause variation or destruction of the resource system. Resource conservation requires effective interaction in all aspects such as innovation, coordination, adaptation, openness, and sharing; Secondly, both types show a high level of openness and sharing, reflecting the high openness and sharing characteristics of marine resources as common pool resources. The cross-border mobility of marine resources also endows the development process with relatively high openness characteristics. Stakeholders share the distribution of the results of marine resource development and conservation; Thirdly, the two utilization types of innovation capabilities are relatively insufficient, indicating that in the process of marine resources utilization, whether it is exploration or conservation, further technology research and development and increased capital investment are required. Fourthly, the score of adaptability indicator for two utilization types are at a low level among the five factors, indicating that all types of marine resources development needs to deal with the complex and huge risk of the natural environment system and its changes.

5. Conclusion

Based on the description and quantitative evaluation of classic cases of global regional marine resource development, preliminary research has drawn the following conclusions:

(1) Regional marine resources development activities bear basic features of a regional resource economic system. But if compared with the general terrestrial resource economic system, the marine resource development is more difficult in coordinating development subject and the development target. The large scale variability of the environment, highly dynamics of marine resources, widely opening boundaries, and discrete roles of stakeholders, make the evaluation of the marine resource development more sophisticated than general regional resource development systems.

(2) From the perspective of human-ocean spatial relationship, the order of marine resource development state in coastal zones, islands and surrounding waters, offshore waters, high seas as well as international seabed areas is in line with the research results of relevant authoritative documents, and also in line with the common-sense judgments of the regional economics professionals, laying the foundation for further relevant evaluation research.

(3) The complex system characteristics of the regional marine resource system challenge the basic development model of the resource system, based on the relatively fixed and closed land area. This research needs to be improved and upgraded in at least the following aspects: (a) a standardized and refined indexes system suitable for regional marine resources development; (b) expansion in scales and categories of cases, i.e. traditional marine resources development and new marine resources development, single marine resources development and integrated marine resources development, etc.

Acknowledgement
Foundation item: Study on the impact of sea level rise on the development of key coastal areas in China (No. 15ZDB170). The author would like to thank Wang Fang of China Ocean Mineral Resources R & D Association and Chen Jiliang of the Greenovation Hub for their guidance and evaluation. Thanks for their help in the writing of this article. We take sole responsibility for our views.

References

[1] Steffen W, Richardson K, Rockström J, Schellnhuber H J, Dube O P, Dutreuil S, Lenton T M and Lubchenco J 2020 The emergence and evolution of Earth System Science. *Nat Rev Earth Environ* 54–63
[2] Neumann B and Unger S 2019 From voluntary commitments to ocean sustainability. *Science* **363** 35-36
[3] Han X C and Wang Y W 2014 Global commons: Ideological origin, conceptual pedigree and academic reflection. *Social Sciences in China* 188-205
[4] Xenophon 1981 Economics Revenue of Athens translated by Zhang B J and Lu D N *Commercial Press* **63** 66-67.
[5] Kirk T A 2005 Genoa and the Sea: Policy and Power in an Early Modern Maritime Republic *Johns Hopkins University Press*
[6] Beerbühl M S 2012 Networks of the Hanseatic League *Institute of European Hist*ory*, retrieved*
[7] Crosby A W 1987 The Columbian Voyages: the Columbian Exchange and their Historians *American Historical Association*
[8] Ames G J 2008 The Globe Encompassed: The Age of European Discovery 1500-1700 Upper Saddle River: Pearson Prentice Hall
[9] Smith S 1998 British Imperialism 1750–1970 *Cambridge University Press*
[10] Romer P M 1989 What Determines the Rate of Growth and Technological Change World Bank Publications
[11] Ostrom E 2009 A general framework for analyzing sustainability of social-ecological systems *Science* **325** 419-422
[12] Tafon R V 2018 Taking power to sea: Towards a post-structuralist discourse theoretical critique of marine spatial planning *Environment and Planning C-Politics and Space* **36** 258-273
[13] Shafer, C S, Inglis G J and Martin V 2010 Examining Residents' Proximity, Recreational Use, and Perceptions Regarding Proposed Aquaculture Development *Coastal Management* **38** 559-574

IOP Conf. Series: Earth and Environmental Science **865** (2021) 012019 doi:10.1088/1755-1315/865/1/012019

[14] Hadjimichalis C and Hudson R 2014 Contemporary Crisis Across Europe and the Crisis of Regional Development Theories *Regional Studies* **48** 208-218
[15] Ferrol-Schulte D, Wolff M and Ferse S 2013 Sustainable livelihoods approach in tropical coastal and marine social-ecological systems: a review *Marine Policy* **42** 253-258
[16] Liu S G, Xu Y J and Wang J Y 2020 An Approach to the Relationship between the River Basin Economic System Opening-up and Sustainable Development: International Experiences and Its Instruction to Yellow River Basin High-quality Development *Resources Science* **042** 433-445
[17] Slob A F L, Geerdink T R A, Rockmann C and Voege S 2016 Governance of the Wadden Sea *Marine Policy* 325-333
[18] Egberts L and Hundstad D 2019 Coastal heritage in touristic regional identity narratives: a comparison between the Norwegian region Sørlandet and the Dutch Wadden Sea area *International Journal of Heritage Studies* 1-15
[19] Swart J A A and Windt H J V D 2012 *Knocking on Doors: Boundary Objects in Ecological Conservation and Restoration* Sustainability Science: the Emerging Paradigm and the Urban Environment Springer
[20] Vincent M A et al 2004 *Marine nature conservation and sustainable development – the Irish Sea Pilot*. Peterborough: Report to DEFRA by the Joint Nature Conservation Committee
[21] Yates K L 2014 View from the wheelhouse: Perceptions on marine management from the fishing community and suggestions for improvement *Marine Policy* **48** 39-50
[22] Danovaro R, Aguzzi J and Fanelli E, et al 2017 An ecosystem-based deep-ocean strategy *Science* **355** 452-454
[23] Thomson R, Fine I and Rabinovich A et al 2011 Observation of the 2009 Samoa tsunami by the NEPTUNE Canada cabled observatory: Test data for an operational regional tsunami forecast model *Geophysical Research Letters* **38** 1-5
[24] Barnes C R, Best M M R and Johnson F R et al 2013 Challenges benefits and opportunities in installing and operating cabled ocean observatories: perspectives from NEPTUNE Canada *IEEE Journal of Oceanic Engineering* **38** 144-157
[25] Pala C 2009 Scientists laud Bush's Blue Legacy but want more *Science* **5911** 192-193
[26] Sekoe K A, Halpern B S and Toonen R J 2008 Aquatic conservation: marine and fresh water ecosystems *Aquatic Conserve: Marine Freshwater Ecosystem* **18** 1149–1165
[27] Gleason K 2014 *A monumental distance: education and outreach from the most remote archipelago on earth* In: D.A. Scott-Ireton (ed.) New York. Between the Devil and the Deep: Meeting Challenges in the Public Interpretation of Maritime Cultural Heritage Springer Science & Business Media
[28] Chan H L 2020 Economic impacts of Papahanaumokuakea Marine National Monument expansion on the Hawaii longline fishery *Marine Policy* **115** 103869
[29] Lee C and Dover V 2011 Mining seafloor massive sulphides and biodiversity: what is at risk *ICES Journal of Marine Science* **68** 341–348
[30] Childs J 2019 Greening the blue? Corporate strategies for legitimising deep sea mining *Political Geography* **74** 1-12
[31] Childs J 2020 Performing 'blue degrowth': critiquing seabed mining in Papua New Guinea through creative practice Sustainability *Science* **15** 117–129
[32] Dutton I M and Resosudarmo B P 2005 If only fish could vote: the enduring challenges of coastal and marine resources management in post-reformasi Indonesia *politics & economics of Indonesia's natural resources* 162-178
[33] Gurney G G, Cinner J and Ban N C et al 2014 Poverty and protected areas: An evaluation of a marine integrated conservation and development project in Indonesia *Global Environmental Change* **26** 98-107
[34] Sumaila U R, Cisneros-Montemayor A M and Dyck A et al 2012 Impact of the Deepwater Horizon well blowout on the economics of US Gulf fisheries *J. Fish Aquat Sci* **69** 499–510

[35] Oudhuis M and Tengblad S 2018 *BP and Deepwater Horizon: a catastrophe from a resilience perspective.* In Tengblad S Oudhuis M. (eds.) The Resilience Framework Work Organization and Employment Springer Nature Singapore Pte Ltd
[36] Court C, Hodges A W and Coffey Ket al 2020 Effects of the Deepwater Horizon Oil Spill on Human Communities: Catch and Economic Impacts. In S. A. Murawski et al. (eds.). Deep Oil Spills. Springer Nature Switzerland AG
[37] Joye B 2015 Deepwater Horizon 5 years on Samantha *Science* **349** 592-593
[38] MacKenzie C A, Baroud H and Barker K 2016 Static and dynamic resource allocation models for recovery of interdependent systems: application to the Deepwater Horizon oil spill *Ann Oper Res* **236** 103–129
[39] Tomczak M and Godfrey J 2003 Regional Oceanography: An introduction(2nd Edition). Daya Publishing House
[40] Polacheck T 2012 Assessment of IUU fishing for Southern Bluefin Tuna *Marine Policy* **36** 1150–1165
[41] Ellis D, Kiessling I and Benetti D D et al 2016 Ranching of Southern Bluefin Tuna in Australia Chapter-9: Advances in Tuna Aquaculture, Elsevier
[42] Mobsby D, Steven A H and Curtotti R 2020 Australian fisheries and aquaculture outlook 2020 Research by the Australian Bureau of Agricultural and Resource Economics and Sciences
[43] Davies A M and Xing J 2003 The Influence of wind direction upon flow along the west coast of Britain and in the north channel of the Irish Sea *Journal of Physical Oceanography* **33** 57-74
[44] Renewable UK 2019 Export Nation: A Year in UK Wind, Wave and Tidal Exports EMEC online report
[45] Higgins P and Foley A M 2015 *Chapter 73: Power system performance of offshore wind in the UK in 2030* In A. Sayigh (ed.) Renewable Energy in the Service of Mankind Vol I. Switzerland: Springer International Publishing
[46] Nijen K, Steven Van Passel S V and Squires D 2018 A stochastic techno-economic assessment of seabed mining of polymetallic nodules in the Clarion Clipperton Fracture Zone *Marine Policy* **95** 133-141
[47] Markus T and Singh P 2016 Promoting Consistency in the Deep Seabed: Addressing Regulatory Dimensions in Designing the International Seabed Authority's Exploitation Code *RECIEL* **25**
[48] Boetius A and Haeckel M 2018 Mind the seafloor *Science* **359** 34-36
[49] Feichtner I 2020 Contractor liability for environmental damage resulting from deep seabed mining activities in the area *Marine Policy* **114** 103502
[50] Sykora-Bodie S T, Morrison T H, Sykora S T and Bodie Morrison T H 2019 Drivers of consensus - based decision - making in international environmental regimes: Lessons from the Southern Ocean *Aquatic Conserv: Mar Freshw Ecosyst* **29** 2147-2161
[51] Dunn A, Vacchi M and Watters G 2017 The Ross Sea region Marine Protected Area Research and Monitoring Plan SC-CAMLR-XXXVI / 20 Commission for the Conservation of Antarctic Marine Living Resources
[52] CCAMLR 2015 A proposal for the establishment of a Ross Sea Region Marine Protected Area Delegations of New Zealand and the USA CCAMLR-XXXIV/29 Rev
[53] Als T D, Hansen M M and MAES G E et al 2011 All roads lead to home: panmixia of European eel in the Sargasso *Sea Molecular Ecology* **20** 1333-1346
[54] Laffoley Dd' A, Roe, H S J and Angel M Vet al 2011 The protection and management of the Sargasso Sea: The golden floating rainforest of the Atlantic Ocean Summary Science and Supporting Evidence Case Sargasso Sea Alliance 44
[55] Gjerde K, Reeve L L N and Harden-Davies H et al 2016 Protecting Earth's last conservation frontier: scientific, management and legal priorities for MPAs beyond national boundaries' *Aquatic Conservation: Marine and Freshwater Ecosystems* **26** 45-60

IOP Conf. Series: Earth and Environmental Science **865** (2021) 012020 doi:10.1088/1755-1315/865/1/012020

Analysis on the spatial accessibility of parks and green spaces and optimal site selection—Taking Weifang prefecture-level city as an example

Wenqi Yan[1, 2, 3] and Shijie Wang[1, 2, 3, 4]

[1]Faculty of Geomatics, Lanzhou Jiaotong University, Lanzhou, Gansu 730070, China;
[2]National-Local Joint Engineering Research Center of Technologies and Applications for National Geographic State Monitoring, Lanzhou, Gansu 730070, China;
[3]Gansu Provincial Engineering Laboratory for National Geographic State Monitoring, Lanzhou, Gansu 730070, China

[4]Corresponding author's e-mail: wangshijie@mail.lzjtu.cn

Abstract. Based on the ArcGIS platform, we took the parks and green spaces of the improved potential model to analyze the spatial accessibility of local parks and green spaces in Weifang prefecture-level city via combining traffic and population distribution with the distribution of parks, and cuckoo search algorithm to optimize the site selection of park green space. According to the results of the accessibility analysis. The results show that: (1) the distribution of population and green spaces is not balanced. Parks are mainly distributed in the downtown, while they are fewer in the township. (2) The spatial accessibility of the park green space is significantly different between areas, and the southeastern and northwestern areas are higher than other parts. (3) We optimize the location of green spaces to achieve fairness in areas with low accessibility of parks and green spaces. The article intuitively reveals the different accessibility of parks and green spaces in various streets of Weifang prefecture-level city, effectively distinguishes their scarcity, and provides a scientific basis for the space allocation and optimization of park green spaces.

1. Introduction

Parks and green spaces are referred to as service facilities with certain rest functions, open to the public, mainly for playing and relaxation, and vast ample green land with functions such as viewing and avoiding danger at the same time [1]. The rapid development of cities has brought about traffic congestion, environmental pollution, and other problems, and at the same time has accelerated the pace of people's lives. The number, size, and spatial layout of parks and green spaces directly affect the quality of the urban environment and the living standards of residents [2]. Parks and green spaces can meet the life demands of urban residents for parks and green spaces for entertainment and various outdoor sports [3]. Therefore, the accessibility of parks and green spaces has always been the current focus and research object. Moreover, the increase in the number of parks and green spaces is a necessary means to satisfy the everyday contemporary lives of residents.

Spatial accessibility refers to how easy it is for residents to overcome a specific resistance to reach another service facility, and it is an important criterion to measure the reasonable spatial layout of urban service facilities [4]. With the development of 3S technology, scholars began to apply GIS spatial analysis to the accessibility of urban parks and green spaces. Nicholls [5] used spatial analysis methods to evaluate the accessibility and fairness of park green spaces. Xiong Huijin [6] used the technique of network analysis to analyze the accessibility of parks and green spaces in Nanchang. X. Zhou [7] discussed the accessibility of parks for different ethnic groups in six cities in Illinois, combining the distribution of parks and tree canopies. Shi Tuo [8] used the method of combining network analysis and buffer zone to analyze the accessibility of urban parks in Shenyang. Refining urban park system planning and green space system planning through the evaluation of park accessibility will focus on the research and application of accessibility theory. The potential model takes into account the effect of distance attenuation, and accurately measures the difficulty of the residents to the park green space.

Yang [9] proposed a multi-objective cuckoo search algorithm, which applying to multi-objective problem optimization. Hui Qian [10] measured the Xi'an Big Wild Goose Pagoda based on walkability and used a particle swarm algorithm to obtain the optimal solution for the newly added park. Zhou Yuan [11] used GIS and a multi-target area location configuration model to select the location of Shenyang parks. In the current study, genetic algorithm, ant colony algorithm, particle swarm algorithm are widely used, still these algorithms have the characteristics of slow convergence, easy to fall into local optimal, search is easy to stagnate or premature, etc., it is necessary to find a way to solve the optimization location problem. The Cuckoo search algorithm has the advantages of few parameters, simple operation, high search efficiency, random search path optimization, strong optimization ability. It is widely used in various fields and has become a new heuristic algorithm after GA and PSO [12-13]. The Cuckoo search algorithm builds a location model to make the park layout achieve spatial fairness, improves efficiency, and reduces the right justice of residents using public services differences.

Since most of the current researches focus on first-tier and second-tier cities, and lack of research on ordinary third-tier cities. There are many rural areas in Weifang, the analysis of the accessibility of parks and green spaces in this area and the optimization of site selection can better respond to the country's rural revitalization, and policies to build a civilized village. This study takes Weifang's parks and green spaces as the research object. With the help of an improved potential model, the accessibility analysis is carried out from the spatial layout of the parks and green spaces in the city, which enriches the evaluation methods of parks and green spaces. It conducts the park green spaces according to the results of the accessibility analysis. The spatial layout and site selection of Weifang prefecture-level city provides guidance and a scientific basis for the reasonable layout and construction of green space in Weifang prefecture-level city's parks.

2. Overview of the study area and data sources

Weifang prefecture-level city, Shandong Province, is located in the middle of the Shandong Peninsula. It has jurisdiction over four districts, six county-level cities, and two counties, with 58 sub-districts and 62 townships. Figure 1 shows the location map of county-level administrative divisions. According to POI, there here are parks and green spaces in Weifang prefecture-level city, and the distribution of parks is shown in Figure 2.

Figure 1. Study area.

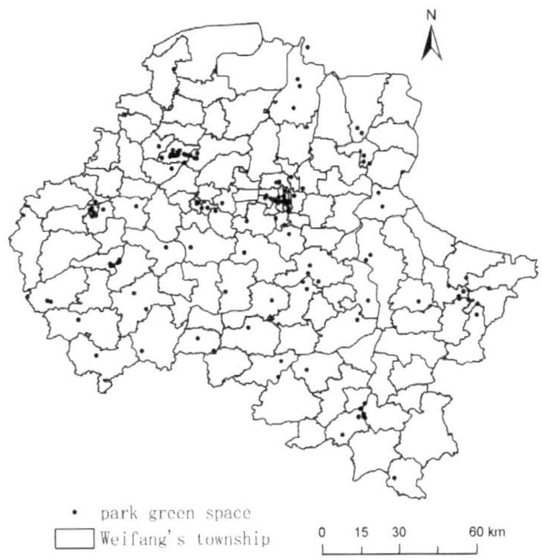

Figure 2. Distribution of park green space.

The population data is obtained from the 2016 Weifang Statistical Yearbook and the statistical yearbooks of counties and districts in Weifang. The data of parks have come from the profile of each park's official website and telephone consultation. The road network data is obtained from OSM (Open Street Map: www.openstreetmap.org) and the official website of the Shandong Provincial Geographic Information Public Service Platform. Due to the complexity of the road network, it is necessary to preprocess the road network and using ArcGIS to check the topology of the road network to optimize the existing road area to ensure the connectivity of the road network, using ArcGIS to digitize relevant information to establish a database for accessibility analysis.

3. Research methods

3.1. Improved potential model

The potential model only considered the demand of the supply side and ignored the competition between the demand side. Joseph [14] added the population size influencing factor to the general formula of the potential model, as shown in equation (1).

$$A_i = \sum_{j=1}^{n} \frac{M_j}{V_J D_{ij}^{\beta}} V_j = \sum_{k=1}^{m} \frac{P_k}{D_{KJ}^{\beta}} \tag{1}$$

In the formula, V_j represents the impact factor of the population scale; P_k represents the population of the township street; M_j represents the activity scale at points j, expressed by the area of park green space; D_{ij} represents the travel impedance factor at point i and point j, expressed by distance; n, m represents the number of parks and green spaces and township. β represents the friction coefficient. Research [15-17] believed that the value of β mainly refers to the range of [0.9-2.9]. After comparison, the value of β is 1 for the accessibility of park green space, so β=1 in the text.

3.2. Cuckoo search algorithm

Site selection for parks and green spaces is a process that is affected by multiple factors. There are more restrictive conditions in the site selection process, and it is impossible to consider every aspect of the planning process. This paper takes spatial fairness as the standard and distance as the main factor to construct the objective function. The unit distance cost of residential areas and parks and green spaces is only related to reach and has nothing to do with the geographical location of parks; that, is the length of reach, the higher the cost. The cuckoo search algorithm is basing on the cuckoo's nest-finding parasitic habit. The basic assumptions of the algorithm are as follows:

(1) Each cuckoo lays only one egg at a time and is randomly stacked in a bird's nest to hatch;

(2) The eggs in the nest of the highest quality are retained to the next generation;

(3) The number of nests in which cuckoos can place eggs is a fixed value n, and the probability of eggs being found is Pa (0<Pa<1).

According to the above three assumption rules, the cuckoo search algorithm updates the search location and path as formula (2)

$$x^{(t+1)} = x_i^t + \alpha \oplus L(\beta) \tag{2}$$

In the formula, x^t_i is the individual i in the nest of the t generation; ⊕ is the quantity product; α is the step control amount, usually identified as 11; L(β) is the random search path. The construction model is showing in formula (3)

$$T = \sum_{i=1}^{NG} \sum_{j=1}^{NH} d_{ij}\, NP_j \tag{3}$$

In the formula,d_{ij} is the distance between the residential area and the green space in the park; NP_j is the number of residents using the green space in the garden; NG is the number of townships, and NH is the number of the park.

4. Spatial analysis and optimal site selection

4.1. Analysis of the spatial accessibility of Park Green Space

The spatial accessibility of the existing parks in the Weifang prefecture-level city is imbalanced, with significant differences in area and the uneven distribution. Parks are mainly distributed in the urban area, while less spread in the township. It can be seen from Figure 2 that parks and green spaces are concentrated in the city center, while towns and some small parks are only distributing in the township. As shown in Figure 3, there are more parks and green spaces in areas with higher population density. It can be seen from Figure 4 that the center, northwestern, and southeastern parts of the city have better accessibility, while the east and west parts are relatively poor. Regions with higher accessibility are distributing in areas with lower population density. It can be seen from that there is a specific particular misplaced phenomenon between the accessibility distribution of parks and the distribution of population density, especially in the north and southwestern of the city. The higher accessibility of park green space does not match the lower population density distribution. The distribution of green space in the park is unreasonable, increasing residents' difficulty gaining the green space resources.

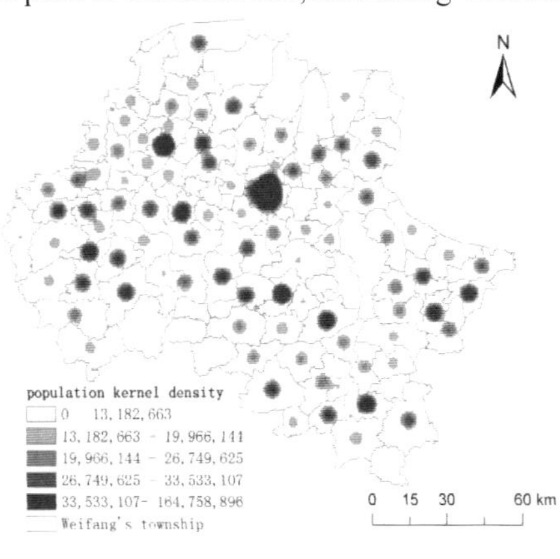

Figure 3. Population kernel density.

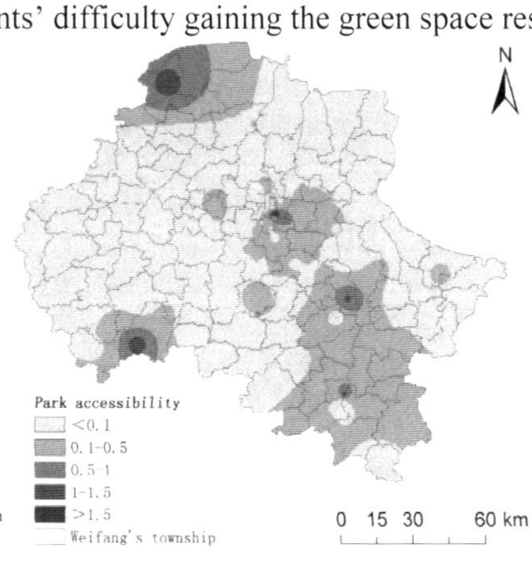

Figure 4. Park accessibility.

4.2. Optimal site selection for parks and green spaces

Aiming at the existing park and green space in line with the actual overall distribution, this paper used the cuckoo search algorithm to optimize the location of the parks and green spaces in Weifang prefecture-level city, so that residents can enjoy the services of the parks and green spaces in a timely and effective manner. Based on the original park green space, a certain amount of park green space n is added. Under normal circumstances, the number of new additions n is usually set to 15-40.

Therefore, this paper uses different parameters to conduct experiments and compare the results of optimized parameters. In this paper, we will analyze the results with n=15, n=30, and n=40, and the number of iterations is 5000 [18-19]. The shortest average distance between each residential area and the newly-added park green space is calculated as shown in Table 1. The result shows that when n=40, the average travel distance of residents is the shortest, when n=15, the average travel distance of residents is the longest, it is between 15 and 40 when n=30, so when n=40 The location of the park and green space is optimal at that time. After adding these parks, making the distribution of the park green space in Weifang more balance. As shown in Figure 5, after the site selection analysis by the cuckoo algorithm, the park green space of Weifang prefecture-level city effectively reduces the difference in the ability of residents to obtain park green space services. The calculation results are better, most of which are distributed in regions with low accessibility, indicating that the cuckoo search algorithm can solve green space in parks.

Table 1. The average travel distance of residents.

n	average travel distance of residents/m
15	23113.94688
30	12311.01312
40	10969.13325

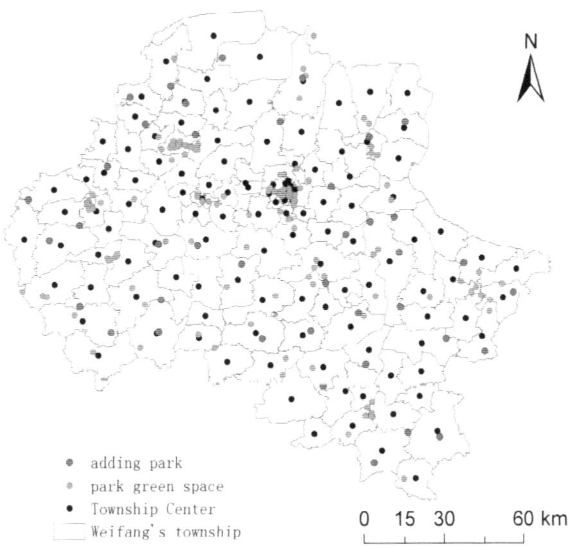

Figure 5. Optimize site selection result.

4.3. Discussion

Based on the improved potential model, this paper comprehensively evaluates the accessibility of Weifang's parks and greens, finds problems and proposes a reasonable optimization strategy for greening, and improves the accessibility of Weifang's parks and green spaces to achieve the result of nearby recreation for current residents. From the results of accessibility, the distribution of green space in parks is uneven, and the overall distribution of parks varies significantly in number and area. Most of the regions with higher population density are less distributed. Studies have shown that the optimization of the layout of parks and green spaces is mainly based on the balance of supply and demand and the perspective of residents' travel [20-21], such as the addition of buses and subways, multiple layouts of community parks, community parks, etc. In this paper, the park green space optimization strategy is mainly based on park accessibility, quantitatively analyzes the park green space accessibility, and provides a scientific basis for the rational planning of park green space. Regarding the addition of green space, only the current new location can be determined, and the size of the new addition cannot be determined, the dynamic study of park green space in time series is not considering. Therefore, accessibility should be considered comprehensively in future research.

5. Conclusion

This paper adopts an improved potential model, starting from the supply-demand relationship between population and park green space, the number, area, spatial distribution of park green space, and the number of residents, density, and spatial distribution of the park green space are studied for accessibility. The cuckoo search algorithm uses to analyze the local optimizing the location of the park green space, providing a reference for the planning of the park green space in Weifang prefecture-level city. (1) the distribution of population and green spaces is not balanced, with large differences in area and uneven population distribution. Parks are mainly distributed in the urban area, while less distributed in townships. (2) There is a significant difference in the spatial accessibility of parks and green spaces. The accessibility of the southeastern and northwestern of the city is higher than other parts. The accessibility of the city center is moderate, and the accessibility of other areas is low. (3) To optimize the site selection of parks and green spaces, it is necessary to coordinate the space allocation and development of parks and green spaces, appropriately expand the area of central parks and green spaces, promote the upgrade of the central parks and green spaces, followed by increasing the distribution of parks and green spaces in the township to improve the life of rural people. The cuckoo search algorithm can effectively optimize the location of urban parks, provides scientific guidance for the spatial layout of urban parks and green spaces, enabling the park green space to form an organic overall in the city space.

Acknowledgments

This article is one of the phased achievements of the National Natural Science Foundation of China Project "Research on the Accuracy of Existing Wired Track Detection Based on UAV" (41861061), supported by the outstanding research platform of Lanzhou Jiaotong University (201806).

References

[1] Ministry of Housing and Urban-Rural Development of the People's Republic of China. Urban green space classification standard (CJJ/T85-2017). Beijing:China Construction Industry Press
[2] Gao Junbo, Yu Bin, Jiang Haiyan 2011 Investigation on the Spatial Distribution of Urban Public Service Facilities——Taking Guangzhou as an Example[J]. *Urban Issues* **(08)** 55-61
[3] Zhou Zhaosen, Lin Guangsi 2018 Review of the Equitable Use of Urban Park and Green Spaces[J]. *Southern Architecture* **(03)** 53-59
[4] Zhang Jinguang, Zhao Bing 2018 Location and layout optimization of urban parks based on accessibility[J]. *Journal of Nanjing Forestry University (Natural Science Edition)* **42(06)** 151-158
[5] Sarah Nicholls 2001 Measuring the accessibility and equity of public parks: a case study using GIS[J]. *Managing Leisure* **6(4)** 201-219
[6] Xiong Huijin, Shen Qinwei, Wei Lingwei, et al. 2021 Accessibility to Urban Park Green Space in Nanchang Based on Network Analysis[J]. *Journal of Chinese Urban Forestry* 19(01) 89-94
[7] Xiaolu Zhou, Jinki Kim 2013 Social disparities in tree canopyy and park accessibility:A case study of six cities in Illinois using GIS and remote sensing[J]. *Urban Forestry&Urban Greening* **12** 88-97
[8] Shi Tuo, Li Junying, Li Ying, et al. 2016 Analysis of urban park accessibility in Shenyang City[J]. *Chinese Journal of Ecology* **35(05)** 1345-1350
[9] Yang X S, Deb S 2013 Multiobjective cuckoo search for design

optimization[M]//Multiobjective cuckoo search for design optimization.Computers and Operations Research.

[10] Xi Qian 2020 Research on Distribution of Urban Parks Green Space Based on Walking Accessibility[D]. *Xi'an University of Architecture and Technology*

[11] Zhou Yuan, Shi Tie-mao, Hu Yuan-man, et al. 2011 Location selection for Shenyang urban parks based on GIS and multi-objective location allocation model[J]. *Chinese Journal of Applied Ecology* **22(12)** 3307-3314

[12] Liu Zifan, Zhang Jianhua 2007 Optimal planning of substation locating and sizing based on refine multi-team PSO algorithm[J]. *Proceeding of the CSEE* **27(1)** 105-111

[13] Cao Fang, Meng Qibin, Miao Peiqing 2015 Optimal substation planning based on improved weighted Voronoi diagram and genetic algorithm[J]. *Power System Technology* **39(2)** 511-516

[14] Joseph Alun E, Bantock Peter R 1982 Measuring potential physical accessibility to general practitioners in rural areas: A method and case study[J]. *Social Science & Medicine* **16(1)** 85-90

[15] Song Zhengna, Chen Wen, Che Qianji, et al. 2010 Measurement of Spatial Accessibility to Health Care Facilities and Defining Health Professional Shortage Areas Based on Improved Potential Model—A Case Study of Rudong County in Jiangsu Province[J]. *SCIENTIA GEOGRAPHICA SINICA* **30(02)** 213-219

[16] Peeters D, Thomas I 2000 Distance predicting functions and applied location allocation models[J]. *Journal of Geographical Systems* **2(2)** 167-184

[17] Tao Zhuolin, Cheng Yang, Dai Teqi 2014 Measuring spatial accessibility to residential care facilities in Beijing[J]. *Progress in Geography* **33(05)** 616-624

[18] Zhang Yi 2020 Research and application of cuckoo search algorithm [D]. *Qufu Normal University*

[19] Zhang Shubo 2019 Evaluation and Optimization of Spatial Layout of Public Service Facilities in the Center District of Beijing[D]. *Capital University of Economics and Business*

[20] Wu Jiansheng, Si Menglin, Li Weifeng 2016 Spatial equity analysis of urban green space from the perspective of balance between supply and demand: A case study of Futian District, Shenzhen, China[J]. *Chinese Journal of Applied Ecology* **27(09)** 2831-2838

[21] Lu Yanchi, Fang Shiming 2019 Study on the Spatial Distribution Optimization of Park Green Spaces in the Main City of Wuhan from the Perspective of Both Equilibrium and Efficiency[J]. *Resources and Environment in the Yangtze Basin* **28(01)** 68-79

Stage characteristics and causes of a mixed severe convective weather process

Hongwu Liu[1,2], Jingyu Xu[1,2], Jie Tang[1,2,3], Rong Yao[1,2] and Yan Hu[1,2]

[1]Hunan Meteorological Observatory, Changsha 410006, Hunan, China;
[2]Hunan Key Laboratory of Meteorological Disaster Prevention and Reduction, Changsha 410006, Hunan, China

[3]Corresponding author's e-mail: 7101768@qq.com

Abstract. Based on the conventional observation data, the intensive automatic weather station data, the 1°×1° Final Operational Global Analysis data and the Doppler radar data, this study analyzes the stage characteristics and causes of a mixed severe convective weather process in Hunan Province from April 23 to 24, 2019. The results show that this severe convective weather mainly has two stages. In the first stage, the disastrous weather is mainly the short-term heavy precipitation. The convective weather in this stage is a severe mixed convective weather process forced by warm advection. The short-term heavy precipitation is mainly caused by the continuous influence of multi-cell storms. The disastrous weather in the second stage is mainly the thunderstorm gale. It is a squall line process forced by upper-level cold advection. Besides, there is apparent mid-altitude radial convergence and high wind core on the radial velocity map. The surface convergence line is the primary dynamic uplift condition triggering the mixed convection in the first stage, and there are several abnormal low centers of three-hour pressure change in the western and southern Hunan. The severe convective weather process in the second stage is mainly triggered by the forward-tilted trough. The dynamic conditions and energy conditions are favourable for the convections at both stages. Compared with the second stage, water vapor is more abundant in the first stage, which is more favorable for the occurrence of short-term heavy precipitation. The three-body scatter spike of hail is not obvious before the formation of the squall line. The early warning signals are mainly the height of the strong echo and vertical integrated liquid water content, both of which increase significantly before the formation of the squall line. The early warning of thunderstorm gale after the formation of the squall line can be issued in advance by monitoring the low-level wind in the downstream areas of the squall line.

1. Introduction

Severe convective weather is featured by small scale and short duration, and occurs suddenly. The forecast and research on severe convective weather have always been the concerns of meteorologists [1-3], and it is also a difficulty in operational weather forecast. Although the temperature gradually increases in spring in Hunan Province, the cold air from the north still frequently affects Hunan, and causes strong convective weather such as gale, hails and rainstorms. In summer, the convective weather occurs more frequently than in other seasons due to the warm-moist atmospheric environment. Many scholars [4-7] have comprehensively analyzed the background, potential energy conditions and triggering mechanisms of severe convective weather in Hunan in spring and summer. They used radar data to rapidly identify heavy precipitation echoes and supercells, discussed the practicability of early warning, and explored the nowcasting signals that could be used for early warning. Cai et al [8] divided the weather situation configuration of thunderstorm gale into low-level warm advection forcing type, baroclinic frontogenesis type, quasi barotropic type and high-level cold advection forcing type.

During April 23–24, 2019, a mixed severe convective weather process occurred in Hunan, with short-term heavy precipitation and thunderstorm gale. It is featured by a wide influence range, high intensity, long duration and overlapping disaster-affected areas. This weather process brought strong wind and hail disasters, resulting in the collapses of river banks, water slopes and dams, house damage and crop damage such as tobacco and vegetables in some areas. In this study, the weather situation analysis and physical quantity diagnosis of the process are carried out by using the conventional surface and upper-level observation data and the National Centers for Environmental Prediction/National Center for Atmospheric Research 1°×1° reanalysis data with an interval of 6 h. Moreover, based on the radar data, we further discuss the causes of the strong convective weather.

2. Overview of the severe convective process

From 08:00 on April 23 to 08:00 on April 25, 2019, a wide range of severe convective weather occurred to the south of central Hunan. The main influence periods are from 08:00 to 15:00 on April 23, from 23:00 on April 23 to 08:00 on April 24, and from 19:00 on April 24 to 06:00 on April 25 (Table 1). As the convection in the daytime of April 23 was weak, this study mainly focuses on the latter two periods, which are called the first stage and the second stage, respectively. Meteorological stations observed 22 thunderstorm gales in the first stage and 58 in the second stage. The maximum wind speed appeared at Xiucailing station in Puman Town, Jiahe County at 00:49 on April 25, reaching 31.1 $m{\cdot}s^{-1}$ and level 11. There were eight counties with local hail, mainly concentrated in the first stage. Among them, four counties in western Hunan had the hail with the largest diameter of 20 mm, namely, Fenghuang County (23:48–23:51 on April 23), Mayang County (23:45–23:55 on April 23), Chenxi County (01:00–01:10 on April 24) and Zhongfang County (01:10–01:15 on April 24). Two counties in southeastern Hunan had small-scale hail, namely, Jiahe County (around 00:00 on April 25) and Guiyang County (06:36–06:41 on April 24). Meteorological stations observed 441 short-term heavy precipitation events. Among them, 234 events occurred in the first stage, with the maximum hourly rainfall of 61.4 mm occurring at Haihua Experimental School station in Chenxi (00:00–01:00 on April 24). Besides, 149 events occurred in the second stage, with the maximum hourly rainfall of 54.2 occurring at Sugu Village, Guanqiao Town, Liuyang City (05:00–06:00 on April 25). In summary, the frequency and intensity of thunderstorm gale in the second stage were higher than those in the first stage. However, the short-term heavy precipitation in the first stage has larger intensity and wider range than in the second stage. In the early morning of April 24, a hail event occurred in western Hunan. Therefore, the strong convective weather in the first stage was mainly short-term heavy precipitation, local hail and thunderstorm gale, but the second stage was dominated by thunderstorm gale (Figure 1).

Table 1. Types and periods of the severe convective weather.

Periods	Thunderstorm gale (station observations)	Hail (county)	Short-term heavy precipitation (station observations)
23:00 on April 23 to 08:00 on April 24 (the first stage)	22	5	234
19:00 on April 24 to 06:00 on April 25 (the second stage)	58	1	149
Total	80	6	383

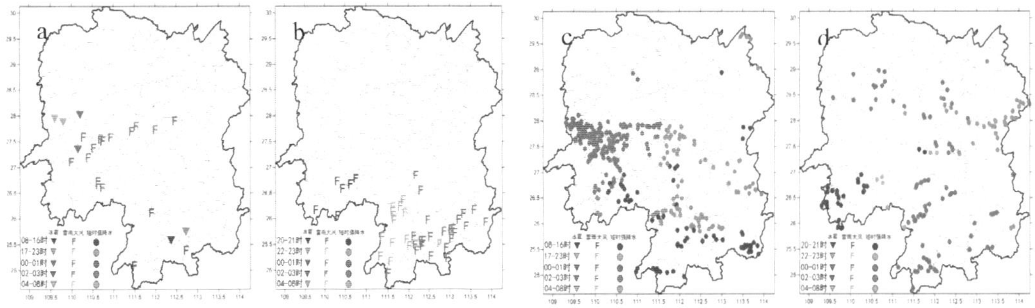

Figure 1. Distribution of thunderstorm gales and hails (a) from 08:00 on April 23 to 08:00 on April 24 and (b) from 08:00 on April 24 to 08:00 on April 25. Distribution of short-term heavy precipitation (c) from 08:00 on April 23 to 08:00 on April 24 and (d) from 08:00 on April 24 to 08:00 on April 25.

3. Characteristics of radar echoes and the early-warning respect

3.1. The first stage

The evolution characteristics of radar echoes observed at Zhangjiajie (Figure 2) show that the echoes were mainly the cumuliform-stratiform-mixed precipitation echoes in the first stage. The area with radar echoes larger than 35 dBZ reached 70 km×50 km, and the maximum reflectivity factor reached 60 dBZ. The echoes moved eastward and southward slowly.

At 22:54 on April 23, a strong storm cell was generated on the boundary between Hunan Province and Guizhou Province, and moved eastward slowly. At 23:36, the three-body scatter spike (the circle in Figure 2a) was detected in the radial direction of the storm cell (the southwest side of the storm cell), but it was not prominent. It is tentatively inferred that there were cumuliform-stratiform-mixed precipitation echoes on the western side of the three-body scatter spike, and thus the long spike was covered. At that time, the maximum reflectivity factor of the storm cell at the elevation of 0.5° reached 60 dBZ and the vertical height extended to 9.4 km (figure omitted). For the vertical integrated liquid water content (VIL) map corresponding to the moving track of the storm cell, the VIL increased to more than 60 $kg \cdot m^{-2}$ at 23:12 (Figure 2b). Thereafter, the high VIL above 60 $kg \cdot m^{-2}$ is sustained for five volume scans (23:12–23:42), with the echo-top height exceeding 14 km (Figure 2c). It is evident that the convection developed vigorously, and strong hail clouds already developed at this stage. The vertical cross-section of the reflectivity factor at Fenghuang County (Figure 2d) shows that the strong echo larger than 55 dBZ extended to the height of the −20℃ layer, with a distinct overhanging structure. According to the disaster investigation, it can be found that strong hail occurred in Fenghuang County at 00:00 on April 24, with the maximum size similar to an egg. Then, the storm affected Mayang County and also brought hail. A bow echo then developed in front of the mixed precipitation echoes, and crossed over Xiangxi Prefecture, northern Huaihua City, Shaoyang City and other places, resulting in regional short-term heavy precipitation.

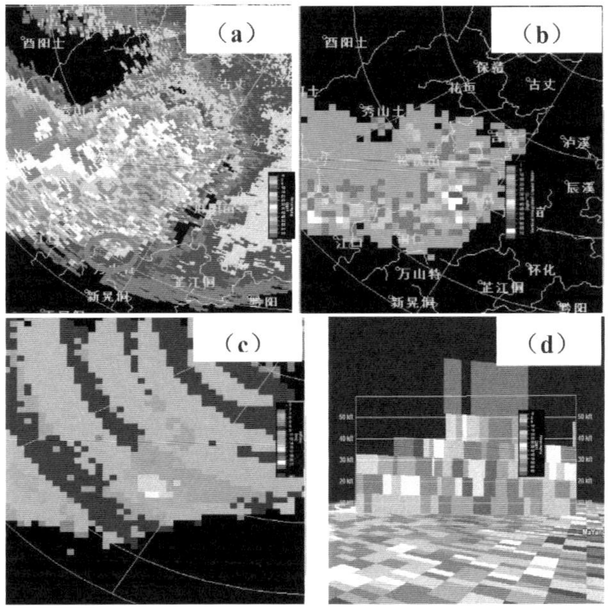

Figure 2. (a) Composite reflectivity factor at 23:36, (b) VIL at 23:12, (c) echo-top height at 23:12 and (d) vertical cross-section of reflectivity factor at 23:36 on April 23 observed by the Zhangjiajie radar.

At 00:48 on April 24, another storm cell began to affect Chenxi County. Meanwhile, an obvious mesocyclone was detected on the radial velocity map (Figure 3a), with a rotation speed of 12.3 $m \cdot s^{-1}$ (medium intensity). Its bottom extended downward to 4.1 km and its top extended to 7.7 km. Generally, the higher the vertical height of the mesocyclone is, the stronger the updraft is, and the more intense the hail collision growth in the air is. The storm cell had obvious bow echo characteristics (Figure 3b). The backward inflow gap and forward inflow gap indicated the existence of strong inflow and outflow. Moreover, multiple storm cells were concentrated on the southwest side of the bow echo, indicating a vigorous convection development. The continuous influence of the bow echo and multiple storm cells caused the short-term heavy precipitation in Huaihua, Fenghuang, Mayang and Xupu successively.

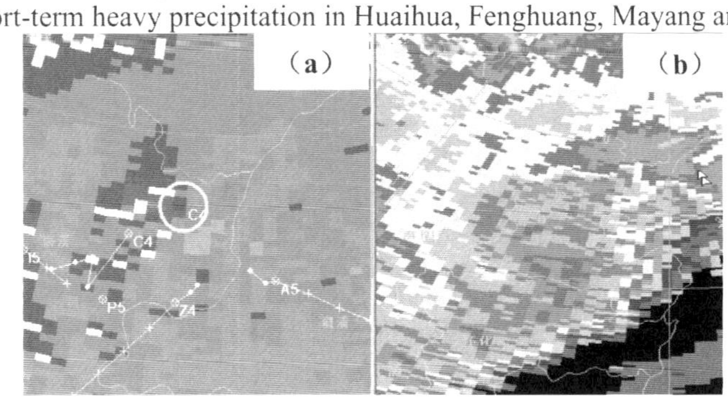

Figure 3. (a) Radial velocity and (b) composite reflectivity factor of Zhangjiajie radar at 1.5°elevation angle at 00:48 on April 24.

In summary, the hail early warning is difficult in the first stage. This is because the three-body scatter spike of strong hail is obscured due to the restriction of the detection conditions and environmental fields. Moreover, Fenghuang and Mayang are located at a radial distance of 230 km from the radar station, and the elevation of the Zhangjiajie radar is as high as 1.48 km. Because of the geographical location, Fenghuang and Mayang are in the purple area on the radial velocity map. Thus, the radial velocity in these areas cannot be judged. So even if mesocyclones occur, it is impossible to observe and identify them. Hence, the early-warning respect is mainly on the height of the strong echo and the abrupt increase of the strong VIL. However, the early warning time is still relatively limited, less than 15 minutes.

3.2. The second stage

The disastrous weather in the second stage was mainly thunderstorm gale. The reflectivity factor figure shows that the bow echo developed significantly in western Hunan at 20:34 on April 24 (Figure 4a), causing thunderstorm gale in Jingzhou County and Tongdao County. The radial velocity map (Figure 4b) shows that the negative velocity area was larger than the positive velocity area in the first range circle of the Yongzhou radar. Meanwhile, there were high-value areas of velocity below the 1 km height within the 25 km range. As can be seen, there was significant convergence in the lower and middle levels in the downstream areas. After ten volume scans, the squall line moved eastward and southward slowly. In addition, new supercells constantly merged at the southern end of the squall line (circle in Figure 4c) and travelled northeastward, resulting in the continuous growth of the southern end of the squall line. On the corresponding velocity map, there was obvious radial convergence in the middle level associated with the squall line echo (circle in Figure 4d) and gale centers on the southern side corresponding to supercells. The squall line broke at 22:34, and the northern part of the broken squall line began to affect western Yongzhou (Figure 4e). There was an obvious gale core (circle in Figure 4f) in the middle layer of the low elevation velocity map. In addition, the wind convergence was more significant in the first distance circle. Therefore, the thunderstorm gale occurred successively in the urban areas of Yongzhou, Dao County, Shuangpai County and Jianghua County (23:15–23:50).

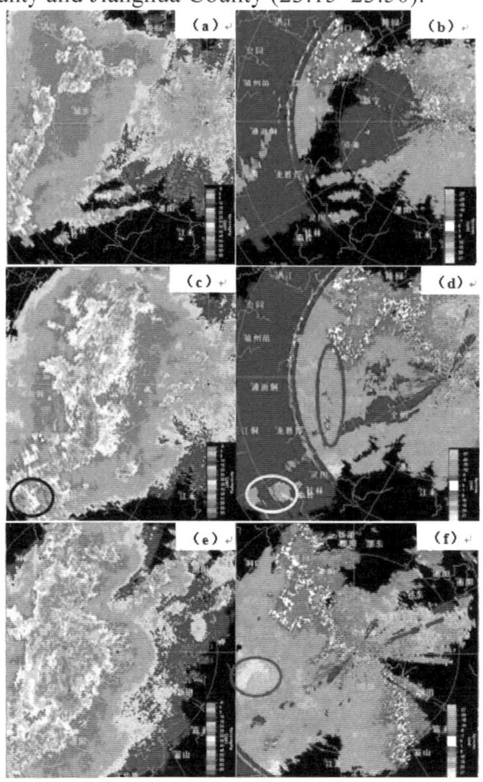

Figure 4. Composite reflectivity (left) and radial velocity at 1.5° elevation angle (right) of the Yongzhou radar at (a, b) 20:34, (c, d) 21:34 and (e, f) 22:34 on April 24.

The vertical wind profile (Figure 6) from Yongzhou radar shows that an upper-level trough passed through in the second stage. Before 23:34 (Figure 5a), the vertical wind profile showed a consistent southwesterly wind from the upper levels to low levels, and the wind speed reached 12 $m \cdot s^{-1}$ above 1.5 km. From 23:50, the southwesterly wind turned to the westerly wind at the height of 4–6 km (Figure 5b). At the same time, the weak southwesterly wind turned to the southeasterly wind at the 0.3 km height, and the wind speed increased. Besides, it can be seen that the vertical wind shear increased significantly and rotated clockwise, indicating the development of warm advection. From 00:22 on April 25, the westerly wind at the height of 4–6 km turned to a northwesterly wind (Figure 5c). The evolution of the wind profile reveals that the passage of the upper-level trough was closely related to the generation and movement of the squall line.

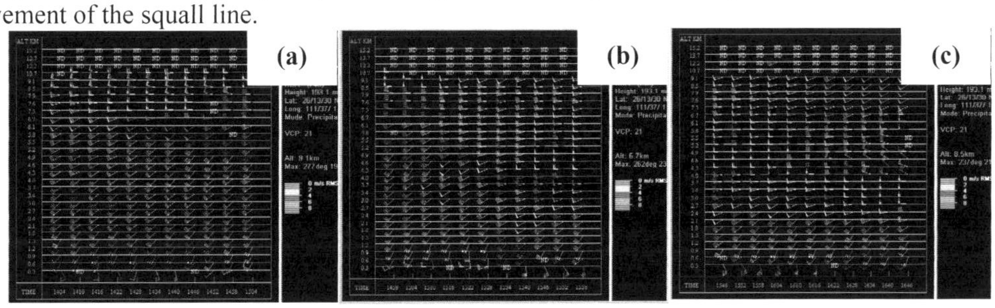

Figure 5. Vertical wind profile of Yongzhou radar (a) at 23:04 and (b) 23:50 on April 24 and (c) 00:46 on April 25.

In summary, the main cause of the thunderstorm gale in southern Hunan is the passage of the squall line. If the squall line is monitored in the upstream area and the gale has also appeared in the corresponding observation, meanwhile there is convergence of wind field in downstream areas, then we could release the thunderstorm-gale early-warning in advance.

4. Potential of severe convective weather

4.1. Large-scale circulation background

From the circulation configuration at 20:00 on April 23 (Figure 6a), it can be found that there was a long and narrow upper level jet crossing over Guizhou-Xiangxi-Yangtze River Delta at 200 hPa. Hunan was located in the strong divergence area on the right side of the upper level jet entrance region. At 500 hPa, the middle-high latitudes were controlled by two troughs and one ridge, and the subtropical high was in blocks in an east-west direction. In addition, there was a trough in the area from Chongqing to southwestern Guizhou, and Hunan was in front of the cold trough. At 850 hPa, the southwest vortex was located in northern Guizhou, and one end of the warm shear in the eastern section of the herringbone shear line extended to the central and western Hunan. Simultaneously, the vortex moved eastward and southward to the warm-humid areas of western and southern Hunan, and was significantly strengthened. To wee hours on April 24, the range of convective cloud cluster expanded and its intensity increased. For the low level wind field, the southwest jet crossed from northern Guangxi to Hunan at 700 hPa and 850 hPa, with the abundant water vapor and unstable energy transported to Hunan. In terms of humidity conditions, central and southern Hunan was in the significant wet zone at 850 hPa and the dry tongue at 500 hPa. Such humidity condition indicated that the dry-cold air intrusion at middle levels was a favorable condition for triggering strong convections. It also indicated that the atmosphere was unstable, with warm-wet air at upper levels and dry-cold air at low levels in the first stage. Moreover, the temperature ridge at 850 hPa was located from eastern Guizhou to western Hunan, with a vertical temperature lapse rate of more than 25℃ in western and southern Hunan. This result indicated that western Hunan was located in the strongest convergence area in front of the warm ridge, which was also an essential cause for hail and other disastrous weather. For the weather situation configuration, the strong convection at this stage belonged to the low-level warm advection forcing type proposed by Xu et al.

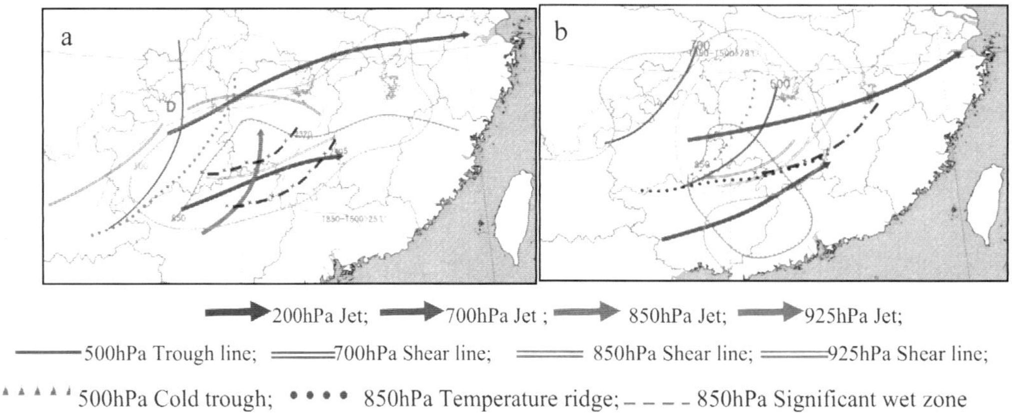

Figure 6. Configuration of main upper-level circulation systems at (a) 20:00 on April 23 and (b) 20:00 on April 24, 2019. Purple, dark red, light red and gray arrows are the jets at 200 hPa, 700 hPa, 850 hPa jet and 925 hPa, respectively. The brown single line denotes the 500 hPa trough, the brown parallel line 700 hPa shear line, the red parallel line 850 hPa shear line, the gray arrow 925 hPa shear line, the blue triangle point line 500 hPa cold trough, the brown dotted line 850 hPa temperature ridge, the green dotted line 850 hPa significant wet zone, and the black solid lines the 588 contours at 500 hPa.

For the main circulation system configuration at 20:00 on April 24 (Figure 6b), the 200 hPa upper-level jet moved southward and was located in the areas from western Hunan to southern Jiangxi Province and Yangtze River Delta. Meanwhile, the southern Hunan was located in the strong divergence area on the right side of the upper-level jet entrance range. The 500 hPa upper-level trough moved to central Hunan, and the temperature trough lagged behind the height trough. The 700 hPa trough was located between Chongqing and northwestern Guizhou, and Hunan was in the positive vorticity advection in front of the trough. At that time, the forward-tilted trough provided more favorable dynamic conditions for the strong convections in the second stage. The 700 hPa southwest jet was located in southern Hunan, with the maximum wind speed of the jet axis reaching 16 m/s. The shear lines at 850 hPa and 925 hPa were both in southern Hunan, and the ultra-low-level shear line was conducive to enhancing the ascending movement. Besides, the terrain of the Nanling Mountains strengthened the convergence. In terms of humidity conditions, southern Hunan was located in the wet zone at 850 hPa and the dry tongue at 500 hPa. The unstable stratification with dry air at upper levels and wet air at low levels provided favorable environmental conditions for convections. The warm temperature ridge at 850 hPa and the cold trough at 500 hPa converged in southeastern Hunan, and the convergence of cold and warm air was beneficial to triggering convections. The strong convections at this stage were consistent with the upper-level cold advection forcing type proposed by Xu et al.

Based on the comprehensive analysis of the 3-hour upper-level chart and the disaster weather observations, it can be found that the 500 hPa trough was located from Huaihua to northern Guangxi at 20:00 on April 24, and the 700 hPa trough was at the boundary of Hunan and Guizhou. For the disaster weather records, the thunderstorm gale with 17.5 $m \cdot s^{-1}$ speed occurred in Jingzhou at 20:00, and the thunderstorm gale with 18.2 $m \cdot s^{-1}$ speed occurred in Tongdao at 21:00. This result indicated that the thunderstorm gale appeared behind the 500 hPa trough and in front of the 700 hPa trough (Figure 7). When the 700 hPa trough passed through ultimately, the weather in southeastern Hunan turned clear.

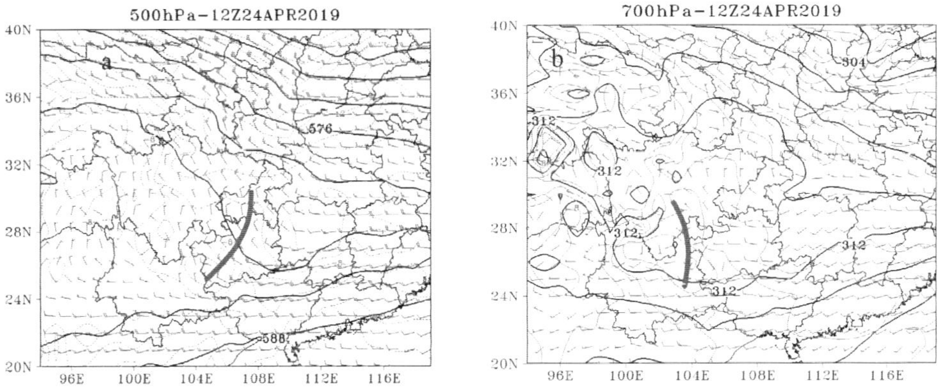

Figure 7. (a) 500 hPa and (b) 700 hPa situations at 20:00 on April 24.

4.2. Favourable environment field

By analyzing the sounding data at two representative stations (Huaihua station and Chenzhou station) in the two stages, it is found that there was a significant inversion in low levels at 20:00 on April 23. After revision the convective available potential energy (CAPE) increased significantly (Table 2). The two stages had several similarities as follows. Heights of the 0℃ level and −20℃ level were suitable, with a slight difference in the two stages, which was beneficial to hail occurrence. There was an obvious dry intrusion in the middle level. The CAPE were large, more than 1000 $J \cdot kg^{-1}$. The Showalter index (SI) was negative, less than −1.8℃. Moreover, there was a jet stream at 500 hPa, which was conducive to the downward momentum transportation. The 0–6 km vertical wind shear reached the strong level, favoring the generation of strong storms (supercells). The K index had little significance in indicating except for that at Huaihua station at 20:00 on April 24.

However, there were still several differences. For example, the CAPE in the first stage was larger, which was more conducive to large-scale hail. At 0–2 km height, the vertical wind shear is significantly larger in the first stage than in the second stage, which was more conducive to severe thunderstorms (supercells). 3) In the first stage, the wind rotated clockwise with the height and then rotated anticlockwise when above 600 hPa. It meant that there was warm advection in the low levels and cold advection in the middle levels in the first stage, with a more unstable stratification than in the second stage. In the second stage, the wind rotated clockwise with the height, and there was warm advection in the whole layer. The relative humidity in low levels was higher in the first stage, more favorable for the occurrence of short-term heavy precipitation. However, the wet layer in southeastern Hunan was shallow in the second stage, and the short-term heavy precipitation was not significant. Moreover, the stratification curve in Chenzhou was characterized by the shape of double-trumpet, more conducive to thunderstorm gales in southeastern Hunan.

Table 2. Comparison of convective indexes between Huaihua station and Chenzhou station at 20:00 on April 23 and 20:00 on April 24.

Huaihua station	K (℃)	SI (℃)	CAPE ($J \cdot kg^{-1}$)	DCAPE ($J \cdot kg^{-1}$)	0–6 km vertical wind shear ($m \cdot s^{-1}$)	0–2 km vertical wind shear ($m \cdot s^{-1}$)	0℃ layer (km)	−20℃ layer (km)
20:00 on April 23	27	−4.7	1612.7	72.2	21	12	4.7	7.6
20:00 on April 24	46.2	−8	1264.2	4.8	10	3.5	4.3	7.2
Chenzhou station	K (℃)	SI (℃)	CAPE ($J \cdot kg^{-1}$)	DCAPE ($J \cdot kg^{-1}$)	0–6 km vertical wind shear ($m \cdot s^{-1}$)	0–2 km vertical wind shear ($m \cdot s^{-1}$)	0℃ layer/ (km)	−20℃ layer (km)
20:00 on April 23	28	−3	1068.3	2.4	21	12	5.0	7.7
20:00 on April 24	31	−2	1551.1	0.1	22	4	4.9	7.5

4.3. Favourable triggering conditions

The surface convergence line was the main dynamic condition triggering the mixed convection in the first stage from the night of April 23 to the early morning of April 24, the surface convergence line north-south migrated in southern Hunan, with the maximum amplitude not exceeding three latitudes. Due to the uplift effect caused by the topography of the northern Nanling Mountains and Xuefeng Mountains, the convective weather mainly occurred on the windward side of mountains. By analyzing the 3-hour continuous pressure change, pressure and wind field data from 23:00 on April 23 to 08:00 on April 24 (figure omitted), it can be found that the 3-hour pressure change was still positive at 23:00 on April 23 and rapidly turned to negative in western and southern Hunan at 02:00 on April 24. The negative pressure change appeared in many areas at 05:00 on April 24. The pressure change of most parts of Hunan turned to positive again at 08:00 on April 24. This result indicated that a small-scale low-pressure system affected western and central Hunan in the wee hours on April 24. By comparing the climatic standard deviation of the 3-hour pressure change at each time in the first stage with those of Xu's statistics (Table 3), it can be found that a large area of abnormal low centers of 3-hour pressure change appeared in western and southern Hunan during the convection development in the wee hours on April 24, which was conducive to promoting the development and strengthening of the mesoscale systems. The surface weather chart also showed that the small-scale low-pressure system moved from Guizhou to Xiangxi Prefecture, and then moved southward to central Huaihua. Note that there was also a small-scale low-pressure in southern Hunan. This result indicated that the small-scale low-pressure system accompanied by the surface convergence line was the main triggering mechanism of the strong convections in the first stage.

On the surface weather chart at 14:00 on April 24 (figure omitted), Hunan as a whole was located in a low-pressure area, with temperature exceeding 30℃. Thus, the thermal instability in the region was significantly enhanced. The thunderstorm gales mainly occurred in the area near 850 hPa warm ridge with high surface temperature, indicating that a large amount of unstable energy was triggered and released in this area. The hourly data from intensive automatic weather station shows that the low-pressure system was generated in southwestern Hunan at 19:00 on April 24 and gradually moved southeastward. From 19:00 to 22:00 on April 24, it was located at the southern foot of the Xuefeng Mountains. Due to the terrain friction, the low-pressure system moved slowly and accelerated after climbing over the mountain. After 23:00 on April 24, the low-pressure system arrived near Nanling Mountains, and its movement slowed down, bringing thunderstorm gale to southeastern Hunan. From 00:00 to 04:00, the low-pressure system moved eastward along the surface convergence line. During this process, the near-surface pressure dropped due to the blocking effect of the mountains and the surface convergence line that strengthened the upward movement.

Table 3. Standard deviation of 3-hour pressure change at each time of April 24.

ΔP_3	**02:00**	**05:00**	**08:00**	**11:00**	**14:00**	**17:00**
Standard deviation	0.7–0.8	0.7–0.9	0.7–0.8	0.7–0.8	0.9–1.0	0.8–1.0

5. Conclusions

Based on the comprehensive analyses of the mixed convective weather in Hunan from April 23 to 24, 2019, the following conclusions are drawn.

The severe convective weather can be divided into two stages. The disastrous weather in the first stage was mainly the short-term heavy precipitation, accompanied by local thunderstorm gale and small-scale hail. This weather process in the first stage belongs to the warm advection forcing type. The severe convective weather in the second stage was mainly thunderstorm gale, with scattered short-term heavy precipitation, which belongs to the upper-level cold advection forcing type.

The first stage was dominated by broad mixed precipitation echoes, while the second stage was dominated by typical linear convective echoes. Besides, new supercells constantly merged into the southern end of the squall line and travelled northeastward, resulting in the continuous growth and development of the squall line on its southern end. The radar figures in the second stage were featured by the bow echo, the mid-altitude radial convergence, the low-level divergence, the high-value area of the velocity and the velocity ambiguity.

The surface convergence line was the main dynamic uplifting condition triggering the mixed convection in the first stage. From the night of April 23 to the early morning of April 24, the surface convergence line north-south migrated in southern Hunan, with the maximum amplitude not exceeding three latitudes. The severe convective weather in the second stage occurred after the upper-level low-pressure trough moved eastward and turned into a forward-tilted trough. Linear convections were generated near the upper-level trough. Simultaneously, in the low-level unstable energy area the linear convections developed into a squall line during the eastward movement of the forward-tilted trough. Then, the thunderstorm gale swept the whole southern Hunan.

The energy conditions and dynamic conditions were favorable for the convections in both stages. In the first stage, the water vapor was more abundant than in the second stage, with the near-surface specific humidity reaching 16 $g{\cdot}kg^{-1}$. The unstable energy area was mainly distributed in western Hunan. In the second stage, the unstable energy in southeastern Hunan was larger, with the CAPE value of more than 2000 $J{\cdot}kg^{-1}$.

Acknowledgments
Fund project: Special project for forecasters of China Meteorological Bureau (CMAYBY2017-050, CMAYBY2020-085), R & D projects in key areas of Hunan Provincial Department of science and technology (2019SK2161); The construction of Hunan Province's business ability (NLJS14).

References
[1] Xu B, Xu A H and Chen Y H et al 2019 Application of severe convection weather conceptual model to forecast of an extreme gale event occurred in Jiangxi on 4 March 2018 [J] *Torrential Rain and Disasters* **8(2)** 144-151
[2] Si F Y, Zhao H Q and Li S Z 2018 Flow Pattern and Physical Characteristics Analysis of Two Categories of Severe Convective Weather Influced by Westerly Trough in Northwest Henan [J] *Meteorological and Environmental Sciences* **41(3)** 57-63
[3] Emanuel K A 1994 Atomospheric Convection [M] New York:Oxford University Press 168-175
[4] Ye C Z, Tang M H and Chen H Z et al 2013 Synthetic discussion on the causes of the first severe convective weather causing disaster over Hunan in 2013 [J] *Torrential Rain and Disasters* **32(1)** 1-10
[5] Wang X L et al 2017 Contrastive Analysis of Two Mixed Convective Weather Processes in Hunan in Spring of 2015 [J] *Journal of Arid Meteorology* **35(2)** 250-259
[6] Yao R, Tang J and Lan M C et al 2017 Analysis of environment condition and evolution characteristic for a mixture strong convective process [J] *Torrential Rain and Disaster* **36(3)** 217-226
[7] Tang M H, Wang Q and Xu J Y et al 2019 Comparative analysis of two squall line events under low-level warm convection forcing in Hunan [J] *Torrential Rain and Disasters* **38(2)** 115-125
[8] Cai R H, Yao R and Huang X Y et al 2017 Synopic Situation Classification and Forecast Analysis of Thunderstorm in Dongting Lake *Meteorological Monthly* **43(5)** 560-572

IOP Conf. Series: Earth and Environmental Science **865** (2021) 012022 doi:10.1088/1755-1315/865/1/012022

The effects of meteorological factors on ozone during summer in Shanghai, China

Shengyang Duan[1], Chenhang Zhao[1], Yonghang Chen[1] and Qiong Liu[1, 2]

[1]Environmental Science and Engineering College, Donghua University, Shanghai, China

[2]Corresponding author's e-mail: liuqiong@dhu.edu.cn

Abstract. Based on the surface observation data from 2011 to 2013, the effects of meteorological factors on ozone are revealed during summer in Shanghai by analyzing the visibility, temperature, and relative humidity. When ozone pollution occurred, visibility and relative humidity decreased significantly, 10 % – 30 % and 15 % – 35 %, respectively. The ozone concentration was negatively correlated with visibility and relative humidity. While the higher the temperature is, the more conducive to the photochemical formation of ozone. Ozone pollution in Shanghai mainly occurred in meteorological conditions with visibility from 6–8 km, temperature from 25 ℃ – 35 °C, and relative humidity from 20 % – 50 %. The temperature made a more outstanding contribution to ozone generation when ozone concentrations were much higher than 200 μg/m^3.

1. Introduction

In recent years, ozone has become the primary pollutant affecting air quality during summer in megacities in China [1,2]. Pairidiguli et al. [3] found that the concentrations of PM2.5 and ozone in the Yangtze River Delta were decreasing and increasing, respectively. Yi et al. [4] found that the most severe ozone pollution in the Yangtze River Delta usually occurred from May to August using the data from China National Environmental Monitoring Centre. Precursors and meteorological factors significantly impact urban summer ozone concentration [5,6]. Using the Weather Research and Forecasting Model-Community Multi-scale Air Quality Model(WRF-CMAQ), Yan et al. [7] found that volatile organic compounds (VOCs) mainly controlled ozone concentrations in Shanghai.

In addition to precursors, meteorological factors also have significant effects on ozone formation. Temperature and relative humidity are two crucial meteorological factors affecting the formation of near-ground ozone [8,9]. Yan et al. [10] analyzed the meteorological data of Jiangsu Province from 2013 to 2017, and the results showed that the ozone mass concentration in Jiangsu Province was generally positively correlated with temperature and negatively correlated with relative humidity. The temperature is higher than 25 ℃, the relative humidity is 30 %-50 %, or the wind speed below 4 m·s-1 is prone to ozone pollution [9]. Cui et al. [11] studied the factors affecting the rapid growth of near-surface ozone concentration in the Beijing-Tianjin-Tangshan region in 2016 and found that the influence of temperature was positively correlated. The most obvious was a specific negative correlation with pressure and a slight positive correlation with evaporation. Wang et al. [12] found that ozone and the daily visibility change showed a significant positive correlation based on the 2013 Shenyang ozone and meteorological data. When particles, especially the secondary fine particulate matter, increase in the air, visibility becomes lower, which can change solar radiation and thus affect ozone concentration [13].

IOP Conf. Series: Earth and Environmental Science **865** (2021) 012022 doi:10.1088/1755-1315/865/1/012022

The main task of air pollution control in China is to control ozone and fine particulate matter concentrations simultaneously. According to the previous studies [14-16] in China, haze is caused by fine particulate matter in the air, when the visibility is less than 10 km, and the relative humidity is less than 95% without any precipitation, dust storm, blowing sands, floating dust, smokes, blowing snows, or snowstorms. The visibility is an essential indicator of air quality and can better reflect the cleanliness of the ground atmosphere [17]. Therefore, visibility can be used as a meteorological parameter reflecting particle concentrations in the atmosphere. In addition to particulate matter, other meteorological factors also affect visibility, with relative humidity having the most significant impact [18-23]. In Shanghai, visibility is mainly affected by PM_{10}, which correlates with temperature and relative humidity [24].

Visibility is a critical meteorological parameter, which can reflect the particulate matter concentrations and affect the formation of near-surface ozone [25]. Therefore, the primary goal is to further understand the influence of meteorological factors on near-surface ozone concentrations by considering visibility in Shanghai.

2. Methods

2.1. Observation site

The Shanghai Meteorological Bureau began to build an atmospheric composition observation network in 2005. The data (2011-2013) were from six stations representing different functional zones (Figure 1) to analyze the impact of different meteorological factors on ozone concentrations. Xuhui is a large commercial, residential zone in the southwest of central Shanghai, not far from the Shanghai Min Viaduct and inner ring viaduct, one of the important urban transportation hubs. Pudong is a large green space, office, and residential zone. Baoshan and Jinshan are steeling industrial zone and petrochemical zone, respectively. Chongming is an ecological island area with a high green coverage rate. Dongtan is a large wetland nature reserve, far away from the urban area and less affected by human activities, representing Shanghai's atmospheric background value of atmospheric pollutants.

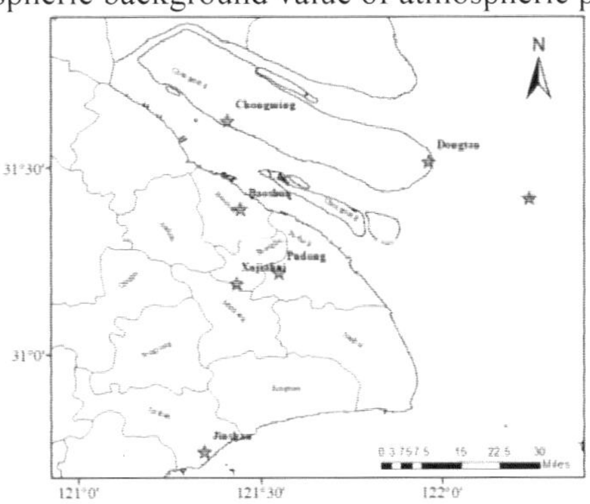

Figure 1. The geographical location of the observation stations.

2.2. Surface observation data

In this paper, the ozone concentrations were obtained from the ML9810B ozone analyzer. When the hourly ozone concentration is higher than 200 μg/m^3, ozone concentration exceeded the standard, and ozone pollution occurred. The meteorological data such as temperature, atmospheric pressure, relative humidity, wind speed, wind direction, visibility, precipitation, etc., are all from the Milos500 seven-element automatic weather station.

2.3. Data processing method

Table 1 shows the statistical data of the hourly average ozone concentrations of the six stations. The lack of data is mainly caused by instrument maintenance, zero range calibration, or data quality control.

Ozone analysis and evaluation mainly use hourly average concentration without precipitation. Since ozone is a product of photochemical reactions, ozone pollution is mostly concentrated in summer and daytime. Thus, the data from May to August and 6 o'clock to 20 o'clock from 2011 to 2013 were used in this paper. The hourly data are divided into two situations: exceeding the standard (ozone concentration higher than 200 μg/m^3) and meeting the standard (ozone concentration less than 200 μg/m^3).

Table 1. Statistics on the characteristics of hourly average concentrations of ozone at six stations in Shanghai.

Station	The total number of valid samples	Data integrity rate	Max (μg/m^3)	Mean (μg/m^3)	Median (μg/m^3)	Standard deviation (μg/m^3)
Baoshan	24576	93.43%	327	52	46	39
Chongming	24188	91.96%	357	70	66	42
Dongtan	24115	91.68%	344	84	82	39
Jinshan	24488	93.10%	418	74	66	52
Putong	24466	93.01%	353	61	55	43
Xujiahui	24663	93.76%	291	42	36	34

3. Results and discussion

3.1. The relationship between ozone and visibility

Figure 2 shows the visibility comparison when the ozone concentrations exceeded or not exceeded 200 μg/m^3 in Shanghai. The visibility of almost all sites decreases when the ozone concentration exceeds the standard. In all the stations, the difference between Baoshan and Jinshan in the industrial park is more prominent, decreasing by 23.84 % and 26.16 %. Moreover, the other stations are not prominent. The reason is that Baoshan and Jinshan are industrial areas, and their PM_{10} concentration is relatively high, which causes a significant change in visibility.

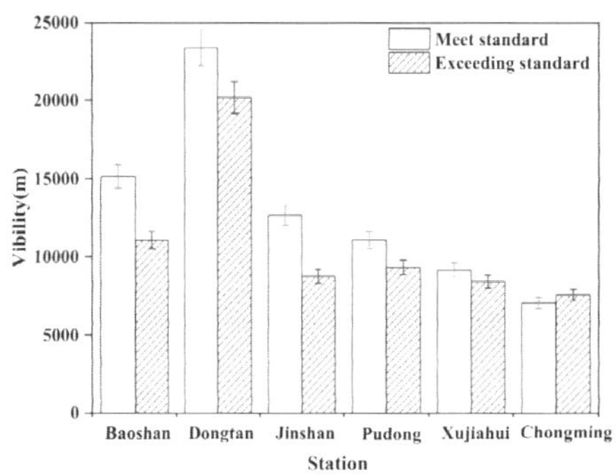

Figure 2. Comparison of average visibility when ozone concentrations meet and exceed the standard in Shanghai.

Figure 3 shows the daily changes in visibility when ozone concentrations meet and exceed the standard in Shanghai. Unlike Figure 2, when the ozone concentration meets the standard, the visibility of each site has a very consistent diurnal variation, showing a single-peak and single-valley structure. Zhou et al. [26] showed that the diurnal variation of visibility in Shanghai in winter is a single-peak feature. The valley appears at about 5 o'clock, the peak appears at about 14 o'clock, and the changes are relatively smooth. When the ozone concentration exceeds the standard, the visibility is generally lower. The visibility trend of Jinshan station from 8 o'clock to 23 o'clock is the same as that when the ozone concentration meets the standard. There is a peak at 16 o'clock, and the visibility at each time is, on average 3 km-4 km lower than when the ozone concentration was less than 200 μg/m^3.

At Dongtan station, except for the time when the fluctuation is significant, the trend from 11 o'clock to 19 o'clock is more similar to that when there was no ozone pollution; Baoshan station is an industrial park site when the ozone concentration exceeds the standard, the average level of visibility is higher than Pudong and Jinshan. However, the visibility of the three stations varies greatly, and the overall decrease is between 28% and 33%, indicating the daily change. The ozone concentration in Shanghai is negatively correlated with visibility, consistent with the study of Wang et al. [27]. When the fluctuation is significant, high ozone concentrations are mostly concentrated at 12:00 - 16:00, and the fluctuation of ozone concentration is slight. At other times, the corresponding visibility samples are also small due to the small number of samples when ozone concentration exceeds the standard, and there will be large fluctuations [28]. For example, the ozone concentrations exceeding the standard at Xujiahui and Baoshan stations at 13:00 were the highest, only accounting for 23.68 % and 20.65 %, respectively [29].

The summer visibility in Shanghai has the most remarkable correlation with PM_{10} [30]. The greater the concentration of PM_{10}, the lower the visibility. Many studies have found that PM_{10} and ozone concentrations show opposite trends, Zhao et al. [29], Shao et al. [31], and other studies have found that ozone and PM_{10} in Shanghai are opposite. It may be that the atmosphere is highly oxidative in summer, and secondary particles are generated at the same time as O_3, which leads to a positive correlation between PM_{10} and O_3 concentrations, which in turn causes a negative correlation between ozone concentration and visibility. In general, urban areas are more prone to strong updrafts in the afternoon of summer, a significant driving force for severe convective weather. The short-time heavy precipitation in the afternoon is essential for the wet deposition of aerosols [32], improving visibility. It may be one of the reasons for the peak visibility at 17:00 when the ozone concentrations in Xujiahui exceeded the standard.

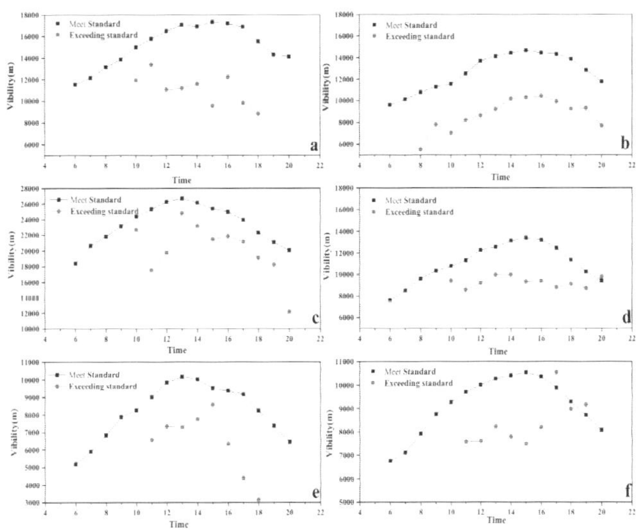

Figure 3. Comparison of daily changes in average visibility when ozone concentrations meet and exceed the standard in Shanghai (a: Baoshan; b: Jinshan; c: Dongtan; d: Pudong; e: Chongming; f: Xujiahui).

3.2. The influence of temperature

As shown in Figure 4, the temperature significantly changes when the ozone reaches the standard and exceeds the standard. When the ozone concentrations were less than 200 μg/m^3, the temperature was about 25 ℃. When ozone pollution occurs, temperatures at all six sites exceed 30 ℃, with Baoshan and Xujiahui reaching 34 ℃ and 35 ℃. The temperature when the ozone concentration exceeds the standard is generally higher than when the ozone reaches the standard.

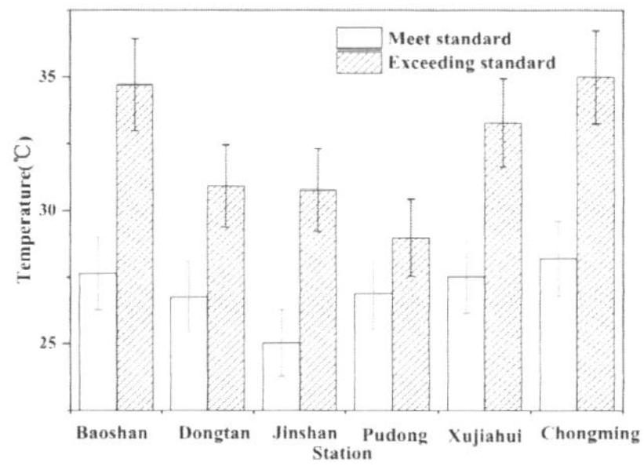

Figure 4. Comparison of average temperature when ozone concentrations meet and exceed the standard in Shanghai.

Figure 5 further analyzes the daily average temperature changes under different ozone concentrations in detail. When the ozone reached the standard, the daily temperature changes at all sites were relatively consistent, peaking around 14:00. Except for the generally low temperature in Dongtan, the daily temperature changes at other sites were between 24 °C and 30 °C. When the ozone concentration exceeds the standard, the daily temperature changes of each station are different, and there is no certain regularity. The temperature range of each station is 33 ℃ to 37 ℃ at Baoshan station, 30 ℃ to 36℃ at Pudong station, 28 ℃ to 34℃ at Dongtan station and 27 ℃ to 31℃ at Jinshan station. As a wetland at the mouth of the Yangtze River, Dongtan has a low base temperature. When the ozone concentration exceeds the standard at Jinshan station, the temperature is significantly lower, especially in the high-temperature area than in other stations. Jinshan is an area with severe ozone pollution, making it many high ozone concentration situations during non-high temperature periods. Compared with other areas, Jinshan is an industrial area, producing more VOCs (ozone precursors), which is more conducive to the photochemical formation of ozone. Therefore, when ozone pollution occurs in Jinshan station, the temperature is lower than in other stations.

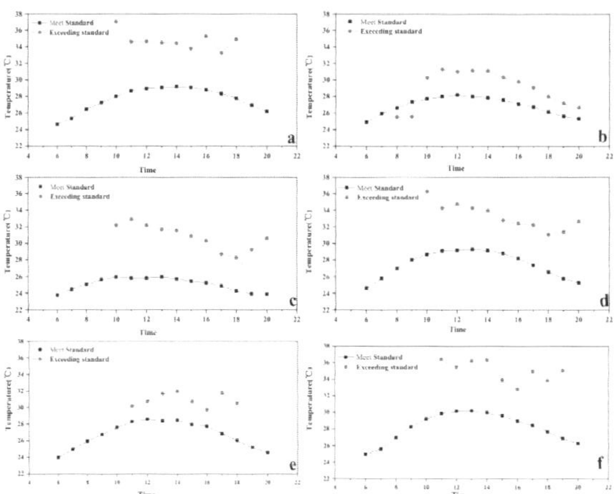

Figure 5. Comparison of daily average temperature changes when ozone concentrations meet and exceed the standard in Shanghai (a: Baoshan; b: Jinshan; c: Dongtan; d: Pudong; e: Chongming; f: Xujiahui).

3.3. The influence of relative humidity

Figure 6 shows that the relative humidity changes significantly when the ozone concentration reaches the standard and exceeds the standard. Relative humidity in the six stations when ozone concentration exceeds standard is lower than when ozone concentration meets standard. Specifically, the percentages

IOP Conf. Series: Earth and Environmental Science **865** (2021) 012022 doi:10.1088/1755-1315/865/1/012022

of relative humidity reduction at the time when ozone concentration exceeds the standard are 40.00%, 38.75%, 30.49%, 25.32%, 38.67%, and 47.83% in Baoshan, Chongming, Dongtan, Jinshan, Pudong, and Xujiahui, respectively.

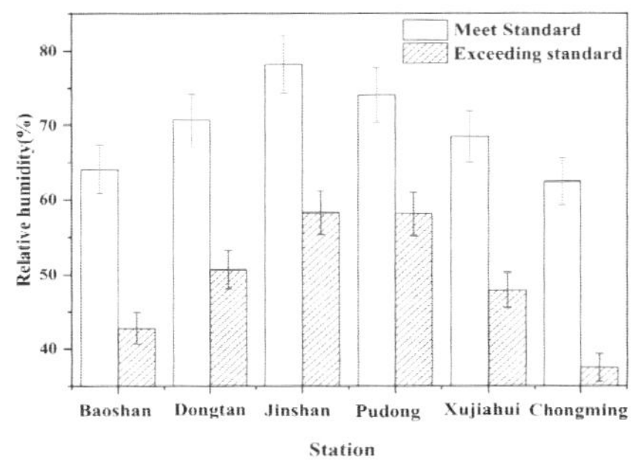

Figure 6. Comparison of the average relative humidity when ozone concentrations meet and exceed the standard in Shanghai.

Figure 7 shows the specific changes in relative humidity when ozone reaches the standard and exceeds the standard. When the ozone reaches the standard, the daily change of humidity is relatively regular, opposite to the daily change of temperature, and the difference between each site is not significant. It is a valley around noon, and the relative humidity is basically between 60% and 70%. When the ozone concentration exceeds the standard, the relative humidity of each station has a significant decrease, and its trend is similar to the daily change of the relative humidity when the ozone reaches the standard, and the decrease ratio is about 15%-35%. The results are consistent with Cui et al. [33], Chen et al. [34], and Li et al. [35]. The "wet scavenging" effect of high relative humidity on O_3 is mainly related to the activity of OH in the atmosphere. OH· is an essential oxidant in the tropospheric atmosphere. According to the photochemical reaction mechanism [36]. The reaction between VOCs and OH to form peroxyalkyl groups is a fast-paced step in the chain photochemical reaction process [37]. Because OH· will affect the photochemical process of O_3 production, it is not conducive to the accumulation of O_3.

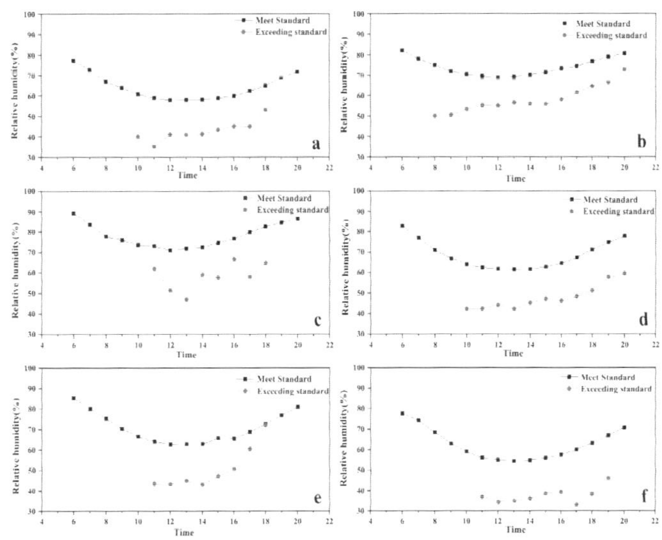

Figure 7. Comparison of daily changes in average relative humidity when ozone concentrations meet and exceed in Shanghai (a: Baoshan; b: Jinshan; c: Dongtan; d: Pudong; e: Chongming; f: Xujiahui).

3.4. The joint influences of temperature, relative humidity, and visibility on ozone

Figure 8 shows that the visibility is mainly 6 km-9 km when the ozone concentrations exceed the standard. When the visibility is 6 km-9 km, the temperature is 25 ℃-36 ℃, and the relative humidity is 25 %-60 % (Figure 9). As the ozone concentration increases, the relative humidity gradually decreases, and the temperature slowly rises. When the visibility exceeds 9 km, the ozone concentration exceeded the standard decreases gradually, especially in the situation of ozone concentration exceeds 225 μg/m^3. The temperature is positively correlated with visibility. When the visibility is less than 6 km, ozone pollution occurs less. The relative humidity is high, from 45 % to 70 %, and the temperature is low. Ozone pollution in Shanghai often occurs under low visibility, high temperature, and low relative humidity [38]. With the decrease of visibility, the temperature is lower, and the relative humidity is higher when ozone pollution occurs. When the visibility is in the range of 6 km-8 km, the temperature is about 30 °C, the relative humidity is 20 % -30 %, ozone concentration is more likely to exceed 250 μg/m^3, and more severe ozone pollution events may occur.

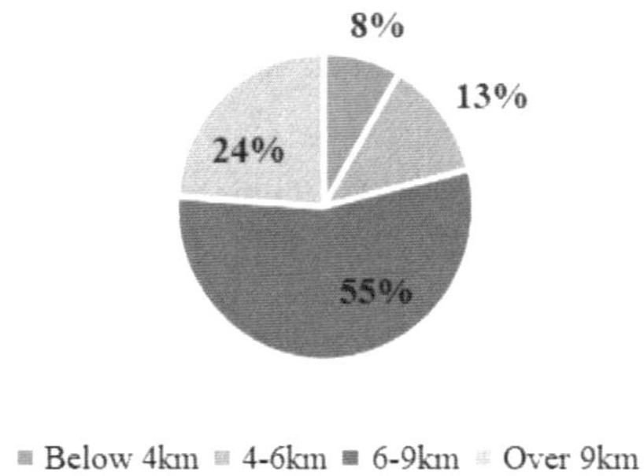

Figure 8. The ratio of samples with different visibility when ozone concentrations exceed the standard during May-August of 2011-2012 at Pudong station.

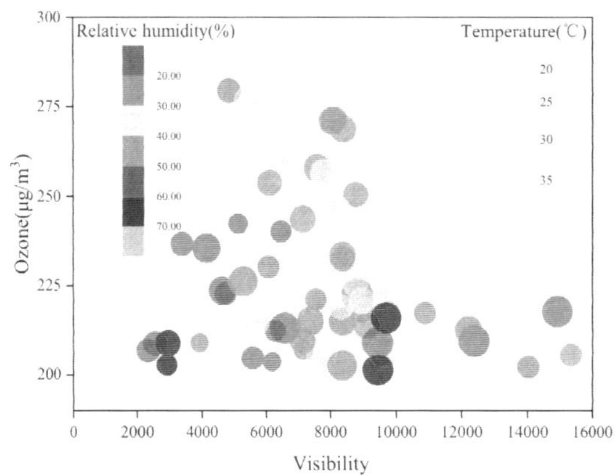

Figure 9. Bubble graphs of temperature and relative humidity under different visibility when ozone concentrations exceed the standard during May-August of 2011-2012 at Pudong station.

4. Conclusion

(1) The ozone concentration was negatively correlated with visibility. When ozone pollution occurs in Shanghai, the visibility of six stations generally decreased. Among them, the visibility in Baoshan, Pudong, and Jinshan decreased most, with an overall decline of 28% to 33%.

(2) When ozone pollution occurs, the temperature range of each site is 27 °C to 37 °C, indicating that the higher the temperature is, the more conducive to the photochemical formation of ozone. When there was no ozone pollution, the diurnal variation trend of temperature in the six stations was consistent, and

IOP Conf. Series: Earth and Environmental Science **865** (2021) 012022 doi:10.1088/1755-1315/865/1/012022

the peak temperature appeared at about 14: 00. The temperature of the other stations was between 24 °C and 30 °C, except that the temperature of Dongtan was generally low.

(3) When the ozone concentrations were less than 200 μg/m^3, the diurnal variation of relative humidity reached the minimum at 12: 00, which was basically between 60 % and 70 %. When the ozone concentrations were higher than 200 μg/m^3, the relative humidity of all stations had a significant decline, and the decline rate was about 15%-35%, indicating that the lower the relative humidity, the higher the ozone concentration.

(4) Ozone pollution mainly occurred in meteorological conditions with visibility from 6 km-9 km, temperature from 25 °C-36 °C, and relative humidity from 25%-60%. The temperature made a more outstanding contribution to ozone generation when ozone concentrations were much higher than 200 μg/m^3.

Acknowledgments
This research was funded by the Chinese Ministry of Science and Technology (Nos. 2018YFC1506305 and 2018YFC1506303) and the Shanghai Municipal Bureau of Ecology and Environment (No. 202102).

References

[1] Ministry of Ecology and Environment of the People's Republic of China. 2020 State of the Environment Bulletin of China [R]. 2021.

[2] Xu X B 2021. Recent advances in studies of ozone pollution and impacts in China: A short review. *Current Opinion in Environmental Science & Health*, **19**.

[3] Pairidiguli A, Yipareguli A and Yu T 2020. Study on Ozone Pollution in Yangtze River Delta [J]. *Journal of Agricultural Catastropholgy*, **10(07)**: 166-168. (in Chinese)

[4] Yi R, Wang Y L, Zhang Y J, Shi Y, Li M S 2015. Pollution characteristics and influence factors of ozone in Yangtze River Delta[J].*Acta Scientiae Circumstantiae*, **35(8)**: 2370-2377 (in Chinese)

[5] Liu P F, Song H Q, Wang T H, Wang F, Li X Y, Miao C H, Zhao H P 2020. Effects of meteorological conditions and anthropogenic precursors on ground-level ozone concentrations in Chinese cities. *Environ Pollut*, **262**, 114366.

[6] Hu C Y, Kang P, Jaffe D A, Li C K, Zhang X L, Wu K, Zhou M W 2021. Understanding the impact of meteorology on ozone in 334 cities of China. *Atmospheric Environment*, **248**.

[7] Dai H B, Zhu J, Liao H, Li J D, Liang M X, Yang Y, Yue X 2021. Co-occurrence of ozone and PM2.5 pollution in the Yangtze River Delta over 2013–2019: Spatiotemporal distribution and meteorological conditions. *Atmospheric Research*, **249**.

[8] Yan R S, Li L, An J Y, Huang C, Huang H Y, Lu Q 2016. Simulation study of ozone generation and its precursors control in Shanghai during summer time[J]. *Environmental Pollution & Control*, **38(01)**: 30-35+40(in Chinese)

[9] Chen Z Y, Li R Y, Chen D L, Zhuang Y, Gao B B, Yang L, Li M C 2020. Understanding the causal influence of major meteorological factors on ground ozone concentrations across China. *Journal of Cleaner Production*, **242**.

[10] Yan W L, Liu D Y, Kang Z M, Xie Z Z, Sun Y 2019. The characteristics of ozone pollution and its relationship with meteorological factors in Jiangsu [J]. *Journal of the Meteorological Sciences*, **39(04)**: 477-487. (in Chinese)

[11] Cui M R, Bai L Y, Feng J Z, Lin X S, Li H L, Gao W W, Li Z W 2021. Analysis of temporal and spatial variations of ozone coupling with dynamics ofmeteorological factors in the Beijing-Tianjin-Tangshan region [J]. *Acta Scientiae Circumstantiae*, **41(02)**: 373-385. (in Chinese)

[12] Wang C, Wang S, Yang B B, Zhang L H, Wang L, Liu M 2015. The Key Laboratory of State Environmental Protection for Organic Pollutants Monitoring and Analysising in Air. Shenyang .*Environmental Monitoring Centre*, **31(03)**: 32-37. (in Chinese)

[13] Zhang J Z, Sun J, Miao Y P 2014. Analysis of the Causes of Smog and Haze Weather and

Countermeasures [J]. *China Emergency Management*, **(01)**:16-21.(in Chinese)
[14] Wu D 2008. Distinct ion between haze and fog in urban metropolitans and hazy weather warnings. *Environmental Science and Technology*, **31 (9)**, 3. (in Chinese)
[15] QX_T113, 2010. Observation and Forecasting Levels of Haze. (in Chinese)
[16] Chen Y H, Liu Q, Geng F H, Zhang H, Cai C J, Xu T T, Ma X J, Li H 2012. Vertical Distribution of Optical and Micro-Physical Properties of Ambient Aerosols during Dry Haze Periods in Shanghai. *Atmospheric Environment*, **50**, 50 - 59.
[17] Zhang Q H, Zhang J P, Xue H W 2010. The challenge of improving visibility in Beijing. *Atmospheric Chemistry and Physics*, **10 (16)**, 7821-7827.
[18] Wang S Y, Zhang X L, Xu X F 2003. Analysis of variation features of visibility and its effect factors in Beijing[J]. *Meteorol Sci Technol*, **(02)**: 109-114.(in Chinese)
[19] Chen K, Dong H Y, Bian W L 2013. Correlation Analysis of Variation Features of Visibility and Its Effect Factors in Tianjin [J]. *Environment and Sustainable Development*, **38(03)**: 85-88. (in Chinese)
[20] Wu Y, Chen R M, Wang J, Liu X F 2015. Analysis of Temporal Variation Characteristics and Meteorological Conditions of PM10 and PM2.5 in the South-Central of Hebei Province in 2013 [J]. *Meteorological and Environmental Sciences*, **38(04)**: 68-75. (in Chinese)
[21] Shen J F, Feng J J, Xie L, Lin Y, Mo C H 2007. Characterization of visibility and its effect factors in Guangzhou [J]. *Ecology and Environment*, **(04)**: 1199-1204. (in Chinese)
[22] Song Y, Tang X Y, Fang C, Zhang Y H, Hu M, Zheng L M, Li C C, Mao T T, Michael B 2003. Relationship between the visibility degradation and particle pollution in Beijing[J]. *Acta Scientiae Circumstantiae*, **(04)**: 468-471. (in Chinese)
[23] Zhang K, Chai F H, Chen Y Z, Wang S L, Hu B Q, Yu J, Zheng Y 2008. Analysis on the Characteristics of Visibility in Wuqing Tianjin [J]. *Climatic and Environmental Research*, **13(06)**: 800-806. (in Chinese)
[24] Shi H, Chen M, Han J J 2008. Analysis of Factors Affecting Visibility and Its Variation Features in Pudong Area of Shanghai [J]. *Atmospheric Science Research and Application*, **(02)**: 1-8. (in Chinese)
[25] Ma M L, Yao G B, Guo J P, Bai K X 2021. Distinct spatiotemporal variation patterns of surface ozone in China due to diverse influential factors. *J Environ Manage*, 2021, **288**, 112368.
[26] Zhou W D, Zhang L Y, Mu H Z 2020. Variation Characteristics and Regional Difference of Winter Visibility in Shanghai [J].*Meteorological and Environmental Sciences*, 2020, **43(4)**: 18-25. (in Chinese)
[27] Wang S B, Zhang H,Yi M J, Wei Y W 2020. Research on the Correlation between Ozone Pollution Distribution and Meteorological Elements in Typical Cities [J]. *Low Carbon World*, **10(12)**: 1-3. (in Chinese)
[28] Xu J L and Zhu Y X 1994. Some characteristics of ozone concentrations and their relations with meteorological factors in Shanghai[J]. *Atmospheric Environment*, **28(20)**:3387-3392
[29] Zhao C H 2015. Study on The Characteristic and Grading Standard of Photochemical Smog in Shanghai [D].*Donghua University*.(in Chinese)
[30] Chen K, Dong H Y, Bian W L 2013. Correlation Analysis of Variation Features of Visibility and Its Effect Factors in Tianjin [J]. *Environment and Sustainable Development*, **38(03)**: 85-88. (in Chinese)
[31] Shao P, Xin J Y, An J L, Wang J X, Wu F K, Ji D S, Wang Y S 2017. An analysis on the relationship between ground-level ozone and particulate matter in an industrial area in the Yangtze River delta during summertime [J]. *Chinese Journal of Atmospheric Sciences*, **41 (3)**: 618−628.(in Chinese)
[32] Wang H Q, He Q S, Chen Y H, Kang Y M 2014. Characterization of black carbon concentrations of haze with different intensities in Shanghai by a three-year field measurement. *Atmospheric Environment*, **99**, 536-545.
[33] Cui M, An X Q, Sun Z B, Wang B X, Wang C, Ren W H, Li Y J 2019. Characteristics and

meteorological conditions of ozone pollution in Beijing[J].*Ecology and Environmental Monitoring of Three Gorges*, **4(3)**: 25-35. (in Chinese)

[34] Chen C, Hong Y Y, Wu M, Zheng Q K, Shen J, Gong J F, Bu Q L, Cheng Y L, He Q R 2020. The changes of ozone pollution and polluted meteorological conditions of Foshan area in 2019[J]. *Environmental Science & Technology*, **43(S2)**:16-22. (in Chinese)

[35] Li K W, et al. Meteorological and chemical impacts on ozone formation: A case study in Hangzhou, China. *Atmospheric Research*, 2017, **196**, 40-52.

[36] Seinfeld J H, Pandis S N 2016. Atmospheric chemistry and physics:from air pollution to climate change [M]. *John Wiley & Sons*.

[37] Yu X F, Cheng P, Gu Y G, Li M, Tian Z L 2018. Formation potential of ozone and secondary organic aerosol from VOCs oxidation in summer in Guangzhou,China[J]. *China Environmental Science*, 2018, **38(03)**: 830-837. (in Chinese)

[38] Dong Y M, Li J, Guo J P, Jiang Z J, Chu Y Q, Chang L, Yang Y, Hong L 2020. The impact of synoptic patterns on summertime ozone pollution in the North China Plain. *Sci Total Environ*, **735**, 139559.

IOP Conf. Series: Earth and Environmental Science **865** (2021) 012023 doi:10.1088/1755-1315/865/1/012023

Accuracy analysis of 3D model of open-pit mine based on tilt photogrammetry

Jinjin Lu[1, 2], Yizhan Zhao[2], Lv Zhou[2, 3, 6], Kejia Wei[2], Jun Ma[4] and Xin Li[2, 5]

[1]Bowen college of Management, Guilin University of Technology, Guilin 541004, China;
[2]College of Geomatics and Geoinformation, Guilin University of Technology, Guilin 541004, China;
[3]Beijing Key Laboratory of Urban Spatial Information Engineering, Beijing 100038, China;
[4]China Railway Siyuan Survey And Design Group Co., LTD, Wuhan 430063, China;
[5]College of Geology Engineering and Geomantic, Chang'an University, Xi'an 710054, China

[6]Corresponding author's e-mail: zhoulv@glut.edu.cn

Abstract. Mineral resources account for a large proportion in the development of national economy in China. The topography conditions around the mine site are complex, and accurate measurement of the mine topography can ensure that the mineral resources are mined safely and reasonably. The traditional mine survey has certain danger. With the advantages of flexibility and no site constraints, the UAV is equipped with a high-resolution lens and combined with tilt photogrammetry technology to obtain high-resolution images of multiple angles of the mine area and fine texture information of the feature surface in a short period of time. A total of 918 optical images were collected as raw data using a consumer-grade UAV equipped with a common optical lens. They were pre-processed and modeled to generate a high-precision, clear-textured 3D model of the study area. The root mean square of the pixel is 0.53 after aerial triangulation and adjustment. Compared with the actual measured value, the mean square error of plane coordinates is ±0.1cm, and the maximum error is 0.2cm, the mean square error of height is ±0.1cm, and the maximum error is 0.2cm. The mean square error of model plane coordinates is ±2.9cm, and the maximum error is 10.2cm. The mean square error of height is ±6.4cm, and the maximum error is 8.6cm. The results show that the production model meets the requirements of relevant specifications and the use of tilt photogrammetry in hazardous areas can improve the safety of surveyors.

1. Preface

Mineral resources provide a great help to the development of our national economy. High accuracy surveying of the terrain around the mine can provide guidance for the scientific programming of the mining area so that the mineral resources can be rationally exploited. Traditional mine surveying takes a long time, the workload is large, and the work of surveyors is at great risk [1]. The new technology of 3D laser scanning also causes little practical application due to its high cost [2]. UAV has the advantages of high mobility, low cost and no site constraints. Combining tilt photogrammetry 3D modeling with mine surveying can effectively solve the tricky problems in mine mapping, and can also provide effective information and data for the environment around the mine [3]. The development

IOP Conf. Series: Earth and Environmental Science **865** (2021) 012023 doi:10.1088/1755-1315/865/1/012023

of tilt photogrammetry in foreign countries began in the 1990s. Pictometry, a US company, was the first to investigate tilt photogrammetry and designed the EFS/POL system [4]. In recent years, tilt photogrammetry has made great progress both at home and abroad. Harwin and Lucieer [5] proposed multi view stereo vision technology in 2012, and produced digital products with point error of 1-3cm to monitor sub decimeter coastal erosion. Grun et al. [6] used high-resolution satellite imagery (GeoEye-1) combined with UAV imagery to generate high-resolution 3D models to restore and reproduce historical buildings in Bhutan. China is a late starter in UAV tilt photogrammetry. In 2014, Wu et al. [7] used close-field photogrammetry to obtain three-dimensional information of the open-pit mine area, but it was difficult to obtain good results due to the complex topographic conditions of the mine area. Xie et al. [8] used fixed-wing UAVs for dynamic monitoring of mine reserves, and the relative error of the calculated mine reserves did not exceed 5%, which proved the reliability of UAV monitoring. Ma et al. [9] compared the UAV photogrammetry technology with GPS-RTK measurement data and concluded that the measurement results of UAV aerial survey technology can meet the demand. Yu [10] showed that the accuracy of the real 3D model made by using the tilt image collected by UAV is reliable, and it can be applied in practical engineering.

2. Tilt photogrammetry and 3D modeling

2.1. Principle of tilt photogrammetry technology

Tilt photogrammetry technology means that on the basis of traditional photogrammetry, it can acquire more complete, specific and complete terrain information by adding sensors and increasing angle of view. Conventional single camera aerial survey systems can only obtain orthophotos at vertical angles, while UAV tilt photogrammetry technology can acquire one set of orthophotos and four sets of tilt images by carrying multiple (up to five) lenses of tilt cameras（Figure 1）. Five groups of images with different angles are collected and processed to extract information such as plane and elevation texture, color and shelter relationship of the terrain, thus creating a high-precision 3D model [11].

Figure 1. Multi-shot image data acquisition.

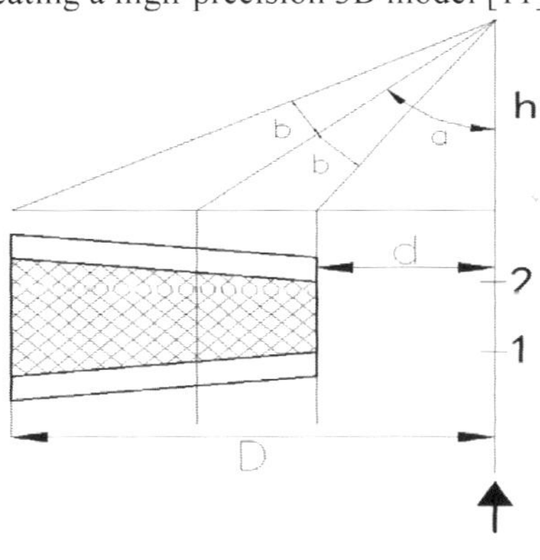

Figure 2. Geometric principles of tilt photography.

The geometry of tilt photogrammetry is shown in Figure 2. The direction of aerial flight is indicated by an arrow, the tilt of the camera is a, the symbol b indicates the viewing angle of the camera, the flight height of the UAV is h, and the maximum and minimum values of the horizontal distance between the UAV and the corresponding feature in the multi-view tilt image are indicated by the symbols D and d respectively.

2.2. Relevant techniques for tilt photogrammetry modeling

2.2.1. UAV tilt photogrammetry system. Tilt photogrammetry system consists of GNSS navigation system, inertial navigation system and tilt camera system. GNSS navigation system can obtain geographic coordinates, inertial navigation system can obtain the attitude angle at the moment of lens exposure, and tilt photogrammetry system can obtain images of ground objects in five different directions.

2.2.2. Image matching. Image matching is a process of establishing a connection between points of the same name within an image through eigenvalue extraction. The main two algorithms are SIFT and ASIF. The basic steps of image matching in the process of UAV tilt photogrammetry are roughly divided into three steps: firstly, screening the image pairs to be matched and extracting the feature values of the images; secondly, coarse matching of the image pairs to obtain the initial homonymous point pairs; and thirdly, exact matching on the basis of the second step, while deleting the mis-matched points in the second step. This process becomes known in many literatures as pyramid image matching.

2.2.3. Tilted multi-view image block combined adjustment. Geometric correction and correct recognition of the dislocation relationship between images are difficult in the data processing of oblique photogrammetry, In order to ensure the matching accuracy and positioning accuracy of the same name points in the image, the exterior orientation elements of the inclined image are selected from the POS data information, and the least square matching method, the matching file method of the same name and the bundle block combined adjustment method are used to solve the problem.

2.2.4. Texture mapping. The fineness of model texture is one of the key indexes to evaluate the quality of a 3D model. The traditional 3D modeling mostly uses the field photos to map the surface of the model after the model is built, which increases the additional workload. The raw data collected by tilt photogrammetry has texture information, Therefore, it is only necessary to correspond the texture in the image to the model, that is, to establish the corresponding relationship between the two-dimensional space and the 3D space in the image. This process is called texture mapping.

3. Example of 3D modeling by tilt photogrammetry in mining area

3.1. Data source

The selected experimental area is an open pit mine near TanTan Village, Dongxing City, Guangxi Zhuang Autonomous Region (as shown in Figure 3). The selected modeling area is 500 meters long and 400 meters wide with an elevation difference of about 160 meters. The terrain undulating greatly can intuitively verify the quality of modeling. 918 tilt photography photos taken with DJI Phantom 4 Pro (It is divided into two groups, each group is 459), The camera model is SONY ILCE-6000, the lens is E 20mm, F 2.8, the image size is 6000*4000, the sensor size is 23.5mm, and the spatial reference system is WGS-84.

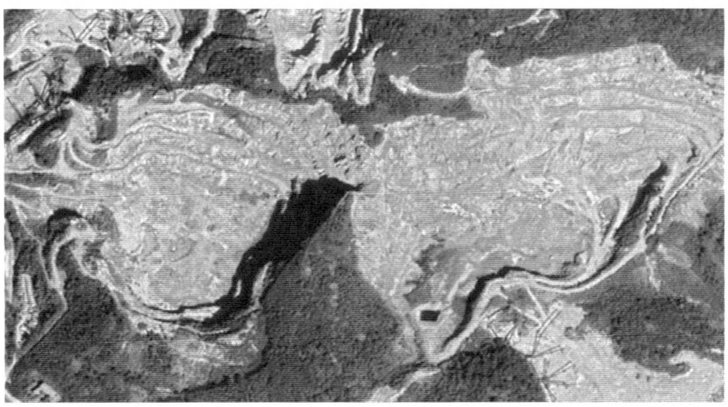

Figure 3. Satellite images of open pit mines.

3.2. Data processed
Firstly, the image control points are laid manually in the study area and the coordinates of the image control points are measured (to ensure accuracy); the flight altitude, side overlap and forward overlap parameters are set correctly after the air route is planned. Check and correct the image, and delete the unqualified image. Next, aerial triangulation was performed to obtain pre-production data, aerial triangulation is performed to obtain pre-production data, and then the image matching correction and image association (called stabbing points) were carried out with the image control point data. Finally, the 3D model with coordinates was obtained. After obtaining the preliminary model, the parts of the production results that are inconsistent with the actual need to be repaired with the help of software, and then return to map to obtain the color 3D model data.

3.3. 3D modeling results
Using high resolution optical image to generate point cloud data and establish a 3D model of the study area (as shown in Figure 4 and Figure 5), the established model can fully obtain the real scene of the surrounding environment of the mining area, and obtain the accurate spatial position, geometric shape and geological characteristics of the mining area, which can provide important parameters for the reasonable mining of the mining area.

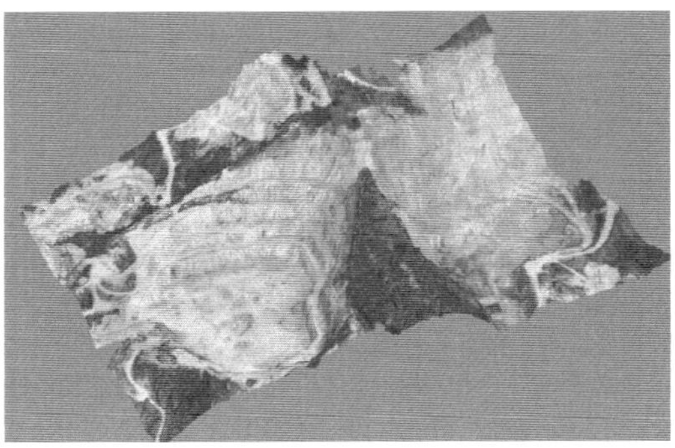

Figure 4. Results of open pit mine 3D model building.

Figure 5. Results of open pit mine 3D model building.

4. Precision analysis
In this paper, the accuracy analysis of the 3D model of the open-pit mine established by tilt photogrammetry technology was carried out from three aspects: 3D model results, aerial triangulation accuracy and model accuracy.

4.1. Mine modeling results
The overall appearance of the mountain model shape undulating, the number of significant errors is very small, and the real 3D world mountain structure consistent. The model as a whole has no cavities, no excess debris, and no suspended pieces, consistent with natural phenomena.

In local detail, the curvature of the road is exactly the same, and the stones on the ground in a real 3D world can be highlighted rather than treated as a plane. From this it can be seen that the 3D model is consistent with the real world in spatial shape (as shown in Figure 6 and Figure 7). From the perspective of texture color, the model color is slightly thicker and the whole picture is sharper, but it is basically the same as the real world, and there is no big change. Mine details are not overlooked and can be fully expressed whether the rock layers at the side of the hill or the scratches on the ground in the pits at the bottom. It shows that the 3D model of open pit can clearly and completely express the true structure of the area with high accuracy.

Figure 6. Open pit mine reality photos.

Figure 7. 3D model photo of open pit mine.

4.2. Accuracy evaluation of aerial triangulation

4.2.1. Pixel accuracy evaluation. The mean square root of the image element can be used to evaluate the results of combined adjustment of aerial triangulation. The root mean square of the image control points of the region studied in this experiment is 0.53 pixel after adjustment, which does not exceed the specified 0.6 pixel, so it meets the requirements.

4.2.2. Accuracy evaluation of image control points. The coordinate values of image control points are changed after the combined adjustment of aerial triangulation, and according to the original external measurement plane and elevation error values to calculate the error table (such as Table 1). Finally, compare with the national maximum limit difference table (such as Table 2) to check the accuracy. The mean square error of plane coordinates is ± 0.3mm, and the maximum error is 1.1mm; the mean square error of height is ± 1.0mm, and the maximum error is 1.1mm. The maximum difference between the two must not exceed the stipulation.

Table 1. Error table of encrypted coordinate values of image control points and original data.

Name	Number of stabbing point photos	Horizontal error /mm	Elevation error /mm
dx1-1	4	1.1	-1.3
dx1-2	6	0.4	0.3
dx1-3	5	0.5	1.0
dx1-4	3	0.6	-0.5
dx1-5	5	0.1	-1.1
RMSE		±0.3	±1.0

Table 2. Check point error maximum difference table.

Mapping scale	plane /m				Altitude /m			
	flat	hilly	mountains	plateau	flat	hilly	mountain	plateau
1:500	0.5	0.5	0.7	0.7	0.4	0.4	0.6	1.2
1:1000	1.0	1.0	1.4	1.4	0.4	0.4	1.0	1.5
1:2000	1.8	1.75	2.5	2.5	1.0	1.0	2.0	2.5

IOP Conf. Series: Earth and Environmental Science **865** (2021) 012023 doi:10.1088/1755-1315/865/1/012023

4.3. Model Accuracy Evaluation

4.3.1. Plane Accuracy Evaluation. The check point data in the model is extracted and then compared with the measured data. The residuals in X, Y and Z directions are calculated to evaluate the accuracy of the model. As can be seen from Table 3, the maximum error in the X-direction among the statistical check points in the area is 5.6 cm, with a mean square error of ± 5.0 cm; the maximum error in the Y-direction is 8.6 cm, with a mean square error of ± 5.0 cm. The plane accuracy of the points on the 3D model produced in this experiment meets the accuracy requirements of the relevant specifications.

Table 3. 3D model control point coordinate residual statistics table.

Name	△X/cm	△Y/cm	△Z/cm	Plane /cm
dx1-1	-0.9	4.0	-4.0	4.1
dx1-2	1.4	-2.3	-3.6	2.7
dx1-3	-6.6	-3.5	-5.0	7.5
dx1-4	5.5	-8.6	-8.4	10.2
dx1-5	-5.6	2.7	8.6	6.2
RMSE	5.0	5.0	6.4	2.9

4.3.2. Model elevation accuracy. In the industry, the mean square error is an important measure when conducting elevation accuracy evaluation. The requirements for elevation error in different scales and in different terrains (flat, hilly, mountainous, plateau) are also different. The maximum error in the Z-direction of the check point for this experiment is 8.6 cm, and the mean square error is ± 6.4 cm. From Table 4, we can see that the elevation accuracy meets the requirements.

Table 4. Topographic accuracy table.

Accuracy level	Grade I	Grade II	Grade III
Mapping scale	1:500	1:1000	1:2000
Flat land mean square error of height /m	0.37	0.37	0.75
Hilly mean square error of height /m	0.75	1.05	1.05
Mountains mean square error of height /m	1.05	1.50	2.25
Plateau mean square error of height /m	1.50	3.00	3.00

In summary, the root mean square of the image element is 0.53 image elements after aerial triangulation and survey adjustment. Compared with the real measurement value, the mean square error of plane coordinates is ± 0.1cm, and the maximum error is 0.2cm. The mean square error of height is ± 0.1cm, and the maximum error is 0.2cm. The mean square error of model plane coordinates is ± 2.9cm, and the maximum error is 10.2cm. The mean square error of height is ± 6.4cm, and the maximum error is 8.6cm. It is obvious that the 3D modeling using tilt photogrammetry has high accuracy and uniform accuracy distribution. Its accuracy is fully compliant with industry reference specifications.

5. Conclusion

In this paper, a 3D model of a mine with accuracy up to centimeter level is obtained through the study of tilt photogrammetry and 3D modeling. The feasibility of combining tilt photogrammetry 3D modeling with mine mapping was verified, and the following conclusions were obtained:

(1) 3D modeling of UAV tilt photogrammetry meets the mapping needs of the mine surface, and the accuracy of the model meets the requirements of relevant specifications.

(2) UAV remote sensing technology can solve the geological disaster investigation restricted by complex terrain conditions. By establishing a high-resolution 3D model of the survey area, it can improve efficiency, reduce the intensity and risk of field work.

(3) Modeling has high requirements for source data. The image needs to have a good texture in order to be better recognized, the same tone, mirror and smooth surface processing are poor, must be post-processing and repair. When modeling, due to occlusion at the bottom or side of the building, feature points are not extracted enough, and the model will be distorted or even empty. These issues are to be discussed by all colleagues in the industry to make progress together.

Acknowledgements

This work was supported by the Guangxi Science and Technology Plan Project (Grant No. GUIKE AD19110107); the Natural Science Foundation of Guangxi (Grant Nos. 2018GXNSFBA050006 and 2020GXNSFBA159033); the Wuhan Science and Technology Plan Project (Grant No. 2019010702011314); the Beijing Key Laboratory of Urban Spatial Information Engineering（Grant No. 2020217; the Guangxi Universities Young and Middle-aged Teachers' Basic Scientific Research Ability Improvement Project（Grant No. 2020KY06032）; and the Foundation of Guilin University of Technology (Grant No. GUTQDJJ2018036).

References

[1] Huang F 2015 RTK Technology and Total Station Instrument in Jinduicheng Mine Surveying in Joint Operations [J] *Geomatics & Spatial Information Technology* **38(02)** 156-158

[2] Chen G and Bai T 2019 Application of 3D laser scanner in open pit survey [J] *World Nonferrous Metals* **(24)** 21+23

[3] Li Y and Lai H 2017 The Application of UAV Photogrammetry Technology in the Law Enforcement Supervision Toward Mine's Extraction [J] *Bulletin of Surveying and Mapping* **(S1)** 95-99

[4] Song B 2012 Research on Key Technology of BuildingOblique Aerial Virtual Imaging System [D] *Henan Polytechnic University*

[5] Harwin S and Lucieer A 2012 Assessing the Accuracy of Georeferenced Point Clouds Produced via Multi-View Stereopsis from Unmanned Aerial Vehicle (UAV) Imagery[J] *Remote Sensing* **4(6)** 1573-1599

[6] Gruen A, Zhang Z and Eisenbeiss H 2012 UAV photogrammetry in remote areas-3D modeling of Drapham Dzong Bhutan[J] ISPRS - International Archives of the Photogrammetry Remote Sensing and Spatial Information Sciences XXXIX-B1 375-379

[7] Wu Qingshen and Yi Ying 2014 Technology of Three-dimensional Reconstruction of Opencast Slope based on Distant Scenes Image [J] *Metal Mine* **(10)** 130-132

[8] Xie Yi, Yi Guixuan, Liang Huanqing and Fan Kunfei 2017 Research on the Application of UAV Photogrammetry in Open-pit Mineral Monitoring [J] *Urban Geotechnical Investigation & Surveying* **(02)**114-117

[9] Ma Guochao, Wang Lijuan, Ma Song and Tang Yao 2018 Application of UAV photogrammetry in construction planning of mine tailings reservoir [J] *Science of Surveying and Mapping* **43(01)**84-88

[10] Yu Jiankang 2020 Application of UAV tilt photogrammetry in 3D reconstruction of overpass [J] *Bulletin of Surveying and Mapping* **(10)**106-109

[11] Miao Yuzhou 2019 Research on Application of UAV Oblique Photogrammetry Technology in Building Real 3D Model of Urban [D] *East China University Of Technology*

IOP Conf. Series: Earth and Environmental Science **865** (2021) 012024 doi:10.1088/1755-1315/865/1/012024

Quick determination of phenanthrene in coloured latexes using graphene-modified glassy carbon

Furong Liu[1, 3], Yu Wang[2], Qiuxiang Liu[1], Fangping Yan[1], Xiangyu Ye[1], QiwangWu[1] and Wei Shi[1]

[1]Zhejiang Light Industrial Product Inspection and Research Institute, Hangzhou 310018, China;
[2]Zhejiang LAINI New Material Technology Co., Ltd, Haining 314413, China

[3]Corresponding author's e-mail: 29607219@qq.com

Abstract. A facile detection system was devised by attaching graphene to glassy carbon, and the electrooxidation of phenanthrene (Phe) was investigated. The linear sweep voltammetry revealed that the modified liquid quantity and electrolyte concentration affected the oxidation of Phe-acetonitrile-$NaClO_4$ whose applicability conditions were optimised. The peak oxidation current in the range of 1-120 nmol L^{-1} has a good linear relationship with Phe concentration (I= 0.1066cpHe +19.04, R^2=0.9975). Furthermore, we reported an electrochemical assay for Phe in coloured latexeson a graphene-modified glassy carbon. It was meaningful for quick determination of Phe, especially in simpleness and detection time compared with GC–MS.

1. Introduction

Polycyclic aromatic hydrocarbons are environmental pollutants with carcinogenic and mutagenic effects [1,2]. They are present in water, soil and latexes, constituting a danger to human beings [3-5]. Phenanthrene (Phe) is a chemical carcinogen with strong carcinogenicity [6,7]. Phe is not intentionally added to the latexes but is produced in the manufacturing process. They are widely found in synthetic latexes, especially in coloured latexes. Electrooxidation supplies available ways to reduce the toxicity of organic compounds [8,9]. Given the issues of the high social popularity rate of latex products, it is important to investigate the Phe in dark coloured latexes.

Existing techniques for determination of Phe include GC–MS [10,11], fluorescence spectrometry [12], HPLC [13] and LC–MS [14]. Although the above methods involve large scale valuable instruments, the complex pretreatment steps make these techniques less attractive [15,16]. The preparation of proper materials on glassy carbon electrodes (GCE) is the main development direction for the quick determination of polycyclic aromatic hydrocarbons.

Here, the sorption of Phe-acetonitrile-$NaClO_4$ on modified GCE and the electrooxidation of Phe were evaluated.Different molar concentrations of Phe (C_{Phe}) were analysed using linear sweep voltammetry (LSV). The electrochemistry (EC) data were compared with those detected by GC–MS. The selective EC method for Phe determination was necessary, establishing an economised and mini-type approach. The EC analysis of Phe in latexes was expected to provide a theoretical basis for the quality control of latexes.

2. Experimental section

2.1. Instruments and materials

The main instruments used were CH Instruments CHI660E electrochemical workstation (Shanghai, China) and Agilent 7890B/5977B gas chromatograph/mass spectrometer (Palo Alto, USA). Additionally, a Hitachi SU3500 scanning electron microscope (Tokyo, Japan), a Grant LMC-3000 centrifuge (Cambridge, UK) and an ID 4-mm diameter GCE (Tianjin, China) were also used.

Graphene (97%), glycol (99.5%)and chromatographic grade acetonitrile were purchased from Macklin Biochemical (Shanghai, China). Phe in methanol (200 μg mL^{-1})was purchased from Biobw (Beijing, China). The other reagents (e.g. hydrochloric acid, sodium perchlorate, ethanol) have at least an analytical grade. Water (resistivity of 18.2 MΩ·cm) was prepared by the Millipore Milli-Q system (Boston, MA, USA).

2.2. Modified GCE fabrication

The surface of GCE was polished with 0.3 μm alumina powder/water suspension on metallographic sandpaper. The three-electrode system consists of GCE, platinum electrode and Ag/AgCl electrode. The GCE was put in 1.0 mol L^{-1} HCl and the oxidation-reduction cycles were studied from −0.4 to 0.7 V. The potential scan rate was repeated as 0.5 V s^{-1} until the cyclic voltammetry (CV) characteristics remained stable and the activation was completed.

Under normal stirring conditions, the modified liquid was obtained by adding 0.03 g graphene to 20 mL glycol under normal stirring conditions. Then, 8 μL of the modified liquid was applied to the surface of the preactivated GCE and dried under an infrared lamp.

2.3. Electrooxidation measurement

For deoxygenation, high purity nitrogen was bubbled above the solution for more than 10 min to separate the reaction from the air. Here, the three-electrode system was employed: graphene-modified GCE, Ag/AgCl electrode and platinum electrode were the working, reference and counter electrodes, respectively. To investigate the electrooxidation of Phe in this system, LSV measurements were performed within the range from 0.7 to 1.3 V at a potential scan rate of 50 mV s^{-1}.

The C_{Phe} of 0, 20, 40, 60, 80, 100, 120, 150, 210 and 250nmol $L^{-1}$was prepared with 0.25 mol L^{-1} acetonitrile-$NaClO_4$ (adjusting the pH of the electrolyte to strong acid condition). The modified GCE was immersed in the Phe standards with open circuit for 20 minutes. The average value of the LSV peak current was expressed as the oxidative current (I). All EC tests were conducted at 25 ℃.

2.4. Sample pre-treatment and chromatographic analysis

For sample pre-treatment, 0.5 g latex in 10 mL acetonitrile was sealed in a thermostatic water tank and extracted at 65 ℃ for 30 minutes. The process was repeated twice with 20 mL acetonitrile, centrifuged at 2000 rpm for 15 min. Then, it was filtered by organic filtration and brought to volume. The extraction solution was added to acetonitrile for LSV.

Thus, 0.8 mL extract was ready for chromatographic analysis and the data were acquired in SIM mode to reach high selectivity (Table 1).The average integral areas of the GC peaks were used to calculate C_{Phe} of the latex samples.

Table 1. Analysis conditions of GC–MS.

Apparatus	Conditions
Gas chromatograph	Column: HP-5MS (30 m × 0.25 mm × 0.25 μm)
	Carrier gas: He (99.999%) at 1.0 mL min^{-1}
	Temperature programme: 150 °C (2 min) at a rate of 10 °C min^{-1} to 280 °C (10 min)
	Injector mode and volume: splitless, 1 μL
Mass spectrometer	Ionisation mode: EI (70 eV)
	Ion source temperature: 250 °C
	MS interface temperature: 280°C
	Solvent delay: 7 min

3. Results and discussion

3.1. Glassy carbon morphology and Phe redox behaviour

GCE was treated with 8 μL of the modified liquid under infrared drying. Scanning electron microscopy was used to observe the microstructures of the modified GCE. After modification, GCE changes and graphene surface depositsto a certain extent (Figure 1).

The modified GCE was placed in 100 nmol L^{-1} Phe-acetonitrile-$NaClO_4$ under open circuit and characterised by CV and LSV analysis. No oxidative peak was observed in 0.25 mol L^{-1} electrolyte (Figure 2*A*); an oxidative peak appeared, which was caused by the electrooxidation of Phe on the electrodes (Figure 2*B*). The oxidative peak at 1.07 V (*vs.* Ag/AgCl) was due to graphene combined with GCE possessing a large specific surface area and outstanding sorption capacity (Figure 2*C*).

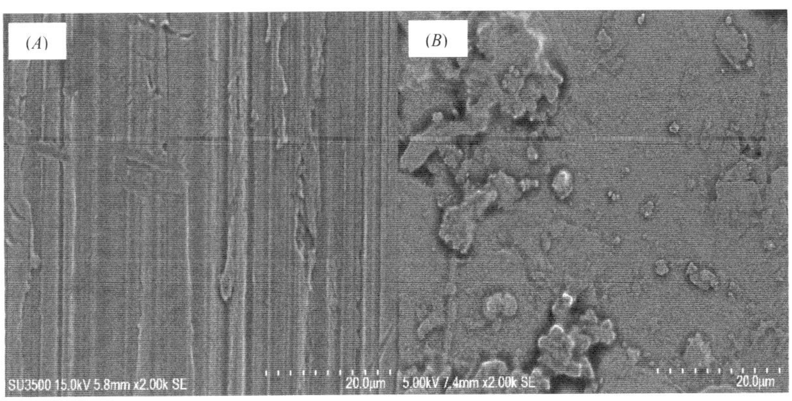

Figure 1. Surface microstructures of the bare GCE (*A*) and graphene-modified GCE (*B*).

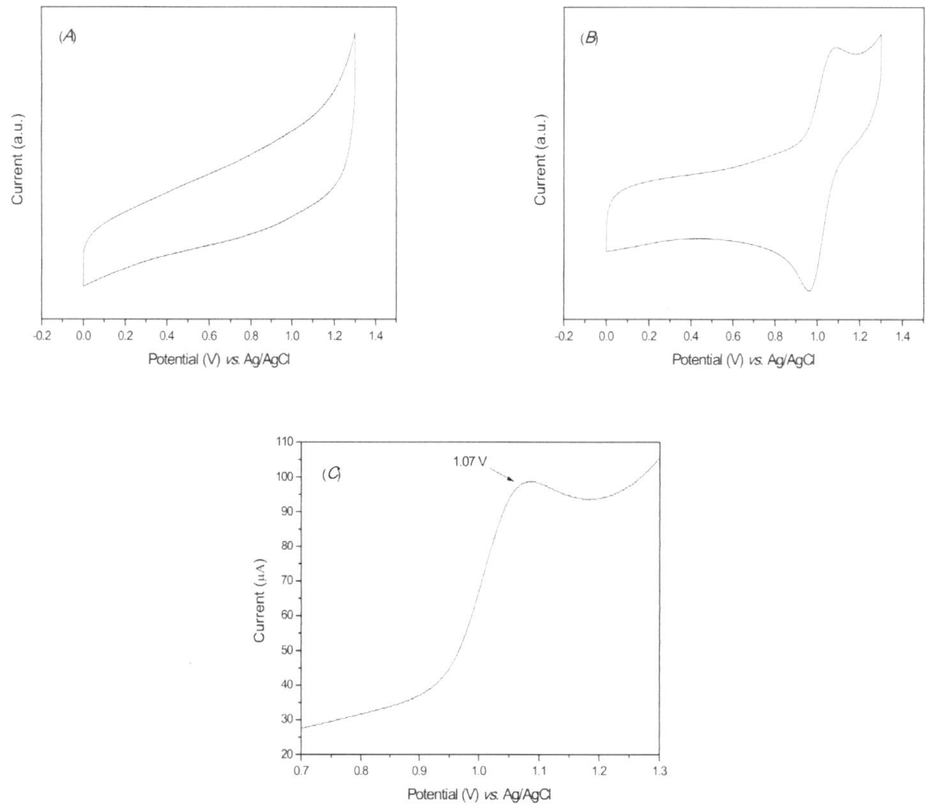

Figure 2. CV of 0.25 mol L^{-1} electrolyte (*A*) and 100 nmol L^{-1} Phe (*B*) on modified GCE within 0–1.3 V; LSV of 100 nmol L^{-1} Phe on modified GCE (*C*) within 0.7–1.3 V.

3.2. Effect of the modified liquid quantity and electrolyte concentration

The modified liquid quantity has a significant effect on the electronic transfer [17]. For this experiment, 3–11 μL graphene suspension was applied to the surface of the preactivated GCE. The *I* of Phe was measured by LSV. Within the volume ranging from 3 to 8 μL, the *I* continued to increase (Figure 3*A*). When the volume was 8 μL, the *I* reached a maximum. By increasing the volume from 8 to 11 μL, the *I* decreased gradually. This was because the modified film was too thick, which hindered the electron transfer.

The *I* under different electrolyte concentrations (0.1–0.4 mol L^{-1}) is shown in Figure 3*B*. In the beginning, the *I* increased significantly. When the electrolyte concentration was 0.25 mol L^{-1}, the *I* reached a maximum. The peak current tended to be stable, exhibiting a final equilibrium. Therefore, the concentration of $NaClO_4$-acetonitrile played an important role in the solution conductivity.

IOP Conf. Series: Earth and Environmental Science **865** (2021) 012024 doi:10.1088/1755-1315/865/1/012024

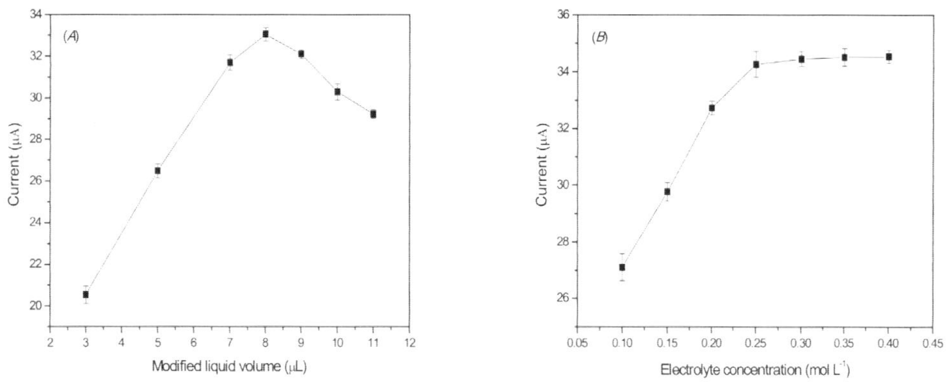

Figure 3. Effect of the modified liquid quantity (*A*) and electrolyte concentration (*B*) on *I* of the graphene-modified GCE.

3.3. EC assay of Phe in dark coloured latex

The modified electrodes were submerged in 1–250 nmol L^{-1} for 20 min and the standard solutions were detected. As shown in Figure 4, the relationship between *I* and C_{Phe} was determined by the equation: $I= 0.1066C_{Phe} + 19.04$, $R^2= 0.9975$. For the LSV analysis of Phe, the linear range was 1–120 nmol L^{-1} and the limit of detection was calculated to be 0.7 nmol $L^{-1}$for S/N = 3. Thus, this EC system has good sensitivity, which can be applied to Phe determination as a fast method.

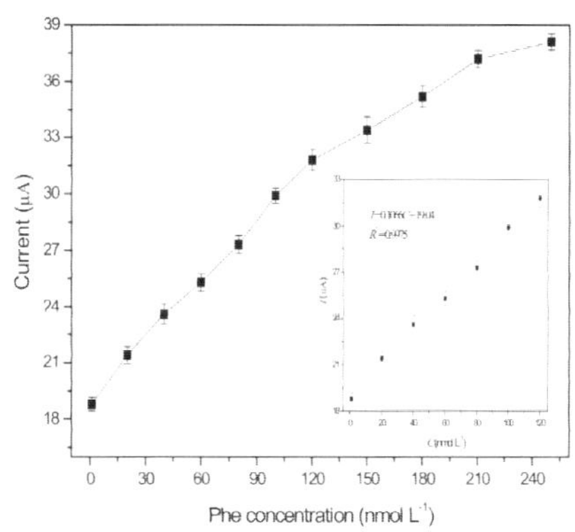

Figure 4. Currentresponse at different concentrations; embedded graph: linear relationship between *I* and C_{Phe} when the concentration ranges from 1 to 120 nmol L^{-1}.

Coloured latexes were pretreated according to Section 2.4 to get the extract. C_{Phe} of the specimens were characterised using EC and GC–MS. The spike-in specimens had the concentrations 0.2, 0.8, 1.4 μ mol L^{-1} in the preparation. The C_{Phe} of the latex was 1.40 μ mol L^{-1} by EC and 1.45 μ mol L^{-1} by GC–MS, of which the deviation was <3.6%. The mean recoveries with EC ranged from 94.3% to 106.4%, while the mean recoveries for the chromatographic method ranged from 98.2% to 105.8% (Table 2).

The relative standard deviations of EC and GC–MS were ≤5.1% and ≤4.5%. The reproducibility and accuracy of the glassy carbon were comparatively good in the field of EC, indicating that this system can be used for quantitative study of Phe in real samples.

Table 2. Mean recoveries of EC and GC–MS forPhe (n = 3).

Specimen	Added (μmol L^{-1})	Found (μmol L^{-1})		Recovery (%)		Relative standard deviation (%)	
		EC	GC–MS	EC	GC–MS	EC	GC–MS
Dark coloured latex	1.4	2.98	2.91	106.4	102.1	4.9	3.8
	0.8	2.33	2.38	105.9	105.8	5.1	4.1
	0.2	1.51	1.62	94.3	98.2	4.3	4.5

4. Conclusion

Electrooxidations of Phe-acetonitrile on the graphene-modified glassy carbon were investigated in this study and a low oxidative peak potential (1.07 V *vs*. Ag/AgCl) was achieved. The modified liquid quantity and electrolyte concentration affected the oxidative peak current. The mean recoveries of Phe in coloured latex by EC ranged from 94.3% to 106.4%, whose relative standard deviations were ≤5.1%. The recovery experiment indicated that the EC method possessed practical applicability to Phe analysis and the oxidising characteristics of Phe showed an excellent selectivity. Hence, the Phe determination in latexes was well developed, and these results can assist future investigations of polycyclic aromatic hydrocarbons.

Acknowledgement

This work was supported by the Zhejiang Supervision and Administration Bureau Research Fund of China (20200313).

References

[1] Li X, Ding L, Song L, Gao W, Wang L and Wang J 2020 Effects of exposure to polycyclic aromatic hydrocarbons combined with high-risk human papillomavirus infection on cervical intraepithelial neoplasia: A population study in Shanxi Province, China *International Journal of Cancer* **146** 2406

[2] Liu X F, Zhou K, Song S, Jiang X and Tian J J 2016 Polyps regeneration experiment forecast of three kinds of polycyclic aromatic hydrocarbons in water pollutant teratogenicity *Biological Chemical Engineering* **2** 52

[3] Hao Z, Wang C L, Feng Z Y and Zou X Q 2020 Review on polycyclic aromatic hydrocarbons and heavy metals in marine environment and organisms *Environmental Ecology* **2** 89

[4] ZhongN F, Yuan R H, He H H, Zhong W H and Wang Y 2021 Electrochemical sensor based on graphene-modified GCE for rapid quantification of benzo(a)pyrene in dark rubber materials *Journal of Physics: Conference Series* **1978**012020

[5] Alegbeleye O O, Opeolu B O and Jackson V A 2017 Polycyclic aromatic hydrocarbons: A critical review of environmental occurrence and bioremediation *Environmental Management* **60** 758

[6] Huang D, Xu B, Wu J Z and Xu J M 2019 Adsorption and desorption of phenanthrene by magnetic graphene nanomaterials from water: Roles of pH, heavy metal ions and natural organic matter *Chemical Engineering Journal* **368** 390

[7] Houshani M, Salehi-Lisar S Y, Motafakkerazad R and Movafeghi A 2019 Uptake and distribution of phenanthrene and pyrene in roots and shoots of maize (*Zea mays* L.) *Environmental Science and Pollution Research* **26** 9938

[8] Xu H, Song Y, Cang L and Zhou D 2020 Ion exchange membranes enhance the electrokinetic in situ chemical oxidation of PAH-contaminated soil *Journal of Hazardous Materials* **382** 121042

[9] Zhu H Q, Lou C Y, Liu F R, Jiang G J, Ye X Y and Yuan R H 2021 Nanosized S-doped TiO_2

with effective visible-light degradation of BTEX from wood-based panels *Key Engineering Materials* **880**139

[10] Santos P M, Campo L, Olgiati L, Polledri E, Sánchez M N and Fustinoni S 2020 Development of a method to profile 2- to 4-ring polycyclic aromatic hydrocarbons in saliva samples from smokers and non-smokers by headspace-solid-phase microextraction-gas chromatography-triple quadrupole tandem mass spectrometry *Journal of Chromatography B* **1152** 122273

[11] Luo K, Hochalter J B, Carmella S G and Hecht S S 2020 Quantitation of phenanthrene dihydrodiols in the urine of smokers and nonsmokers by gas chromatography-negative ion chemical ionization-tandem *Journal of Chromatography B* **1141** 122023

[12] Wei C X, Ren P, Cen Q L, Zhu Y X and Zhang Y 2019 Simultaneous determination of dissolved phenanthrene and its metabolites by derivative synchronous fluorescence spectrometry with double scans method in aqueous solution *Talanta* **195** 339

[13] Mateos R, Vera-López S, Saz M, Díez-Pascual A M and Andrés M P 2019 Graphene/sepiolite mixtures as dispersive solid-phase extraction sorbents for the anaysis of polycyclic aromatic hydrocarbons in wastewater using surfactant aqueous solutions for desorption *Journal of Chromatography A* **1596** 30

[14] Zyakun A M, Kochetkov V V, Zakharchenko V N, Baskunov B P, Anokhina T O and Boronin A M 2019 Application of high-performance liquid chromatography/high resolution mass spectrometry to the investigation of the biodegradation and transformation of phenanthrene by a plasmid bearing rhizosphere bacteria *Pseudomonas aureofaciens Journal of Analytical Chemistry* **74** 1355

[15] Yuan R H, Liu W H, Teng Y J, Nie J and Ma S Z 2015 Detection of ethoprophos using SERS coupled with magnetic Fe_3O_4/Ag composite materials *Spectroscopy and Spectral Analysis* **35** 1276

[16] Wang F, Jia Z, Su W, Shang Y and Wang Z L 2019 Adsorption of phenanthrene and 1-naphthol to graphene oxide and L-ascorbic-acid-reduced graphene oxide: Effects of pH and surfactants *Environmental Science and Pollution Research* **26** 11062

[17] Gan Z, Song N, Zhang H, Ma Z, Wang Y and Chen C 2020One-step electrofabrication of reduced graphene oxide/poly(n-methylthionine) composite film for high performance supercapacitors *Journal of The Electrochemical Society* **167** 085501

Case study on the landscape of Yuhang Entrance of Shanghai-Hangzhou expressway under the background of the park city

Yang Wang[1], Lingzhi Shen[1], Xiaomin Ye[1] and Yun Xie[2, 3]

[1]Power China Huadong Engineering Corporation Limited, Hangzhou 311122, Zhejiang, China;
[2]Jiyang college of Zhejiang A & F University, Zhuji 311800, Zhejiang, China

[3]Corresponding author's e-mail: 247446054@qq.com

Abstract. Constructing the "Park City" has become a new trend in urban construction and development. Under this background, as an essential part of the urban system, the expressway entrance plays a symbolic role in constructing park cities. In this case, building a park-style entrance of a city is the quickest way to exhibit the image of a park city and an important means to realize the concept of the park city. Yuhang Entrance of Shanghai-Hangzhou expressway is the main entrance in the north of Hangzhou city, where we have effectively and sustainably built an ecological compound welcoming park with strong identification and harmony between human and nature. Besides, the park at the entrance of the city blends in the life of citizens, thus human, the city, environment, and industry being harmoniously integrated. Therefore, it is shown that Yuhang Entrance has become a beautiful scenery line under the guidance of the concept of the "Park City", which can be a classic case and a valuable reference for other designers.

1. Introduction

The concept of the "Park City" was proposed by General Secretary Xi Jinping during his visit to Chengdu in February 2018 [1]. So far, the new direction of China's urban construction and development has been determined, and the urban construction has entered an advanced form under the guidance of the new development concept. It is the latest Chinese urban planning formation that inherits and improves from contemporary Garden City.

1.1. Brief on Garden City concept and history

Garden City concept was first developed by Ebenezer Howard in his book *Tomorrow: A peaceful path to realm reform* [2], where the novel combination of urban and rural space was devised. In 1925, the definition of Garden City was further distilled by Purdom as "a town designed for healthy living and industry; of a size that makes possible a full measure of social life, but not only larger; surrounded by a rural belt; the whole of the land being in public ownership or held in trust for the community" [3]. Later in the 1930s, Frederick Law Olmsted actively advocated a large area of park reservations for EastBay Cites development in United States. In 1946, Peter Bredsdorff and Sten Eiler Rasmuussenw developed a significant "Finger Plan", that later became the model of global ecological and sustainable development, for Copenhagen where a large area of natural and rural areas had formed high-quality natural parks and agricultural area, and bred diverse wildlife [4]. Further in 1963, Koenigsberger, an urban planning expert of the United Nations, proposed the "Ring Plan" concept of Singapore. A

IOP Conf. Series: Earth and Environmental Science **865** (2021) 012025 doi:10.1088/1755-1315/865/1/012025

complete pattern and island-wise ecosystem have been formed after 40 years of developments. By 1996, 13 scattered towns were connected by railway, and a large amount of land between the towns has been gradually developed into ecological corridors and diverse parks [5]. In 21st century, when ecologist Richar T. T. Forman carried out ecological planning in the Barcelona metropolitan area, he continued to utilize agricultural parks and natural parks as core tools to promote the harmonious development of humans and nature in urban areas [6]. In 2007, the "Science" magazine published the *Domesticated nature: shaping landscapes and ecosystems for human welfare*, which pointed out: "A more durable stewardship would manage tradeoffs among ecosystem services so that nature and people simultaneously thrive" [7].

1.2. The development of urban greening in China

In 20th century, "Shan-shui City", "Landscape Garden City", "Ecological Landscape Garden City" and "Park City" successively appeared in Chinese urban greening development [1]. The concept of Shan-shui city was put forward by Professor QIAN Xuesen in his letter to Professor WU Liangyong of Tsinghua University on 31st July, 1990, in which he said: Can we integrate China's Shan-shui poetry, Chinese classical landscape architecture and Shan-shui painting together to create the concept of 'Shan-shui city'? People left from the nature and would be back to the nature again. China can construct the urban area with Shan-shui idea [8]. As one of the unique spatial planning concepts in Chinese history, Shan-shui City combines the urban construction and the natural environment which is mainly composed of the mountains ("Shan") and the water ("Shui")[9]. It not only contains classical Chinese garden art and the philosophy of "Harmony between nature and people", remained also is the design concept or vision of integrating the natural environment and cities, which includes lots of aesthetics and human settlement ideas [10].

In 1992, the Chinese Ministry of Housing and Urban-Rural Development formulated the evaluation criteria for landscape garden cities to strengthen the construction of urban landscaping and create garden cities nationwide [1]; "garden city" solely inherited the form of "Garden City" and emphasized more on landscaping and greening beautification; In 2007, The Chinese Ministry of Housing and Urban-Rural Development initiated the development of ecological garden city nationwide, aiming to use ecological principles for afforestation, increasing biodiversity, and improving urban ecological functions on the basis of Garden Cities [1]. In addition to beautification, more ecological requirements were applied, where not only 30%-40% natural areas but also 60%-70% pseudo-natural urban ecosystems were required. In February 2018, General Secretary Xi proposed to build "Park City", which emphasizes more on the "Greening to the Doorway" on the basis of the former concept, and "serve all people". The conception of "Park City" is elaborated in Section 2.

1.3. The research related to expressways and the landscape of expressway entrance(city gateway)

In the post-World War II era, expressways were conceptualized and planned in many cities around the world, especially in China, where the total length of expressway ranks the first in the world. As a symbolic urban infrastructural component, the highway is capable to lead the modernization of cities, and meanwhile contribute to the urban scene. However, at the same time, the expressway was regarded as a force of destructive modernization that displaces communities, degrades the environment, facilitates suburbanization at the expense of the traditional urban environment [11]. Thus, previous researches focused on the ecological impact of expressway [12], yet seldom mentioned about how to seamlessly blend expressways with the surrounding context without creating additional incongruous landscape [13]. Also, few attentions have been paid to the landscape design of expressway entrance, except independent practical case studies [14-19]. Actually, as a tool for promoting global integration and urban gateway zones can attract significant amounts of capital accumulation, and lead the panoramic arrangements of cityscape [20]. In the meantime, the changing climate, degrading ecosystems and the loss of biodiversity, combined with rapid urbanisation and population growth, keep reminding us that new designs of retrofitting and living in cities are critically needed [21]. Therefore, as expressway entrance being a crucial part of urbanization, we advocate more elaborated and in-depth researches are needed for landscape designing of expressway entrance to be more integrated, systematic, multifunctional, identifiable and environmental-friendly.

1.4. The relationship between the entrance of the expressway and the park city

The entrance of the expressway is the gateway of the city, the window for displaying the image of the city, and an essential component of the urban space. To be specific, it not only takes on the role of space definition for material, energy and information exchange, but also presents the social and cultural information and the urban image of the city. The entrance is considered as the fastest window of exhibition, and it is also an extremely active juncture and important growth point in the green urban space system. As the first landscape exhibiting the city image, the expressway entrance plays an important part in both the urban system and the construction of the "Park City". In addition to that, as an indispensable part of the park city, the expressway entrance demonstrates the concept and image of the park city.

1.5. Background of the landscape construction of Yuhang Entrance

As both a famous historical and cultural city in China and a pivotal city in East China and the Yangtze River Delta, Hangzhou pays much attention to the landscape construction at the entrance of the expressway, and launched the entrance renovation project in April 2015 [22]. In 2018, as an important window of Hangzhou's external links, Yuhang Entrance of Shanghai-Hangzhou expressway was profoundly carried out the "Park City" concept, which has not only relieved the poor transportation, beautified the environment, clarified traffic signs and achieved the eight goals of "Freedom, Cleanness, Greenness, Beauty, Brightness, Clearness, Culture, and Excellence" [22], but also built a welcoming park which welcomes incoming people and creates a sense of "the City in the Park". As for the welcoming park, it is of an ecological composite type that highlights the environmentally friendly image of Hangzhou Shan-shui city with the composition of public leisure, cultural and commercial space, and shows the beauty of the cultural landscape and natural ecology. Furthermore, it becomes the benchmark and a new highlight for park city constructions and has great reference value.

2. The connotation and application of the "park city" concept

2.1. Cognition concerning the form of the park city

After the concept of the "Park City" was proposed, domestic scholars have made a lot of discussions about the "Park City". Besides, it is generally believed that the "Park City" is an upgraded version of the "Garden City"[23] and is also the combination of dwelling environment, urban infrastructure construction, urban space and the urban ecosystem. The park-style urban and rural ecological pattern enables citizens to live, work, transport and recreate in the park city. Apart from that, as the city is presented in the form of the park, the city is built in the park and blended into the park. Thus, the city is the park, and vice versa [24-27], or "urbanize in the park; reside in the park" [28]. Therefore, the "Park City" is a full play of the concept of the landscaping city and Chinese garden culture [25], and also an innovated concept of both urban-rural dwelling environment construction and an ideal city construction pattern [29].

2.2. Ecosystem cognition in park cities

The park city is a complex urban ecosystem [30, 31] that possesses rich public wealth, profound historical culture, and good ecological environment. Apart from that, it is also an ecological complex public space built upon the harmonious fusion of people, cities, environment and industry [32]. The "Park City" (1) emphasizes ecology, sustainability and human-nature harmony, (2) plays a role as an sponge city that enhances the tenacity and water-saving and energy-saving effect [33], (3) reserves more urban public leisure, cultural and commercial space while at the same time attaches great importance to the ecological restoration and low carbon design, ecological corridor [34], (4) creates a new growth pole, (5) builds the urban ecological formation, and realizes intensive, efficient and sustainable urban development [1]. Based on the concept of the "Park City", the construction of the expressway entrance should gather wisdom of various industries, break the traditional ideas and methods, maximize the best ecological and social benefits, and build the expressway entrance into a beautiful park with great identification and harmonious coexistence of man and nature.

2.3. Urban construction practice of park city theory

At present, many cities have carried out the practice of park city construction. For example, in 2018, Tianfu District of Chengdu became the first experimental city in China to carry out the concept of "Park City". Chengdu has made an urban plan for a beautiful and livable park with a "whole-region park system", and determined the "ecological pattern of two mountains, two networks, two rings and six sections", as well as the ideas of the "three-level ecological corridor" and the "green traffic system"[35]. Hangzhou designed the urban spatial pattern of "one main city, three deputy cities, six groups, and six ecological belts"[24], and has devoted to creating a city of "green every 300 meters and garden every 500 meters". In addition, Nanning, Guilin, Yangzhou, Zibo, Taiyuan, Shenzhen, Yangquan, Chenzhou and other cities are all putting effort to explore the construction of park cities.

2.4. Application of park city theory to landscape engineering

The concept of the "park city" has been widely adopted in the landscape engineering of waterfronts [36], parks [37], communities [38], roads [39] and other projects, but there is no record concerning the application in the landscape construction of high-speed entrances. Therefore, for the first time, the concept of the park city was integrated into the landscape construction of Yuhang Entrance of Shanghai-Hangzhou Expressway, which has become a classic case of park city landscape.

3. Landscape construction concerning Yuhang Entrance of Shanghai-Hangzhou Expressway based on the "park City"

3.1. Landscape exploration of Yuhang Entrance based on the park city

3.1.1. Design concept. Hangzhou advocates natural landscape. As the window and bond of Hangzhou connecting Shanghai, Yuhang Entrance of Shanghai-Hangzhou Expressway defines itself as an ecological complex welcoming park. With the appearance of a park, it exhibits the form of the city and a picture of the park city with both humanistic landscape and natural ecological beauty. Therefore, people can appreciate "the grace of the forest in the prosperous city" and the highly enjoyable image of the "Park City" – city in the park.

According to the concept of the "City Park", the four elements of "Mountains, Water, Cities and Human" were fused (Figure 1). With the construction requirement of "Global Beauty", "Eight Entries, Eight Lines" in Yuhang District, and the design concept of the "Green Gateway, Urban Forest", Yuhang Entrance was constructed into parks using the "Half Image, Half Park" method, which means not only was the image of the city displayed, but also the multi-functional ecological park complex was built (Figure 2). Taking serving people as the core objective and nature as foundation, and implementing the concept of "Sponge City", ecological remediation and low carbon design, Yuhang Entrance was also built into the starting point of the city's ecological corridor and the junction of the urban-rural ecological picture with further increasing spaces for leisure, social life, activities, fitness, culture and commerce. Yuhang Entrance has become the identifiable, human-nature harmonious scenery gateway, and carrying out the organic unity of ecology and the urban form, intensification, efficiency and sustainability of the city have been realized [15].

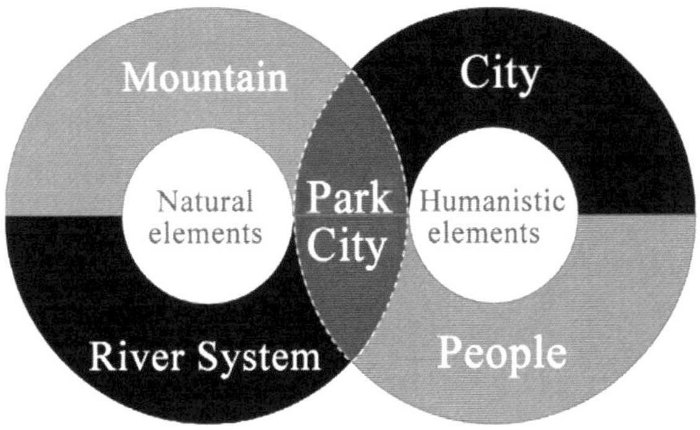

Figure 1. The relationship between the park city and the entrance of the expressway integrating "mountains, water, the city and human".

Figure 2. The definition of the entrance park which is "Half Images and Half Parks".

3.1.2. Design objectives and strategies. Yuhang Entrance of Shanghai-Hangzhou Expressway took the multi-functional ecological complex park as the main form, when its goal was to build an ecological complex park that highlights the "Shan-shui City" image of Hangzhou, featured with high recognition, public participation, low carbon design, and natural science unification. Yuhang Entrance Park integrated the ecological resources, namely mountains, forest, water and lakes, fused urban design strategy of sponges, exhibited the objective of "atmosphere, livability, culture, intelligence and leisure" in Yuhang District and the ecological pattern of "introducing mountains into the landscape, water diversion into the city", showed the temperament, culture and characteristics of Linping city, and created the scenery line that possessed the beauty of Hangzhou's humanistic landscape and natural ecology.

The main design strategies included:

(1) Integrity: connected insides and outsides, emphasized the entirety, reasonably arranged spatial levels and spatial sequences, and paid attention to the collocation of foreground, medium and distant perspectives [40] (Figure 3).

(2) Characteristic: developed the distinct theme, and showed the city culture, city image and city characteristics.

(3) Ecological aspect: took greening as the main feature, developed with low-impact, integrated function of the sponge city, constructed underwater forest, constructed, maintained and utilized existing vegetation, enriched vegetation from the perspective of biological habitat, and maximized the retention and reuse.

(4) Humanistic aspect: with the people-oriented objective and overall consideration, created more leisure, communication, activity, fitness, culture and commerce spaces.

(5) Compound: achieves multi-functional landscape, sustainable development, innovation consciousness, operability, ecological education and display, fixed-point display and integration of activities, and stimulates the environmental awareness of the public.

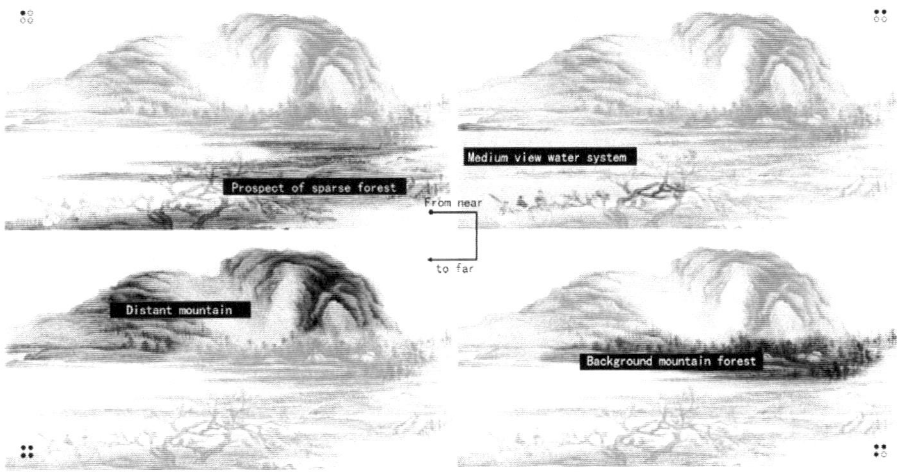

Figure 3. Reasonable arrangement of spatial hierarchy and spatial sequences, when paying attention to the collocation of foreground, middle-ground and background scenes.

3.2. *Design practice*

3.2.1. Analysis of driving sight: image exhibition and sight range analysis. Spatial scenery provides an optimal ornamental surface or viewing angle for riders, when the best view is related to speed and view distance. Besides, pre-arranged view distance and view spots offer the best viewing effect to visitors. In the landscape architecture, when the car speed is 40km/h, the appropriate viewing distance is 150m [41]. Therefore, the gentle slope was designed in the front of the best viewing area, leaving more grassland space and providing a reasonable viewing distance for the landscape nodes and the overall environment.

3.2.2. Layer development: landscape integrity and spatial layer. Through different layers of sensory experience, three spaces were formed in Yuhang Entrance space design: "Identification– City Image"; "Welcome–Yuhang Impression"; "Entering City– Poetic New Town". Besides, different landscape was also created through sequentially changed layers.

The rhythm of spatial level was progressively and cleverly organized (Figure 4). Specifically, the three landscape sequences made full use of urban background, topography and vegetation collocation, created large-scale plant levels, formed foreground, middle-ground, background three dimensional spatial layers, sorted the existing phytoecommunity, reshaped the terrain and the relationship between plant community, built a total of three spatial layers with lawn in the foreground, scarce forest, grassland and the belt of flowers in the middle, forest and urban constructions in the background, and enriched and improved crown canopy lines and forest edge lines. Thus, a harmonious relationship and mutual interlacing with the urban crown canopy line were formed, thus highlighting the unique charm of Linping New Town as a "Shan-shui City".

3.2.3. Terrain treatment-enhance mountains and water. (1) On the basis of retaining a large background forest belt, two separate lands were combined into a whole by taking advantage of the current terrain of the site and making gentle slope grassland through micro-topographic treatment, which simultaneously solved the drainage problem inside the site.

(2) To ensure the optimal sight line at Yuhang Entrance, the current terrain was raised as a whole, which also provides space for the foreground.

(3) Taking advantage of the depression, trimming mountains and water, following the natural law of "Mountains Have Veins; Water Has Sources", the urban landscape space framework of "light smoke, flowing water and quiet city scenery" was constructed with enriched spatial layers (Figure 5).

Figure 4. Get straight to the point-step by step, and organize a hierarchical rhythm.

Figure 5. Terrain treatment-the elevation analysis chart of mountains and water.

3.2.4. Symbol: The expression of regional characteristics. The symbol is an important part of urban characteristics and images [42]. In the design of Yuhang Entrance, the landmark features were mainly reflected in the entrance square and the theme sculpture.

(1) Entrance square: a magnificent and elegant vegetative landscape was created through the specific design of vegetation and symbols. Between two forest belts, a corridor of sight lines was built to enhance the deep feeling of landscape. Besides, welcoming atmosphere was created with broad and grand vegetative space and ethereal in-season flowers.

(2) Entrance theme sculpture: the large-scale green sculpture with the themes of "Millennium Yuhang" and "Fashion Yuhang" artistically presented the beautiful face of Yuhang, and created a warm, joyful and grand welcoming atmosphere.

3.2.5. Beautiful color: The expression of regional characteristics. Effective color application has strong visual impacts of attracting attention and leaving a deep impression. In the vegetation space of Yuhang Entrance, flower belts were creatively laid out, and grand theme sculpture with rich characteristics was also introduced. Consecutive and characteristic colorscape was then formed, thus creating impassioned welcoming atmosphere. Apart from that, the portion of flowering tree species with colorful leaves was deliberately increased, and therefore, there is mass application of "ginkgo woods", "sakura woods", "sequoia woods", "Purple leaf plum woods" and "Malus and Chaenomeles woods". In this case, the infinite charm of natural seasonal change was unreservedly shown.

3.2.6. Plants: Four seasons and four levels. Adapting to the terrain and landforms, trees were widely planted. Therefore, people can enjoy the blooming flowers in spring, the cool shades of the trees in summer, the sunglow diluted by the mist in autumn, and the sunset through sparse forest in winter. Furthermore, it should be noted that plant configuration was divided into four levels: evergreen background forest, the forest belt with various leaves and flowers, lawn space with the plant community, and the characteristic flower belt with marking landscape.

3.2.7. Improve facilities-people-oriented. In accordance with the people-oriented principle and the so-called "City Face, Garden Heart" idea, and with ensuring transportation safety, park ancillary facilities were enhanced. Moreover, the organization of park roads was improved. Especially near the north side of the public buildings, a place was provided for people to wander. Thus, Yuhang Entrance has become a real "Park City" with recreation function.

3.3. New technology application and its innovation and advantage

3.3.1. Underwater forest construction. Yuhang Entrance was featured with a design that combines with the original low depression terrain for terrain reconstruction, forming a lake area of 8500 square meters. In addition to that, beautiful scenery was lined along the shore, and underwater overflows with vigour. Underwater forest construction technology and techniques including the water detection and treatment, soil improvement, and ecology-beneficial mineral were introduced. By planting the submerged plants such as Vallisneria spiralis and then putting in fish and benthic organisms, the underwater ecological balance system was constructed, and the water body was clear and transparent all the way to the bottom throughout the years.

3.3.2. Sponge city technology application. The project made use of the regulating and storing function of the landscape lake to effectively solve the problem of rainwater runoff discharge on the site, relieve the pressure of the surrounding municipal network, and eliminate the phenomenon of serious water accumulation on the road at Wangmei intersection. Besides, the regional "green sponge" will be established to recycle and utilize water resources for greening and watering of the whole garden, so that rainwater can fulfil a comprehensive service function (Figure 6, 7).

Figure 6. Analysis on Sponge City.

Figure 7. Technology of Sponge City.

3.3.3. Compound function and strategy. In this design, the best conjunction point of urban image display and function utilization is put forward, when following the intensive and efficient land usage, revisiting the humanistic care, constructing the pluralist activity space, blending the park into the citizen's life, promoting the industry integration with some small commercial spaces, and realizing the maximization of garden comprehensive benefit.

4. Conclusion

4.1. The necessity of the application of the concept of park city at the expressway entrance

Xi Jinping's call for "park at the doorstep" - the park city Coincides with Kaplan's call for "nature at the doorstep" three decades ago, which were supported by strong evidence for a positive association between the size of green space around people's living area and perceived mental health and mortality due to all causes in general adult populations. There are links between contemporary greening practice and study on related health benefits, wherein proximal greening emerges as a distinct form, and possible norm, for twenty-first-century urban design [43]. Park City is good for people's health, also with the parks at the expressway entrance, which provides repetitive and long-term exposures for people to greenery that can provide cumulative benefits through "micro-restorative opportunities." Also, Parks at expressway entrance raise people's awareness to greening and health.

4.2. The success of the entrance park

Yuhang Entrance project adapted to the "Park City" concept well, and at the same time, a "City Face, Garden Heart" feature of the welcoming park was obtained. After the completion of the project, the four elements-the people, the city, the environment and the industry-were integrated into a harmonious communion. Considering that, the first impression of Hangzhou was directly exhibited in the form of parks at Yuhang Entrance, as well as efficient, intensive and sustainable construction of an ecological complex space (Figure 8 and Figure 9). Yuhang Entrance also applied low carbon design such as the sponge city and underwater forest. While building the ecological corridor, the city also provided people with living, work, transportation, recreation, culture and entertainment space, and created an ideal environment of "the city in a garden", known as "Building a Shan-shui City; Enjoying All-season Landscape". Thus, it can be seen that the most vivid and poetic picture of the city's dwelling was exhibited. Then, it was concluded that Yuhang Entrance was built into an identifiable, human-nature harmonious park and became a classic case of the expressway entrance landscape project that follows the concept of the "Park City".

Figure 8. The Entrance Park of "City Face, Garden Heart"-a space of ecological complex.

Figure 9. A close-up view of the Entrance Park of "City Face, Garden Heart".

4.3. The deficiencies of the entrance park and future study

Although the attempt to build a park at the Yuhang Entrance provide a "natural opening" for people, but the cost of construction and daily maintenance is high. The artificial traces are heavy and undesirable. How to Increase biodiversity, and to be more ecological and sustainable are the next problems to be solved. Agricultural parks and agricultural landscape is potentially one of the means to solve these problems, promoting the industry integration with agricultural production space and experiential commercial space, and "a more durable stewardship would manage tradeoffs among ecosystem services so that nature and people simultaneously thrive" [7].

References:

[1] Shi Cheng and Yuning Cheng 2018 Form Garden City to Park City Design–The Dialectic Relationship between Urban Ecology and Morphology *Chinese Landscape Architecture* **34** 41
[2] Howard, Ebenezer 1898 Tomorrow: A Peaceful Path to Real Reform London: Swann Sonnenschein
[3] Purdom, CB 1925 *The building of satellite town* London: J.M. Dent & Sons
[4] Vejre H, Primdahl J and Brandt J 2007 The Copenhagen Finger Plan: keeping a green space structure by a simple planning metaphor Europes living landscapes Chapter 19 310
[5] Abrams, C., Kobe, S. and Koenigsberger, O. 1963 Growth and urban renewal in Singapore: Part I-Action planning Editorial introduction *Habitat International* **5** 63
[6] Forman R T T 2008 Urban regions: ecology and planning beyond the city *Cambridge: Cambridge University Press*
[7] Kareiva P, Watts S, McDonald R, et al 2007 Domesticated nature: shaping landscapes and

ecosystems for human welfar *Science* **316** 1866
[8] Renwei Wu and Xi-e Fu 2009 On the Origin and Meaning of "Shan-shui City" *Chinese Landscape Architecture* **06** 39
[9] Chen Yulin 2010 Shan-shui City: A Chinese Spatial Planning Tradition *46th ISOCARP Congress*
[10] Fan Fu, Hong Li and Caijun Zhao 2020 From Shan-shui City To Park City-The Way of City Development in China *Chinese Landscape Architecture* **36** 12
[11] Neta Feniger and Roy Kozlovsky 2020 Expressway urbanism: highway planning and the reimagining of Tel Aviv-Jaffa *Planning Perspectives* **36** 259
[12] Jiayin Gao, Mingfei Zhang, Zhaoguang Hu, et al 2020 Influence of Expressway Construction on the Ecological Environment and the Corresponding Treatment Measures: A Case Study of Changyu (Changchun-Fuyu Lalin River) Expressway, China *Nature Environment & Pollution Technology* **19** 1195
[13] Arun George, Anita Sinha and Varkki George Pallathucheril 2005 The Goshree Road Link, Kochi, India: A Design Approach towards Creating a Legible Streetscape *Journal of Urban Design* **10** 275
[14] Haiyan Dong, Minli Jin and Jianyan Wang 2011 The Reflection of the Forest Artistic Conception in the Design of Hangzhou-Qiandao Lake Expressway Entrance Area Shelter Forest *Jour of Fujian Forestry Sci and Tech* **38** 144
[15] Rui Zhang, Jun yan and Cuiping Lu 2017 The Research on the Landscape Design Based on the Regional Culture of City Entrance-A Case Study of MaAnShan CaiShiJi High-Speed Entrance *Natural Sciences Journal of Harbin Normal University* **33** 119
[16] Shuming Duan and Jing Mao 2021 Practice of "Road to Nature" in Landscape Design of Expressway Interchange *The Chinese and foreign road* **41** 5
[17] Juan Yu 2017 Landscape Planning and Design of Expressway Entrance-Taking Wucheng Section (Yingbin Avenue-Chisong Town Section) as an Example *China Housing Facilities* **3** 32
[18] Ruiqi Li and FanYang 2015 Landscape Planning and Design of Expressway Entrance of Danfu, Shenyang *Beautiful Garden* **S1** 146
[19] Jincheng Chen 2019 Research on Landscape Design of Expressway Entrance Based on Regional Culture-A Case Study of Landscape Design of Expressway Entrance Square in Lianhuo, Kaifeng *Modern Horticulture* **8** 58
[20] Olguc and Alis, Kan 2010 Urban Gateway: Just a Symbol, or More? (Reappraising an Old Idea in the Case of Ankara) *Journal of Urban Design* **15** 91
[21] Maibritt Pedersen Zari 2018 Regenerative urban design and ecosystem biomimicry (vol 11) *Routledge*
[22] Qiang Yan 2016 The changed city–Record of comprehensive rectification of Hangzhou entrances welcoming G20 *Summit Hangzhou (Weekly)* **09** 42
[23] Xiucheng Liu 2018 Establishing Confidence on Landscape Architecture and Building Park City *Chinese Landscape Architecture* **34** 7
[24] Xinyue Wu and Xueyan Li 2020 Study on Park City Design Strategy Bassed on Hangzhou *Art Education Research* **08** 83
[25] Xiong Li and Yunlu Zhang 2018 A New Approach in Urban Green Development for the New Era–Strategies for Building Park Cities *Chinese Landscape Architecture* **34** 38
[26] Jia Liu 2020 The application of low carbon design in the concept of park city *Design* **7** 45
[27] Jinlu Li 2018 Discussion "Park City" under the Background of the New Era *Chinese Landscape Architecture* **34** 26
[28] Hao Wang 2018 Natural landscape Locates Cities in Parks, Artificial Landscape Builds Parks in Cities–Discussion on the Planning and Construction of Park City *Chinese Landscape Architecture* **34** 16
[29] Yan Wu, Zhongjie Wang, Chenyang Su, etc 2018 Discussion on the Concept & Practice

Approach of "Park City" *Chinese Landscape Architecture* **34** 30
[30] Yungui Shi and Xiaojun Liu 2019 Green Governance: A Rational Path to Park City *Journal of Sichuan University (Philosophy and Social Sciences Edition)* **3** 38
[31] Yang Liu, Xiaorui Zhang and Qizhi Zhang 2020 Research Status and Future Prospects of Park City *Journal of Hunan City University (Natural Science)* **29** 44
[32] Qi Zhang 2020 City Design under the Thinking of Park City and Landscape ecology *Intelligent City* **19** 29
[33] Shi Cheng and Yuning Cheng 2020 On Ecology and Survival Wisdom-Also on the Ecological Wisdom of Sponge City *Chinese Landscape Architecture* **36** 13
[34] Binyi Liu 2018 Methodology of Park City Research and Construction *Chinese Landscape Architecture* **34** 10
[35] Jiuli Zeng, Peng Tang and Geng Peng 2020 Exploration and Practice of Planning and Building Park City in Chengdu City Planning *Review* **08** 112
[36] Qi Zhang 2019 Urban waterfront space design under the concept of park city *Sichuan Construction* **6** 37
[37] Lina Zhai 2020 Urban mountain park reconstruction design under the background of park city-A case of the landscape reconstruction of Longcheng Park in Longgang District, Shenzhen *Business & Luxury Travel* **8** 57
[38] Qinyu Zhang, Yuan Liu and Pingping Li 2020 Micro-update of Community Landscape based on the Concept of Park City-Taking Nanjing Fuhougang Community as an Example *Landscape Architecture* **6** 24
[39] Yikang Zhang 2019 Thoughts on Road Landscape Renovation Design Under the Background of "Park City"-A Case of Shennan Blvd *Urbanism and Architecture* **15** 143
[40] Junling Zhang 2012 The artistic principles of landscape architecture China Forestry Publishing House
[41] Xiansheng Xu and Li Feng 2017 *Preliminary landscape design* China Forestry Publishing House
[42] Kevin Lynch 1960 *The Image of the City* Mit Press
[43] Rachel Kaplan 1985 Nature at the Doorstep: Residential Satisfaction and the Nearby Environment *Journal of Architectural and Planning Research* **2** 115-27

IOP Conf. Series: Earth and Environmental Science **865** (2021) 012026 doi:10.1088/1755-1315/865/1/012026

Analysis of the stability of Zhangjiapo landslide

Qian Hou[1], Lixin Liu[2], Jinxin Chong[1] and Xueping Li[1, 3]

[1]China University of Geosciences, Wuhan, China;
[2]China Construction Industrial & Energy Engineering Group Co., Ltd., Nanjing, China

[3]Corresponding author's e-mail: 1201920245@cug.edu.cn

Abstract. The Zhangjiapo landslide in Xinling Town; Badong County has obvious deformation and a great hidden danger of safety in the past two years. In order to better treat the Zhangjiapo landslide, I quantify the stability of Zhangjiapo landslide using transfer coefficient method. And this method is compared and verified with the method of strength reduction stability analysis after using Rhino to establish geological model into FLAC3D. The results show that the stability coefficient is basically the same, and the Zhangjiapo landslide is extremely unstable under rainstorm condition. It proves the reliability of the paper's numerical simulation method in landslide research and provides a reference for the analysis of the landslide. According to the engineering geological characteristics of Zhangjiapo landslide and the influence of deformation area on landslide body, combined with the present situation of landslide, the comprehensive treatment scheme of landslide body "anti-slide pile+reinforcement of landslide front+drain" is put forward from the point of view of protecting environment.

1. Introduction

Landslide is a serious geological hazard. People pay more and more attention to the casualties and economic losses caused by landslide. The landslide disaster in the Three Gorges Reservoir area has the characteristics of wide distribution, high frequency, heavy disaster and strong sudden occurrence [1-3]. Zhangjiapo landslide is located in 12 groups of Loess Slope Community in Xinling Town, Badong County, Three Gorges Reservoir area. In the past two years, the landslide has obvious deformation and further deterioration trend. Once the disaster is caused, it will seriously affect the safety of the existing people's lives and property, and limit the development around the city line. Effective prevention and control work is needed [4].

The stability coefficient of landslide can be obtained by engineering geological survey and quantitative analysis of landslide stability, but it is not accurate enough to determine landslide stability from only one aspect [5-9]. Therefore, the present situation of landslide is simulated by numerical simulation method, and the stability coefficient is obtained. The accuracy of landslide stability coefficient can be verified by comparing the two methods [10, 11].

Taking Zhangjiapo landslide in Xinling Town, Badong County as the research object, the three-dimensional geological model of landslide is established. Based on finite difference software FLAC 3D, the deformation characteristics and stability coefficient of landslide in natural and rainstorm state are studied, and the results of transfer coefficient method are compared and verified with the commonly used transfer coefficient method in engineering, which provides the basis for engineering analysis and prevention [12-14].

2. Geological conditions of landslides

Zhangjiapo landslide is located in the middle and upper part of the fan-shaped slope in Xincheng District of Badong County. Elevation 410-600 m for slope distribution. The geographical location is shown in Figure 1.

Figure 1. Location of Zhangjiapo landslide.

The outcrop strata in Badong County are mainly Lower Triassic Jialingjiang Formation, Middle Badong Formation and Quaternary.Because of the construction of 209 national highway around the city line, the original geomorphology is changed greatly, the stability of slope body is reduced, and landslide deformation and failure are easy to occur.

3. Quantitative analysis of landslide stability

3.1. Transfer coefficient method

The transfer coefficient method is simple and easy to grasp. The influence of rainfall on landslide stability can be considered, and it can be used as the preferred method for stability analysis in engineering practice. In this paper, the safety factor is calculated according to the general formula of transfer coefficient method. The problem of landslide stability is regarded as plane strain problem, the sliding force acts on the sliding surface by the shearing strength τ parallel to the sliding surface and the normal stress σ perpendicular to the sliding surface, the landslide is regarded as ideal rigid plastic material. Figure 2, 3 show calculated profiles 2-2' and 3-3' selected according to drilling data. Profile 1-1' is similar to profile 2-2'.

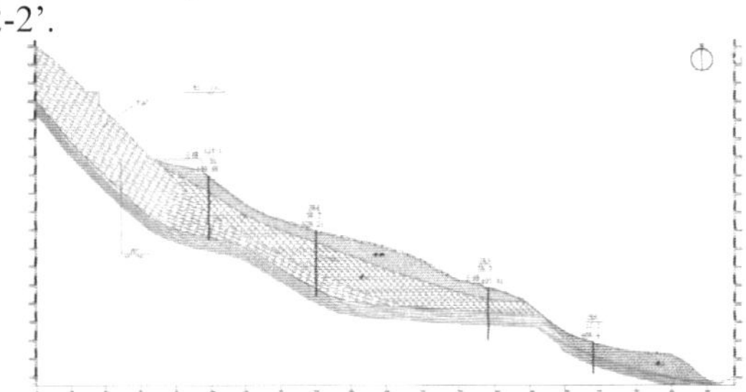

Figure 2. Engineering geological profile 2-2'.

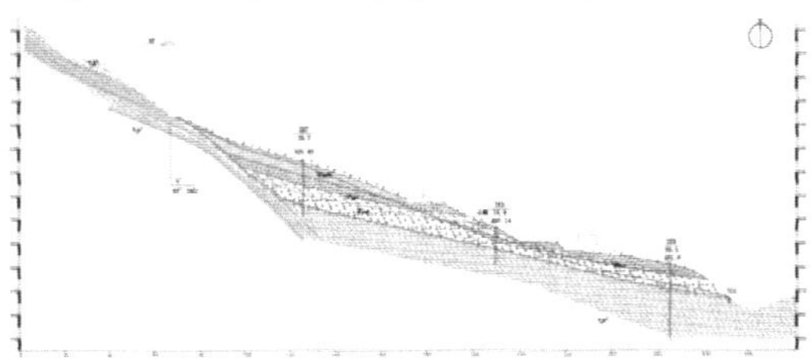

Figure 3. Engineering geological profile 3-3'.

3.2. Calculation parameters
According to the test index, inversion parameters and engineering analogy, the shear strength index of Zhangjiapo landslide stability calculation is determined. The sliding surface C =20 kPa, φ_C =12 in saturated state and 25 kPa, φ_C =15 in natural state. According to the drilling data, the natural conditions in the 1-1',2-2'sections are no water, and the natural groundwater level in the 3-3'section is very high.

3.3. Calculation conditions
The sliding zone and lower voluminous rock of Zhangjiapo landslide are relative water barrier, the material of landslide is broken, the surface rainfall is easy to infiltrate, and the groundwater level in landslide will rise soon after rainfall infiltration. The landslide is a non-wading landslide, and the seismic load is not considered. The calculation conditions are divided into the following two types, as shown in Table 1.

Table 1. Stability calculation load combination.

Working condition combination number	Content of load combination
I	Dead load+Surface load+Current water level
II	Dead load+Surface load+Heavy rain levels in 20 years

3.4. Computing result
According to the Technical Requirements of Geological Disaster Prevention and Control Engineering Survey in the Three Gorges Reservoir Area, the stability state is determined according to the stability coefficient of landslide. As shown in Table 2.

Table 2. Classification of landslide stability.

Stability coefficient of landslide	Fs<1.00	1.00<Fs≤1.05	1.05<Fs≤1.10	Fs≥1.10
stable state	unstabitily	understable	basic stability	stabilize

The results of the transfer coefficient method are shown in Table 3. For profile 1-1', 2-2', because there is no groundwater in the natural state or the groundwater level is low, so the rainstorm state has a great influence on its stability coefficient;

Table 3. Results of calculation of stability coefficient by transfer coefficient method.

Profile \ Working condition	I	II
1-1	1.085	0.864
2-2	1.028	0.819
3-3	1.002	0.99

4. Numerical simulation of landslide stability

4.1. Modeling process
Rhino is professional 3D modeling software with high graphics accuracy, it can input and output dozens of file formats. Griddle is Rhino plug-in for generating advanced meshes, which can accurately generate and divide surfaces into different types and sizes of surface meshes in Rhino. Flac3D is a powerful discrete element numerical simulation software, but limited by complex geological modeling. Through Griddle, the 3D modeling function of Rhino can be combined with the numerical calculation

and analysis of FLAC3D. The technical route of this paper is shown in Figure 4. The rhino modeling effect is shown in Figure 5.

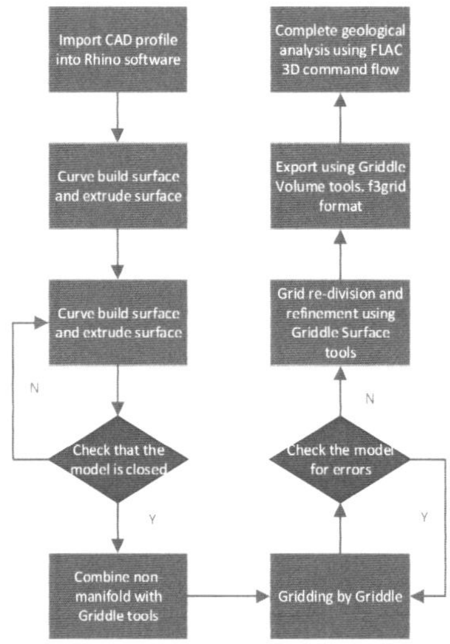

Figure 4. Model pre-processing flow diagram.

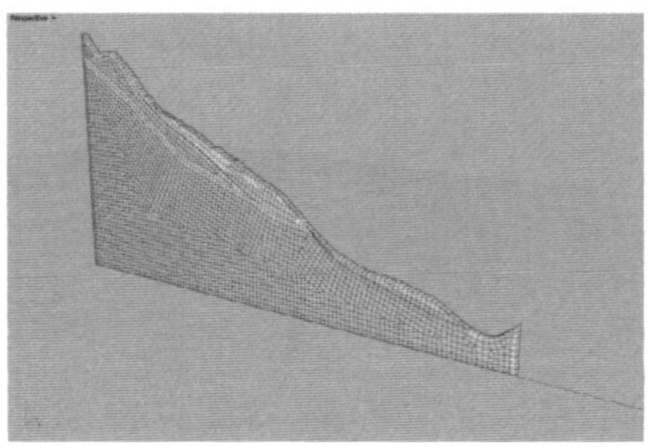

Figure 5. Profile grid in Rhino.

4.2. Building models

4.2.1. Basic principle of strength reduction method. The strength reduction method defines the slope safety factor as the reduction of the strength parameters when the slope reaches the critical failure state.When the slope adopts the Mohr-Coulomb criterion, the strength parameters affecting its stability are c and ϕ. Divide the original cohesion and internal friction angle of the slope body by reduction coefficient K, then numerical analysis. By increasing K, repeated analysis until the slope reaches critical failure state. Suppose the cohesion and angle of internal friction are c^{cr} and ϕ^{cr} at this time, because the slope is critical, the corresponding safety factor is $K^{cr}=1$, the original safety factor corresponding to the slope can be obtained as:

$$F=\frac{K}{K^{cr}}=K=\frac{c^{0}}{c^{cr}}=\frac{\tan\phi^{0}}{\tan\phi^{cr}}$$

In this paper, we assume that all non-empty regions of the numerical model adopt the Mohr-Coulomb constitutive model, the command Solve fos is used to solve the safety factor in the FLAC3D.

IOP Conf. Series: Earth and Environmental Science **865** (2021) 012026 doi:10.1088/1755-1315/865/1/012026

4.2.2. Definition of coordinate system and model scope. The model boundary condition is one-way constraint: along both sides of the aspect is a displacement constraint in the X direction, the bottom of the model is the displacement constraint in the model, the both boundaries perpendicular to the profile are Y direction displacement constraints. Calculation model range: profile 1-1' X axis direction 333 meters, Z axis direction 160 meters; profile 2-2' X axis direction 395m, Z axis direction 193 meters; profile 3-3' X axis direction 340 m, Z axis direction 150 meters. Y axis direction thickness of all profiles is 4 m. The grid consists of 4*4*4 hexahedron.

4.2.3. Determination of calculation parameters. The strata are grouped into L1~L6, as shown in Table 4 and assigned values in the Flac3D according to the stratigraphic characteristics.

Table 4. Parameters of rock and soil.

Stratigraphic grouping	Name of soil layer	Density (g/cm^3)	Cohesive strength /kPa	Internal friction angle /(°)
L1	Surface tillage or artificial fill	1850	5	15
L2	Silty clay or silty clay with gravel	1900	23	14
L3	Gravel soil or broken stone	2250	2	36
L4	Block stone	2300	3	38
L5	Argillaceous siltstone	2100	4	34
L6	Marl, limestone	2490	30	40

4.3. Analysis of landslide stress

Figure 6 shows the maximum principal stress cloud of the model:

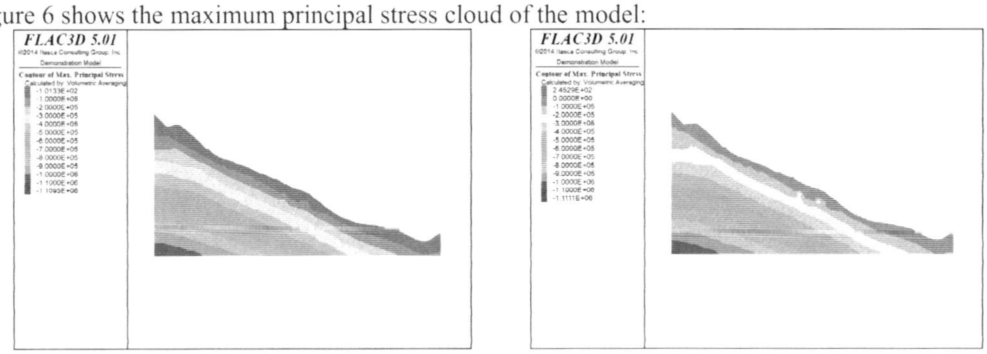

6(a) profile 1-1'(I) 6(b) profile 1-1'(II)

Figure 6. Maximum principal strain.

Similar to most slopes, the maximum principal stress of the model is "strip distribution", which decreases from the bottom to the surface, and most areas of the model are basically compressive stress.

4.4. Displacement characteristics of landslide

Figure 7 shows the horizontal displacement of the model landslide:

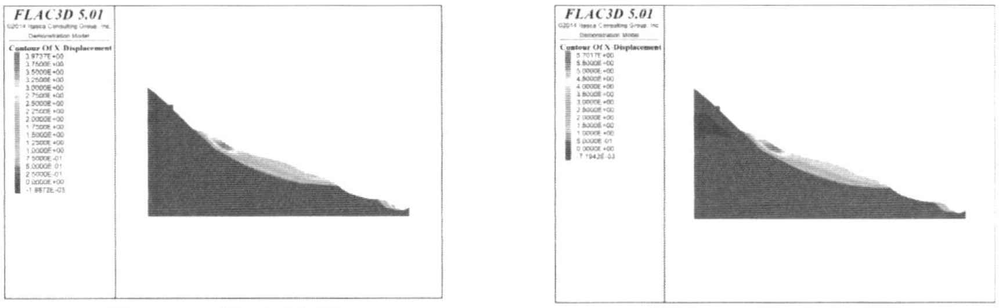

7(a) profile 2-2' (I) 7(b) profile 2-2' (II)

Figure 7. Horizontal displacement of landslide.

From the horizontal displacement map of the landslide, we can see that in both conditions the maximum horizontal displacement of the profile 1-1' is at the foot of the landslide, the profile 2-2' is at the back edge of the landslide, the profile 3-3' is in the middle and rear part of the landslide.

4.5. Maximum shear strain increment

The maximum shear strain increment cloud is shown in Figure 8:

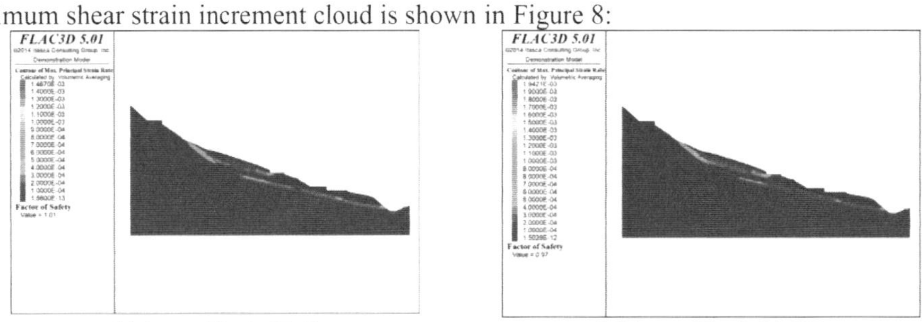

8(a) profile 3-3' (I) 8(b) profile 3-3' (II)

Figure 8. Maximum shear-strain increment cloud.

Profile 1-1' has the largest shear strain increment at the foot of slope, profile 2-2' has the largest shear strain increment at the back edge of the landslide, profile 3-3' has the largest shear strain increment at the front edge of the landslide. The maximum shear strain increment of landslide reflects the area of shear deformation in the sliding body, and its penetration can indicate the characteristics of landslide instability. Figure 8 shows that the region with the largest change in shear stress variability within the slope body is arc-shaped, most likely a potential sliding surface.

4.6. Displacement analysis of monitoring points

In order to further grasp the landslide deformation stability, monitoring points were set up at the front, middle and back of the three sections of the landslide, to provide the basis for the analysis of the landslide stability. The location of the monitoring points is arranged as shown in Figure 9.

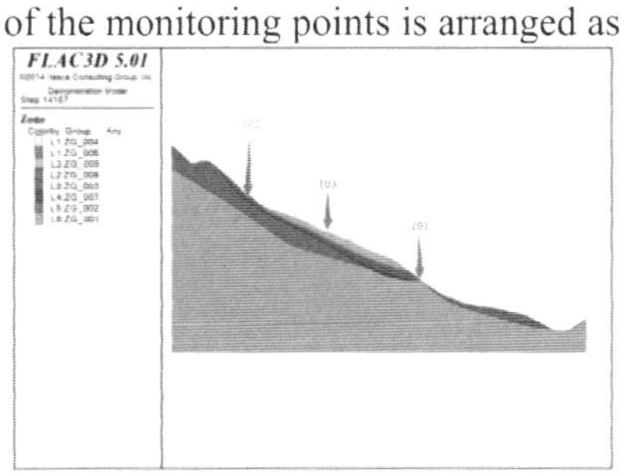

9(a) profile 1-1'

Figure 9. Layout of displacement monitoring points.

The displacement curve of the monitoring point is shown in Figure 10.

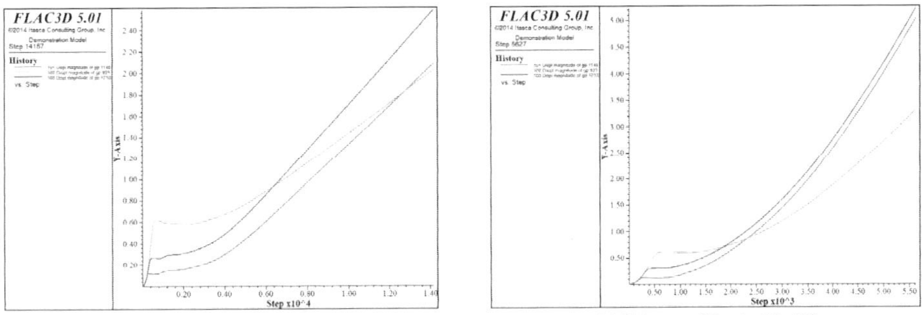

10(a) profile 1-1' (I) 10(b) profile 1-1' (II)

Transverse axis-time step (s), longitudinal axis-displacement (m)

Figure 10. Displacement monitoring curve.

The results show that the front and middle of the landslide are the most displacement, and the rear part of the landslide is the first displacement.

4.7. Evaluation of simulation results

(1) Under the rainstorm condition, the shear strain increment of landslide appears obviously through compared with the natural condition, and the shear strain increment through area of the simulation result is consistent with the sliding surface revealed by drilling. Because the shear strain increment basically runs through the whole slope body, the landslide is in an unstable state under rainstorm condition.

(2) According to the location of the initial slide, Zhangjiapo landslide belongs to the rear thrust load caused landslide. This is consistent with the conclusion of engineering geological survey and provides the basis for putting forward targeted landslide prevention and control measures.

(3) The safety factor of three profiles under two working conditions is calculated, which is basically consistent with the safety factor calculated by the transfer coefficient method. The landslide safety coefficient pair is shown in Table 5.

Table 5. Results of the calculation of stability coefficients.

Method \ Profile(working condition)	1-1' (I/II)	2-2' (I/II)	3-3' (I/II)
Transfer coefficient method	1.085/0.864	1.028/0.819	1.002/0.990
Numerical simulation	1.09/0.87	1.04/0.84	1.01/0.97

5. Conclusion and suggestion

This paper constructs a complex three-dimensional numerical model of Zhangjiapo landslide in Flac3D based on Griddle in Rhino. On this basis, the stability of Zhangjiapo landslide under two working conditions is analyzed by comparing strength reduction method and transfer coefficient method. The study concluded that:

(1) The stability coefficient of the three profiles under the rainstorm condition of Zhangjiapo landslide is less than 1. In view of the possibility of instability under the rainstorm condition, it is necessary to adopt comprehensive prevention and control project.

(2) Zhangjiapo landslide is the rear thrust load caused landslide. The propositional project plan of landslide prevention and control is as follows: the anti-slide pile is arranged outside the steep ridge in

the front of the landslide, the drainage measures are adopted on the periphery of the landslide, and the anti-slip retaining wall is arranged at the front edge of the landslide.

The three-dimensional geological model established by Rhino can directly and truly reflect the actual situation of landslide, but how to divide the grid more accurately, improve the accuracy of numerical simulation calculation, and make the Rhino model provide more accurate pre-processing model for Flac3D is the focus of future research.

References

[1] Dejun Zhao 2017 An Effective Investigation and Prevention of Landslide in West Hubei-Dongyu Padang Formation [D]. China University of Geosciences.

[2] Guangdian Zou and Rulong Wei 2006 Theoretical and Method of Numerical Solution of Universal Limit Equilibrium Method for Stability Analysis of Soil Slope[J]. *Journal of Rock Mechanics and Engineering* **(02)** 363-370

[3] Xiangbin Zhang 2018 Study on landslide stability analysis and prediction technology[D]. China University of Geosciences (Beijing)

[4] 2017 Rock slope stability analysis according to Eurocode 7, discussion of some dilemmas with particular focus on limit equilibrium analysis, 1 Norwegian University of Science and Technology (NTNU), Bulletin of Engineering Geology and the Environment Vol. 76 (4) pp.1229-1236 Bjørn Nilsen

[5] Y.M. Cheng, D.Z. Li, L. Li, Y.J. Sun, R. Baker and Y. Yang 2011 Limit equilibrium method based on an approximate lower bound method with a variable factor of safety that can consider residual strength[J]. *Computers and Geotechnics* **38(5)**

[6] Yumin Chen and Dingping Xu FLAC\FLAC3D Foundation and Engineering Examples[M]. China Water Conservancy and Hydropower Press China

[7] Yang Ming 2018 Numerical Analysis of Slope Stability of a Gold Mine in Inner Mongolia[J]. *World Nonferrous Metals* **(17)** 277-278

[8] Ying Zhang, Zhaotao Sun and Zhongwen Pei 2018 Numerical Analysis of Slope Stability of a Gold Mine in Inner Mongolia[J]. *Prospecting Engineering* (Geotechnical Drilling Engineering) **45(03)** 57-61

[9] Effect of Intermittent High Frequency Magnetic Field on Initial Solidification in Continuous Casting[J]. *Acta Metallurgica Sinica(English Letters)* 1997 **(02)** 151-156

[10] Yu-chuan Yang, Hui-ge Xing, Xing-guo Yang, Kang-xin Huang and Jia-wen Zhou 2015 Two-dimensional stability analysis of a soil slope using the finite element method and the limit equilibrium principle[J]. *The IES Journal Part A: Civil & Structural Engineering* **8(4)**

[11] Wang Min, Li Yi, Shang Qin, Zheng Huakang and Zhou Jiaqing 2018 Analysis on Stability of Xituo landslide in Three Gorges Reservoir Area[J]. *Jiangxi Water Conservancy Science and Technology* **44(04)** 275-279

[12] Xu Feng 2010 Study on the Relationship between the Mechanism and Prevention Engineering of Typical Giant Landslide in Southwest China[D]. Chengdu University of Technology

[13] Weifu Teng, Yiping Wu and Dongying Yang 2005 Analysis on the Stability of West Slope Landslide in Badong County[J]. *Geological science and technology information* **24(s1)** 142-144

[14] Qiulin Liao, Qian Gang Zeng and Tong Liu 2005 The Automatic Generation of Complex Geological Body FLAC3D Model Based on ANSYS Platform[J]. *Journal of Rock Mechanics and Engineering* **024(006)** 1010-1013

IOP Conf. Series: Earth and Environmental Science **865** (2021) 012027 doi:10.1088/1755-1315/865/1/012027

A novel reduced-order method for analysis of hyperspectral images

Yanming Zhang[1, 2] and Lijun Jiang[1]

[1]Department of Electrical and Electronic Engineering, the University of Hong Kong, Pokfulam, Hong Kong

[2]Corresponding author's e-mail: ymzhang@eee.hku.hk

Abstract. Hyperspectral imaging technology has been broadly applied in remote sensing because it collects echoed signals from across the electromagnetic (EM) spectrum and provides fruitfully helpful information. However, the processing or transformation of high-data-volume hyperspectral images, also viewed as snapshots varying with the EM spectrum, burdens the hardware resources, especially for the high spectral resolution and spatial resolution cases. To tackle this challenge, a novel reduced-order method based on the dynamic mode decomposition (DMD) algorithm is presented here to analyze hyperspectral images. The method decomposes the spatial-spectral hyperspectral images in terms of spatial dynamic modes and corresponding spectral patterns. Then, these spatial-spectral patterns are utilized to recover the raw hyperspectral images. Our proposed approach is benchmarked by the actual hyperspectral images measured at the Salinas scene. It is demonstrated that the proposed approach can represent the hyperspectral images with a low-rank model in spectral dimension. Our proposed approach could provide a useful tool for the model order reduction of hyperspectral images.

1. Introduction

Hyperspectral images refer to a set of images of the same observation scenario, each of which is measured at a specific EM spectrum. Moreover, it has been demonstrated that hyperspectral images have wide application in classifying objects, detecting targets, and extracting terrain's features as it offering plenty of information not only in the spatial dimension but also the spectral dimension. Generally, the collected images form three dimensional hyperspectral data cube, (x, y, λ). Hereinto, the x and y denote the two spatial dimensions, and λ refers to the spectral dimension, which corresponds to the EM wavelength. However, while the development of hyperspectral images with high spectral resolution and spatial resolution can provide extremely rich and diverse spatial-spectral information, it also causes various issues. In particular, it is still challenging for hardware processor to deal with high spectral and spatial resolution hyperspectral images that are often very high-volume dataset. To address this challenge, several model order reduction (MOR) technologies have been developed. Hereinto, the principal components analysis (PCA) is explored to realize dimension reduction, and the principal components decomposed by PCA are evaluated to robustly detect the difficult targets in hyperspectral image datasets [1]. Afterwards various spatial dimensionality reduced methods have been developed and applied to MOR of hyperspectral images, such as graph embedding (GE) method [2], locally linear embedding (LLE) approach [3], and the low-rank learning [4]. It is worth mentioning that, with the recent development of machine learning and deep learning, the dimensionality reduction methods based on convolutional neural networks have been proposed.

However, in order to achieve great results, a large amount of hyperspectral data and consumption are required during the training step, which also puts a processing burden on the hardware. While the above-mentioned studies have achieved significant developments, it is still an open and challenging issue for the dimensionality reduction of hyperspectral imagery.

In this paper, we investigate the DMD approach to reduce the dimension of hyperspectral imagery. Through the DMD method, the spatial-spectral hyperspectral data cube is decomposed into the spatial dynamic modes and corresponding spectral patterns. These decomposed patterns can represent the raw hyperspectral data via the low-dimensional linear combination. The actual hyperspectral images measured at the Salinas scene are used to verify the proposed method. The advantages of this newly proposed approach are listed as followed. (1) The DMD-based method is a spectral dimension reduction and could be hybridized with traditional spatial dimension reduction methods. (2) Unlike supervised learning, this method does not demand a training process and then could be executed more economically and efficiently. (3) The output of this approach offers a spatial-spectral correlated subspace, which could be helpful for feature extraction and target classification. The remainder of the article is formed as follows. Section II presents the methodology of the proposed approach. The verification based on the actual hyperspectral imagery dataset is exhibited in Section III. Finally, the conclusions are drawn.

2. Methodology

2.1. Background

Often, the spatial-spectral hyperspectral images can be regarded as a three-dimensional data cube. To be specific, we assume that $d(x_i, y_j, \lambda_k), i = 1, \dots, O, j = 1, \dots., P, k = 1, 2, \dots l, \dots$N denotes a pixel observed at a specific EM spectrum. If the vector $\boldsymbol{r} = (x, y)$ is defined to represent the spatial domain, one can obtain a set of snapshots varying with the EM spectrum, which is expressed as followed.

$$\mathbf{D} = \begin{bmatrix} \mathbf{d}(\mathbf{r_1}, \boldsymbol{\lambda_1}) & \cdots & \mathbf{d}(\mathbf{r_1}, \boldsymbol{\lambda_l}) & \cdots & \mathbf{d}(\mathbf{r_1}, \boldsymbol{\lambda_N}) \\ \vdots & \cdots & \vdots & \cdots & \vdots \\ \mathbf{d}(\mathbf{r_M}, \boldsymbol{\lambda_1}) & \cdots & \mathbf{d}(\mathbf{r_M}, \boldsymbol{\lambda_l}) & \cdots & \mathbf{d}(\mathbf{r_M}, \boldsymbol{\lambda_N}) \end{bmatrix} \in \mathbb{R}^{\mathbf{M*N}} \tag{1}$$

Where $\mathbb{R}$ denotes real analytic space, M is the number of pixels in each image, and lth column of **D** means the observation sequence $\boldsymbol{d}_l$, also named as the lth snapshot. Then, the raw hyperspectral images can be expressed as: $\mathbf{D} = [\boldsymbol{d}_1, \boldsymbol{d}_2, \cdots, \boldsymbol{d}_l, \cdots, \boldsymbol{d}_M]$.

2.2. Dynamic mode decomposition

Traditionally, the DMD is used to decompose the spatial-temporal signal into the term of oscillation decay/growth rate and frequency [5]. This approach has been applied in many fields, such as EM vortex beam demultiplexing [6], EM radiation analysis [7], financial trading strategies [8], and ship wake detection [9]. If the spectral domain of hyperspectral images is analogous to the temporal domain, the spatial-spectral hyperspectral images can be analysed by this decomposition method. In particular, we first separate the data into two sets of snapshots as

$$\boldsymbol{D_1} = [\boldsymbol{d}_1, \boldsymbol{d}_2, \cdots\cdots, \boldsymbol{d}_{M-1}] \tag{2}$$

$$\boldsymbol{D_2} = [\boldsymbol{d}_2, \cdots\cdots, \boldsymbol{d}_{M-1}, \boldsymbol{d}_M] \tag{3}$$

We assume a local approximation: $\boldsymbol{d}_{l+1} = \boldsymbol{F}(\boldsymbol{d}_l)$, so the connection between equation (2) and equation (3) can be established, which is expressed as follows.

$$\boldsymbol{D_2} = \boldsymbol{P}\boldsymbol{D_1} \tag{4}$$

Where $\boldsymbol{P}$ means the DMD mapping matrix. DMD is to compute the eigenvectors and eigenvalues of the mapping matrix $\boldsymbol{P}$. Hereinto, we adopt the singular value decomposition (SVD) based approach to realize the DMD algorithm. The steps are shown as follows. First, $\boldsymbol{D_1}$ is represented by the decomposed results obtained by SVD.

$$\boldsymbol{D_1} = \mathbf{T\Sigma}\boldsymbol{V}^* \tag{5}$$

Where * refers to the conjugate transpose. Then the mapping matrix $\boldsymbol{P}$ can be expressed by

$$\boldsymbol{P} = \boldsymbol{D}_2\boldsymbol{V}\boldsymbol{\Sigma}^{-1}\boldsymbol{T}^* \tag{6}$$

Finally, through the eigen decomposition of $\boldsymbol{P}$ in low-rank system that is calculated via computing proper orthogonal decomposition (POD) on the mapping matrix, one can obtain the approximate solution of the hyperspectral snapshots, which is given by:

$$d(\lambda) = \sum_{k=1}^{n} \psi_k exp(w_k\lambda) b_k = \boldsymbol{\Psi} exp(\boldsymbol{\Omega}\lambda)\boldsymbol{b} \tag{7}$$

where ψ_k and w_k refer to the eigenvalue and eigenvector of the mapping matrix $\boldsymbol{P}$, which correspond to the spatial dynamic mode and spectral pattern respectively, b_k refers to the initial intensity of kth mode. The physical meaning of this decomposition is straightforward. The spatial-spectral hyperspectral images are represented by the combination of spatial dynamic mode and low-rank spectral pattern. Hence, MOR for high-volume hyperspectral datasets could be realized with this architecture.

3. Results

To verify the proposed method, we adopt the actual measured hyperspectral data at the Salinas scene. The hyperspectral images dataset is measured by the Airborne Visible Infrared Imaging Spectrometer (AVIRIS) sensor. Moreover, this hyperspectral observation is collected across 224 EM bands. Also, the spatial resolution of hyperspectral data measured at Salinas Valley is about 3.7 meter in each pixel. Figure 1 shows the hyperspectral images with different EM bands. It can be seen that the observed area has different geomorphic features. These geomorphic features can be distinguished well in some EM wavebands (such as number of band 100), but it is difficult to distinguish them in several EM wavebands (such as number of band 1). After obtaining the hyperspectral image data, we first reshape the two-dimensional spatial matrix into a spatial vector. Then, the three-dimensional hyperspectral data cub is arranged into a matrix according to the index of the EM spectrum. By inputting this narrow and tall deformation matrix into the DMD algorithm, we can obtain the spectral pattern and spatial modes. As shown in Figure 2, there are 11 eigenvalues in the decomposition. Since the hyperspectral data is in real analytic space, the extracted eigenvalues with non-zero imaginary part are all conjugate pairs to ensure that the reconstructed hyperspectral data is also in the real analytic space. It is worth noting that the extracted spectral pattern can be viewed as the dynamic characteristics of spectral direction.

Figure 3 plots the extracted spatial modes. We can see that all the geomorphic features are well represented in these modes. In different spatial modes, the intensity of the geomorphic features is also different, which provides a good foundation for classification problems in the hyperspectral images. Notably, unlike the classification is explored based on the original hyperspectral data, the classification in this scheme only use a streamlined amount of hyperspectral data, which is similar to the noise cancellation technology. In order to verify the effectiveness of the dimensionality reduction technology, we reconstruct the hyperspectral images according to the combination relationship shown in equation (7). Figure 4 shows the comparisons of the actual and reconstructed hyperspectral images under different EM bands. It is clear that the raw hyperspectral data is well represented in the reduced dimensionality obtained by the DMD approach. Also, the differences among the geomorphic features are also well portrayed. That is to say, the hyperspectral data of 224 EM bands are compressed and well expressed through 11 the spectral pattern and spatial modes. Hence, the dimensionality of the hyperspectral images can be effectively reduced by the proposed DMD approach.

To further validate the proposed reduced-order approach, we quantitatively analyzed the error by calculating the normalized mean absolute error (NMAE) of the reconstructed results. Table 1 shows the NMAE versus the index of EM bands. It can be seen that NMAEs have small value at different EM bands, and all NMAE of the EM bands is less than 10%. This error is acceptable from the viewpoint of engineering. Performance of error analysis further demonstrates the effectiveness of the proposed approach.

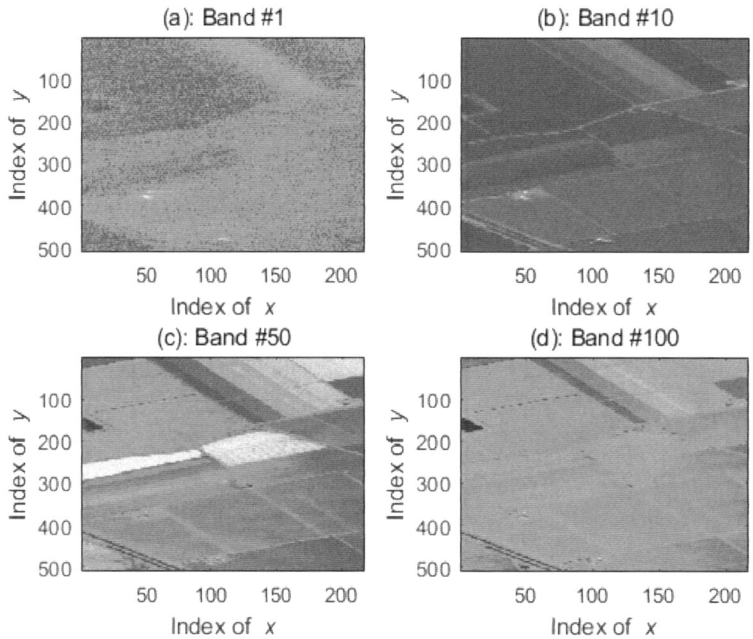

Figure 1. The hyperspectral images at the Salinas scene with different EM bands.

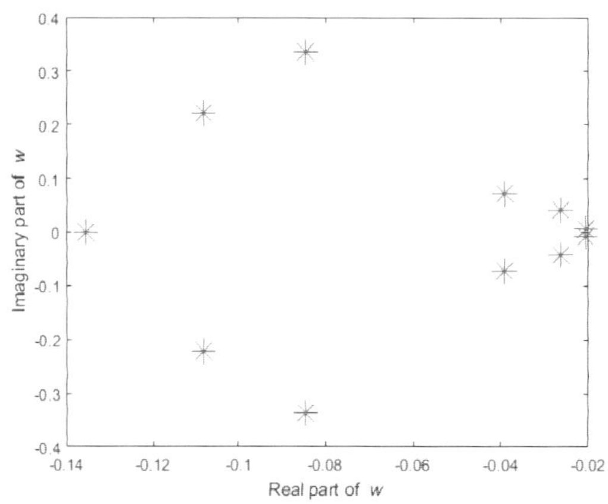

Figure 2. The extracted spectral pattern.

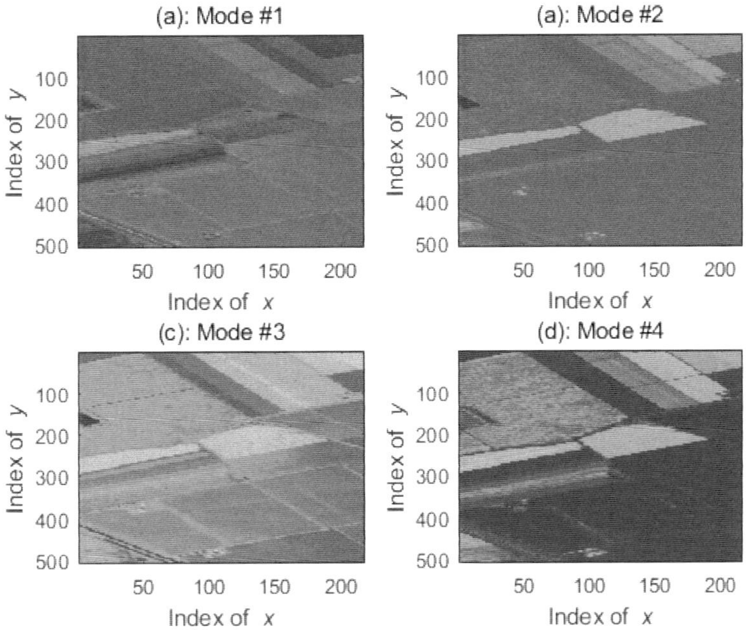

Figure 3. The extracted spatial modes.

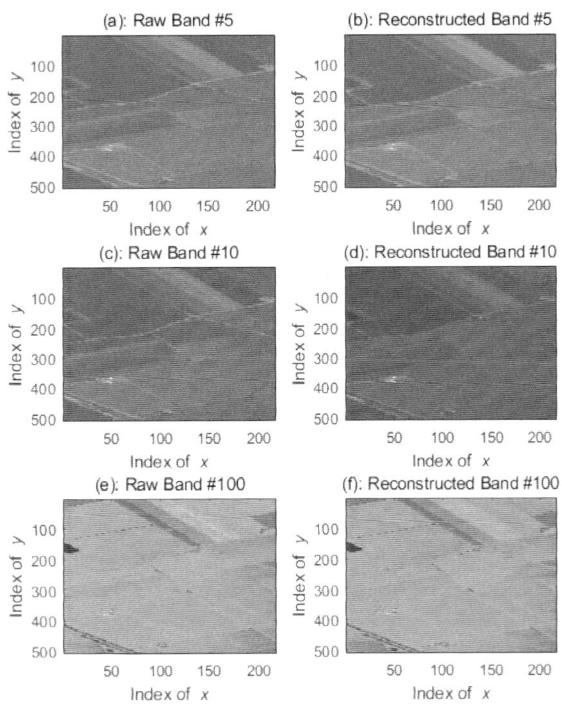

Figure 4. The comparisons of the actual and reconstructed hyperspectral images with different EM bands.

IOP Conf. Series: Earth and Environmental Science **865** (2021) 012027 doi:10.1088/1755-1315/865/1/012027

Table 1. NMAE versus the index of EM bands.

Number of EM bands	NMAE
#5	0.0087
#10	0.0158
#100	0.0402
#160	0.0581

4. Conclusion

In conclusion, we demonstrated the DMD-based approach to investigate and decompose the spatial-spectral hyperspectral images. The proposed data-driven approach deals with the hyperspectral image data as a two-dimensional correlated real signal. It decomposes hyperspectral image data into the DMD spatial modes and corresponding spectral patterns, which can be utilized to reconstruct the hyperspectral images. We verified the proposed based on the actual measured hyperspectral data at the Salinas scene. A two-dimensional spatial-spectral subspace can be obtained for feature classification through the extraction of these two relevant domains. Therefore, our paper offers a practical reduced-order technique for hyperspectral images.

References

[1] Farrell M D and Mersereau R M 2005 On the impact of PCA dimension reduction for hyperspectral detection of difficult targets *IEEE Geosci Remote Sens Lett,* Vol. 2, no. 2, pp 192-195

[2] Huang H, Shi G, He H, Duan Y and Luo F 2020 Dimensionality Reduction of Hyperspectral Imagery Based on Spatial–Spectral Manifold Learning *IEEE Trans Cybernetics,* vol. 50, no. 6, pp 2604-2616

[3] Chen G and Qian S E 2007 Dimensionality reduction of hyperspectral imagery *J Appl Remote Sens* **1(1)** 013509

[4] Gao L, Hong D, Yao J, Zhang B, Gamba P and Chanussot J 2021 Spectral Superresolution of Multispectral Imagery With Joint Sparse and Low-Rank Learning *IEEE Trans Geosci Remote Sen,* vol. 59, no. 3, pp 2269-2280

[5] Tu J, Rowley C, Luchtenburg D, Brunton S, and Kutz J 2013 On dynamic mode decomposition: Theory and applications *J Comput Dynamic,* vol. 1

[6] Yanming Zhang, Menglin L. N. Chen, and Li Jun Jiang 2019 Analysis of electromagnetic vortex beams using modified dynamic mode decomposition in spatial angular domain *Opt Express* **27** 27702-27711

[7] Zhang Y and Jiang L 2020 A Novel Data-Driven Analysis Method for Electromagnetic Radiations Based on Dynamic Mode Decomposition *IEEE Trans Electromagn Compat,* vol. 62, no. 4, pp 1443-1450

[8] Mann J and Kutz J N 2016 Dynamic mode decomposition for financial trading strategies *Quant Financ*, vol. 16, no. 11, pp 1643–1655

[9] Zhang Y and Jiang L 2020 A Novel Data-Driven Scheme for the Ship Wake Identification on the 2-D Dynamic Sea Surface *IEEE Access* vol. 8 pp 69593-69600

Classification of terrestrial laser scanning data based on relative projection density

Xinyi Zhang[1], Maolin Chen[1, 2], Xiyu Wang[1] and Xiangjiang Liu[1]

[1]School of Civil Engineering, Chongqing Jiaotong University, No.66, Xuefu Avenue, Nan'an District, Chongqing, China;

[2]Corresponding author's e-mail: maolinchen@cqjtu.edu.cn

Abstract. Point cloud classification is a critical step in ground 3D scene analysis. The density of large-scale terrestrial laser scanning data will decrease rapidly with the increase of distance, which will affect features extraction. Focusing on this problem, we propose a grid features based on the relative projection density for point cloud classification in this paper. Geometric features are constructed based on the neighborhood covariance eigenvalue of each point. In grid feature extraction, the relative projection density is used to replace the number of projection points as grid density feature directly. In an outdoor scene obtained by the panoramic scanning of Reigl-VZ400 scanner, the grid features based on the relative projection density and the traditional projection density features are compared and analyzed. Based on the Random Forest for classification, the result shows that the relative projection density features with overall accuracy of 96.51%. Compared with traditional projection density feature, it is more accurate in classification, and it also performs relatively well in the extraction of cars, pedestrian and pole.

1. Introduction

Terrestrial laser scanning(TLS) technology is a new measurement technology in the field of surveying and mapping. It can capture accurate 3D information of the target object efficiently. In recent years, TLS has been widely used in many areas, such as building reconstruction [1],vegetation detection [2], roads extraction [3] ,and forest resource survey [4]. Since original scanning data is a discrete point cloud and does not contain semantic information, point cloud classification is a necessary processing before TLS data application, which specifies a specific semantic label for each 3D point.

Feature extraction is a significant step in point cloud classification, and many features have been proposed in previous studies, including geometric features, RGB colors, echo-based features etc. Geometric features are calculated from the 3D coordinates of all points. To describe the shape of the neighborhood, dimensionality features can be constructed by the eigenvalues of the local structure tensor [5]. The eigenvalues are extended in both 3D and 2D space by Weinmann, and he combined with projection density, height difference and height standard deviation for classification [6][7][8]. The grid projection density is greatly affected by the point cloud density, and it is difficult to distinguish the geometric features of ground objects themselves. Focusing on this problem brought by point density, Chen *et al.* [9] presented polar grid features, it can possibly take the place of the regular grid to adapt to density change, and further improve the classification accuracy of buildings from a large distance. A single point description method based on the point feature histogram (PFH) can provide multi-value features with good robustness at the overall scale, but the computational complexity is high [10][11].In order to solve this problem, a similar approach called Point Feature Histograms [12] has been proposed.

Li *et al.* [13]used hundreds of dimensional features including spin image, RGB colors, normal vector, point elevation to describe a single point.

RGB color is usually obtained by the built-in camera of the scanner and assigned to each point by the registration of image and point cloud. In [14], taking advantage of the different colors of objects in the mining area, RGB colors of the point cloud are converted to the HSV space to achieve the classification of the surface point cloud data. Both RGB color and geometry features are involved in constructing 23-dimensionality features, and classify with random forest [15]. In [16], RGB colors were combined with intensity information modified by angle and distance for classification.

Echo features mainly include the number of echoes and echo intensity. Pirotti *et al.* [17] proposed a method, that can identify non-ground points and classify TLS data based on the number of echoes. Echo intensity is one of the most commonly used features of TLS point cloud, and its value is the maximum amplitude of the returned pulse [18].The intensity of artificial object such as buildings and roads is often higher than natural features such as vegetation, therefore, the intensity can be used to extract buildings or roads directly [19]. In [20], the optimal neighborhood is selected based on the intensity feature, the outliers of intensity will be eliminated. Then the geometric features such as normal vector and dimensional feature are combined to form multi-dimensionality features, participating in classification with SVM.

As RGB colors and echo information are not available for all datasets, the use of non-geometric features will limit the universality of a classification method. Because the calculation of geometric features only relies on the coordinate information of the points, we only involve geometric features in our investigations. As a part of geometric features, grid features have also been studied in many studies[7][21], among which rectangular grid is the most commonly used grid. Grid feature is influenced by the local structure and point density of the target object. The number of projected points in each sub-grid commonly decreases with the increase of scanning distance. Accordingly, it may not be suitable for large-scale TLS to use the number of projection points as grid density feature directly. Focusing on the problem brought by density variation, we propose the relative projection density to optimize the classification accuracy of TLS data.

2. Methodology

2.1. Neighborhood selection

In point description, it is necessary to find a subset that can represent the characteristics of the point cloud best. However, point cloud of different sizes and densities will affect the definition of neighborhood and the selection of corresponding scale parameters. In order to overcome the disadvantages of fixed neighborhood. In this paper, the method in [6] is used to determine the optimal neighborhood size based on K-nearest Neighbor (KNN) by gradually increasing the number of neighboring points k ($k\epsilon[10,100]$, Δk=10).

2.2. Feature extraction

Most public datasets contain geometric information of spatial 3D coordinates, so this paper mainly involves geometric features in our investigations. Firstly, the ground points are eliminated by the method of Cloth Simulation Filter(CSF) [22], and secondly the features are calculated from the remaining points. Finally, the geometric features (3D features, 2D features and grid features) of single points are generated by the optimal neighborhood selection method.

2.2.1. 3D features. In this paper, we use the method in [6] to construct 3D features. Linearity L_λ, planarity P_λ, scattering S_λ, Shannon entropy DE, eigenentropy EE, Omnivariance O_λ, anisotropy A_λ, change of curvature C_λ, the local point density D and the verticality V of single point are added to describe the 3D geometric properties of point cloud. The 10-dimensonal 3D vectors can be obtained as

$$F_{3D} = \left[L_\lambda, P_\lambda, S_\lambda, DE, EE, O_\lambda, A_\lambda, C_\lambda, D, V\right] \quad (1)$$

2.2.2. 2D features. Similarly, the method in [6] can also construct 2D features including 2D density D_{2D}, linearity $L_{\lambda,2D}$, planarity $P_{\lambda,2D}$, Omnivariance O_{2D}, Shannon entropy DE_{2D}, eigenentropy EE_{2D}. Therefore, a 6-dimensonal 2D vectors can be obtained as

$$F_{2D} = \left[D_{2D}, L_{\lambda,2D}, P_{\lambda,2D}, O_{2D}, DE_{2D}, EE_{2D} \right] \tag{2}$$

2.2.3. Grid features. 2D grid feature extraction takes 2D grid as the processing object, which is usually based on rectangular grid. Grid features are extracted by projecting the 3D points onto a 2D grid and analyzing the points in each sub-grid. Projection density is a crucial grid feature. In large-scale scene, the point density will decrease rapidly with the increase of scanning distance. The grid near the scanner usually contains more points than the grid farther away from the scanner, as shown in Figure 1. The projection density based on the number of points is not only related to the height difference in one sub-grid, but also directly affected by point density. For example, the projection density of tall buildings at a distance may be lower than that of low vegetation or tree canopies at a close distance, which affects the classification accuracy. Therefore, directly taking the number of projection points as the grid density characteristic may not be suitable for large-scale laser scanning data. In this paper, the relative projection density is proposed as grid density feature based on rectangular grid.

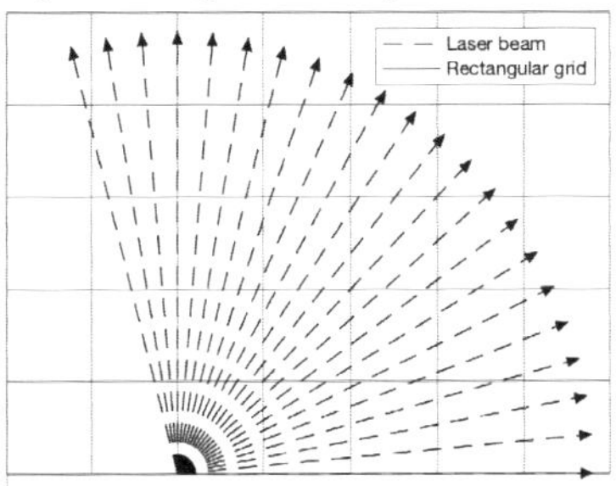

Figure 1. Rectangular grid point density varies with scanning distance.

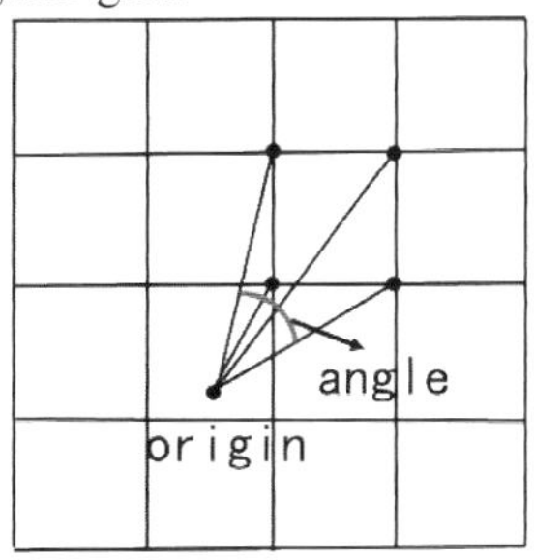

Figure 2. The angle range of a grid.

The relative projection density(Den_r) is defined as the number of points in a grid(N_{Grid}) divided by the horizontal number of scan lines in a grid ($N_{Grid,h}$).

$$Den_r = N_{Grid} \,/\, N_{Grid,\mathrm{h}} \tag{3}$$

From the angular resolution and the angle range of each grid (the maximum angle difference of four vertex coordinates in the grid, as shown in Figure 2), the number of the scanning lines in the horizontal plane can be obtained

$$N_{Grid,\mathrm{h}} = (180° \times angle) \,/\, (\alpha \times \pi) \tag{4}$$

In addition, the parameters of the projected density grid feature also include the maximum height difference and height standard deviation of elevation values within the grid. Hence, the grid feature of relative projected density can be expressed as

$$F_{\mathrm{Grid}} = \left[Den_r, \Delta Z, \sigma_Z \right] \tag{5}$$

Therefore, a 19-dimensonal geometric vectors can be obtained as

$$F = \left[F_{3D}, F_{2D}, F_{\mathrm{Grid}} \right] \tag{6}$$

2.3. Classification method

Reference label of each point is marked manually, while ground points can be obtained by CSF filtering. This paper randomly selects an equal number of sample points from each category as training samples, and uses Random Forest(RF) [23] for classification. We use the package in [24] and set the number of trees in the forest to 500.

3. Experiments and analysis

3.1. Data

The experimental data in this section are obtained in an urban scene by the REIGL-VZ400, and the angular resolution is 0.057°. The information of each category in the scene is shown in Table 1. The ground points are filtered by CSF filtering and do not participate in feature extraction and classification. Therefore, 5,723,196 non-ground points participate in feature extraction and classification, and 1,000 points are randomly selected from each of the six categories as samples.

Table 1. Basic information about per class of objects in the scene.

Class	Point number	Sample
Buildings	3322440	1000
Vegetation	1000590	1000
Pedestrian	214888	1000
Cars	83023	1000
Pole	10145	1000
Others	1092110	1000
Ground	2958727	1000
Total	8681923	7000

3.2. Classification results

Table 2. Overall accuracy (OA, %) under different features, mean ± standard deviation.

classification method	OA/%
RPDGF	96.51±0.11
PGF	93.31±0.21

Table 3. Recall, precision and F1-score under different features, mean ± standard deviation.

Class	classification methods	Recall/%	Precision/%	F1/%
Buildings	RPDGF	96.92±0.15	99.62±0.04	98.25±0.07
	PGF	94.12±0.34	99.39±0.08	96.68±0.16
Vegetation	RPDGF	96.32±0.31	96.86±0.28	96.59±0.14
	PGF	94.91±0.36	93.89±0.42	94.40±0.19
Cars	RPDGF	98.58±0.29	49.59±1.40	65.98±1.22
	PGF	95.50±0.54	29.50±0.84	45.07±0.96
Pole	RPDGF	98.23±0.42	23.63±1.79	38.06±2.31
	PGF	96.00±0.56	18.57±2.21	31.06±3.10
Pedestrian	RPDGF	98.33±0.24	96.65±0.74	97.48±0.41
	PGF	92.39±0.46	85.84±1.19	88.99±0.67
Others	RPDGF	94.90±0.38	96.90±0.41	95.88±0.21
	PGF	89.38±0.47	96.46±0.73	92.78±0.35

The relative projection density is compared with the projection density based on polar grid which is also proposed to adapt to the density variation in TLS data in [9]. Besides the grid features, the 3D and 2D features in [9] are the same in our method. As the 3D and 2D neighborhoods of the two methods are both generated adaptively by the method in [7], parameters requiring manual setting are related with the grid features. The main parameters of grid feature proposed in this paper are grid size, which are set as 1m*1m in this test. For polar grid, its main parameters are angular width $\theta_G$and polar grid size Δr.

The angular width is set to N times of the angular resolution (0.057°) with $N=10$ and Δr is set to 0.5. Due to the randomness of the selection of samples, the result of each run may be different. Thus, 100 experiments are repeated for each method. The evaluation metrics are composed of overall accuracy (OA), recall, precision and F1-score. The experimental results are shown in Table 2-3.

It can be seen that our method obtain better overall accuracy than polar grid features and the corresponding overall accuracy value are larger than 90%. The best overall accuracy of a single run is 96.75% for relative projection density grid features, and the average overall accuracy of relative projection density grid features is 3.2% higher than that of the polar grid features.

For the class level, the relative projection density grid features of each class are superior to the polar grid features. Both two methods have similar precision on buildings class, but in terms of recall and F1-score, relative projection density grid features still remain a higher level. In addition, as shown in Figure 3, a large number of internal structures of a close distance buildings and facades of distant buildings are identified as car and vegetation in the polar grid features. In the vegetation class, recall, precision and F1-score are approximate to each other in relative projection density grid features, and the false detections are mainly from the car and pole classes. However, in the polar grid features, more false detections are from the classes of car and building, which will have a greater impact on the extraction of buildings. Moreover, in vegetation class, there is an obvious gap between two methods of accuracy indicators. In the classes of cars and pole, the recall value of both methods are higher than 95%, which indicate that most of the surface points of cars and poles could be correctly identified, but the precision is lower than 50%, and the precision of pole under the polar grid features is only 18.57%, lower than any values in all classes. What's more, the two classes also have lower F1-score than other classes. In pedestrian and others classes, the relative projection density grid features are superior to polar grid features. Our method performs better on all classes extraction than polar grid features, but it improves more in car, pole and pedestrian.

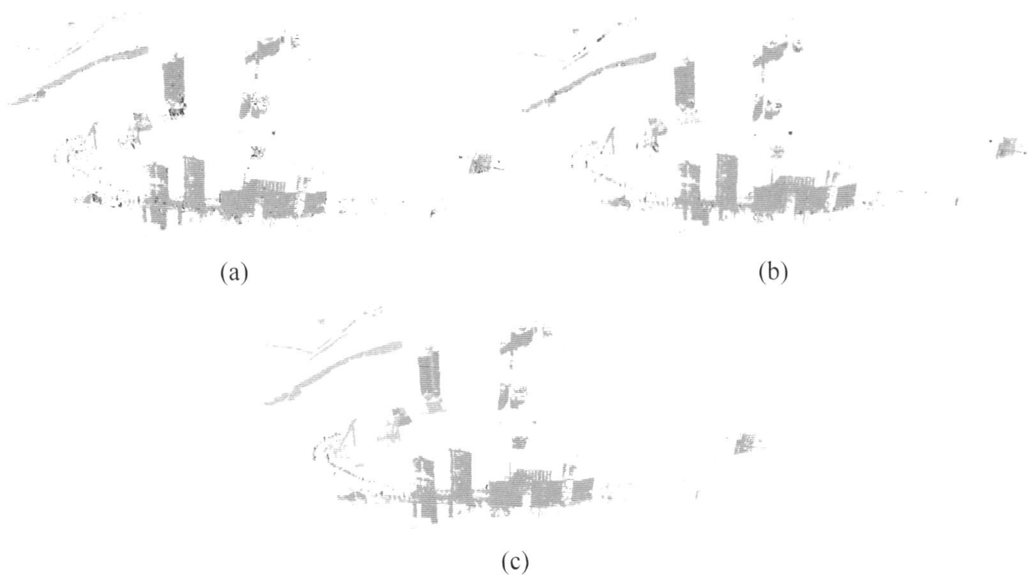

(a) (b)

(c)

Figure 3. Classification results. (a) Result of our method. (b) Result of polar grid features. (c)ground truth. All the points are colored by building(grey), vegetation(green), car (yellow), pole (blue), pedestrian(red), and others(aqua).

3.3. Feature Importance

Random Forest can be used to evaluate the feature importance. In this paper, feature importance of 19 features is evaluated with the help of random forest classification, as shown in Figure 4. It can be seen that the three grid features improved by relative projection density grid features proposed are much more important than the other two types of features. Among the rest of features, the feature importance of single point verticality V is the highest. Among the dimensionality features, the importance of planarity P_λ is the highest. The eigenentropy EE and EE_{2D} have higher discriminability than the Shannon entropy DE and DE_{2D}.

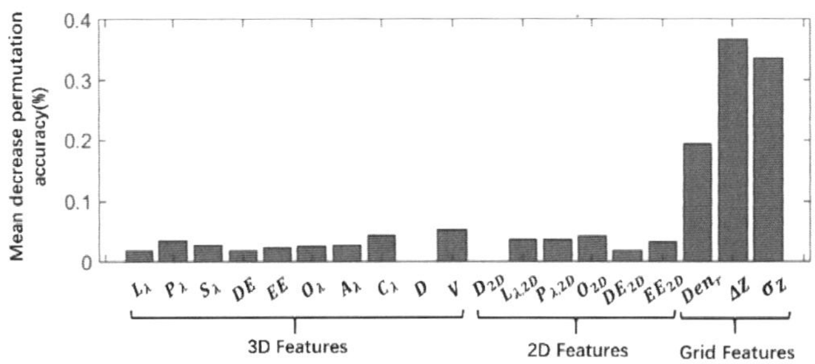

Figure 4. Feature importance of the relative projection density grid features.

4. Conclusion

This paper presents a method of grid feature extraction based on relative projection density. The results show that, the proposed features can obtain better classification result than polar grid features. Compared with the previous density adaptive feature extraction methods based on polar grid, the method in this paper can achieve similar building extraction accuracy, and better improvements in car, pole, pedestrian and others classes at the same time. In future work, the method can be further optimized by exploring more appropriate feature selection methods combined with label smoothing methods.

Acknowledgments

This work is supported in part by National Natural Science Foundation of China under grant no. 41801394, in part by Chongqing Natural Science Foundation under grant no. cstc2019jcyj-msxmX0370, in part by the Science and Technology Research Program of Chongqing Municipal Education Commission under grant no. KJQN201900729 and in part by Innovative Research Program for Graduates of Chongqing under grant no. CYS21339.

References

[1] Li Z, Zhang L, Mathiopoulos P T, et al. 2017 A hierarchical methodology for urban facade parsing from TLS point clouds[J]. *ISPRS Journal of Photogrammetry and Remote Sensing*, 123(JAN.): 75-93.

[2] Monnier F, Vallet B, Soheilian B 2012 TREES DETECTION FROM LASER POINT CLOUDS ACQUIRED IN DENSE URBAN AREAS BY A MOBILE MAPPING SYSTEM[J].

[3] Boyko A , Funkhouser T 2011 Extracting roads from dense point clouds in large scale urban environment[J]. *Isprs Journal of Photogrammetry & Remote Sensing*, **66(6)**: S2-S12.

[4] Bisheng Y , Wenxia D , Zhen D , et al. 2016 Automatic Forest Mapping at Individual Tree Levels from Terrestrial Laser Scanning Point Clouds with a Hierarchical Minimum Cut Method[J]. *Remote Sensing*, **8(5)**: 372.

[5] Jérôme Demantké, Clément Mallet, David N , et al. 2011 DIMENSIONALITY BASED SCALE SELECTION IN 3D LIDAR POINT CLOUDS[J]. *ISPRS - International Archives of the Photogrammetry Remote Sensing and Spatial Information Sciences*, **38(5)**: 97-102.

[6] Weinmann M, Jutzi B, Hinz S, et al. 2015 Semantic point cloud interpretation based on optimal neighborhoods, relevant features and efficient classifiers[J]. *ISPRS Journal of Photogrammetry and Remote Sensing*, 105(jul.): 286-304.

[7] Weinmann M, Urban S, Hinz S, et al. 2015 Distinctive 2D and 3D features for automated large-scale scene analysis in urban areas[J]. *Computers & Graphics*, 49(jun.): 47-57.

[8] Weinmann M, Jutzi B, Clément Mallet 2013 Feature relevance assessment for the semantic interpretation of 3D point cloud data[J]. *Isprs Annals of Photogrammetry*, II-5/W2:313-318.

[9] Chen M, Pan J, Xu J 2018 Classification of Terrestrial Laser Scanning Data With Density-Adaptive Geometric Features[J]. *IEEE Geoscience and Remote Sensing Letters*, **15(11)**: 1-5.

[10] Rusu R B, Blodow N, Marton Z C, et al. 2008 Aligning Point Cloud Views using Persistent Feature Histograms[C]// *2008 IEEE/RSJ International Conference on Intelligent Robots and Systems*, September 22-26, 2008, Acropolis Convention Center, Nice, France. IEEE.

[11] Rusu R B, Marton Z C, Blodow N, et al. 2008 Learning informative point classes for the acquisition of object model maps[C]// *International Conference on Control*. IEEE.

[12] Rusu R B, Blodow N, Beetz M. 2009 Fast Point Feature Histograms (FPFH) for 3D registration[C]// *IEEE International Conference on Robotics & Automation*. IEEE.

[13] Li Z, Zhang L, Tong X, et al. 2016 A Three-Step Approach for TLS Point Cloud Classification[J]. *IEEE Transactions on Geoence and Remote Sensing*, **54(9)**: 1-13.

[14] Song D Y, Cai L L, Gu S D 2019 Color point cloud classification of mine surface based on color space[J]. *Bulletin of Surveying and Mapping*, **2019(S2)**: 56-60.

[15] Cheng X L, Cheng X J, Guo W, Xie D 2015 Point cloud classification and feature extraction of building facades with the corrected Laser intensity value[J]. *Journal of Tongji University* (Natural Science), **43(9)**: 1432-1437.

[16] Hu H Y, Hui Z Y, Li N 2020 Airborn LIDAR point cloud classification based on multiple-entity eigenvector fusion[J]. *Chinese Journal of Lasers*, **47(08)**: 237-247.

[17] Pirotti F, Guarnieri A, Vettore A 2013 Ground filtering and vegetation mapping using multi-return terrestrial laser scanning[J]. *Isprs Journal of Photogrammetry & Remote Sensing*, 76(FEB.): 56-63.

[18] Reitberger J, Schnoerr C, Krzystek P, et al. 2009 3D segmentation of single trees exploiting full waveform LIDAR data[J]. *Isprs Journal of Photogrammetry & Remote Sensing*, **64(6)**: 561-574.

[19] Chehata N, Li G, Mallet C 2009 Airborne lidar feature selection for urban classification using random forests[J]. *Geomatics & Information Science of Wuhan University*, **38**: 207--212.

[20] Yang B, Dong Z 2013 A shape-based segmentation method for mobile laser scanning point clouds[J]. *ISPRS Journal of Photogrammetry and Remote Sensing*, **81(7)**: 19-30.

[21] Fan H, Yao W, Tang L 2013 Identifying Man-Made Objects Along Urban Road Corridors From Mobile LiDAR Data[J]. *IEEE Geoscience & Remote Sensing Letters*, **11(5)**: 950-954.

[22] Zhang W, Qi J, Wan P, et al. 2016 An Easy-to-Use Airborne LiDAR Data Filtering Method Based on Cloth Simulation.[J]. *Remote Sensing*.

[23] Leo Breiman 2001 Random Forests[J]. *Machine Learning*, **45(1)**.

[24] Liaw A, Wiener M, Liaw A 2002 Classification and Regression with Random Forest[J]. *R News*, **23(23)**.

Evaluation of marine green economy efficiency and its influencing factors in China

Jianhua Tang[1, 2], Shujuan Wu[1] and Jianhua Xiao[1]

[1] School of Economics and Management, Wuyi University;

[2]Corresponding author's e-mail: 1909214436@qq.com

Abstract. In order to measure the sustainable development ability of marine economy and provide reference for improving the sustainable development ability of marine economy in China, after constructing a resource input index based on the entropy method, the paper uses the SBM-DEA model with undesirable output to evaluate the marine green economy efficiency of 11 coastal provinces(cities or districts) in China from year 2006 to 2016. Further, Tobit regression model is used to analyze the marine green economy efficiency of influencing factors. It is found that according to the sustainable development ability, the 11 coastal provinces(cities or districts) can be divided into 4 levels: Tianjin, Guangdong, Shandong, Shanghai, and Jiangsu are of highest level, Zhejiang and Fujian are of the high level, Liaoning and Hebei are of the median level, and Guangxi and Hainan are of the low level; the marine industry structure and marine science and technology level have significant positive impacts on the marine economy's sustainable development ability, even though the impact of marine industry structure is weak, while the environmental regulation tools have a weak insignificant negative impact on the sustainable development of the marine economy. Based on the above conclusions, this article provides some suggestions for the sustainable and healthy development of China's marine economy.

1. Introduction

Since the 1990s, marine economy in China has entered a stage of rapid development and become a new driving force for the development of the national economy. The report of the 19th National Congress of the Communist Party of China proposed that the development of the marine economy needs to adhere to land and sea coordination, so as to promote the construction of a marine power and promote the high-quality development of the marine economy. However, the current marine economic development in China is under the extensive developing mode, which is of high input but low output. With the rapid development of the marine economy, this mode of marine economy development has caused a series of problems, such as low utilization of marine resources and damaged marine ecological environment and so on. It runs counter to the response of the "Strategic Proposal for Implementing Marine Negative Emissions and Practicing Carbon Neutrality" initiated at the 2020 International Forum on Marine Ecological Economy. It has become a stumbling block for the sustainable development of the marine economy. In this context, it is very important to use scientific evaluation indicators to measure the sustainable development of the marine economy. How to improve the sustainability of marine economic development and to explore a low-energy, high-efficiency marine economic development method has become a challenge for our country. Therefore, this paper evaluates the marine green economy efficiency and analyses the influencing factors in coastal provinces (cities or districts), aiming to provide basis approaches for promoting the sustainable development ability of marine economy.

IOP Conf. Series: Earth and Environmental Science **865** (2021) 012029 doi:10.1088/1755-1315/865/1/012029

In recent years, marine green economy efficiency is generally defined as a ratio of inputs i.e. human resources, capital, technology and other production factors in marine economy to outputs i.e. GDP and environment loss, under the premise of limited marine resources and marine environment. It aims to measure the sustainable development of marine economy [1-3].

In recent years, relevant domestic researches mainly include: Lin Xu [4] evaluated the efficiency of financial support of 11 coastal provinces(cities, districts) in China by adopting the panel SFA model, and used gray correlation analysis to verify the financial development and the correlation between the marine economy; Tianyi Gao [5] measured the scientific and technological innovation efficiency of its marine economy based on the Qingdao City by adopting data envelopment analysis(DEA), and analyzed the relationship between the marine economy and technological innovation from both dynamic and static perspectives; Lingui Qin [6] measured marine green TFP of the 11 coastal provinces(cities, districts) by adopting the SBM-DEA model, and emphasized the impact of technological innovation on marine green TFP; Mei Gai [3] measured marine economy efficiency of 11 coastal provinces(cities, districts) by adopting stochastic frontier analysis (SFA) model, and studied its spatial-temporal differences and influencing factors; Wei Zou [7] measured marine economy efficiency of 17 cities in the Bohai Sea region by adopting bootstrap-DEA model, and used the standard deviation ellipse and barycenter coordinates to analyze the spatial pattern characteristics. XiXi Zhang [8] measured the green efficiency of marine carbon sink fishery of major coastal marine carbon sink fishery production areas in China by adopting the Super-SBM model 9, and analyzed its influencing factors and spatial spillover effects.

In addition to Chinese scholars, others have been engaged in related research, mainly including: Jamnia [9] adopted SFA model to evaluate the technical efficiency of fishery production in Bahar region in southern Iran; Wanke [10] study the cargo handling efficiency of Brazilian ports and evaluate the integration efficiency of internal production and operation equipment by adopting network DEA model.

Summarizing the previous literature reviews, we can know that the amount of research on marine economy efficiency evaluation has been quite rich, but there are still some shortcomings: first, most of the existing studies used the traditional DEA model. However, this model only consider the desirable output of marine economy, and it can not resolve the problem of input-output slack; Second, only a small number of scholars carried out the study of green efficiency around the marine economy, and failed to fully consider the impact of the undesired output of the marine environment and the environmental resource input elements. In view of this, this paper takes 11 coastal provinces(cities or districts) in China as the research object, on the basis of fully considering the elements of resource input, uses SBM-DEA model considering undesirable output to calculate their marine green economy efficiency. This not only solves the problem of input-and-output slack, but also improves the input index system to make efficiency measurement results more accurate; Second, this article constructs a Tobit regression to study the influencing factors of its green efficiency; Finally, useful suggestions in this paper are provided for accelerating the construction of a maritime power, improving the efficiency of the use of marine resources, and realizing the high-quality development of the marine economy.

2. Model construction and index system

2.1. Construction of index system

In view of the complexity of marine economic activities and the availability of data, it is impossible to take all the input-output production factors into account when carrying out the empirical research on the marine green economy efficiency. Therefore, on the basis of fully considering the definition of marine green economy efficiency and the characteristics of input-output factors of marine economic production activities, this paper conducts an indicator system that can reflect the changing trend of marine green economy efficiency of coastal province (city, district).

2.1.1. Input factor Index. Based on the definition of marine green economy efficiency and the character of marine economy activities, this paper selects labour input factors, capital input factors and resource input factors as investment indexes.

In terms of labor input factors, it is measured by the number of sea related employees in each province(city, district).

On the basis of fully considering the three main industries of marine economy, this paper uses the entropy method to construct the resource input index for the length of coastal large-scale production ports, the number of star hotels and the marine fishing output, so as to serve as the resource input factors.

As far as the capital input factors are concerned, it is measured by the number of sea related employees in each province (city, district). Since there is no data related to the stock of marine fixed assets in the Marine Statistical Yearbook, combined with the research results of Jun Zhang [11], runsong Jia and Guangshun he, this paper uses the perpetual inventory method to calculate the stock of fixed assets in coastal provinces, i.e.:

$$K_{it} = (1 - \delta_i) \times K_{it-1} + I_{it}/P_t \quad (1)$$

Where, K_{it} indicates the capital stock of i province(city, district) in t year, K_{it-1} indicates the capital stock of the province(city, district) in the $t-1$ year. I_{it} represents the fixed capital formation of i province (city, district) in t year. P_t represents the fixed asset price index in year t. Taking into account that the faster the regional economic development, the higher the capital depreciation rate, so this paper learn from Runsong Jia's [12] practice of using different depreciation rates for different provinces, that is, choose i depreciation rates δ_i for i province (city, district).

The actual capital stock of each province is measured with the stock of fixed assets in 2005 as the base period, and then the calculation method of Guangshun He [13] is used to calculate marine capital stock of coastal provinces (cities, districts). The formula is:

$$Y_{it} = K_{it} \times (GOP_{it}/GDP_{it}) \quad (2)$$

Where, Y_{it} represents the marine capital stock of i provinces (city, district) in the t year. GOP_{it} represents the gross marine economic product of i province (city, district) in the t year, GDP_{it} represents gross domestic product of i province (city, district) in the t year.

2.1.2. Output factor Index. The gross ocean product and industrial wastewater discharge are selected as output indicators to respectively represent the desirable output and undesirable output of the marine economy.

2.2. SBM-DEA model of undesirable output

Data envelopment analysis (DEA) is analytical tool which uses mathematical programming to evaluate the relative efficiency of decision-making units (DUM) with multiple inputs and outputs. Based on the Slack-Based Measure(SBM) model considering undesirable output proposed by tone[14], this paper introduces the slack variable into the objective function, which not only overcomes the slack problem of input-output index in the traditional DEA model, but also overcomes the impact of undesirable output on the accuracy of efficiency measurement. Therefore, this paper uses the SBM-DEA model considering the undesirable output to measure the marine economy green efficiency of 11 coastal provinces (cities, districts). Not only that, this index value is used to evaluate the sustainable development ability of marine economy.

2.3. Data sources

The research period of this study is 2006-2016, and the research objects are 11 coastal provinces (cities and districts). Relevant data comes from China Marine Statistical Yearbook, China Statistical Yearbook and China Urban Statistical Yearbook, which finally are compiled into panel data of 11 coastal provinces (cities, districts) from 2006 to 2016.

3. Empirical results of marine green economy efficiency

Samples in this study included 11 coastal provinces (cities, districts) in China, each of which is treated as a DUM under DEA-SBM analysis. The MATLAB 2017(b) software is adopted for calculating results displayed in Table 1. According to the research results of Zhanxin Ma [15], the score of marine green economy efficiency is divided into four level: if the efficiency value is in the interval of $0.8 \leq \rho^* \leq 1$, it

IOP Conf. Series: Earth and Environmental Science **865** (2021) 012029 doi:10.1088/1755-1315/865/1/012029

indicates that the sustainable development ability of marine economy is of highest level; if the efficiency value is in the interval of $0.5 \leq \rho^* < 0.8$, it indicates that the sustainable development ability of marine economy is of the high level; if the efficiency value is in the interval of $0.4 \leq \rho^* < 0.5$, it indicates that the sustainable development ability of marine economy is of the median level; if the efficiency value is in the interval of $0 \leq \rho^* < 0.4$, it indicates that the sustainable development ability of marine economy is of the low level.

Generally speaking, the empirical results show that Tianjin, Guangdong, Shandong, Shanghai and Jiangsu are of highest level. Compared with other coastal provinces, the marine economic production in these provinces (cities, districts) consumes the least resources and causes minimal damage to the marine environment; Zhejiang and Fujian are of high level; Liaoning and Hebei are in the median efficiency value interval. This shows that the sustainable development ability of this provinces (cities, districts are in the median level; Guangxi and Hainan are in the low efficiency value range. Compared with other coastal provinces (cities, districts), their marine economy development has the greatest damage to marine resources and environment. This shows that the sustainable development ability of marine economy in Guangxi and Hainan are of low level.

The reasons why the sustainable development ability of marine economy in coastal provinces (cities, districts) have the above characteristics are:

(1) The marine industrial structure of Tianjin, Guangdong, Shandong and Shanghai is "3 2 1". These coastal provinces (cities and districts) rely on the development of the tertiary industry to extend the industrial chain of marine products and services, improve the added value of products, and enhance the sustainable development ability of marine economy. Not only to optimize the industrial structure, Tianjin, Guangdong, Shandong, Shanghai and Jiangsu vigorously relies on marine science and technology innovation to improve the utilization efficiency of marine resources, reduces the amount of invalid investment in production and reduces the degree of environmental pollution.

(2) Although the marine industrial structure of Zhejiang and Fujian has achieved "3 2 1", there is still a certain degree of waste in traditional industries, such as shipping industry and marine salt industry. The technical level of marine industry is low and there is a phenomenon of more invalid investment, which aggravates the deterioration of the marine environment.

(3) The marine primary industries of Liaoning, Guangxi and Yunnan account for a large proportion of GOP, which indicates that their marine industrial structure is unreasonable. Moreover, insufficient investment in marine science and technology has led to more low-end seafood and failed to increase the added value of output. In recent years, Hainan Province long-termly presents the "3 2 1" industrial structure, and its marine green economy efficiency shows a rising state. However, its marine tourism resources development fails to achieve scale efficiency, and the investment scale is small, which leads to the marine resources not being fully utilized.

Table 1. Marine green economy efficiency of coastal provinces (cities and districts).

	TianJin	HuBei	LiaoNing	ShangHai	JiangSu	TianJin	ZheJiang	FuJian	ShangDong	GuangDong	GuangXi
2006	1.00	1.00	0.45	1.00	0.86	0.50	0.52	1.00	1.00	0.38	0.28
2007	1.00	1.00	0.45	1.00	1.00	0.55	0.56	1.00	1.00	0.38	0.34
2008	0.84	1.00	0.41	1.00	1.00	0.50	0.50	1.00	1.00	0.30	0.30
2009	1.00	0.49	0.41	1.00	1.00	0.53	0.52	1.00	1.00	0.31	0.30
2010	1.00	0.57	0.43	1.00	1.00	0.55	0.54	0.72	1.00	0.32	0.36
2011	1.00	0.47	0.43	1.00	1.00	0.55	0.49	1.00	1.00	0.27	0.36
2012	1.00	0.47	0.41	1.00	1.00	0.55	0.49	1.00	1.00	0.27	0.35
2013	1.00	0.47	0.42	1.00	1.00	0.55	0.48	1.00	1.00	0.28	0.38
2014	0.74	0.56	0.43	1.00	1.00	0.72	1.00	1.00	1.00	0.41	0.33
2015	1.00	0.46	0.36	1.00	1.00	0.49	0.53	1.00	1.00	0.27	0.37
2016	1.00	0.41	0.29	1.00	1.00	0.49	0.54	1.00	1.00	0.27	0.43

IOP Conf. Series: Earth and Environmental Science **865** (2021) 012029 doi:10.1088/1755-1315/865/1/012029

4. Analysis on the factors influencing the efficiency of marine green economy efficiency

4.1. Selection of influencing factors and model setting of marine green economy efficiency

The rationalization of marine industrial structure can not only show the coordination degree of marine industry, but also show the reasonable utilization degree of marine resources. With the transfer of marine production factors from low productivity to high productivity in marine industries, the marine industrial structure tends to be rationalized, the utilization rate of marine resources is greatly improved, the pollution of the ocean is constantly reduced, and the marine green economy efficiency is enhanced; Investment in marine pollution control can slow down the deterioration of marine resources, but there are different views on whether it can improve the sustainability of marine resources; Under other conditions unchanged, advanced marine science and technology can improve the production efficiency of various industries, enhance the added value of products and services, and enhance the sustainable development ability of marine economy.

Therefore, in the comprehensive consideration of the above situation, this paper chooses marine green economic efficiency(Y) as the dependent variable to measure the sustainable development ability of marine economy in China's coastal provinces(cities, districts); chooses marine industrial structure(X_1), marine environmental regulations(X_2) and the level of marine science and technology (X_3) as explained variable to study the factors affecting the sustainable development of the marine economy. The structure of the marine industry(X_1) is expressed by the proportion of the gross output of the marine tertiary industry in the gross marine production. Marine environmental regulation(X_2) is measured by investment in marine environmental industrial pollution control. The level of marine science and technology(X_3) is measured by the number of patents accepted by marine scientific research institutions.

After calculating the marine green economy efficiency, this paper explores the factors that affect the marine green economy efficiency. Since the efficiency score calculated in the previous article is between 0 and 1, the data type is merged data. If the least squares method is used to explore the factors affecting the marine green economy efficiency, the results will be biased. Therefore, this paper uses the Tobit model to construct a regression model. The model is as follows:

$$Y_{it} = \beta_0 + \beta_1 X_{1it} + \beta_2 X_{2it} + \beta_3 \ln X_{1it} + \varepsilon_t \quad (3)$$

Where β_0 represents a constant; $\beta_{1\sim3}$ respectively represents the parameter value of each variable; ε_t represents the error term. According to this model, this paper uses STATA16.0 for empirical analysis, and the empirical results are shown in Table 2.

Table 2. Regression analysis on influencing factors of marine green economy efficiency in coastal provinces.

Y	$Coef.$	$P > \|t\|$	[95% $Conf.$	$Interval$]
X_1	0.0167	0.0050	0.0052	0.0282
X_2	-0.0034	0.2530	-0.0092	0.0024
X_3	0.1541	0.0000	0.1060	0.2022
$\_cons$	0.2392	0.0180	0.0417	0.4366
$var(e.Y)$	0.1529		0.1012	0.2311

4.2. An empirical analysis of the factors influencing the marine green efficiency economy

According to the empirical results:

(1) Marine industrial structure has a significant positive impact on marine green economy efficiency, and its estimated value of the parameter is 0.0167. This shows that reasonable marine industrial structure can improve the sustainable development ability of marine economy, but its driving force is relatively weak. The possible reasons are as follows. with the continuous development of marine industrial structure adjustment and the continuous optimization of marine industrial structure, 11 coastal provinces (cities, districts) in China have basically achieved the "3 2 1" industrial structure, and their influence on the sustainable development ability of marine economy has gradually weakened.

(2) Environmental regulation tools have a negative impact on marine green economy efficiency. The estimated value of the parameter is -0.0034, but the estimated result does not pass the significance test. This shows that the negative effect of environmental regulation tools on the sustainable development ability of marine economy is not significant. Even so, the empirical results show that the estimated value of the parameter is negative, which indicates that the marine economy development mode of "pollution before treatment" is not desirable, which is not conducive to the sustainable development ability of marine economy.

(3) The scientific and technological level of marine industry has a significant positive impact on marine green economy efficiency, the estimated value of the parameter is 0.1541. This shows that advanced marine science and technology can improve the sustainable development ability of marine economy. This proves that the advanced marine science and technology can improve the production efficiency of marine industry, that is, to obtain more output with less input of resources and more green production mode.

5. Conclusion and Enlightenment

5.1. Conclusion

Based on the resource input index constructed by entropy method, this paper uses the SBM-DEA model of undesirable output to evaluate the marine green economy efficiency of 11 coastal provinces (cities and districts) from year 2006 to 2016. Tobit regression model is established to analyze its influencing factors. the main conclusions are as follows:

(1) The sustainable development ability of marine economy of 11 coastal provinces(cities and districts) can be divided into four levels: Tianjin, Guangdong, Shandong, Shanghai and Jiangsu are of highest level, and the input and output are in the ideal state; Zhejiang and Fujian are in the high level; Liaoning and Hebei are in the median level; Guangxi and Hainan are in the low level.

(2) Marine industrial structure has a significant positive impact on marine green economy efficiency. However, its driving force to marine green economy efficiency is relatively weak, the estimated value of the parameter is 0.0167. The scientific and technological level of marine industry has a significant positive impact on the marine green economy efficiency, with an estimated value of 0.1541. Environmental regulation tools have a negative impact on the marine green economy efficiency, but they have not passed the significance test.

5.2. Inspiration

According to the evaluation results of marine green economy efficiency and the analysis conclusions of its influencing factors, in terms of enhancing the sustainable development ability of marine economy, we get the following enlightenment:

At the theoretical level, this paper uses SBM-DEA model to evaluate marine green economy efficiency in coastal provinces, which not only improves the accuracy of efficiency evaluation, but also provides a case material for the existing theoretical research of efficiency evaluation and makes the existing theoretical research more abundant. Moreover, on the basis of the existing efficiency evaluation index system, this paper improves the resource input factors to make the results more in line with the definition of marine green economic efficiency.

At the practical level, through the empirical research on the influencing factors of marine green economy efficiency, it provides guidance for improving the sustainable development ability of China's marine economy in the future. First of all, scientific and technological innovation of marine economy is the main focus to improve the sustainable development ability of marine economy. In order to accelerate the transformation of marine industrial production from scale efficiency to technical efficiency and improve the sustainable development ability of marine economy, we should increase investment in marine science and technology innovation, create a good atmosphere for marine science and technology innovation. Then, we should develop low energy consumption and environment-friendly marine industry and eliminate backward and low value-added production capacity. We will promote the adjustment of marine industry to a more advanced "three two one" industrial structure and develop coastal tourism services with higher added value. Finally, the marine environmental protection mode that pollution end treatment can not be an effective way to enhance the sustainable development of marine economy.

Acknowledgments
This research was supported by the Specialty and Innovation Projects for Colleges and Universities of Guangdong Province (Research of Database Construction and Evaluation Technology in China's Regional Marine Economy, Project No. 2020WTSCX097). This research was supported by "13th Five-Year Plan" 2020 Research Projects for Guangdong Province Educational Science (Research on the Dynamic Evolution Law and Influencing Factors Identification of Intelligent Efficiency in Guangdong-Hong Kong-Macao Greater Bay Area, Project No. 2020GXJK105)

References

[1] Zhao Xin, Peng Yong, Ding Lili 2016 Spatiotemporal evolution and influencing factors of marine green economic efficiency in China [J]. *JOURNAL OF HUNAN AGRICULTURAL UNIVERSITY* (SOCIAL SCIENCE EDITION), **17 (05)**: 81-89
[2] Wu Shujuan, Luo Shaoyu, Xiao Jianhua 2015 Measurement of green efficiency of China's marine economy and its influencing factors [J]. *Industrial technology economy*, **2015 (11)**: 105-112
[3] Gai Mei, Liu Dandan, Qu Benliang 2016 Spatial temporal differences and influencing factors of green marine economic efficiency in coastal areas of China [J]. *Ecological economy*, **32 (12)**: 97-103
[4] Xu Lin, Lai Qianru, Yan Cheng 2019 Measurement of financial support efficiency for China's marine economic development: An Empirical Study Based on three marine economic circles [J]. *Statistics and information forum*, **34 (03)**: 64-75
[5] Gao Tianyi, Chang Fei, Gao Guoqi 2018 Research on transformation and upgrading of industrial structure of marine economy in Qingdao -- Analysis and evaluation based on scientific and technological innovation efficiency [J]. *Management review*, **30 (12)**: 42-48
[6] Qin Linggui, Shen Tiyan 2020 Does Technological Innovation Promote the High Quality Development of China's Marine Economy-Empitical Test based on Effect of Technological Innovation on GTFP [J]. Scientific *and technological progress and countermeasures*, **37 (09)**: 105-112
[7] Zou Wei, Sun Caizhi, Qin Xionghe 2017 Spatial evolution and influencing factors of marine economic efficiency in Bohai Rim region based on bootstrap DEA model [J]. *Geographic science*, **37 (06)**: 859-867
[8] Zhang Xixi, Zheng Shan, Yu Lianghong 2020 Measurement of green efficiency of China's ocean carbon sink fishery and its spatial spillover effects[J]. *China Rural Economy*, **2020(10)**: 91-110
[9] Jamnia A R, Mazloumzadeh S M, Keikha A A 2015 Estimate the technical efficiency of fishing vessels operating in Chabahar region, Southern Iran[J]. *Journal of the Saudi Society of Agricultural Sciences*, **14(1)**: 26-32.
[10] Wanke P F 2013 Physical infrastructure and shipment consolidation efficiency drivers in Brazilian ports: A two-stage network-DEA approach[J]. *Transport Policy*, **29**: 145-153.
[11] Zhang Jun, Wu Guiying, Zhang Jipeng 2004 Estimation of China's provincial physical capital stock: 1952-2000 [J]. *Economic research*, **2004 (10)**: 35-44
[12] Jia Runsong, Zhang Sican 2014 [J]. *Statistical research*, **31 (11)**: 35-42
[13] He Guangshun, Ding Lili, Song Weiling 2014 Theory, method and practice of marine economic analysis and evaluation [M]. Beijing: Ocean Press
[14] Tone K 2004 Dealing with undesirable outputs in DEA: A slacks-based measure (SBM) approach[J]. *Presentation At NAPW III*, Toronto, **2004**: 44-45.
[15] Ma Zhanxin 2010 Data envelopment analysis model and method [M]. Science Press

IOP Conf. Series: Earth and Environmental Science **865** (2021) 012030 doi:10.1088/1755-1315/865/1/012030

The analysis of impact factor on temporal and spatial distribution characteristics of petroleum pollution in surface sediments of Laizhou Bay

Fan Gao[1, 2, 4], Qinbang Sun[1, 2, 4], Pengfei Zhao[1], Yumiao Zhang[3], Wenbo Jiang[1] and Chong Zhang[1]

[1]National Marine Environmental Monitoring Center, Dalian 116023, PR China;
[2]State Environmental Protection Key Laboratory of Coastal Ecosystem, Dalian 116023, PR China;
[3]College of Civil Engineering and Architecture, Zhejiang University, Hangzhou 310058, PR China

[4]Corresponding author's e-mail: gaofan@nmemc.org.cn

Abstract. According to the monitoring data of five cruises in Laizhou Bay from 2013 to 2018, the content, pollution status and spatial distribution characteristics of petroleum hydrocarbons in the surface sediments were analyzed. The results showed that the content range of petroleum hydrocarbons in the sediments was from 0.00~ 377.00 mg / kg, with an average of 63.34 mg/kg, which indicated that the pollution extent was light. The obtained monitoring data were analyzed through DPS18.10 software and PRIMER 7 software. The analysis results showed that the spatial dispersion extent of inter-annual petroleum hydrocarbon content at the monitoring stations was high, the difference was not significant, the clustering similarity was low, and the grouping was not obvious. The results showed that the spatial distribution trend of petroleum hydrocarbon contents in the sediments was uniform, and there were high value areas in some areas.The spatial distribution of petroleum hydrocarbons in Laizhou Bay was dominated by the distribution of ports and the input of human activities. The inter-annual variation of petroleum hydrocarbon contents in the sediments was not obvious. The area where the decrease or increase of petroleum hydrocarbon content in each monitoring station was obvious, was in the SW-NE direction.

1. Introduction

Oil pollution has become one of the prominent problems of the global Marine ecological environment, which is easy to become the focus of social attention [1]. Marine ecological environment pollution may be caused by all links of offshore oil exploration, exploitation, transportation and use [2, 3]. Petroleum hydrocarbons are homogeneous mixtures composed of alkanes, olefin, aromatics, heterocyclic aromatics and other complex molecular structures with multiple components. Once the offshore oil spill occurs, it will not only affect the photosynthesis of phytoplankton, but also reduce the primary productivity of the waters [4]. Moreover, it will reduce the Marine biodiversity index, cause the decline of Marine biological resources and weaken the Marine ecosystem. Furthermore, it affects human health through biological enrichment and food chain transmission [5, 6]. Petroleum hydrocarbons in seawater may partially enter the sediments through physical processes and pollute the Marine sedimentary environment, which may do harm to benthic organisms [7].

Laizhou Bay, one of the three major bays of the Bohai Sea, is a semi-closed shallow sea with weak water exchange capacity and poor self-purification ability to pollutants. It is an important spawning and feeding ground for the three major Marine organisms of the Yellow Sea and the Bohai Sea, as well as a famous fishing ground in China [7-9]. The shipping industry of Laizhou Bay is well-developed. Laizhou Port, Longkou Port, Weifang Port and many other ports are distributed along the coast. And there are rich submarine oil and gas resources. With the rapid development of regional society and economy and the increasing intensity of human activities, the pressure of Marine ecological environment protection has become increasingly prominent. At present, ecological environment studies of Laizhou Bay mainly involve benthic fauna community, ecological environment quality, nutrient content, persistent pollutants and seawater intrusion [10-14]. However, there are few reports on petroleum hydrocarbons in the surface sediments of Laizhou Bay. The content and distribution of petroleum hydrocarbons in sediments can not only reflect the pollution status of the bay, but also have great significance for the study of the influence of petroleum hydrocarbons on benthic organisms [8,9,15-17]. In this paper, the content and spatial distribution of petroleum hydrocarbons in surface sediments were analyzed based on the monitoring data of five voyages in Laizhou Bay from 2013 to 2018. The single factor index method and coefficient of variation method were used to analyze and evaluate the pollution status of petroleum hydrocarbons, in order to provide scientific support for the control of petroleum hydrocarbon pollution, the analysis of pollution sources and the management of ecological environment protection in the surface sediments of Laizhou Bay.

2. Materials and methods

2.1. Sample source and analysis

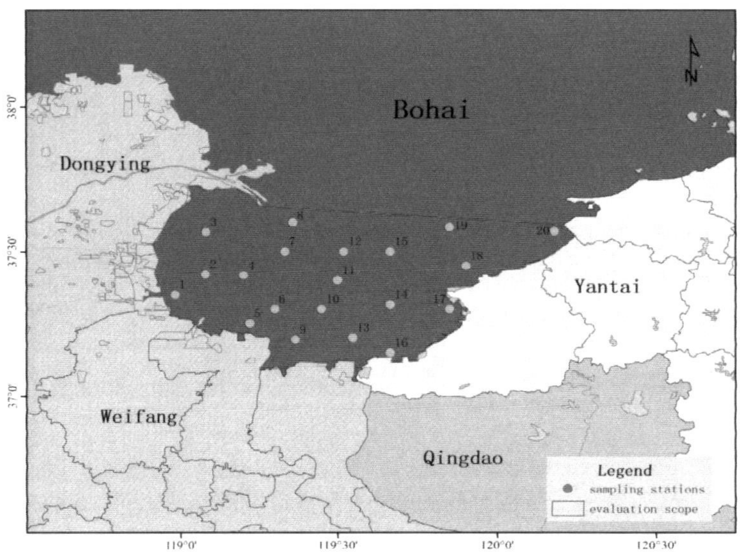

Figure 1. The distribution of sampling stations in Laizhou Bay.

In this study, data came from the national Marine sediment quality monitoring work organized and implemented by the former State Oceanic Administration from 2013 to 2018 and a total of 20 surface sediment monitoring stations in Laizhou Bay were selected (Figure 1). The monitoring factor was petroleum hydrocarbon, and the sampling depth was 0~10cm. Sample collection and analysis was carried out in accordance with Marine Monitoring Specifications (GB17378-2007) of China, Marine Monitoring Technical Specification Part 2: Sediments (HY/T147.2-2013) and "Basic Technical Requirements for Annual Marine Ecological Environment Monitoring" of China, and the quality control was carried out in accordance with "Annual National Marine Ecological Environment Monitoring Quality Assurance Work Plan" of China.

2.2. Data evaluation and analysis methods

2.2.1. Evaluation method. In order to evaluate the environmental quality of the monitoring station in the simplest and most intuitive way, the single factor index method is used to obtain the results of the oil hydrocarbon content in the monitoring station, such as the rate of reaching the standard or exceeding the standard, and the multiple of exceeding the standard. The single factor index method is used to evaluate each pollution factor separately. By comparing the measured value of the evaluation factor with the standard reference value, the result of reaching the standard rate or exceeding the standard rate and exceeding the standard multiple can be obtained through statistics.The mathematical expression of single-factor index calculation is as follows [9,18]:

$$I_i = C_i/S_i \tag{1}$$

Where I_i is the pollution index of the i factor, C_i is the measured value of the i factor and S_i is the standard reference value of the i factor. When $I_i > 1$, it shows that the evaluation factor in sediment has exceeded the standard and the environment has been polluted. The greater the I_i value, the higher the pollution level. And when $I_i \leq 1$, the evaluation factor in sediment has not exceeded the standard and the environment has not been polluted.

Meanwhile, in order to quantitatively reflect the difference of the fluctuation degree of petroleum hydrocarbon at the spatial scale among the monitoring stations and inter-year, and objectively evaluate the important influence degree of the evaluation factors, this paper used the coefficient of variation method to evaluate the variation degree of petroleum hydrocarbon content at each station and inter-year. The mathematical expression of coefficient of variation is as follows [15]:

$$CV = SD/\bar{X} \tag{2}$$

Where CV is the coefficient of variation, and SD、$\bar{X}$ is the standard deviation and average value of the evaluation factors of each monitoring station.

2.2.2. Data analysis. ArcGIS10.5 software was used to map the spatial distribution of petroleum hydrocarbons in sediments, in which the inverse distance weight method was used for data interpolation. DPS18.10 software was used to carry out mathematical statistical calculation on the monitoring data, and Kruskal-Wallis non-parametric test method was used to analyze the significance of the differences of petroleum hydrocarbons between different monitoring stations and between different years. And then the statistical results were compared. When $P \geq 0.01$, the difference was not significant.

PRIMER 7 software was uesd for Cluster Analysis and Multi-Dimensional Scaling. The results are tested by Stress. When Stress<0.20, the analysis results have certain credibility; when Stress<0.10, the order of the analysis results is better; when Stress<0.05, the analysis results are very representative.

3. Results and discussion

3.1. The contents and changes of petroleum hydrocarbons in the surface sediments

The survey and monitoring data between 2013 and 2018 in Laizhou Bay were statistically analyzed, and the results were as shown in Table 1. The petroleum hydrocarbon content in the surface sediments of Laizhou Bay from 2013 to 2018 ranged from 0.00 mg/kg to 377.00mg/kg, with a mean of 63.34 mg/kg. According to the Quality of Marine Sediments (GB18668-2002) of China, the petroleum hydrocarbon content during 2013-2018 conformed to the Class I standard for Marine sediments (500 mg/kg or less). The order of the average content of petroleum hydrocarbons was: 2015>2013>2017>2018>2016. Due to the petroleum hydrocarbon contents fluctuated little between 2013-2018, showing a slight downward trend overall (Figure 2). The flow of rivers around Laizhou Bay arrived a peak in 2015, thus the average value of petroleum hydrocarbons contents in 2015 was the highest in the five years. The results further illustrated that the rivers were the main sources of petroleum hydrocarbons pollution. The average concentration per unit water mass of Laizhou Bay was 0.0118 ~0.0585 mg /L between 2013-2018, which met the national class I (class II) seawater standard of China. From Table 2, the I_i values of minimum, maximum and average petroleum hydrocarbon contents were all less than 1, indicating that the pollution degree of petroleum hydrocarbon in the surface sediments of Laizhou Bay was slight. Between 2013-2018, the coefficient order of annual variation of petroleum hydrocarbons contents was: 2018>2016>2013>2017>2015. The mean value was 1.12 and the variation level was high, indicating that the petroleum hydrocarbon content of each monitoring station and each year fluctuated greatly on the spatial scale, the spatial dispersion degree was high, the spatial distribution of petroleum hydrocarbon content was not uniform, and there might be some abnormal values in the local areas, which was basically consistent with the performance in Figure 2.

Table 1. The statistical characteristic values of petroleum hydrocarbons in the surface sediments of Laizhou Bay from 2013 to 2018.

Year	Content range (mg/kg)	The average (mg/kg)	The median (mg/kg)	The standard deviation (mg/kg)
2013	0.00~366.00	68.93	48.25	82.13
2015	7.11~199.00	72.63	49.28	57.71
2016	3.14~250.10	50.13	12.60	64.35
2017	6.95~268.00	65.03	40.10	60.54
2018	5.30~377.00	59.99	32.90	85.71

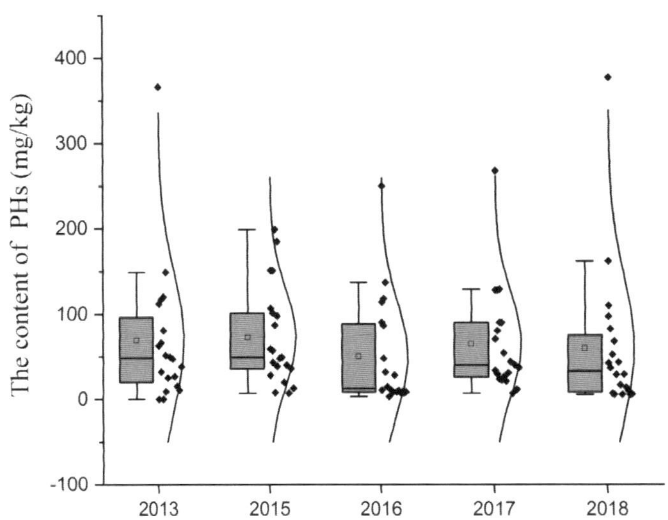

Figure 2. The content change of petroleum hydrocarbon in the surface sediments of Laizhou Bay from 2013 to 2018.

Table 2. The single factor index and coefficient of variation of petroleum hydrocarbon in the surface sediments of Laizhou Bay from 2013 to 2018.

Year	Single factor index (I_i)			Coefficient of variation (CV)
	minimum value	maximal value	average value	
2013	0.00	0.73	0.14	1.19
2015	0.01	0.40	0.15	0.79
2016	0.01	0.50	0.10	1.28
2017	0.01	0.54	0.13	0.93
2018	0.01	0.75	0.12	1.43

3.2. The differences and cluster analysis of petroleum hydrocarbons in the surface sediments

The difference of petroleum hydrocarbon content in the surface sediments of 20 monitoring stations from 2013 to 2018 was analyzed. The significance test of approximate chi-square distribution showed that the P value is approximately equal to 0.19, and the P value of Monte Carlo sampling probability analysis was also approximately equal to 0.19. Two kinds of inspection results values were higher than 0.01, indicating that the difference of petroleum hydrocarbon content between years was not obvious [19, 20]. This was consistent with the analysis conclusion in Section 2.1 that petroleum hydrocarbon content fluctuated little between 2013 and 2018.

The results of cluster analysis showed that the overall clustering similarity of petroleum hydrocarbon content among monitoring stations and inter-years was low, and the grouping was not obvious (Figure 3). In order to further verify the cluster analysis results, the Bray-Curtis similarity index was used to calculate MDS graph (Figure 4). The stress coefficient value in the Figure was 0.03, indicating that the Figures in the results was representative. Figure 4 showed that the distribution of petroleum hydrocarbon content in the surface sediments of 20 monitoring stations in the five voyages from 2013 to 2018 was relatively uniform on the whole, and the overall grouping was not obvious, which further indicated that it was consistent with the results of cluster analysis.

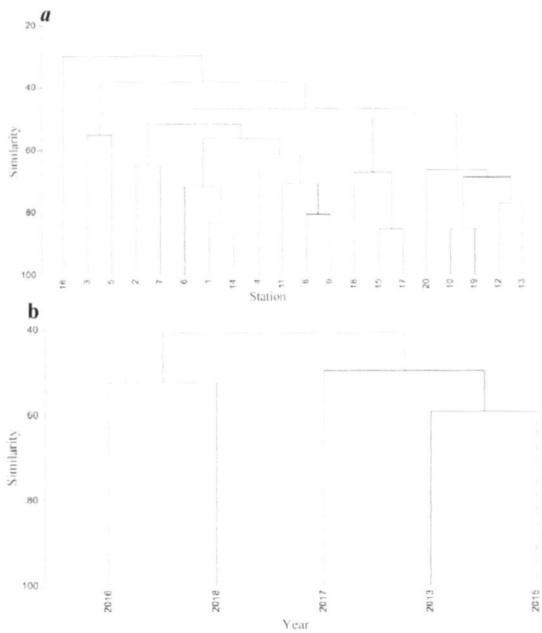

Figure 3. The cluster analysis of petroleum hydrocarbon in the surface sediments of Laizhou Bay: (*a*) the stations cluster analysis results; (b) the annual cluster analysis results.

Figure 4. MDS analysis of petroleum hydrocarbon in surface sediments.

3.3. The spatial distribution of petroleum hydrocarbons in the surface sediments

The spatial distribution trend of petroleum hydrocarbon content in surface sediments of five voyages from 2013 to 2018 was generally consistent (Figure 5), which showed a uniform spatial distribution. The level of spatial difference and variation was basically the same, but there were differences in local areas, mainly manifested as certain abnormal values in local areas. There were two high values of petroleum hydrocarbon content in sediments in the evaluation area, one was station No. 4 of 2013 and 2018 (366 mg/kg and 377 mg/kg, respectively) and the other was station No. 20 of 2016 (250.10 mg/kg). In general, the two areas of high petroleum hydrocarbon content were basically distributed with the monitoring station as the graph of the center of the circle, with the highest value near the center of the circle and the outer edge gradually decreasing. The spatial distribution of petroleum hydrocarbon content in surface sediments of the monitoring station was consistent with the distribution of ports in Laizhou Bay. Weifang Port was near the high value area of the monitoring station No. 4, and Longkou Port was near the high value area of the monitoring station No. 20. Therefore, the results of the paper were consistent with the research results that petroleum hydrocarbon pollutants were mainly derived from human activity input.

The annual variation chart (2013-2018) in Figure 5 indicated that the inter-annual variation of petroleum hydrocarbon content in surface sediments in the study area was generally small. The variations range were from -151.12mg/kg to 141.84mg/kg, and the vast range of the evaluation area varied from -50 mg/kg to 0mg/kg. The areas with large changes in the reduction amount were mainly in the coastal waters of Weifang Port and the waters around the No. 4 monitoring station, while the areas with large changes in the increase amount were mainly in the waters around the No. 5 and No. 10 monitoring stations. The areas with obvious changes in the reduction amount or increase amount at each monitoring station were all in the SW-NE direction.

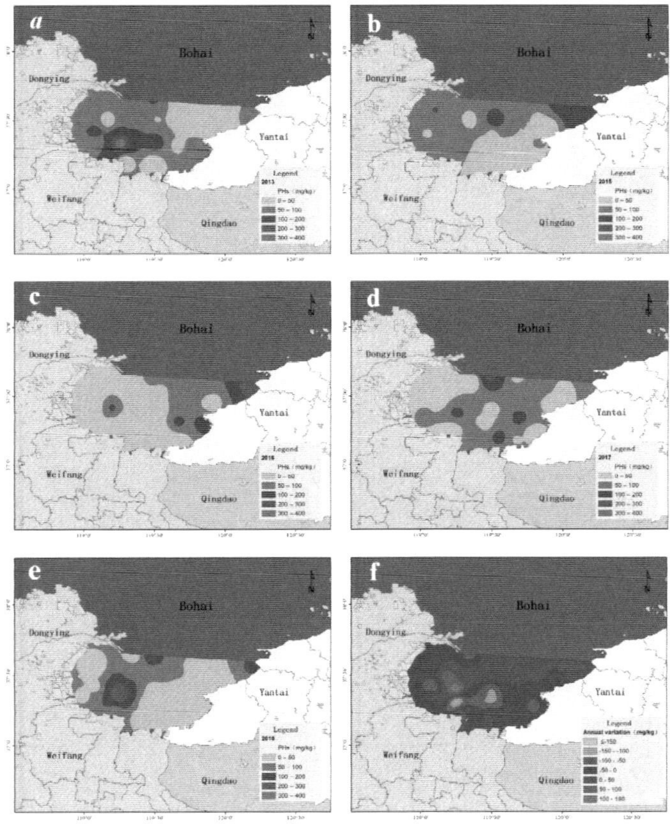

Figure 5. The spatial distribution of contents and annual variation of petroleum hydrocarbon in the surface sediments of Laizhou Bay: (*a*-e) the spatial distribution of petroleum hydrocarbon in the surface sediments of Laizhou Bay; (f) the annual variation of petroleum hydrocarbon contents in the surface sediments of Laizhou Bay.

4. Conclusion

The petroleum hydrocarbon content in the surface sediments of Laizhou Bay from 2013 to 2018 ranged from 0.00 mg/kg to 377.00mg/kg, with an average value of 63.34mg/kg. The pollution degree was slight, and the spatial dispersion degree of petroleum hydrocarbon content among monitoring stations and inter-years was high.

There was no significant difference in petroleum hydrocarbon content between different years. The overall clustering similarity of petroleum hydrocarbon content was low and the grouping was not obvious. The results of MDS analysis were consistent with those of cluster analysis.

The spatial distribution trend of petroleum hydrocarbon content in sediments was generally consistent, showing a uniform spatial distribution, and there were high content areas in some areas. The inter-annual variation of petroleum hydrocarbon content in sediments from 2013 to 2018 was generally small, and the areas with obvious changes of decreasing or increasing amount at each monitoring station were all in the SW-NE direction.

Acknowledgement

The work was supported by the National Key Research and Development Program of China（No. 2018YFC1407604）and Dalian Youth Science and Technology Star Project(2018RQ77).

References

[1] Ozigis M S, Kaduk J D and Jarvis C H et al 2020 Detection of oil pollution impacts on vegetation using multifrequency SAR, multispectral images with fuzzy forest and random forest methods *Environmental Pollution* **256** 113360

[2] Samelak I, Balaban M and Antić M et al 2020 Geochromatographic migration of oil pollution from a heating plant to river sediments *Environmental Chemistry Letters* **18(2)** 459-466

[3] Suneel V, Rao V T and Suresh G et al 2019 Oil pollution in the Eastern Arabian Sea from invisible sources: A multi-technique approach *Marine Pollution Bulletin* **146** 683-695

[4] Shai Y, Rubin-Blum M and Angel D L et al 2021 Response of oligotrophic coastal microbial populations in the SE Mediterranean Sea to crude oil pollution; lessons from mesocosm studies *Estuarine Coastal and Shelf Science* **249** 102-107

[5] Karlapudi A P, Venkateswarulu T C and Tammineedi J et al 2018 Role of biosurfactants in bioremediation of oil pollution-a review *Petroleum* **4(3)** 241-249

[6] Okafor Y I 2018 The effects of oil pollution on the marine environment in the Gulf of Guinea—the Bonga Oil Field example *Transnational Legal Theory* **9(3-4)** 254-271

[7] Zhu H, Yu Y and Liu X 2020 Macrobenthos Community and diversity in Bohai Sea *Marine Science Bulletin* **22(01)** 32-42

[8] Yuan W, Wang J and Zuo T et al. 2020 Macrobenthos Community Structure and its dynamic change in Laizhou Bay *Acta Oceanographica Sinica* **42(06)** 52-61

[9] Qin H, Liu X and Zhang J et al 2016 Distribution characteristics of oil in Laizhou Bay *Marine Environmental Science* **35(05)** 739-742

[10] Zou Y 2020 Pollution characteristics source apportionment and ecological risk of petroleum hydrocarbons and heavy metals in three Bohai bays *University of Chinese Academy of Sciences*

[11] Liu Y 2010 Distribution and source analysis of persistent organic pollutants in Bohai Sea sediments *Dalian Marine University*

[12] Zhuo W 2019 Response of fouling animal community to heavy metal stress in Haikou Bay *Hainan University*

[13] Chen M, Kang P and Li Y 2017 Effect of nitrogen and phosphorus pollution load on the water quality of the east coast of Laizhou Bay *Hydropower and energy science* **35(02)** 58-60

[14] Wu J, Wu Y and Lin J et al 2018 Study on groundwater management and seawater intrusion prevention in coastal areas of the Yellow Sea and Bohai Sea *China environmental management* **10(02)** 91-92

[15] Luo W, Shao M and Zhou L 2004 Application of cluster analysis function in water environmental quality assessment of Dalian port *Journal of Dalian Maritime University* **04** 51-54

[16] Zhang X 2017 Research on decision support technology for marine oil spill emergency response in Liaodong Bay *Dalian Marine University*

[17] Li X 2020 Ecological impact of reclamation on Macrobenthos Community in Bohai Sea *University of Chinese Academy of Sciences*

[18] Xu Y, Qu K and Zhao J et al 2017 Comprehensive assessment of summer environmental quality in the Central Bohai Sea *Progress of fishery science* **38(02)** 8-15

[19] Li L, Jiang M and Wang Y et al 2014 Pollution characteristics of petroleum hydrocarbons in sediments from the Yangtze River Estuary and adjacent waters *China Environmental Science* **34(03)** 752-757

[20] Li L, Jiang M and Shen X et al 2014 Distribution characteristics and pollution assessment of petroleum hydrocarbons in the Yangtze River Estuary and adjacent waters *Environmental Chemistry* **33(08)** 1366-1372

IOP Conf. Series: Earth and Environmental Science **865** (2021) 012031 doi:10.1088/1755-1315/865/1/012031

Assessment of volatile organic pollution control technologies in automotive industry

Chonglei Ren[1], Xuefeng Liu, Jiabao Ren and Lei Wang

China Automotive Technology and Research Center, Tianjin, China

[1]Corresponding author's e-mail: renchonglei@catarc.ac.cn

Abstract. Volatile organic compounds (VOCs) are one of the main factors causing air pollution in China. In the automotive industry, spraying, drying, and other processes will produce a large amount of VOCs pollution. How to select suitable treatment techniques is the major problem for VOCs control. In this paper, a treatment technology evaluation index system was constructed based on the characteristics of VOCs emission in automobile manufacturing. Using the Analytic Hierarchy Process (AHP) model to comprehensively evaluate VOCs governance technologies in the automotive industry. The results showed that the priority of the six VOCs treatment techniques is as follows: runner + RTO > activated carbon + TO > activated carbon + RTO > runner + RCO > activated carbon + CO > activated carbon + RCO.

1. Introduction

Volatile Organic Compounds (VOCs) are defined as organic compounds that participate in atmospheric photochemical reactions, or as organic compounds defined in accordance with relevant regulations.In the process of atmospheric chemical reaction, VOCs can significantly change the physical and chemical properties of the atmosphere, and harm the atmospheric environment through long-distance transmission across the border.Through the analysis of air pollution sources, VOCs are important precursor substances that affect the generation of secondary organic aerosols, PM2.5 and ozone [1]. The photochemical smog caused by VOCs is harmful to human body and plant health. Benzene, toluene, xylene and other benzene series of the VOCs are irritating to the human respiratory tract, sensory system and skin, have harmful effects on the human viscera and nervous system, and can even produce three effects on the human body [2].

In recent years, VOCs have become the focus of China's air pollution prevention and control work, The "13th Five-Year" Volatile Organic Matter Pollution Prevention and Control Work Plan "proposes that during the" 13th Five-Year "period, the total VOCs emissions in key regions and key industries should be reduced by more than 10%. As a key industry, the automobile manufacturing industry is required to promote VOCs emission control in vehicle manufacturing, auto parts manufacturing and other fields, and to build efficient terminal treatment facilities to achieve emission standards. At present, VOCs terminal treatment technologies is mainly divided into recovery technologies and destruction technologies. Recovery technologies includes activated carbon adsorption-desorption, solvent absorption, condensation method, etc. Destruction technologies includes combustion method, plasma flame method, biodegradation method, and concentration-combustion technologies. Different technologies have certain differences in governance effect, application object, and scope of application. Different enterprises have different characteristics of VOCs emission sources, so it is necessary to choose targeted VOCs governance technologies. Therefore it is necessary to establish an evaluation system of VOCs governance technologies in the automotive manufacturing industry and provide a scientific technology selection method for VOCs governance in the industry.

The Analytic Hierarchy Process (AHP) is one of the most commonly used and effective group decision-making methods for technology selection, which is used to evaluate complex multi-standard decision-making methods involving subjective judgment [3][4]. AHP has been applied in the selection of VOCs treatment technologies in oil storage and transportation, packaging, and printing industries [5,6]. In this paper, VOCs treatment technologies evaluation index is selected according to the emission characteristics of the automobile manufacturing industry, and the weight of the evaluation index of VOCs treatment technologies is calculated by using AHP, and a complete evaluation model is constructed, which provides a reference for the selection of VOCs treatment technologies of automobile manufacturing enterprises in the future.

2. Materials and Methods

2.1. Evaluation methods

The AHP method has been widely used in safety science, environmental science, and other fields [7]. The AHP method builds the process of decision problem into an analytic hierarchy model that includes decision target layer, evaluation criterion layer (i.e., first-level indicator layer), second-level indicator layer and the alternative scheme layer to be evaluated. By solving the eigenvector of the judgment matrix, the weight of each element in each level of the model is obtained for the corresponding elements in the next level. The index weight is normalized and the weight of the lowest index relative to the decision target is obtained. Determining the pros and cons of each index of the alternative scheme, and determining the weight of each scheme under each index through the judgment matrix, and calculating the final score of each scheme for the decisive goal through the way of weighted sum, the highest score is identified as the optimal technology. The basic steps are as follows [8]: (1) To establish an analytic hierarchy model with three levels, including decision target layer, index layer, and alternative technology layer. (2) Construction of judgment matrix. (3) Calculate the ranking weight vector of the first-level index and the second-level index and carry out a consistency check. (4) Calculate the final score of the alternative plan for the decisive goal. (5) Evaluate the scheme according to the final calculation results.

2.2. Evaluate technical route

With the ultimate goal of pollutant emission reaching the standard in a long term and stably, a pollution control technology evaluation model was established that comprehensively considered the pollution treatment capability, input and operation cost, technologies maturity, and other indicators of different technology paths. The main steps are shown in Figure 1.

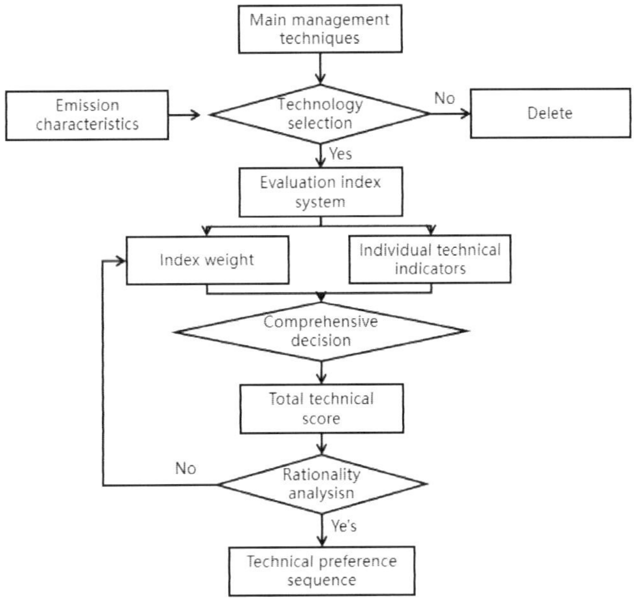

Figure 1. Technologies assessment roadmap.

A) Technical preliminary screening. Through data, literature research, and field investigation, the available pollution control technologies is selected, and some unsuitable VOCs treatment technologies is eliminated through the characteristic parameters of enterprise pollution emission, and the list of alternative VOCs treatment technologies is formed.

B) Technical index system determination. Selected technical index, economic index, environmental index, and other first-level evaluation indexes, and refined the second-level evaluation index. Build pairwise judgment matrix, invite experts to score the importance of the index according to the order of the first level index and the second level index, and normalize the importance of the second-level index to get the weight of the decision goal, to determine the evaluation index system of VOCs governance technologies.

C) The technical index score is determined. In this study, through the preliminary investigation and expert evaluation of VOCs governance technologies, combined with pairwise judgment matrix to calculate the secondary index scores of each alternative technologies.

D) Technical score calculation. Input each VOCs governance technology index weight and score, use linear weighted sum method to calculate the score of alternative technology for the decision goal, according to the final score, determine the priority of VOCs governance technology.

3. Results and discussion

3.1. VOCs emission characteristics of automotive manufacturing industry

VOCs pollution in the automotive industry is mainly caused by the use of solvent-containing products such as coatings and cleaning agents. According to the report released by China Coatings Industry Association, in 2017, China produced 28 million cars and 1.7 million tons of vehicle coatings, among which the proportion of paint used by car manufacturers (original OEM paint) was 53-54%, about 900,000 tons. In the production of various coatings, the proportion of water-based paint with a solvent content of 10% is 2/3, and that of oil-based paint with a solvent content of 50% is 1/3. The total consumption of solvent in the automobile manufacturing industry is about 210,000 tons. In the process of vehicle manufacturing, primer spraying, middle coating, top paint spraying, leveling, drying, and other processes will cause a large number of VOCs emissions. In the process of automobile

IOP Conf. Series: Earth and Environmental Science **865** (2021) 012031 doi:10.1088/1755-1315/865/1/012031

manufacturing, spraying, and drying VOCs pollution production links and their pollution production characteristics are shown in Table 1.

Table 1. VOCs pollution production link and its characteristics.

process		methods	Air volume(m^3/h)	concentration	VOCs emission ratio	characteristics
Spraying	Primer	Closed, 75% internal circulation, 15% fresh air	150000-700000	Low concentration	60-85%	Moisture content is high
	Primer surfacer					
	Base coat					
	Clear coat					
Drying	Electrophoresis drying	Closed, centralized discharge through the exhaust system	3000-60000	The relatively high	10-15%	High temperature (about 140℃)
	Primer surfacer drying					
	Clear coat drying					

3.2. Evaluation of VOCs control technologies in the automotive manufacturing industry

According to the technical assessment route, this section evaluates the optimal control technology of VOCs treatment in the spraying process of the automotive manufacturing industry.

3.2.1. The technology of primary. According to the preliminary investigation, the scope of application of VOCs governance technologies is sorted out, and the specific scope of application of VOCs is shown in Figure 2.

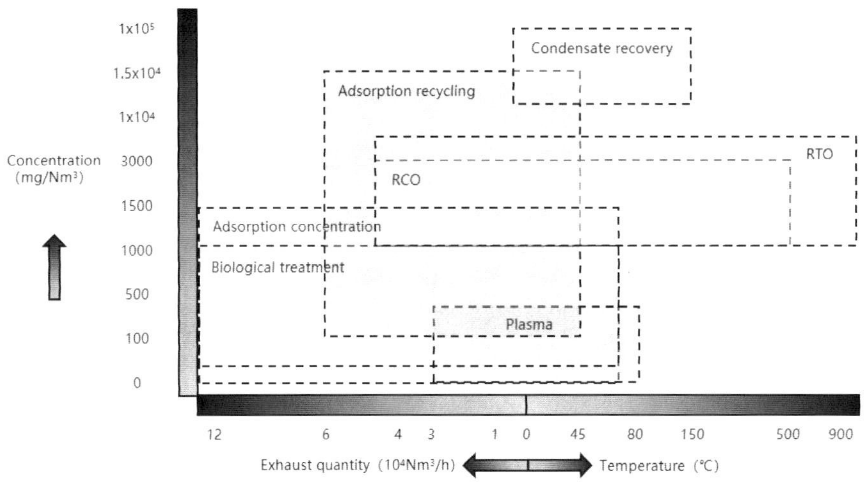

Figure 2. VOCs governance technologies scope of application.

Due to the complexity of VOCs waste gas composition and nature and the limitations of a single treatment technology, in many cases, the use of a single technology is often difficult to meet the treatment requirements, but also uneconomic. Using the advantages of different treatment technologies, the combined treatment process can not only meet the emission requirements, but also reduce the operating cost of purification equipment. Therefore, in recent years, the combination of two or more purification technologies in the treatment of organic waste gas has received great attention, and has been rapidly developed. For example, the combined process of condensation and adsorption can be adopted for the treatment of high concentration organic waste gas. Low concentration organic waste gas can adopt the combined process of adsorption concentration and condensation recovery or incineration, etc. Usually, we can choose the appropriate end treatment technology according to the composition of pollutants, the nature of pollutants and the discharge conditions (air volume, concentration, temperature, particulate matter content, etc.).

Combined with the characteristics of production and pollution of large air volume and low concentration of VOCs exhaust gas produced by automobile manufacturing spraying process, the candidate VOCs treatment technologies was screened and eliminated. After the elimination, Suitable for spraying process treating technologies belong too the adsorption of VOCs concentration + thermal combustion type device, containing activated carbon concentration + TO, activated carbon concentration + CO, activated carbon concentration + RTO, activated carbon concentration + RCO, wheel enrichment + RTO, wheel enrichment + RCO. The preliminary screening results and descriptions of VOCs technologies evaluation are shown in Table 2.

Table 2. VOCs technologies evaluation preliminary screening results.

Processing technologies	Feasibility	Instructions
Adsorption recycling	×	Subsequent combustion treatment is required
Adsorption concentration	×	Subsequent combustion treatment is required
Biological treatment	×	Operating temperature is too low
Low temperature Plasma technology	×	Operating exhaust volume and temperature range is too narrow
Condensate recovery	×	Operating concentration is too high
RTO	×	Spraying exhaust gas concentration is too low
RCO	×	Spraying exhaust gas concentration is too low
Adsorption concentration + thermal combustion	√	Adsorption includes activated carbon and runner concentration, combustion includes RTO, RCO, etc

3.2.2. AHP evaluation model construction. According to the decision target and the VOCs technologies evaluation index system composed of the first and second level indicators of VOCs governance technologies, an AHP analysis hierarchy structure model is constructed, including the decision target layer, the first level indicator layer, the second level indicator layer and the alternative technologies layer. The analysis hierarchy model is shown in Figure 3.

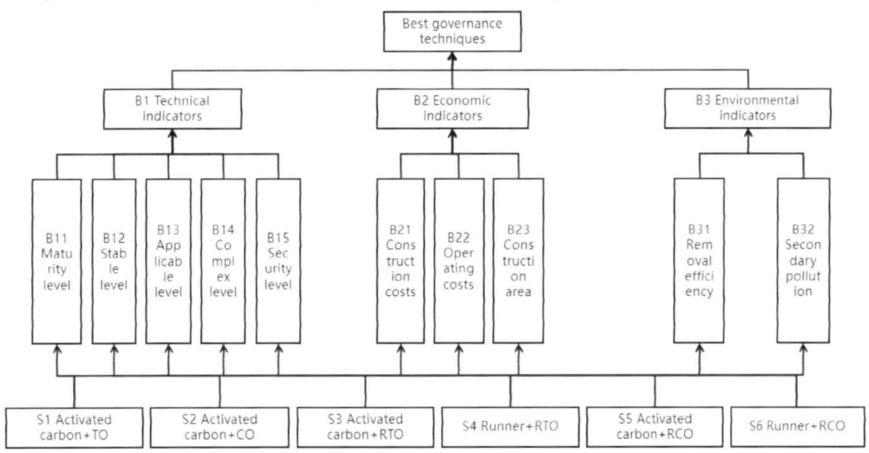

Figure 3. VOCs governance technologies evaluation analysis hierarchy model.

According to the expert assignment results of index weight, combining the research of VOCs treatment technologies, in view of the primary index layer (B1, B2, B3), secondary index layer (B11, B12, bl3, B14, B15, B21, B22, B23, B31, B32) as well as the solution layer (S1, S2, S3, S4, S5, S6) two judgment matrix structure, after the completion of the need to construct judgment matrix with consistency check, calculate the weight of each layer elements relative decision-making target. Take the first-level index weight judgment matrix R1 as an example.

$$R1=\begin{bmatrix} 1 & 3 & 1/5 \\ 1/3 & 1 & 1/7 \\ 5 & 7 & 1 \end{bmatrix} \tag{1}$$

The maximum characteristic root of R1 λ_{max}=3.0658, and its characteristic vector W= (0.1932, 0.0833, 0.7235), of which three components are the weight values of three elements B1, B2, and B3 in the first-level indicator layer B relative to the decision target. According to the consistency test formula of the judgment matrix:

$$C_i = (\lambda_{\max} - n)/(n-1) \tag{2}$$

$$C_I=0.0329 \tag{3}$$

Find the random consistency index R_i, when n = 3, R_i = 0.58, $C_R = C_i/R_i$ =0.0567 is less than 0.1, so the matrix conforms to the consistency test, and the index weight value in W can be applied.

Under the first-level index B1, λ_{max} = 5.3965, corresponding eigenvector W_1 = (0.0822, 0.1674, 0.2766, 0.0483, 0.4256). Among them, the five components are the weight values of the five elements in the second-level indicator layer B: B11 maturity, B12 stability, B13 applicability, B14 complexity and B15 security. The calculated consistency test index C_r = 0.085, less than 0.1.

Under the first-level index B2, λ_{max} = .0183, corresponding eigenvector W_2 = (0.3202, 0.5571, 0.1126). Among them, the three components are the weight values of the three elements in the second-level indicator layer B: B21 construction cost, B22 operation cost, B23 floor area. The calculated consistency test index C_r = 0.0176, less than 0.1.

Under the first-level index B3, λ_{max} = 2, corresponding eigenvector W_3 = (0.75, 0.25). Among them, the two components are the weight values of the two elements in the second-level indicator layer B: B31 removal efficiency, B32 secondary pollution. The calculated consistency test index Cr = 0.0002, less than 0.1.

The weight of all indexes in the second-level index layer is normalized to get the weight value of each second-level index relative to the decision target. The specific weight values of secondary indicators are shown in Table 3.

Table 3. The weight of each secondary index relative to the decision target.

First level indicator	Secondary indicators	Weight
Technical indicators	Maturity level	0.0159
	Stable level	0.0323
	Applicable level	0.0534
	Complex level	0.0093
	Security level	0.0822
Economic indicators	Construction costs	0.0267
	Operating costs	0.0464
	Construction area	0.0102
Environmental indicators	Removal efficiency	0.5426
	Secondary pollution	0.1809

Combined with experts on the VOCs governance technologies of each secondary index evaluation, the use of judgment matrix VOCs governance technologies index score. The scores of activated carbon concentration + TO, activated carbon concentration + CO, activated carbon concentration + RTO, activated carbon concentration + RCO, runner concentration + RTO, runner concentration + RCO are shown in Table 4.

Table 4. VOCs governance technologies index scores.

Alternative technologies	Maturity level	Stable level	Applicable level	Complex level	Security level	Construction costs	Operating costs	Construction area	Removal efficiency	Secondary pollution
Activated carbon + RTO	0.1667	0.1244	0.2455	0.1073	0.0723	0.2217	0.1461	0.0405	0.2356	0.1005
Runner + RTO	0.1667	0.3767	0.2693	0.4388	0.391	0.103	0.1461	0.0824	0.2356	0.1005
Activated carbon+ RCO	0.1667	0.0534	0.0761	0.044	0.1034	0.0744	0.3045	0.2193	0.0755	0.2389
Runner + RCO	0.1667	0.2112	0.0818	0.1191	0.2264	0.0415	0.3045	0.2193	0.0755	0.2945
Activated carbon + TO	0.1667	0.1576	0.2455	0.2069	0.089	0.3776	0.0365	0.2193	0.2541	0.0600
Activated carbon + CO	0.1667	0.0768	0.0818	0.0840	0.1179	0.1819	0.0623	0.2193	0.1237	0.2056

The VOCs governance technologies index score and the index weight through the linear weighted sum method to get the total score of VOCs governance technologies, the final automotive manufacturing industry spraying process VOCs governance technologies score ranking is: wheel concentration + RTO > activated carbon concentration + TO > activated carbon concentration + RTO > wheel concentration + RCO > activated carbon concentration + CO > activated carbon concentration + RCO. The specific evaluation results are shown in Table 5.

Table 5. VOCs Management technologies evaluation results.

Alternative technologies	Weight	sequence
Runner + RTO	0.2219	1
Activated carbon + TO	0.1929	2
Activated carbon + RTO	0.1859	3
Runner + RCO	0.1453	4
Activated carbon + CO	0.1343	5
Activated carbon + RCO	0.1199	6

Through the analysis of the scores of alternative technologies, it is found that the runner concentration + RTO technology has more outstanding advantages in technical indicators, and the comprehensive evaluation score is higher. However, its economic index has a great disadvantage compared with the runner concentration + RCO. Therefore, based on meeting the VOCs emission standard of exhaust gas emission concentration, the technologies with an operating cost advantage can be selected: runner concentration + RCO technology. The results show that the evaluation of VOCs governance technologies needs to make decisions from many aspects. Enterprises can adjust the weight parameters of each index in the model according to their needs and select the optimal VOCs governance technologies in line with their actual situation.

4. Conclusion

With the gradual tightening of China's environmental standards, it is necessary to establish a systematic and scientific-technical evaluation index system to support the screening of VOCs governance technologies. In this study, the VOCs treatment technology evaluation index system of the automotive manufacturing industry was constructed and combined with AHP to evaluate the VOCs treatment technology of the spraying process. Wheeler concentration + RTO technology has advantages in technical indicators and environmental indicators and is selected as the optimal treatment technology. From the perspective of practical application, the result is more reasonable. The technology evaluation model in this study can be adjusted according to the actual needs of enterprises, which provides strong scientific support for the selection of VOCs governance technology in the industry. At the same time, it is necessary to continue to study the VOCs governance technologies evaluation model in the automotive industry in this paper. The next step can be carried out from the following aspects:

(1) The VOCs governance technologies index database is constructed, the evaluation index value is revised according to the technologies development, and the evaluation system is updated.

(2) Expand the evaluation index system, build the characteristic evaluation model in different situations, and customize the evaluation model of VOCs governance technology for enterprises with different needs.

(3) Starting from the pollution control of the whole process of the automobile manufacturing industry, the evaluation model of combined pollution treatment technologies is constructed to provide the best pollution control scheme of the whole process for enterprises.

References

[1] Chou K T 2019 Analysis of VOCs Harm and Treatment Technologies, *J. Environmental Research and Monitoring*. **32** 13-15.
[2] Zhu X, Liu Y 2019 Characterization and Risk Assessment of Exposure to Volatile Organic Compounds in Apartment Buildings in Harbin, *J. Bulletin of Environmental Contamination & Toxicology*, **92(1)** 96-100.
[3] Zhu J l, Wang G, Huang M H 2019 Research and Practice On Third-Party Evaluation of Environmental Remediation of Village-Level Industrial Zones Based On Analytic hierarchy process, *J. Environment and Sustainable Development*, **44(6)** 132-136.
[4] Huang C C, Chu P Y, Chiang Y H 2008 A fuzzy AHP application in government-sponsored R&D project selection, *J. Omega*. **36** 1038-1052.
[5] Xue P L, Zhang D D, Sun X F 2019 Evaluation and Screening of Volatile Organic Compounds Control Technology in Packaging and Printing Industry, *J. Environment & Sustainable Development*. **44** 79-82.
[6] Wu M E, Cai S Y, Zhao B 2015 Evaluation of Volatile Organic Compound Emission Control Technologies in Oil Storage and Transportation Industry, *J. Chemical Environmental Protection*. **35** 64-68.
[7] Wang H L, Wang J H, Zhu C L 2014 Evaluation and Screening of Volatile Organic Compounds Control Technologies in Packaging and Printing Industry, *J. Environmental Science*. **1** 2503-2507.
[8] Guo J Y, Zhang Z B, Sun Q Y 2008 Research and Application of AHP, J. **18** 148-153.

IOP Conf. Series: Earth and Environmental Science **865** (2021) 012032 doi:10.1088/1755-1315/865/1/012032

Comparative analysis of two hazard rainstorm processes in Hunan affected by the Southwest China vortex

Yan Hu[1,2], Hongwu Liu[1,2,5], Huanqian Liu[3], Minghui Tang[1,2] and Yuting Zeng[4]

[1]Hunan Meteorological Observatory
[2]Hunan Key Laboratory of Meteorological Disaster Prevention and Reduction
[3]Hunan Meteorological Observation Technology Support Center 4
[4]Meteorological Center of Southwest Air Traffic Administration

[5]Corresponding author's e-mail: 45877780@qq.com

Abstract. Based on surface rain-gauge data, Doppler radar products, the National Center for Environmental Prediction (NCEP) reanalysis data and other multi-source datasets, two regional rainstorm processes caused by the Southwest China vortex (SCV) in June, 2015 are diagnosed and analyzed from the aspects of water vapor condition, dynamic and thermodynamic mechanisms, and the causes of the regional rainstorms are discussed. The results show that the two processes belong to the SCV rainstorm, and there is the southwest jet in the middle and lower levels. The high-value area of water vapor flux and the water vapor convergence center can well indicate the rainstorm area. In the first process, two water-vapor-flux transport bands from the Bay of Bengal and the South China Sea converge. In the second process, the water vapor mainly comes from the Bay of Bengal. The extreme values of dynamic and thermodynamic factors all appear near 850 hPa in the rainstorm development stage, and the latitudes of the extreme values are close to the severe rainfall area. The first process has a typical rainstorm vorticity configuration with negative vorticity at upper layer and positive vorticity at lower layer. The second process is mainly driven by the strong positive vorticity at the lower layer. There are strong energy frontal zones near the ground, and the rainstorm area is consistent with the convergence region of cold and warm air. The characteristics of radar echo show that there are mid-level and low-level jet streams in the two rainstorm processes. The first process is dominated by the cumulus precipitation, and the adverse wind region promotes the development of the rainstorm. The second process is dominated by cumuliform-stratiform-mixed cloud precipitation, and the train effect is the main cause of the regional rainstorm.

1. Introduction

Hunan Province is located in the subtropical monsoon region where warm and humid southerly wind prevails in summer, with abundant water vapor. The interannual change of precipitation is large, and the average annual precipitation can reach 1200–1700 mm [1]. Regional extreme rainstorms often occur here, which bring a great threat to people's lives and property. There are many mountains, hills, rivers and a lake in Hunan, the distribution of rainfall center is similar to that of terrain. According to the statistics of extreme precipitation in Hunan from 1951 to 2009, the northeastern Hunan and northwestern Hunan are both the centers of extreme precipitation [2]. The statistical results from 1961 to 2016 have shown that the occurrence frequency of persistent rainstorm in June is the highest, and the regional rainstorm in the western and northern Hunan is the most, showing an increasing trend year by year [3]. Other studies have shown that the rainstorm in the Yangtze River Basin is closely related to the southwest jet [4–5], and the large wind speed zone near the exit of the low-level jet is consistent with the rainstorm area in the Yangtze River Basin [6]. Many scholars have also carried out the cause analysis of the regional rainstorm process. It is found that the convective energy is triggered by the mid-level and low-level shear lines in the initial stage of regional rainstorm [7–8], and the warm shear is one of the main formation mechanisms. During the occurrence and maintenance of rainstorms, the wind shear line has a single direction and the low-level jet often appears in the middle and lower levels [9]. In addition, the surface convergence line is also very easy to lead to mesoscale convective regional rainstorm [10], which is an important factor triggering short-time severe precipitation [11].

In June 2015, there were four regional rainstorm processes in Hunan, and the severe rainfall areas were almost overlapped, mainly concentrated in the central and northern Hunan, resulting in many casualties and huge economic losses and inducing some geological disasters such as serious landslides.The accumulated rainfall was more than twice of the normal years. In process I, the affected population reaches 1.774 million, nine people died or missed due to the disaster, and the direct economic loss reaches 2.29 billion Yuan. In this study, the most severely two regional rainstorms (from June 1 to June 4 and from June 6 to June 8 in 2015) are selected, and the thermodynamic and dynamic mechanisms of the two rainstorms are diagnosed, to provide technical reference for the hazard rainstorm forecasting in the future.

2. Materials and methods

The four kinds of data in this study includes surface rain-gauge data and the Hunan Doppler radar products provided by the China Meteorological Data Network, the spatial resolution of the reanalysis data 1°×1°, the reanalysis data which is provided by the National Center for Environmental Prediction (NCEP), the Global Data Assimilation System (GDAS) data from the National Oceanic and Atmospheric Administration (NOAA) , the temporal resolution is 6h, spatial resolution is 1°×1°.

Using the HYSPLIT4 model from the National Oceanic and Atmospheric Administration (NOAA), the backward trajectory is calculated by Lagrangian method. The model is used to track the 7-day trajectory of the rainstorm center in the two regions, determine the source and transport trajectory of water vapor, and quantitatively calculate the contribution rate of different water vapor channels to the rainstorm area by clustering method. The simulation area selected that is 27 – 31°N, 108 – 114°E, and the initial vertical simulation heights are 1000 m, 1500 m, and 3000 m.

3. Characteristics of the two regional rainstorm processes

From June 1 to June 4 in 2015, there was a regional rainstorm process in the north of central Hunan (hereinafter referred to as process I, Figure 1a), with characteristic of strong locality, long duration and strong rainfall intensity. The severe rainfall began from the night of June 1 and moved from north to south. Specifically, the rainfall was the severest from the night of June 1 to June 3, and the severe rainfall was mainly concentrated in the north of central Hunan. The rainfall significantly weakened and moved southward rapidly on June 4. From June 6 to June 8 in 2015, there was another regional rainstorm process in central and northern Hunan (hereinafter referred to as Process II, Figure 1b). The rainfall began from the west of Hunan on June 6, and there was a large area of severe rainfall on the north of central Hunan on July 7. On June 8, the main rain band moved slightly to the south and steadily maintained in the central part of Hunan.

The statistical analysis (Table 1) shows that the average rainfall of process I is 53.0 mm in the whole province. The accumulated rainfall exceeds 100 mm in 653 townships, of which the accumulated rainfall in 84 townships exceeds 200 mm, and the maximum hourly rainfall is 97.1 mm. The average rainfall of process II is 45.5 mm in the whole province. The accumulated rainfall exceeds 100 mm in 189 townships, of which the accumulated rainfall in three townships exceeds 200 mm and the maximum hourly rainfall is 65.9 mm. The comparison shows that in process I, the rainstorm area is wider, the maximum rainfall and hourly rainfall intensity are stronger and the disaster-causing ability is stronger than that of the process II. In the two processes, there are geological disasters such as debris flow caused by landslide, and the disaster degree of process I is more serious than that of process II.

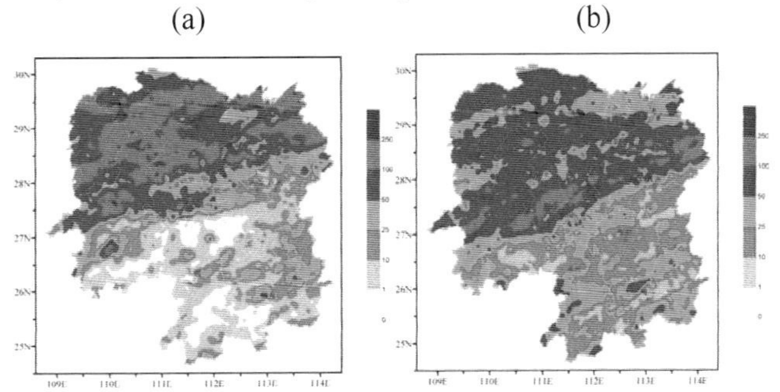

Figure 1. Accumulated rainfall in Hunan (a) from 08:00 BJT (Beijing time, the same below) on June 1 to 08:00 on June 5 and (b) from 08:00 on June 6 to 08:00 on June 9 in 2015 (unit: mm).

Table 1. Rainfall and damage situation from 08:00 on June 1 to 08:00 on June 5 and from 08:00 on June 6 to 08:00 on June 9 in 2015.

Comparison of rainfall and damage situation between the two regional rainstorm processes		
	Process I	Process II
Rainstorm stations (station)	653	1038
Average rainfall (mm)	53	45.5
Daily maximum rainfall (mm)	163.8 (Linxiang County, Yueyang City)	113.8 (Taoyuan County, Changde City)
Maximum hourly rainfall intensity ($mm \cdot h^{-1}$)	97.1	65.9
Duration (d)	4	3
Affected population (thousand)	1774	1471
Affected crop area (hectare)	106.2	74.5

4. The difference of large-scale circulations

The 500hPa upper-level charts in the strongest rainfall periods of the two processes were presented in Figures 2a and 2b. In Process I, the western pacific subtropical high was distributed in blocks. The west ridge point of 588dagpm was 104 ° E, which was more than 20 longitudes westward than that in the same period of the normal year. The northern boundary was 22°N, and the maximum positive anomaly of geopotential height reached 80–100 dagpm. The circulation was a trough-ridge type in mid-high latitude, the upper-level trough moved eastward from Sichuan Basin to the boundaries between Hunan and Hubei. Hunan was located on the northwest edge of the subtropical high and in the warm and humid unstable airflow in the front of the upper-level trough. In Process II, the circulation meridional is large in middle and high latitudes, which is two troughs and one ridge. The central intensity of the northeast cold vortex reached 555dagpm, and the maximum positive distance of the cold vortex reached 80-100dagpm, which was powerful. The northerly airflow behind the cold vortex was conducive to the southward movement of cold air to affect Hunan. The western Pacific subtropical high was distributed in zonal from east to west. The west ridge point of 588dagpm was 95° E, which was 24 longitudes westward than that in the same period of the normal year. The northern boundary was 26°N. the maximum positive anomaly of geopotential height reached 80–100 dagpm, and the intensity was strong. The southwest flow of the trough drived the vortex to the north of Hunan, and the low-level jet transported water vapor to the rainstorm area. The two processes were both influenced by the SCV or the eastern section of the herringbone shear in the middle and low levels, which were the main dynamic driving factors. In Process I, the SCV lasted longer in Hunan, and the intensity of SCV was strengthened under the influence of the upper-level forward-tilted trough, which lasted more than 60 hours from 20:00 BJT (Beijing time, the same below) on June 1 to 08:00 on June 4. Some scholars have found that the SCV is the main cause of rainstorms in the Yangtze River Basin, and it often causes rainstorms in spring and summer in Hunan Province [12].

There are significant differences in circulation background between the two processes. In Process I, the west ridge point of subtropical high is more easterly, the northern boundary is more southerly , which is beneficial for the long wave trough to drive the vortex eastward. Addition to, the strength of the western north pacific subtropical high is more stronger, which hinders the vortex eastward, slows down the eastward movement of the trough, and causes the stagnation of the southwest vortex to be longer. It is also one of the main reasons for a large range of heavy rainfall in the process. In Process II, it is mainly affected by the high-altitude shortwave trough and the southward cold air behind the cold vortex.

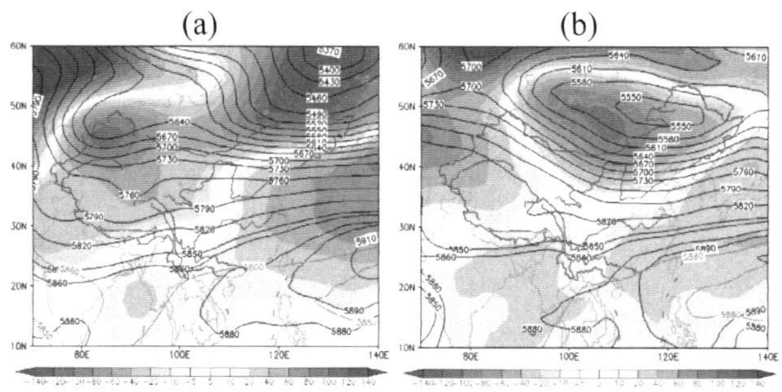

Figure 2. 500-hPa mean geopotential height (black isoline, unit: dagpm) and the anomaly (color spot area, unit: dagpm) at (a) 20:00 on June 1 and (b) 20:00 on June 7. The red isolines represent the climatological mean (unit: dagpm).

5. Difference of water vapor

The summer rainstorm area often distributes around the high-energy tongue and the high specific humidity tongue. The physical diagnostic analysis of water vapor conditions, dynamic and thermodynamic factors is helpful to find out the causes and predictability indexes of disaster-causing rainstorms.

5.1. Water vapor conditions

The water vapor flux vector is a physical quantity that measures the intensity of water vapor transport, and the water vapor flux divergence is a physical quantity that quantitatively analyzes the inflow and outflow of water vapor in the rainstorm area. According to the vertical profile of the water vapor flux in June on average from 1948 to 2020 in Hunan (Figure 3), the water vapor from the ground to 850hPa is growing rapidly, reaching the maximum at 850hPa ($12*10^{-7} g \cdot cm^{-1} \cdot hPa^{-1} \cdot s^{-1}$). As the height increases, the water vapor flux decreases. The lower layer is the main transport layer of rainfall in Hunan.

Before the occurrence of rainstorm in Process I (Figure 4a), 850-hPa water vapor mainly originated from the Bay of Bengal, and there was a significant transport band of water vapor flux from the southern coast of Guangxi to the western Hunan. When the rainstorm 4 developed (Figure 4b), the transport passages from the Bay of Bengal and the transport band from the South China Sea converged over Hunan, maintaining a long time. The strong southwesterly airflow transported water vapor and momentum to the territory of Hunan and Jiangxi, and the water vapor band lasted for more than 36 hours. At 08: 00 on June 3, there was a strong northeasterly inflow, which intersected with the southwesterly airflow in the west and north of Hunan. At the weakening stage (Figure 4c), the transport band of water vapor flux gradually moved southward and weakened, and the main transport pathway was from the Bay of Bengal. From the divergence of water vapor flux(Figure 4a), the strong water vapor sink was located in the southern Chongqing at 08:00 on June 1, with the center of maximum negative value of water vapor flux divergence, and there was a weak water vapor convergence zone in northern Hunan. At 08:00 on June 2(Figure 4b), the cold air from the north moved southward and converged with the strong water vapor band moving northward from the South China Sea. There was an obvious water vapor convergence band on the north of central and western Hunan. At 08:00 on June 3(Figure 4c), the water vapor convergence band moved from northern Hunan to central Hunan. There was a high-value area of water vapor flux divergence in both the northwestern and northeastern Hunan, with the maximum value of -8×10^{-7} g·cm^{-2}·hPa^{-1}·s^{-1}. On June 4, the water vapor flux significantly weakened, and then the water vapor convergence weakened in Hunan. The precipitation gradually weakened from the north to south in Hunan.

In Process II, the water vapor mainly came from the Bay Bengal. Before the rainstorm occurrence (Figure 4d), the strong water vapor flux band distributed northeast-southwestward and was located in central Hunan. In the rainstorm development stage (Figure 4e), there was a weak southward water vapor flux vector in northern Hunan and a significant water vapor flux transport band in central and southern Hunan. The intersection of the two water vapor flux transport bands produced strong water vapor convergence. By 02:00 on June 9 (Figure 4f), the water vapor flux decreased rapidly, and the high-value area retreated to the north of Guangdong and Guangxi. From the analysis of water vapor flux divergence (Figure 4d–f), a water vapor convergence center was formed in Xiangxi Prefecture and Yueyang City at 02:00 on June 8, with the maximum value of -10×10^{-7} g·cm^{-2}·hPa^{-1}·s^{-1}. Until 20:00 on 8 June , the high-value area of water vapor flux divergence stably remained in the central Hunan and its northern area for 18 continuous hours, which was consistent with the heavy rainfall area. After 20:00, the high-value area of water vapor flux and the water vapor convergence center rapidly weakened and moved southward.

The comparative analysis shows that there were the convergence of two water vapor transport bands in the two processes, which resulted in the intersection of cold and warm air and the occurrence of the two regional rainstorm processes. The difference was that the transport band of water vapor flux in Process I maintained longer, and the northerly airflow was also stronger. In Process I, the two water vapor flux transport bands from the Bay of Bengal and the South China Sea converged, which provided sufficient water vapor source for the large-scale rainstorm and resulted in greater accumulated rainfall and hourly rainfall intensity. While in Process II the water vapor mainly came from the Bay of Bengal. In summary, the transport band of water vapor flux vector and the water vapor convergence center can well indicate the rainstorms area. The establishment of water vapor channel in the lower troposphere and the large concentration of water vapor can provide favorable water vapor and unstable conditions for the development of convective systems.

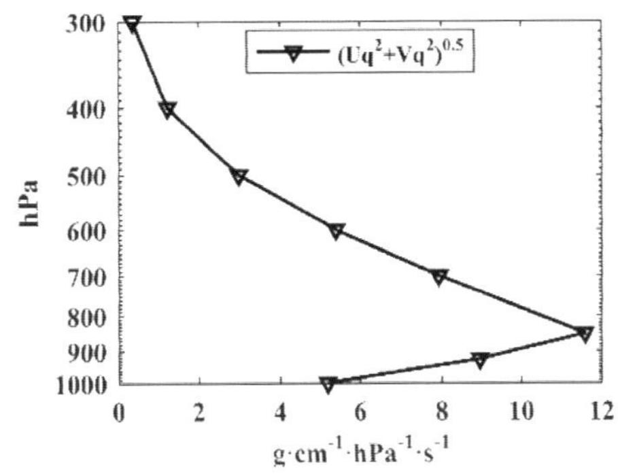

Figure 3. Vertical profile of multi-year average water vapor flux from 1948 to 2020.

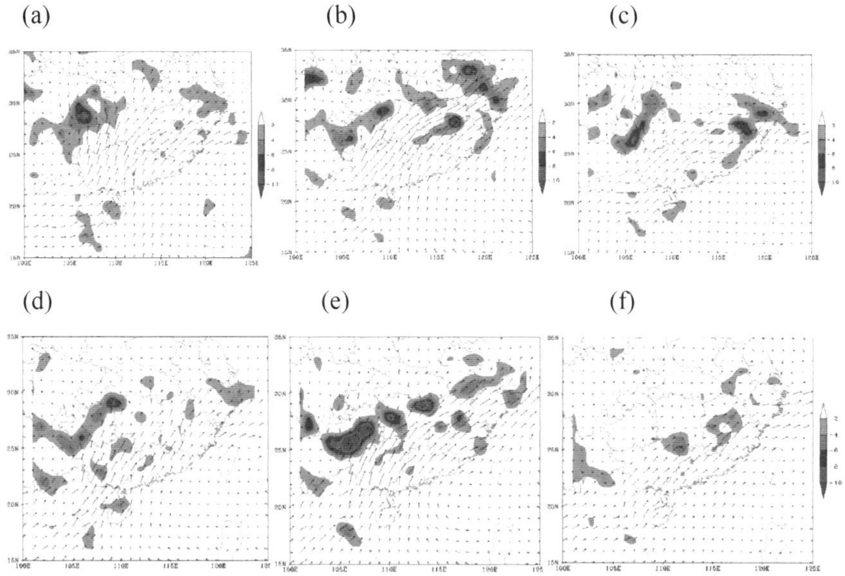

Figure 4. Distributions of water vapor flux and water vapor flux divergence at 850 hPa. The color spot area represents the negative large value of water vapor flux divergence (unit: 10^{-7} g·cm^{-2}·hPa^{-1}·s^{-1}). The arrow is the water vapor flux vector (unit: g·cm^{-1}·hPa^{-1}·s^{-1}). at (a) 08:00 on June 1, (b) 08:00 on June 2, (c) 08:00 on June 4, (d) 02: 00 on June 7, (e) 02: 00 on June 8, (f) 02: 00 on June 9.

5.2. Source and contribution rate of water vapor

The HYSPLIT model is used to track the backward trajectory of the air block in the center of the maximum rainstorm in the two processes(111.0°E, 28.2°N) (111.1°E, 27.8°N) .It is selected as the starting point, and the simulation starting time is 08:00 on June 3rd and 20:00 on June 9th. According to the water vapor transport of multi-year average at 925hPa, 850hPa, 700hPa level, which are corresponding to the altitude of 1000m, 1500m, 3000m, and the backward 168h water vapor trajectory is tracked. The number of cluster water vapor channels was determined by TVS, and the contribution rate of water vapor was analyzed.

Through the backward trajectory tracking comparison (Figure 5), the water vapor of the two processes at 700hPa (3000m) is from the Bay of Bengal. In Process I, the southwest airflow is transported across the central and southern peninsula to the South China Sea, and the southerly airflow is transported northward through the Guangdong and Guangxi. The warm and humid airflow after the Vietnam Ridge Mountains is uplifted and transmitted to the rainstorm center. In Process II, the water vapor is transported through the Yunnan-Guizhou Plateau by the southwest airflow, and the mountains promote the forced uplift of the airflow to the westward airflow to the rainstorm area. During the two processes, the 850hPa vapor mass originated from the rainstorm area in the Yangtze River Basin, and the short-range transmission was the main path. In Process I, the height of the vapor mass fluctuated greatly. In the near-surface process one, there is a water vapor from the East China Sea to the easterly flow through the Yangtze River Basin to the rainstorm center of Hunan. The height of the water vapor mass changes little, the air block is below 500m in the early stage, and rises slightly near the rainstorm area. In Process II, the source and transport trajectory of water vapor are consistent with 850hPa.

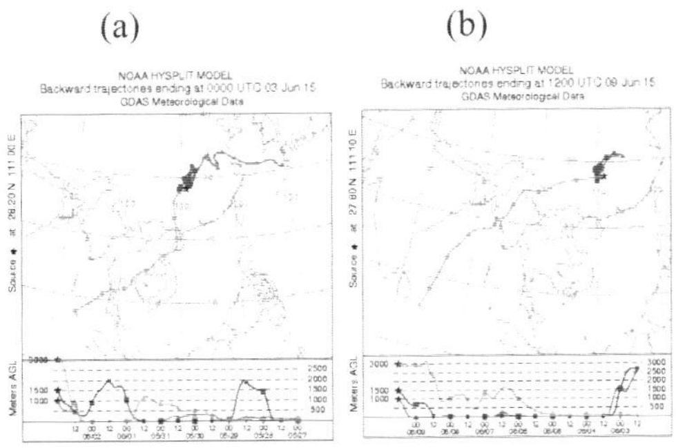

Figure 5. The 168-h back tracking of the air parcel at (a) 08:00 3 June, at (b) 20:00 9 June 2015. (Heights of the air parcel are represented by different colors lines, red line:1000, blue line: 1500, green line: 3000, unit: m)

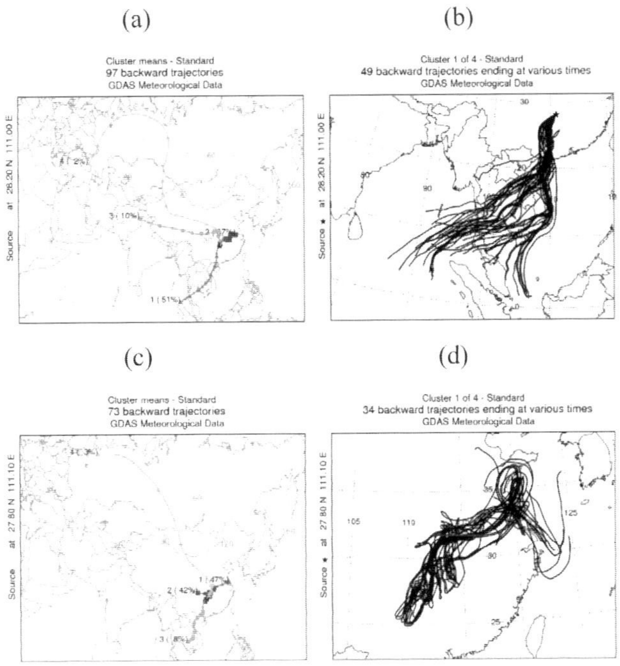

Figure 6. Water vapor transportation paths at (a) 850hPa in Process I, the first water vapor channel in Process I (b), water vapor transportation paths at (c) 850hPa in Process II, the first water vapor channel in Process(d).

Based on the characteristics of 850hPa rainfall in Hunan, the contribution rate of different water vapor transport paths at 850hPa, we analyzed the air mass of the two rainstorm centers. Simulation time of water vapor backward paths is 168h, interval step 12h. The HYSPLIT4 model was clustered to obtain the spatial variance change rate of the two processes (Figure omitted). The TSV all changed abruptly when they were clustered to the four trajectory. The fourth trajectory increased sharply in Process I, and decreased rapidly at the second trajectory. Therefore, four cluster vapor trajectories were selected to analyze the contribution to the rainstorm area.

In Process I (Figure 6a, b), there are four water vapor transport paths at 850hPa with 97 trajectories, which originate from the Viet Nam Peninsula in the Bay of Bengal, and the contribution rate is 51% from the South China Sea. The transportation trajectory from the middle and lower reaches of the

Yangtze River to the rainstorm area reaches 37%, and the two air streams are superimposed to provide abundant water vapor. There are 73 trajectories for the process II(Figure 6c, d), and the southward trajectory from the Yellow Sea through Jiangsu, Anhui, and Hunan to the center of the rainstorm has a contribution rate of 47%. It is the main water vapor transport channel, which may be the northeast cold vortex. The water vapor from the southwest vortex in the eastern part of Sichuan accounted for 42% of the water vapor brought by the northeast airflow from the back rotating southward through the Yellow Sea, and the water vapor transmitted from the South China Sea through the Guangdong and Guangxi regions accounted for 8%. In summary, in Process I, the main water vapor transport sources are the Bay of Bengal and the western Pacific, while in process II, it is mainly from the confluence of the warm and humid airflow in front of the southwest vortex and the northerly airflow.

6. Comparison of dynamic and thermodynamic conditions

6.1. Dynamic conditions

Dynamic uplift is the triggering condition for the rainstorm, and the systematic ascending motion is conducive to the occurrence and maintenance of rainstorm [5]. The cross-sections along 112°E of 850-hPa vorticity in Process I and process II are given in Figure 7a and Figure 7b, respectively. In the initiation and development stages of Process I (Figure 7a), there were two positive vorticity centers in 27°N and 30°N at 20:00 on June 2 and 08:00 on June 3, and the locations and time were corresponded to those of the severe rainfall centers in western and northeastern Hunan. The maximum value of positive vorticity center was $10\times10^{-5}\ s^{-1}$. There was stably weak positive vorticity before 20:00 on June 7 in Process II (Figure 7b). After that time the vorticity increased significantly in Hunan, reaching a maximum value of $12\times10^{-5}\ s^{-1}$ at 14:00 on June 8, and then decreased. Compared with the inappropriate express, the stage of maximum vorticity corresponded to the intense development of precipitation in central Hunan. The analysis shows that the strong positive vorticity appeared at low levels in both processes.

In order to better understand the coordination between the upper-level and low-level vorticities and the relevance with rainstorms, the vertical cross-sections of vorticity are made along 112°E when rainfall was intensively in the two processes (Figure 8). In the two processes the positive vorticity centers both appeared below 500 hPa, and the maximum positive vorticity center in process II was $8\times10^{-5}\ s^{-1}$, which was stronger than that in Process I. The large positive vorticity zone in Process II extended from 850 hPa to 600 hPa, which presented a "Bull's eye" situation with double centers and the thickness was larger than that in Process I. The negative vorticity intensity in the upper troposphere of process II was obviously weaker than that of Process I. The configuration of negative vorticity at upper levels and the positive vorticity at low levels was a typical upper-level and low-level vorticity situation that is conducive to the occurrence and development of severe rainfall. Process II was mainly driven by strong positive vorticity at low levels.

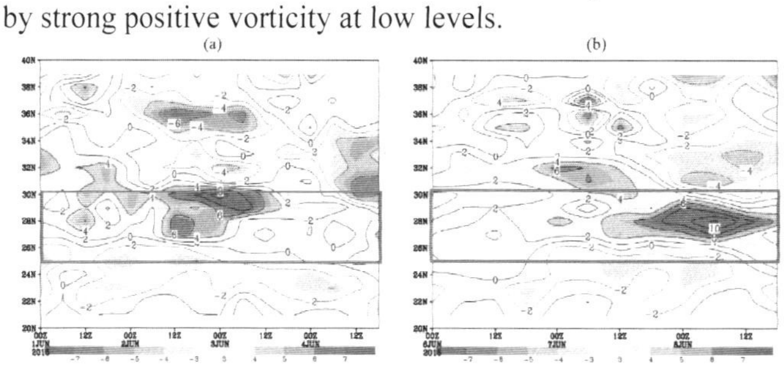

Figure 7. 850-hPa time section of vorticity along 112°E (unit: $10^{-5}s^{-1}$) (a) from June 1 to June 4 and (b) from June 6 to June 8.

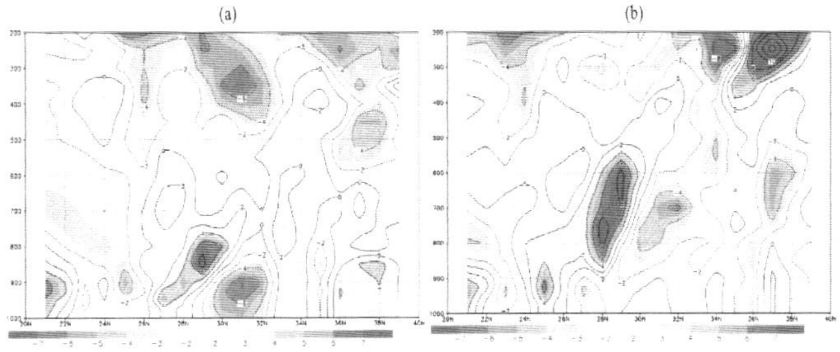

Figure 8. Vertical cross-sections of vorticity along 112°E (unit: $10^{-5}\,s^{-1}$) (a) from June 1 to June 4 and (b) from June 6 to June 8.

6.2. Thermodynamic conditions

Thermodynamic condition is one of the main factors affecting the formation of convective rainstorms, and regional rainstorms often have convective properties. The vertical distribution characteristics of pseudo-equivalent potential temperature (θ_{se}) can characterize the temperature-humidity distribution and vertical stability of the atmosphere. The spatial distribution can reflect the frontal activity, and the vertical velocity is corresponding to the intensity of ascending motion. Therefore, the vertical cross-sections of pseudo-equivalent potential temperature and vertical velocity along 110°E in the two processes are given in Figure 9, so as to analyze the similarities and differences of thermodynamic conditions in the two processes.

From Figure 9, the 350-K high-energy center appeared at 700hPa (25°N–27°N) in the north of central Hunan during the whole precipitation period of Process I. In the rainstorm initiation stage (at 08:00 on June 1), the high-energy tongue above 346 K bulged upwards, extending to 600hPa, and the atmosphere had a convective unstable stratification. The isolines of θ_{se} below 900hPa were dense in the rainstorm development stage (at 08:00 on June 2), and the near-surface θ_{se} reached the maximum of above 360 K, indicating the θ_{se} dense zone for frontal zone, instead of maximum area of θ_{se}..The extension height of the high energy tongue above 350K reached 600 hPa, and the convective instability conditions significantly strengthened. By 08:00 on June 4, the cold air moved southward, and the northern region of central Hunan (25–27°N) was in the rainstorm formation stage (at 20:00 on June 6) in process II (Figure 8). The 346 K isolines of θ_{se} bulged upward, and the expansion height of high-energy tongue reached 500hPa. The atmospheric convection instability was greater than that in Process I. The maximum pseudo-equivalent potential temperature in the near surface was above 366 K in the rainstorm development stage. The density of θ_{se} strengthened, and the energy frontal zone in the near surface was stronger than that in process I.

From the perspective of vertical velocity (Figure 9-10), there were large areas of negative vertical velocity in the rainstorm initiation and development stages on the north of central Hunan (25°N–27°N) during the two processes. The vertical extension height was above 300hPa, and the ascending motion was intense. At the same time, the center of maximum negative value was located at 850hPa. The negative value center in Process I was –2 $pa \cdot s^{-1}$ near 26°N, and the high-value area of negative vertical velocity in Process II had two regional centers, with –1.7 $pa \cdot s^{-1}$ and –1.4 $pa \cdot s^{-1}$ in 26°N and 28°N, respectively. Therefore, the ascending motion in the middle and lower troposphere of Process I was stronger than that of Process II. In the rainfall weakening stage, the vertical velocity of Process I began to weaken and the vertical velocity center of Process II rose to 700hPa, and the velocity remained at a large value, which indicated the rainstorm development in Process II was mainly affected by the strong ascending motion at low levels.

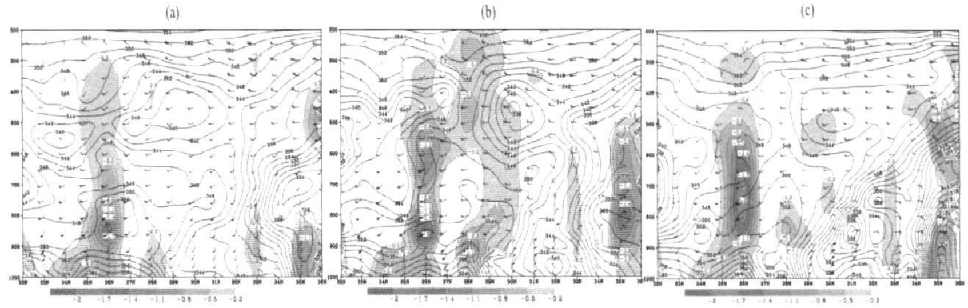

Figure 9. Meridional cross-sections of θ_{se} (isoline, unit: K) and vertical velocity (color spot, unit: $pa\cdot s^{-1}$) along 112°E at (a) 08:00 on June 1, (b) 08:00 on June 2, (c) 08:00 on June 3 and (d) 08: 00 on June 4.

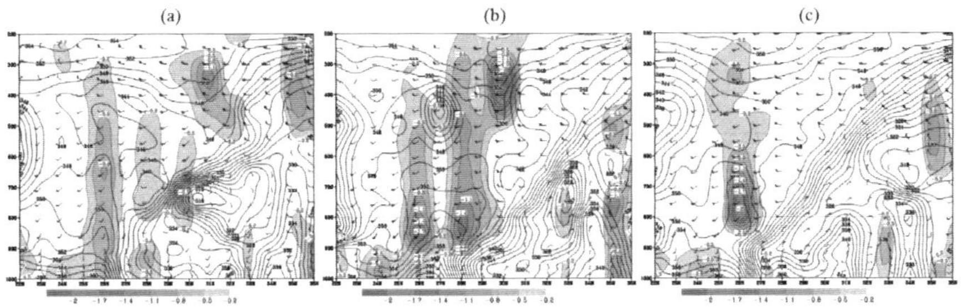

Figure 10. The same as Figure 9, but at (a) 20:00 on June 6, (b) 20:00 on June 7 and (c) 20:00 on June 8 for Process II.

7. Conclusions

Through the comparative analysis of two disaster-causing regional rainstorm processes under the influence of SCV in 2015, the following conclusions are drawn.

Both of the two heavy rainstorm were caused by SCV. The eastward movement of upper-level trough drives the SCV to move eastward and affects Hunan. The unstable warm-humid airflow on the edge of the subtropical high transports water vapor and unstable energy to the rainstorm area.

In the two processes, there is a convergence of two transport bands of water vapor flux, resulting in the intersection of cold and warm air. The intensity of water vapor convergence was enhanced, and caused the regional rainstorm processes. The transport band of water vapor flux in Process I maintains longer, and the northerly airflow is also stronger, resulting in the greater accumulated rainfall and hourly rainfall intensity in Process I. The main water vapor transport sources are the Bay of Bengal and the western Pacific, while in process II, it is mainly from the confluence of the warm and humid airflow in front of the southwest vortex and the northerly airflow.

In the two processes, the maximum values of dynamic and thermodynamic factors all appear near 850hPa in the rainstorm development stage, and the latitude of the extreme value is close to the severe rainfall area. The distribution characteristics of low-level convergence and upper-level divergence appear in both processes. Process I has a typical rainstorm vorticity configuration of negative at upper layer and positive at lower layer. Process II is mainly driven by the strong positive vorticity at low levels.

References

[1] Zhang W, Shou S W, Yang J H 2007 Analysis of anomaly characteristics of extreme precipitation in flood season in the Mid-lower reaches of Yangtze River, *Meteorological monthly*[J], **33(3)**: 61-67

[2] Zhang J M, Liao Y F, Duan L J, et al. 2011 Climatic characteristics of extreme rainstorm events

in Hunan Province during 1960-2009, *Progress in geography*[J], **30(11)**: 1395-1402.

[3] Dai Z J, Cai R H , Peng L L, et al. 2019 Climatic characteristics of regional persistent heavy ran and heavy rainfall pattern over Hunan, *Plateau Meteorology*[J], **38(3)**: 573-582

[4] Tao S Y 1980 *Rainstorm in China*, [M].

[5] Zhu Q G, Lin J R, Shou S W, et al. 2007 *Pinciple of synoptic meteorology*[M].

[6] Wu Z J, Bao Y X, Zhu T, et al. 2019 Study on the mechanism of low-level jet in a heavy rainfall in the Yangtze river basin. *Journal of tropical meteorology*[J], **35(3)**: 409-422.

[7] Zhao Q, Wang N, Li P Y, et al. 2017 Diagnosis of thermal and dynamic mechanisms of two rainstorm processes in Northern Shanxi. *Journal of applied meteorological science*[J], **28(3)**: 340-356.

[8] Yang X X,Wu W, Jiang P, et al. 2013 Contrastive analysis on three extreme heavy rainfalls triggered by warm shear line in Shandong Province, *Meteorological monthly*[J], **39(12)**: 1550-1560.

[9] Fang D X, Dong X N, Zhou G B, et al. 2017 Comparison and application of two kinds vertical wind profile data part II, *Plateau Meteorology*[J], **36(4)**: 971-983.

[10] Yang X B, Dai Y T, Cui M, 2015 Comparison analysis of two shearline-caused rainstorm process in Shandong. *Meteorological science and technology*[J], **43(4)**: 688-696.

[11] Liu X, Ma Y, Lin Y, et al. 2019 Analysis of two cyclone heavy rain processes with defferent Low-level humidity conditions in Qingdao, *Meteorological science and technology*[J], **47(5)**: 818-829

[12] Liu H W,Deng C P, Li G P, et al. 2016 Statistical analysis of the Eastward Southwest vortexes influenced Hunan, *Meteorological and Environmental Sciences*[J], **39(1)**: 59-65.

Effects of hydraulic uncertainty on slope stability in an embankment dam

Xiaoying Chen[1, 4], Haoqing Yang[2], Xiaohui Lu[1], Anling Lu[3]

[1]School of Earth Sciences and Engineering, Hohai University, 8 Focheng West Road, Nanjing, 211106, China;
[2]School of Civil and Environmental Engineering, Nanyang Technological University, 50 Nanyang Avenue, 639798, Singapore
[3]Faculty of Construction and Environment, The Hong Kong Polytechnic University, 11 Yuk Choi Road, Hong Kong.

[4]Corresponding author's e-mail: 1742059852@qq.com

Abstract. It is well-recognized that hydraulic properties are highly variable and heterogeneous suffering from great uncertainty. In this study, the effects of uncertainty of saturated hydraulic conductivity, k_s on pore water pressure head, displacement, and factor of safety of an embankment dam are studied by Monte Carlo simulation based on 500 random fields. The uncertainty of displacement and pressure head will increase as the uncertainty of k_s increases. The standard deviation of the factor of safety will increase with the increase of the uncertainty of k_s, whereas the mean value is almost unchanged. As the coefficient of variation of k_s increases, the factor of safety gradually tends to a normal distribution.

1. Introduction

It is well-recognized that hydraulic properties are highly variable and heterogeneous suffering from great uncertainty [1]. This hydraulic spatial variability may affect preferential flow paths and pore water pressure, and finally change the deformation and stability of slopes [2, 3]. Slope stability analysis is an essential procedure for predicting the safety of soil embankment due to various loading and environmental conditions. Therefore, it is necessary to study the effects of hydraulic uncertainties on seepage, deformation, and stability of embankment dams.

Several scholars have studied the influence of spatial variability of hydraulic parameters on embankment dams of unsaturated soil. Gui et al employed a probabilistic approach to investigate the effects of seepage through an embankment, and the hydraulic conductivity was modeled as a stationary random field [4]. Cho studied the effect of spatial variability of the saturated hydraulic conductivity on the seepage flow in a two-layered embankment [5]. Tan et al assessed the effects of spatial variability of hydraulic parameters on seepage behavior in an earth dam combining Monte Carlo simulation and random field theory [6]. Calamak & Yanmaz investigated the effects of uncertainty in hydraulic conductivity and van Genuchten parameters on transient seepage through the embankment by stochastic analyses [7]. Yang et al proposed an efficient probabilistic back estimation method for characterization of spatial variability by integration of the Karhunen-Loève expansion method, the Polynomial Chaos Expansion method, and the Markov Chain Monte Carlo method [8]. Liu et al investigated the steady seepage through an embankment on a soil foundation with non-stationary random fields [9]. However, most of the studies generally do not involve slope stability analyses or use simple methods such as simplified Bishop method [9]. The influence of hydraulic uncertainty on slope stability is poorly understood.

In this study, the numerical model of an embankment dam is firstly established by Darcy's Law and Mohr-Coulomb model. Random field theory is adopted to account for the spatial variability of hydraulic parameters. The strength reduction method is developed to compute the factor of safety (FOS) of the slope. The effects of uncertainty of hydraulic conductivity, k_s on pore water pressure head, displacement, and FOS are studied by Monte Carlo simulation based on 500 random fields of k_s.

2. Methods

2.1. Random field

In this study, k_s is viewed as a spatial variable to represent hydraulic uncertainty. k_s can be modeled by random field theory following a log-normal distribution. The mean value $\mu_{\ln k_s}$ and the standard deviation $\sigma_{\ln k_s}$ to simulate log-normal distribution can be calculated as:

$$\sigma^2_{\ln k_s} = \ln(1+\frac{\sigma^2_{k_s}}{\mu^2_{k_s}}) \tag{1}$$

$$\mu_{\ln k_s} = \ln \mu_{k_s} - \frac{1}{2}\sigma^2_{k_s} \tag{2}$$

where μ_{k_s} and σ_{k_s} are the mean value and the standard deviation of k_s, respectively. In this study, the mean value of k_s is assumed to be 5×10^{-5} m/s (silty soil).

The correlation length is a parameter to measure the distance over which points of soil samples are significantly correlated in a domain. If the distance of two points is greater than the correlation length, the soil samples at these two points are largely uncorrelated. In probabilistic analysis, the effects of correlation length need to be considered by random field theory, which provides a convenient measure of relevance of soil properties in space [10]. It is reported that the correlation length of lnk_s varies from less than 1 m to around 10 m [1]. Therefore, the horizontal and vertical correlation lengths (l_x and l_z) of the natural logarithm of k_s, Ink_s are assumed to be 10 m and 2 m, respectively. The exponential covariance function $C(\mathbf{x})$ is used to simulate the spatial variability of k_s:

$$C(\mathbf{x}) = \sigma^2_{\ln k_s} \exp\{-[\frac{(x_1-x_2)^2}{l_x^2} + \frac{(z_1-z_2)^2}{l_z^2}]^{\frac{1}{2}}\} \tag{3}$$

where $\mathbf{x} = [(x_1, z_1), (x_2, z_2)]$ represent the coordinates of the two points in the domain.

2.2. Slope stability analysis by strength reduction

Three steps are used to analyze the slope stability. First, the pore pressure in the soil is modeled by Darcy's law. In the unsaturated zone, the hydraulic conductivity is the function of soil suction (Figure 1(a)). Second, the Mohr-Coulomb yield function is used to simulate the soil deformation. The angle of internal friction Φ is given as:

$$\Phi = \text{atan}\left(\tan\phi_u\right)(p<0) + \text{atan}\left(\tan\phi_s\right)(p \geq 0) \tag{4}$$

where ϕ_u and ϕ_s are angles of internal friction for unsaturated and saturated soils, and p is the pore pressure given by Darcy's law.

To compute the factor of safety (FOS), the strength reduction method is developed in the third step. The Mohr-Coulomb material parameters are functions of FOS [11]:

$$c' = \frac{c}{\text{FOS}} \tag{5}$$

$$\Phi' = \text{atan}\left(\frac{\tan\phi_u}{\text{FOS}}\right)(p<0) + \text{atan}\left(\frac{\tan\phi_s}{\text{FOS}}\right)(p \geq 0) \tag{6}$$

IOP Conf. Series: Earth and Environmental Science **865** (2021) 012033 doi:10.1088/1755-1315/865/1/012033

where c is the cohesion of soil; c' and Φ' are the cohesion and angle of internal friction used in Mohr-Coulomb yield function. Decreasing the material parameters results in the reduction of the shear strength of the soil, which eventually leads to the collapse of the embankment. At this moment, the value of FOS is the safety factor of the embankment. In this study, c, ϕ_{u}, and ϕ_s are assumed to be 25 kPa, 30°, and 20°, respectively [12]. The Young's modulus, Poisson's ratio, soil density, and porosity are 5×10^5 kPa, 0.4, 2000 kg/m^3, 0.3, respectively.

2.3. Simulation of embankment dam

The numerical model of an embankment dam is simulated by COMSOL Multiphysics. The geometry of the embankment is shown in Figure 1(b), the height of the embankment is 12 m and the width is 55 m. The water level is 10 m and the possible seepage height is 4 m. Therefore, the left and right sides are constant head boundaries and other boundaries are zero-flux boundaries. In the next step to simulate soil deformation, the hydraulic pressure on the left side of the saturated zone of the embankment is used as boundary load combined effect of gravity. No external stresses are applied in regions of unsaturated soil since pores are considered interconnected and at constant atmospheric pressure.

(a) (b)

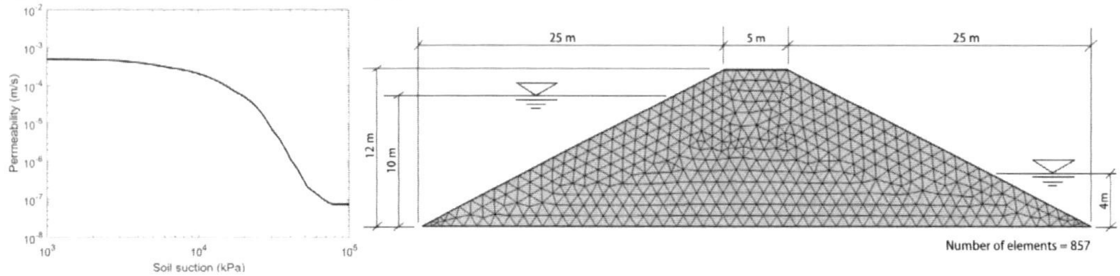

Figure 1. Numerical model of an embankment dam (steady-state): (a) Permeability function; (b) Initial and boundary conditions.

3. Results and discussions

3.1. Deterministic results

The deterministic results are shown in Figure 2. Figure 2(a) displays the displacement of the embankment. The maximum displacement is over 0.34 m located on the side of the reservoir level. The pressure head in the embankment dam is shown in Figure 2(b). In the saturated zone, the pressure head value varies from top to bottom is 0 m to 10 m. In the unsaturated zone, the maximum matric suction head is around 3 m (negative pressure head).

(a)

(b)

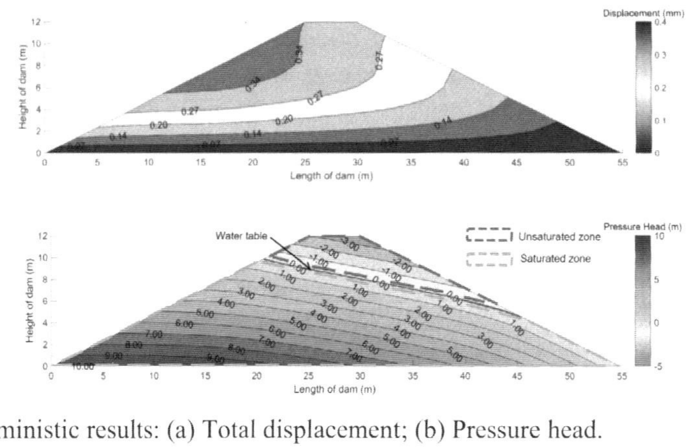

Figure 2. Deterministic results: (a) Total displacement; (b) Pressure head.

The analysis of slope stability by the strength reduction method is shown in Figure 3. The maximum displacement versus FOS is shown in Figure 3(a). The FOS increases from 1.00 to 1.92 with a step of 0.01. The maximum displacement increases significantly at FOS = 1.86 indicating the onset of the collapse of the slope. The plastic strain of FOS = 1.86 is shown in Figure 3(b). The slip surface is well reflected in the figure and the arrows show the direction of the landslide based on the displacement.

(a) (b)

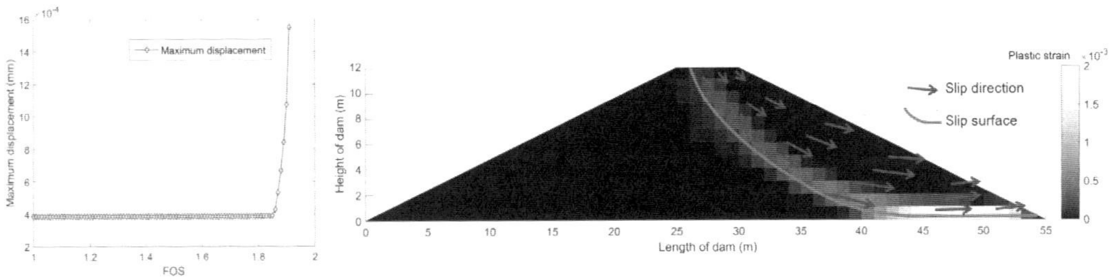

Figure 3. Slope stability by strength reduction method: (a) Factor of safety (FOS); (b) Plastic strain and landslide.

3.2. Effects of k_s uncertainty

The coefficient of variation (COV) and standard deviation (SD) are two typical indices to measure the uncertainty of variables. It is generally recognized that k_s is highly heterogeneous and its COV can vary from 2% ~ 160% [1]. Therefore, in this study, the COVs are chosen to be 5%, 40%, and 80% for parametric study. SD is used to illustrate the uncertainty of model responses.

(a)

(b)

(c)

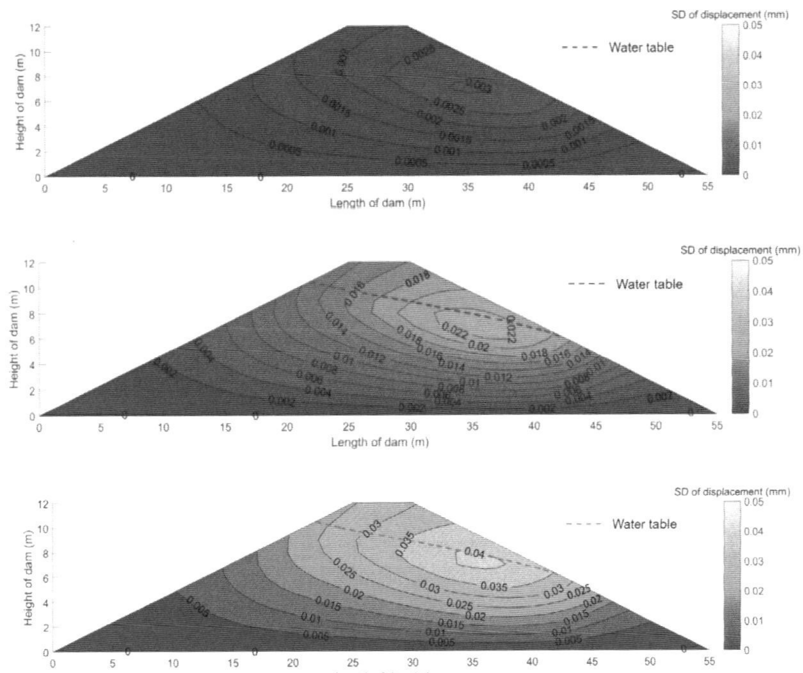

Figure 4. Effect of COV of k_s on uncertainty of displacement: (a) COV = 5%; (b) COV = 40%; (c) COV = 80%.

Figure 4 shows the effect of COV of k_s on the uncertainty of displacement. As the COV increases, the SD of displacement increases significantly. When the COV of k_s is 80%, the maximum SD of displacement is approximate 0.05 mm, which is around 16% of the deterministic result. In contrast, when the COV is 5%, the maximum SD is only around 0.003 mm. It is indicated that the variation of k_s has a significant influence on the uncertainty of displacement. The uncertainty of displacement will increase as the uncertainty of k_s increases and the uncertainty around the water table is significant.

Figure 5 shows the effect of COV of k_s on the uncertainty of pressure head. The SD of pressure head increases rapidly as the COV increases. When the COV of k_s is 80%, the maximum SD of pressure head is around 0.8 m. It is also proved that the variation of k_s has a significant influence on pressure head and the uncertainty of pressure head will increase as the uncertainty of k_s increases. In addition, it can be seen that the uncertainty of pore pressure is mainly distributed on the right side of the embankment. This may be because the right side is less affected by the boundary conditions.

(a)

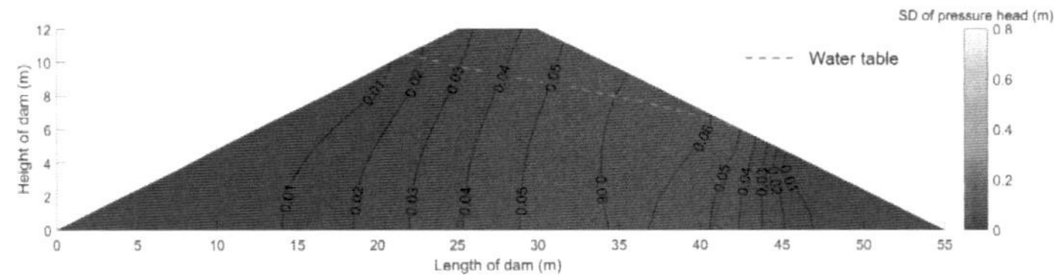

(b)

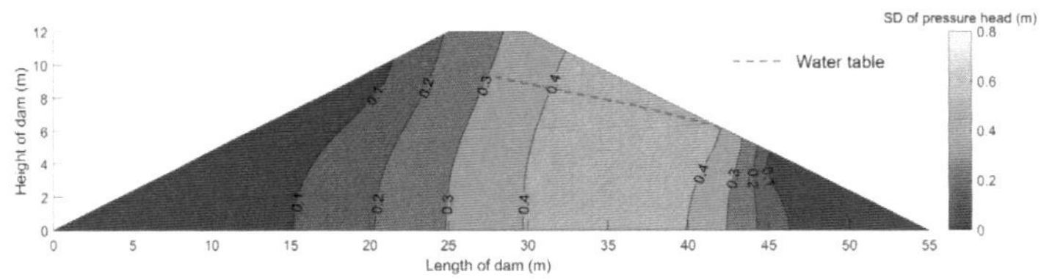

(c)

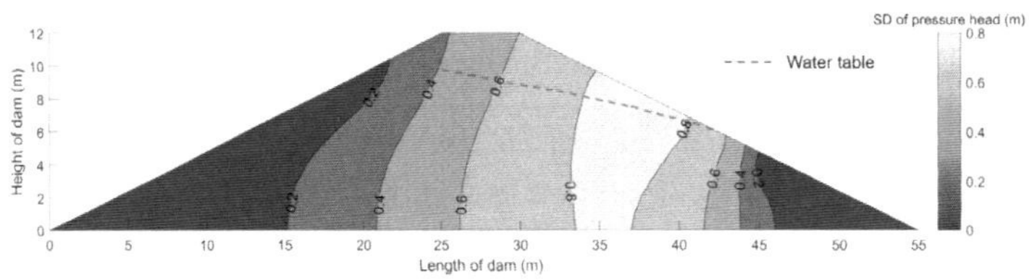

Figure 5. Effect of COV of k_s on uncertainty of pressure head: (a) COV = 5%; (b) COV = 40%; (c) COV = 80%.

(a) (b) (c)

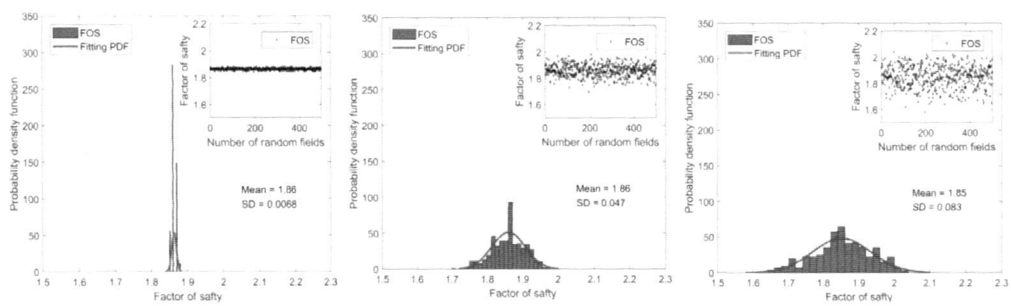

Figure 6. Effect of COV of k_s on distribution of FOS: (a) COV = 5%; (b) COV = 40%; (c) COV = 80%.

Figure 6 shows the effect of COV of k_s on the distribution of FOS. As the COV grows from 5% to 80%, the distributions of FOS become dispersed, which could be illustrated by the SD value of FOS. It ranges from 0.0068 to 0.083. However, the mean value of FOS is around 1.86 constantly. In addition, as the COV increases, the distribution of FOS could be better fitted by a normal distribution. To conclude, the uncertainty of FOS will increase with the increase of the uncertainty of k_s, whereas the mean value of FOS is almost unchanged. Besides, as COV of k_s increases, FOS gradually tends to a normal distribution.

4. Conclusions

In this study, random field theory is adopted to account for the spatial variability of hydraulic parameters. The effects of uncertainty of saturated hydraulic conductivity, k_s on pore water pressure head, displacement, and FOS are studied by Monte Carlo simulation based on 500 random fields of k_s. However, this study only uses the pore pressure of the seepage field as the boundary condition of the stress field to realize the interaction of water and soil, and it is not a fluid-solid coupling in the true sense. The fluid-solid coupling problem of embankment stability analysis needs further study. Major conclusions were summarized below:

1. The uncertainty of displacement and pressure head will increase as the uncertainty of k_s increases. When the COV of k_s is 80%, the maximum SD of displacement is approximate 0.05 mm, and of pressure head is around 0.8 m. It is suggested that the design of embankment needs to be more conservative when the uncertainty of k_s is significant.

2. The uncertainty of FOS will increase with the increase of the uncertainty of k_s, whereas the mean value of FOS is almost unchanged. As COV of k_s increases, FOS gradually tends to a normal distribution. Therefore, the distribution of FOS can be determined by low-order statistics to be understood by engineers when the variation of k_s is high.

References

[1] Zhang L L, Li J, Li X, Zhang J and Zhu H 2018 Rainfall-induced Soil Slope Failure: Stability Analysis and Probabilistic Assessment. *CRC Press*

[2] Polemio M and Lollino P 2011 Failure of infrastructure embankments induced by flooding and seepage: a neglected source of hazard *Natural Hazards and Earth System Sciences* **11** 3383

[3] Thorel L, Ferber V, Caicedo B and Khokhar I M 2011 Physical modelling of wetting-induced collapse in embankment base *Géotechnique* **61** 409

[4] Gui S, Zhang R, Turner J P and Xue X 2000 Probabilistic slope stability analysis with stochastic soil hydraulic conductivity *Journal of Geotechnical and Geoenvironmental Engineering* **126** 1

[5] Cho S E 2012 Probabilistic analysis of seepage that considers the spatial variability of permeability for an embankment on soil foundation *Engineering Geology* **133** 30

[6] Tan X, Wang X, Khoshnevisan S, Hou X and Zha F 2017 Seepage analysis of earth dams considering spatial variability of hydraulic parameters *Engineering Geology* **228** 260

[7] Calamak M and Yanmaz A M 2017 Uncertainty quantification of transient unsaturated seepage through embankment dams *International Journal of Geomechanics* **17** 04016125

[8] Yang H Q, Zhang L and Li D Q 2018 Efficient method for probabilistic estimation of spatially varied hydraulic properties in a soil slope based on field responses: A Bayesian approach *Computers and Geotechnics* **102** 262

[9] Liu L L, Cheng Y M, Jiang S H, Zhang S H, Wang X M and Wu Z H 2017 Effects of spatial autocorrelation structure of permeability on seepage through an embankment on a soil foundation *Computers and Geotechnics* **87** 62

[10] Fenton G A and Griffiths D V 2008 Risk Assessment in Geotechnical Engineering *John Wiley & Sons*

[11] Griffiths D V and Lane P A 1999 Slope stability analysis by finite elements *Geotechnique* **49** 387

[12] Briaud J L 2013 Geotechnical Engineering: Unsaturated and Saturated Soils *John Wiley & Sons*

IOP Conf. Series: Earth and Environmental Science **865** (2021) 012034 doi:10.1088/1755-1315/865/1/012034

The effect of sedimentary facies on coalbed methane in Binchang mining area

Kai Shao[1, 2,4], Dongmin Ma[1], Laixin Li[2], Dongdong Wang[3], Lin Zhang[2] and Gang Yang[2]

[1]Xi'an University of Science and Technology, Xi'an 710054
[2]Shanxi Coalbed Methane Development Company Limited, Xi'an 710119
[3]Shandong University of Science and Technology, Qingdao 266590

[4]Corresponding author's e-mail: sncbmc@163.com

Abstract. Research on geology data, drilling data, logging data of the coalbed methane drilling data in Binchang mining area, 6 sedimentary facies have been found. These sedimentary facies are: braided river facies, meandering river facies, meandering river delta facies, alluvial fan facies, fan delta facies and lake facies. By study of the CBM drilling, well cementing and hydraulic fracturing, Optimization parameter selection on the later CBM well engineering in different sedimentary facies had been summed up.

1. Background

Binchang mining area is located in Changwu county and Binzhou City, Shaanxi Province (Figure 1) and on the Yishan slope southwest of the Ordos Basin, its structural framework is wave-shaped monoclinic structure. The main structure is 3 ~ 5 ° small dip oblique structure. The faults in this area are smaller than 5 meters with relatively simple structure, and the development of strata is mainly affected by the original sediment [1, 2]. The region is rich in coal resources, coal reserves of 5 billion tons, coal bed methane reserves of 7.3 billion cubic meters.

Since 2008, 63 coalbed methane wells have been drilled by Shaanxi Coalbed Methane Development Company Limited, and the peak production reached 30323.78 m^3 per day for the horizontal wells and 3300 m^3 per day for the vertical wells. The total output over 50000 m^3 per day.

Coalbed methane is a clean energy source. The development of this new energy is conducive to reducing the occurrence of coal mine gas accidents, and also provides an example for the development of low-rank coalbed methane in Northwest China. Gas field engineering is an important part of coalbed methane development.

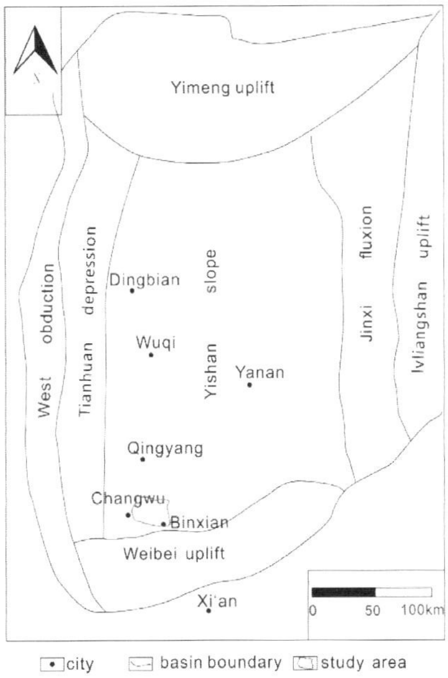

Figure 1. Location of study area.

However, in the process of coalbed methane development, due to drilling quality control, pipe sticking, well deviation and poor fracturing effect, the production reduction is caused. This is a serious problem affecting the engineering quality and the production of coalbed methane wells. This paper introduces the construction process and geological conditions of coalbed methane wells, and tries to find a solution to the common problems of engineering quality.

2. Stratigraphy and sedimentation

The lithological units of the formation include, from base to top, Hujiacun Formation, Anding Formation, Zhiluo Formation, Yanan Formation, Fuxian Formation, Luohe Formation, and Yijun Formation (Figure 2).

The sedimentary characteristics of Hujiacun Formation are gray to dark gray mudstone, silty sandstone mixed with thick fine-grained feldspathic sandstone, and there are current bedding and horizontal layers in the area. Hujiacun Formation is mainly delta front sedimentary facies; The quantitative analysis results show that Anding Formation and Zhiluo Formation are unconformable in contact, and the lithology is red sandy sandstone, silty sandstone and light gray sandstone [3]. From bottom to top, two sets of sediments are formed and returned to the corresponding delta facies, and delta front and front delta sedimentary facies are developed.

The Zhiluo Formation can be divided into two parts. The lower lithology is purple and violet sandy shale mixed with violet fine sandstone. The upper part is gray fine-grained sandstone mixed with violet sandy shale [4]. This formation is in unconformity contact with the underlying Yanan Formation, and the vertical lithologic changes show positive rhythm and river-dual structure. River courses and natural levees are developed by siltation in meandering rivers, and cross bedding and trough cross bedding can be seen at boreholes.

Yanan Formation is the main coal-bearing strata, and its lower lithology is grayish brown bauxite mudstone, argillaceous siltstone and fine sandstone, which is rich in pyrite, oolitic siderite and plant root fossils. Above is a thick No.4 coal seam, overlying a set of shallow dark gray mudstone and sandy mudstone, rich in plant fossils. No.4 coal seam can be mined locally in No.4, 4-1 and 4-2 layers. The upper lithology is gray mudstone, sandy mudstone, siltstone and carbonaceous mudstone, and the bottom is thick sandstone layer, which is connected with the lower part. No.3 coal seam can be found in some areas. The results show that the delta zone is the most favorable coal accumulation area, and the coal seam is deposited in a stable place where the creation rate of accommodation space and peat/coal accumulation rate are balanced [5].

The lithology of Fuxian Formation is represented by gray, sage green mudstone and porphyry siltstone, and the bottom of breccia is often sandy mudstone of Triassic. The strata are in unconformity contact with the underlying strata. The sedimentary facies are meandering river sedimentary facies, and the deposits in Binchang mining area are mostly floodplain deposits [6].

The Luohe Formation can be divided into two parts. The upper part is represented by red thick conglomerate, and purple grain feldspar sandstone with thick conglomerate, is mainly braided river sedimentary facies, about 60-80m thick [7]. Lower part is represented by red fine - grain feldspar quartz sandstone, feldspar sandstone with thick layer of coarse grain and dark brown thin layers of mudstone, and general thickness is about 70 -100 m. It is the main aquifer with large scale cross-bedding and conformable contact with underlying YiJun Formation, representing the fan delta and braided river sedimentary facies [8].

Yijun Formation: The lithology is thick conglomerate containing gravel coarse sandstone lens. The gravel is mainly composed of granite and gneiss, with an average thickness of about 20 m, which is unconformably contacted with the underlying strata. The sedimentary facies are mainly alluvial facies, and the borehole can be divided into root fan and inner-middle fan.

3. Coalbed methane drilling and production

Coalbed methane projects mainly include drilling, logging, cementing and fracturing projects. The main function of logging method is to find out the distribution and lithologic changes of coal and coalbed methane, so as to achieve the purpose of traditional coalfield geological drilling [9, 10].

3.1. Drilling engineering quality

The drilling operation methods of coalbed methane wells in different coalbed methane fields may be quite different. Drilling technology and equipment usually depend on coal seam geological conditions. According to the available data, the main problems of drilling in this area are sticking, deviation and completion.

3.2. Stuck pipe

The main part of the stuck pipe of Luohe Formation is in the upper part. Sedimentary environments are mainly braided rivers, alluvial and fluvial, the lithology beingquartz sandstone, feldspathic quartz sandstone, and gypsum minerals, which have good permeability. During drilling, the expansion of gypsum minerals in water will cause wellbore contraction, and the good permeability of sandstone will cause the loss of drilling fluid. Both will cause wellbore contraction, pipe sticking and completion.

3.3. Well deviation

Deviation will affect subsequent cementing, fracturing and production of coalbed methane wells, and serious deviation will cause failure of coalbed methane wells. The strata in the study area experienced Indosinian movement and Yanshanian movement, and the geological background is an important reason for well deviation. There are three remarkable characteristics of unconformity between Yijun Formation and Anding Formation, Anding Formation and Yanan Formation, Fuxian Formation and Hujiacun Formation. The lithology, dip angle of strata and layered structure of lower strata of unconformities are different. For example, the underlying surfaces of Yijun Formation and Anding Formation are unconformable, the dip angle of strata changes rapidly, and the lithology changes from conglomerate to argillaceous siltstone, resulting in the problem of well deviation.

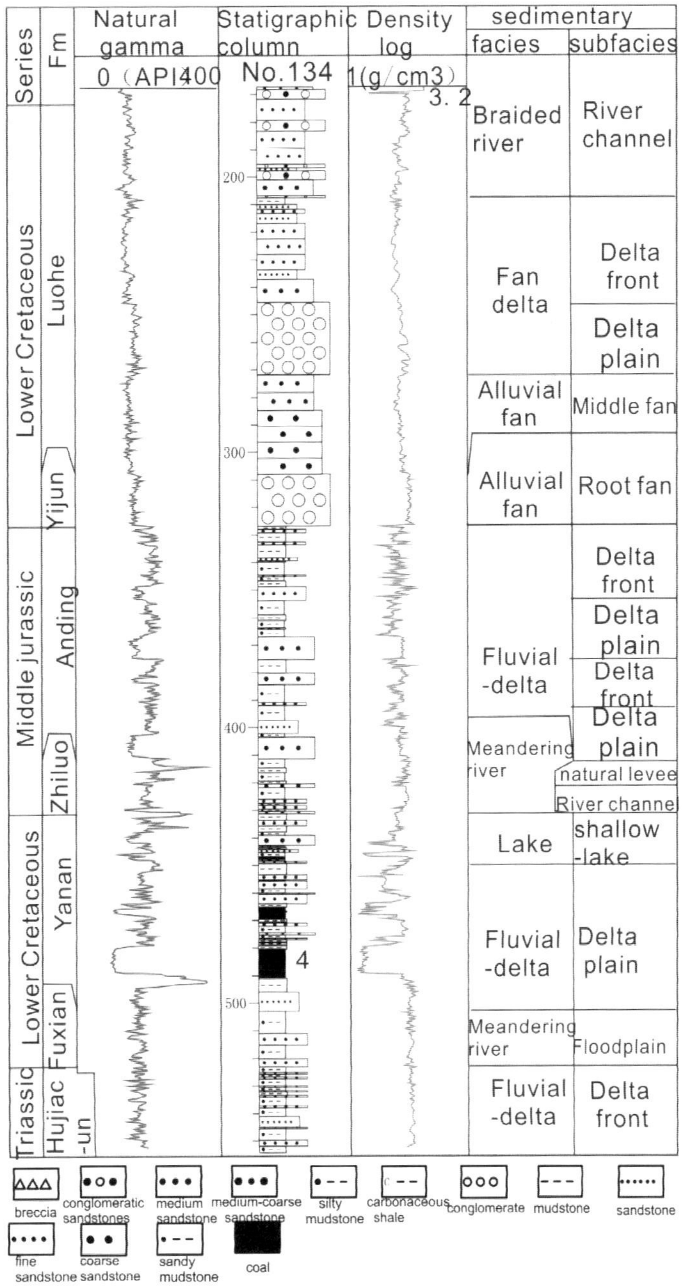

Figure 2. Stratigraphic log showing the position of the formations and sedimentary facies on CBM well engineering in Binchang mining area.

3.4. Completing the Well

According to the completion experience of traditional oil wells, the completion technology of coalbed methane wells has made great progress, but it needs to be further improved to adapt to the characteristics of coalbed methane reservoirs. As stated in the existing statistics, coal seams such as the lower member of Luohe Formation, the middle and lower member of Anding Formation and the lower member of

Zhiluo Formation of Yijun Formation are the main hidden dangers of cementing quality. The sedimentary facies of the middle and lower members of Luohe Formation and Yijun Formation in the study area are alluvial fan deposits with thick glutenite and large grain size, which have typical alluvial fan sedimentary characteristics. The sedimentary facies of the middle and lower members of Yijun Formation and Luohe Formation are alluvial deposits with thick sedimentary layers and large grain size; The lithology of the lower part of the stable formation is mainly sandstone and argillaceous sandstone, and the sedimentary facies is mainly distributary channel microfacies (Figure 2).

After completion, the cementing performance, filtration performance and permeability of alluvial fan and distributary channel sediments are worse than those of other sediments. Due to these characteristics, mud and water quickly enter the stratum, which affects the smooth completion of the project. In order to further improve the completion engineering, the compatibility between cement and formation should be studied.

Due to the engineering problems of drilling, the construction period of coalbed methane project becomes longer and the cost increases. For example, the cementing of Well 05 and Well 06 occurred in alluvium. In order to improve the quality of drilling engineering, the study area should improve the drilling fluid, drilling equipment and technology according to the characteristics of sedimentary facies, and adopt waterborne drilling fluid to achieve the balance between wellbore stability and coalbed methane production.

3.5. Fracturing coal seams and production

In the reservoir, methane is adsorbed on the surface of coal seam, so hydraulic fracturing of coal seam is needed to generate economic gas flow. Hydraulic fracturing is designed to make fractures as long as possible. Although the fracturing equipment and design are the same, the fracturing effect is different under different sedimentary environments [11, 12].

The coal of Yanan Formation is deposited in delta swamp area. Distributary channel has little influence on swamp environment in some stable areas, and can provide sufficient accumulation places for thick coal seams (Figure 3). However, under the action of distributary channel and estuary dam, the coal seam in river delta changes under the influence of river movement. Coal seams are well developed in DFS-133 well area, and the sedimentary environment is more stable than that in DFS-151 well area. Because the Young's modulus and shear strength of hydraulic fracturing coal seam are lower than those of sandstone and mudstone, it is easier to form wider fracture propagation in coal area [13, 14, 15].

The results show that the output of well DFS-133 is 3300 m^3, which is much higher than output of well DFS-151 is only 450 m^3. It shows that sedimentary facies analysis plays an important role in coalbed methane fracturing. Therefore, in the future work, sedimentary facies are an important factor that must be considered in fracturing coal seams. The meandering river environmental well is an important way to optimize the design of fractures, and shale gas may exist in the shale of estuary dam(Figure 3).

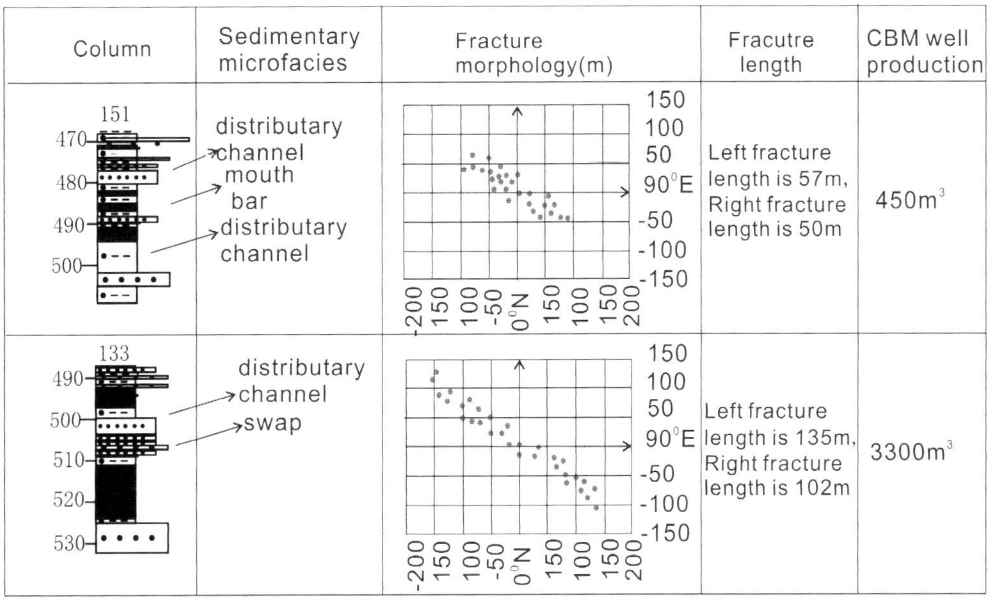

Figure 3. The effect of sedimentary facies on CBM Fracturing in Binchang mining area.

4. Conclusion

1) The main sedimentary environments of drilled strata during coalbed methane well in Bingchang mining area are continental sedimentary environments, braided stream, meandering river, meandering river delta, alluvial fan, fan delta. Lake 6 types of sedimentary facies are found in the area, and the coal seam is mainly formed in the distributary bay facies at meandering river delta.

2) For coalbed methane wells, sedimentary facies affect well deviation, completion, cementing quality and fracturing radius. Therefore, the analysis of sedimentary facies of each coalbed methane well is very important. Attention should be paid to the influence of sedimentary facies in drilling fluid, drilling equipment and fracturing technology.

3) It ultimately affects productivity and economic efficiency. This study provides technical support for the follow-up coal-bed methane engineering, especially for the low-rank coal-bed methane engineering in northwest China.

References

[1] Xin X, Peihe Z, Zaibing J, et al 2014 CBM reservoir parameters and development potential in Binchang mining area [J]. *China coal geology*, **26 (3)**: 19-22

[2] Long L, Xing W 2015 Occurrence characteristics and resource utilization of coalbed methane in Binchang mining area, Shaanxi Province [J]. *Coal geology of China (formerly coalfield geology of China)*, **17(5)**: 44-46

[3] Jiwei L 2017 Research on Jurassic sedimentary system and sequence stratigraphy in Ordos Basin. *Doctoral dissertation of Northwest University*

[4] Dongdong W 2009 Sequence paleogeography and coal accumulation law of Middle Jurassic Yan'an formation in Ordos Basin [D]. *China University of mining and Technology (Beijing). Doctoral dissertation*

[5] Junxing Z, Hongde C,suotang F, et al. 2007 Coupling relationship between Jurassic sequence stratigraphy and basin filling evolution in Ordos Basin [J]. *Journal of Chengdu University of Technology (NATURAL SCIENCE EDITION)*, **3(5)**: 511-517.

[6] Junfeng Z, Chiyang L, Lin Y, et al. 2008 Mesozoic sedimentation and migration of accumulation center in Ordos Basin and its geological significance [J]. *Acta geologica Sinica*, **82(4)**: 540-

IOP Conf. Series: Earth and Environmental Science **865** (2021) 012034 doi:10.1088/1755-1315/865/1/012034

552
[7] Longyi S, Zhenghui X, Wang H, et al. 2008 High resolution sequence stratigraphy and coal accumulation model of Carboniferous Permian coal bearing strata in Qinshui Basin [J]. *Geological Sciences*, **43(4)**: 777-791.
[8] Dongdong W, Longyi S, Zhixue L, et al. 2013 Sequence stratigraphic framework and coal seam formation of Yan'an formation in Ordos Basin [J]. *Journal of Jilin University (Geoscience Edition)*, **43 (6)**: 1726-1739.
[9] AADNOY B S 1987 CHENEVERTME. Stability of highly inclined bore holes .SPE Drilling Engineering, **2(4)**: 364-374
[10] LEE H, ONG S H, AZEEMUDDIN M et al. 2012 A wellbore stability model for formation with anisotropic rock strengths[J]. *Journal of Petroleum Science and Engineering*, **96(97)**: 109-119.
[11] Kui L, Deli G,Yanbin W & Yongsheng L 2016 Effects of local load on shale gas well casing deformation[J]. *Natural Gas Industry*, **36(11)**: 76-82.
[12] Lihui Z, Panfeng W, Xuanqing L, Sun Hao, Fu Yuwei & Nie Shuaishuai 2016 Laboratory experiments on the effect of KCl concentration on desorption capacity of reservoir rocks[J]. *Drilling Fluid & Completion Fluid*, **33(3)**: 117-122.
[13] Yingchun W, Chao L, Daiyong C, et al. 2018 Effect of pulverized coal dispersant on coal in the CBM well-washing technology[J]. *Journal of China Coal Society*, **43(7)**: 1951-1958. doi: 10. 13225/j. cnki. jccs. 0364
[14] Xueqi C, Xiaodong W, Wei L, et al. 2014 Analysis of factors affecting productivity of coalbed gas wells[J]. *Science Technology and Engineering*, **14(4)**: 201-204.
[15] Yong L, YanbinW, Xiaoming N, et al. 2020 Study on identification and control system construction of low efficiency coalbed methane wells [J]. *Coal Science and Technology*, **48(2)**: 185-193. doi: 10. 13199/j. cnki. cst. 02. 024.

Analysis on the geophysical characters of Qiyueshan fault

Yunsheng Liu[1], Jun Chen[2,4], Ziguang Zhang[1], Meiheng Zhu[3] and Aiqiong Chen[3]

[1]Exploration management Department, Jianghan Oilfield Company, SINOPEC,
[2]School of Ocean and Earth Science, Tongji University,
[3]Geophysical Research Institute, Jianghan Oilfield Company, SINOPEC

[4]Corresponding author's e-mail: junxianchen@tongji.edu.cn

Abstract. Qiyueshan fault is not only a topographic boundary, but also a tectonic boundary. It is of great significance to the structural research, oil and gas exploration and engineering application in this area. It is well displayed in geophysical data. All kinds of geophysical data show that the fault has obvious segmentation, and different sections have different characteristics. Gravity and seismic data have a good response to the whole fault, but the response of magnetic anomaly to the fault is better in the north of Shizhu than in the south of Shizhu. Through the processing and analysis of the gravity and magnetic anomaly data around the fault, and the interpretation of some seismic profiles along the fault, response characteristics of faults on different geophysical data are presented. It shows that the Qiyueshan fault not only has typical zoning characteristics in plane, but also has obvious segmentation in vertical direction. From north to south, the dip angle is steeper and steeper. From the south to the north, the difference between the deep and the shallow parts of the fault is increasing.

1. Introduction

Qiyueshan fault (QYF) is distributed intermittently in the axial part of Qiyueshan anticline with NE strike. It intersects with Fangdoushan fault to the south, which is now and then disappearing, and becomes Jinfoshan fault. Then it continues to extend to Loushan, Guizhou province. Geographically, it spans four provinces and cities, that is Chongqing, Sichuan, Hubei and Guizhou.

Qiyueshan fault is basin- mountain transition zone between Xuefeng orogenic belt and Sichuan Basin [1]. It is the boundary fault between the comb like fold belt in eastern Sichuan and the trough like fold belt in Western Hunan and Hubei [2], It is not only the eastern boundary of Sichuan Basin [3], but also Traditional boundary fault of upper and middle Yangtze [4]. In the same time, it is one of the main regional faults in Three Gorges reservoir area [5-9]. The surface exposed strata and faults (Figure 1) on both sides of the fault are different. The quality distribution of source rocks around Qiyueshan is stable. The study of fault characteristics is beneficial to the further study of hydrocarbon accumulation and preservation conditions.

2. Geological setting

In terms of regional geological structure, Qiyueshan fault is located in the Yangtze plate area. It is the boundary fault of upper and middle Yangtze [10]. And it is also the boundary of comb like fold belt in the west and trough like fold belt in the east. The tectonic deformation of sedimentary caprock on both sides of the fault is quite different.

Figure 1 shows the distribution of surface faults. On the west side, the surface faults are less exposed. On the contrary, there are more surface faults in the east side. Especially in the southeast area. The exposed faults are mainly NE trending. The NS and NW trending faults mainly distributed in the south of Wulong, Chongqing municipality [11].

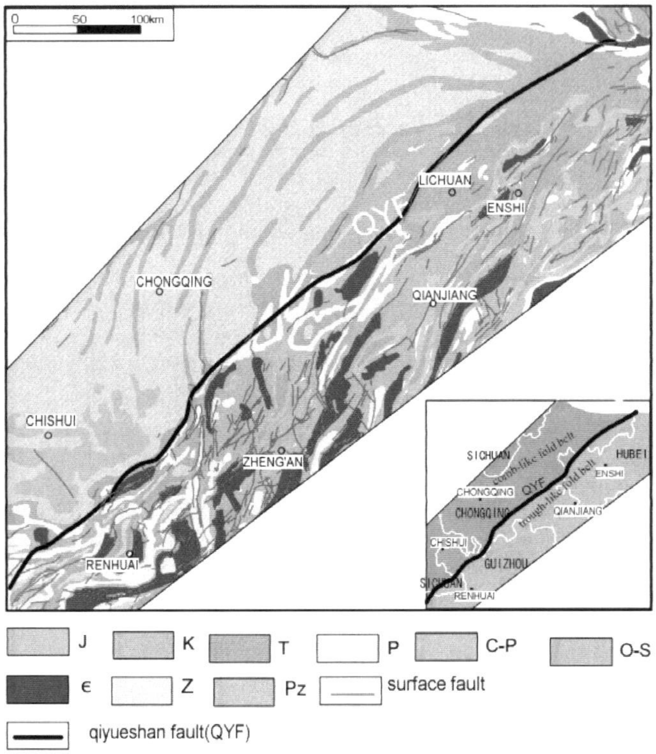

Figure 1. Geological sketch map around Qiyueshan fault.

On the west side, Jurassic is the mainly strata. The Triassic strata occur alternately in NE direction. A few Cretaceous strata are distributed near Chishui, Guizhou province. However, the age of the exposed strata in the East is older than that in the West. There are more stratigraphic ages and obvious regional characteristics. Surrounding Renhuai, Guizhou province, the Cambrian and Triassic strata are mainly exposed. In addition, there are a few Sinian and Permian strata. Between Zheng'an, Guizhou province and Enshi, Hubei province, the main exposed strata are Ordovician and Silurian. In the north of Enshi, Hubei province, the mainly distributed strata is Triassic. The Lower Paleozoic strata are mainly in the east of Enshi. The main strike of the strata is NE, which is obviously controlled by faults. The change trend of stratigraphic age is growing old gradually from west to East.

3. Gravity characters

Figure 2 shows the bouguer gravity contour. All the gravity anomalies are negative. The value is between -160mGal to -50mGal. From west and east sides to the middle, the values become larger. Qiyueshan fault is well reflected in the gravity field. A series of high gravity traps are distributed in beaded shape [12]. There are gradient zones on both sides of the trap. The western gradient zone is the display of Qiyueshan fault. The main trend of anomaly is northeast. But NE anomalies are of discontinuity. Its are cut off by NW trend fault.

Figure 3 is the remnant gravity anomalies after upward 5 km. The positive and negative anomalies are in interphase arrangement. Qiyueshan fault is distributed on the boundary between positive and negative anomalies. The west side are positive anomaly traps. And the east side are negative traps.

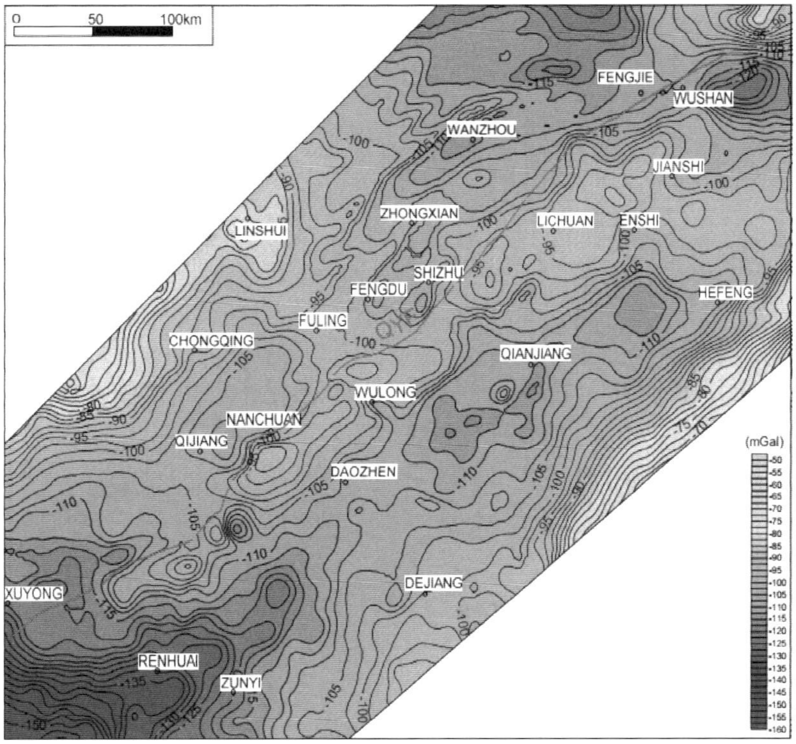

Figure 2. Bouger gravity anomalies around Qiyueshan fault (interface 2.5mGal).

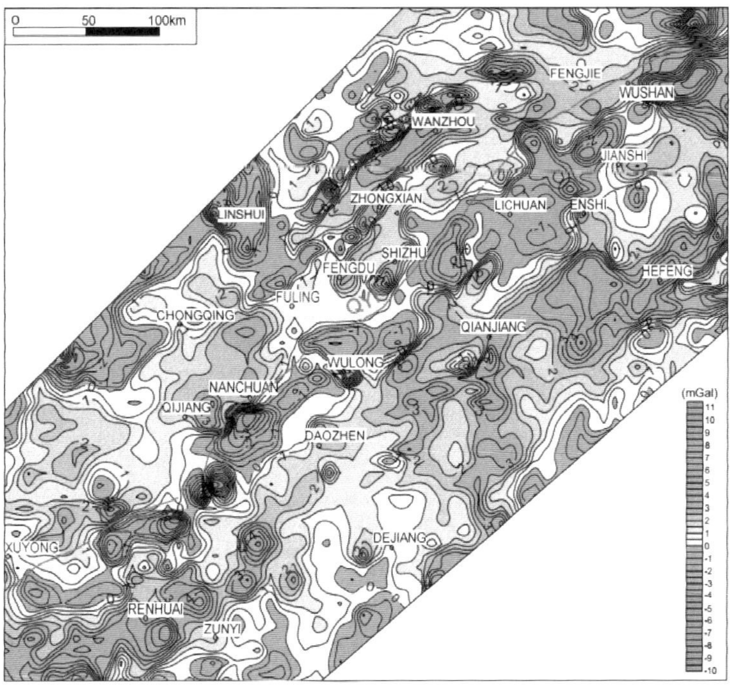

Figure 3. The remnant gravity anomalies of upward 5km around Qiyueshan fault.

4. Magnetic characters

Magnetic anomalies around Qiyueshan fault are showed in Figure 4. The values are from -160nT to 260nT. Taking Nanchuan to Enshi as the boundary, the characters on both sides are different. The values are negative in the southeast side. But in the northwest side, the main values are positive. And the negative values located between Zhongxian and Wanzhou.

Anomalies show typical segmented feature along Qiyueshan fault. The anomaly is negative in the southern part of Nanchuan. The gradient of anomaly is not obvious. The anomaly trap is truncated near Xuyong. The anomaly values are positive between Nanchuan and Lichuan. In this section the fault character is not obvious. The obvious distortion and dislocation of contour indicate the existence of fault. The values are on both sides positive or negative at the same time. This shows that the basement fall on both sides of Qiyueshan fault is not big [13]. Generally speaking, the magnetic anomaly in the south of Shizhu is less obvious than that in the north. It shows that the change of basement rock properties in the south of Shizhu is less than that in the north. The basement along the fault is mainly composed of Banxi group of Upper Proterozoic, while the basement from Shizhu to Zhongxian is composed of medium basic volcanic rocks, and the basement from Fengjie to Wuxi is composed of granite.

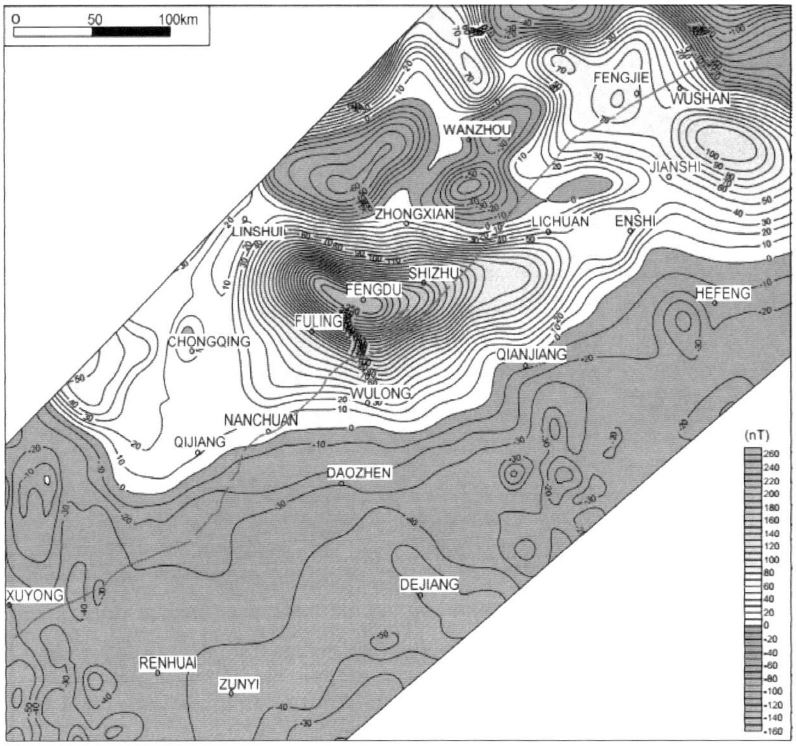

Figure 4. Magnetic anomalies around Qiyueshan fault (interface 10nT).

5. Seismic interpretation

Seismic data can well explain the vertical structural model of fault. The geometrical and kinematic characteristics and fault evolution history of the Qiyueshan Fault were analyzed systematically using seismic and geological data. Figure 5 shows the interpreted characteristics of differential structural deformation on both sides of fault [14].

Figure 5 (a) presented the combination style of Qiyueshan fault and secondary fault is inverse "y". It is characterized by thrusting to hidden thrusting decollement fold, strong thrusting and faulting.

Figure 5 (b) shows that Qiyueshan fault is a number of thrust faults with the same dip and nearly parallel. It's zigzag. And the regional structural deformation is characterized by thrusting to fault spreading decollement fold.

Figure 5 (c) shows that Qiyueshan fault is a high and steep thrust structure with shovel shape, and its bottom end disappears above the basement detachment fault.

Figure 5 (d) shows that the Qiyueshan fault is shovel shaped and the fault breaks out of the surface. The regional structural deformation is characterized by basement thrusting, fault spreading fold and detachment fold, with strong thrusting and faulting, mainly thrust structure, fault spreading anticline and thrust up structure.

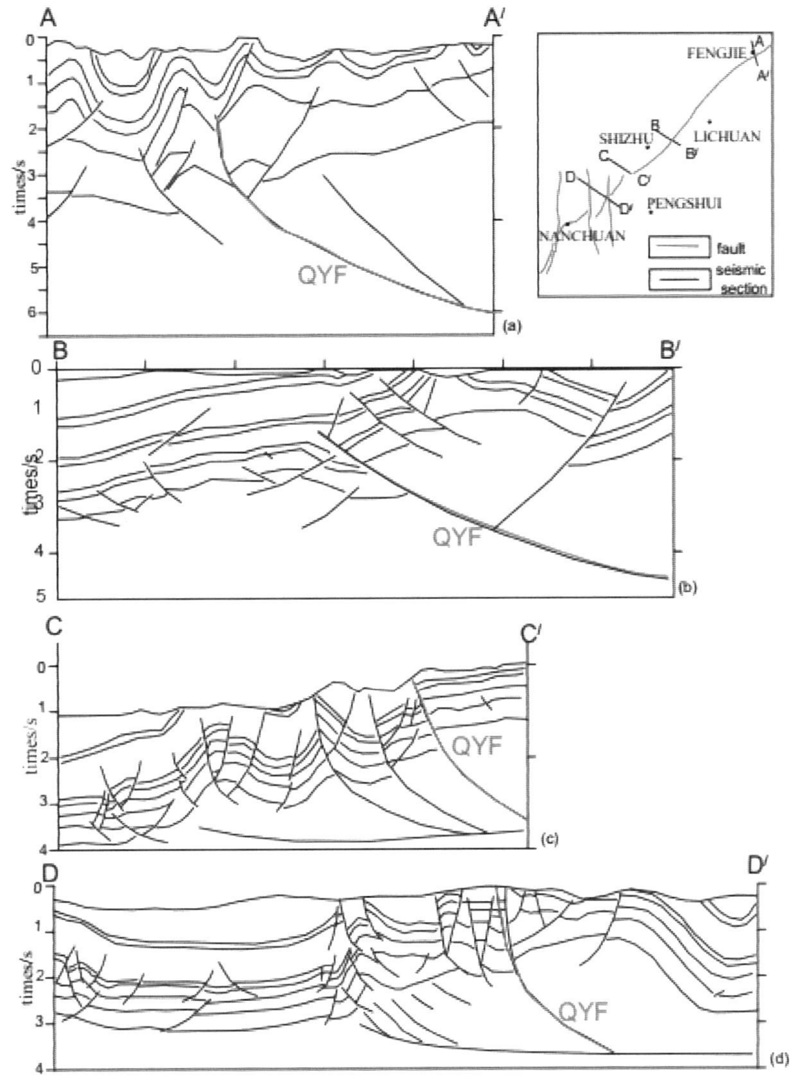

Figure 5. The seismic interpretation results of Qiyueshan fault [3].

Figure 6 presents the interpretation results of another seismic profile near Lichuan, Hubei province. The distribution pattern and depth of strata on both sides of Qiyueshan fault are quite different. Because of the existence of detachment structure layer, the deep and shallow structure model is different.

This fault has different characteristics not only in plane but also in longitudinal direction. From north to south, the dip angle is steeper and steeper. The main fault in the north is deep, which is not consistent with the shallow fault. The shallow part inclines to the northwest, and the deep part inclines to the southeast. When the dip angle of the southward fault becomes steeper, it also penetrates to the surface until it is exposed.

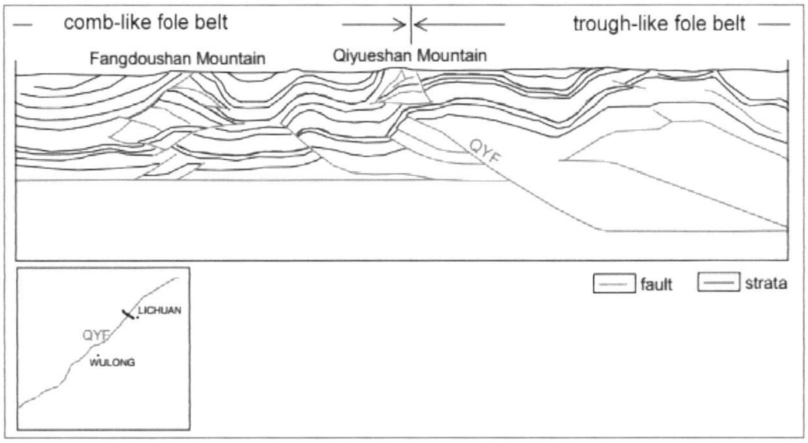

Figure 6. The seismic interpretation results of Qiyueshan fault.

6. Conclution

The Qiyueshan Fault is characterized by a sectional structure not only in plane but also in vertical direction. From north to south, the dip angle is steeper and steeper. From the south to the north, the difference between the deep and the shallow parts of the fault is increasing. Both inclination and angle have great changes.

The magnetic anomaly mainly reflects the structural changes in the basement level, while the gravity reflects the changes from the surface to the deep lithosphere. Therefore, the variation of gravity anomaly on both sides of the fault is more obvious than that of magnetic anomaly. In particular, the magnetic anomaly in the south of Shizhu is less obvious than that in the north. The formation velocity is related to the density, so the seismic velocity profile also has a good response to the fault.

Acknowledgements

This study was supported by Sinopec project of research on characteristics of main faults in the middle Yangtze and their effects on oil and gas preservation conditions (No. ZKW4018001). The authors would like to thank professor Hai Tang, Zhengqiang Li and other members of the project team. At the same time, many thanks should be given to professor Yong Tang for his discussion with him.

Reference

[1] Song H B, Luo Z L 1995 The study of the basement and deep geological stuctures of Sichuan basin, *China Earth Science Frontiers (China University of Geo sciences , Beijing)* **2(2-3)** 231-237

[2] Bai D Y, Xiong X, Yang J, et al 2015 Characteristics and dynamic mechanisms of the folds in the basin-mountain transition zone, east of the Qiyueshan fault Geotectonica et Metallogenia **39(6)** 1008-1021

[3] Deng D F 2014 Study on the intracontimental structure of the enrichment of marine paleo-reservoirs in the northern margin of Jiangnan-Xuefeng uplift, Southern China *China University of Geosciences (Wuhan) (PhD dissertation)*

[4] Wang L Z, Tian Y, Tu B, et al 2012 Paleostress analysis of the Qiyue Shan high-angle nnticline in

the Lichuan Area, western Hubei Province, China *Geotectonica et Metallogenia* **36(4)** 490-503

[5] Tang G Z, Tao M 1991 The study of the neo-tectonic movement of Yangze Gorges region and its relation to the damage of the project construction *Bull. Yichang Inst. Geol. Mineral Resources, CAGS* **19** 1-63

[6] Chen M S, Zhang S H 1998 Characteristics and forming mechanism of opposite protruding actuates structural zones in the Three Gorges area of the Yangtze River, South China *Geology and Mineral Resources of South China* **1** 47-55

[7] He M Y 2019 Investigation and evaluation of neo-tectonic movement and crustal stability in the Three Gorges area *Modern Mining* **598** 1674- 6082

[8] Zhao P 2009 Study on the reservoir induced seismicity in Chongqing under the running of the Three Gorges Reservoir *Chongqing Jiaotong University* (Master's dissertation)

[9] Wu S R, Wu G G, Hu D G, et al 1999 Segmentation and fractal dimension of the NE trending active fault population in the upper reaches of the Qingjiang River, Western Hubei *Bulletin of the Chinese Academy of Geological Sciences* **20(2)** 142-149

[10] Fan Y F 2016 Interrelationship between tectonism and hydrocarbon accumulation in marine strata at critical tectonic movement in west Hubei-east Chongqing, South China *China University of Geosciences (Wuhan) (PhD dissertation)*

[11] Wang Z J, Wang H C, He H L, et al. 2016 Dating of the last active age for Wulong ridge section of Qiyaoshan-Jinfoshan faults *EARTHQUAKE RESEARCH IN SICHUAN* **1** 25-29

[12] Jing X H 2009 Study on Michang mountain and Daba mountain's deep textures and structures *Northwest University (Master's Dissertation)*

[13] Wang X D 2008 The Study of magnetic structure over Xuefeng Tectonic Belt and its adjacent areas *Northwest University (Master's dissertation)*

[14] Wei F, Chen K Q, Tuo X S 2019 Differential tectonic deformation in the northern Qiyueshan Fault, eastern Sichuan Basin *PETROLEUM GEOLOGY & EXPERIMENT* **41(3)** 348－354

Research on glacier elevation changes in Southeast Tibet based ICESat and CryoSat-2 Data

Lanyu Li[1, 2]

[1]No. 8 Guorui Road, Yuhuatai District, Nanjing, Jiangsu Province

[2]Corresponding author's e-mail: Leelionyu1991@163.com

Abstract. Based on the comprehensive analysis of ICESat/GLAS14 data from 2003 to 2009 and CryoSat-2/L2I data from 2010 to 2016 in summer season. This paper revealed the elevation changes of mountain glaciers in small watershed scale in southeastern Qinghai-Tibet Plateau by using outliner elevation of footprints, spatial matching analysis of DEM and elevation trends simulation methods. The result showed, the data of effective footprints of ICESat is significantly higher than CryoSat data on mountain glaciers. Glaciers in the study area has been regressed significantly from 2003 to 2016, with the average regression rate in the study area is $-1.05 \pm 0.53 ma^{-1}$, and a correlation coefficient of 0.72. Compared with the result of other similar studies in southeastern Tibet, the mountain glacier regression rate is credible. Meanwhile, the feasibility of mountain glacier elevation change monitoring in small watershed by CryoSat-2 data has been proven.

1. Introduction

Owing to its high coverage, high precision, periodic nature, and wide range, satellite altimetry has been widely applied to the altitude detection of glaciers and their periodic changes [1]. With onboard instruments, such as microwave radiometer, radar altimeter, and synthetic aperture radar, altimeter satellites can measure the distance from the earth's surface to the sensor, thus providing relevant data for geodetic surveys of surface elevation variation, geodesy, and applications in geophysics [2]. ICESat laser altimetry data are currently preferred for studying surface elevation changes in large-scale glaciers. However, because of the strong characteristics of laser particles, ICESat data are affected by clouds, particularly floating snow particles. For example, in the southeastern part of the Qinghai–Tibet Plateau, where cloud and rain coverage are high, measurement data are likely to be discontinuous [3]. As a representative case of radar altimetry data, CryoSat-2 data have been widely used for studying polar glaciers and sea ice but rarely for the monitoring of inland glaciers. CryoSat-2 data are highly universal owing to the high-precision parameter design and good performance for polar glacier measurement [4]. Unlike laser-based height measurement, surface exploration by radar altimetry can be conducted all-day and in all-weather. Therefore, for mountain glaciers distributed in southeast Tibet, CryoSat-2 data products have more application prospects. They can effectively supplement the temporal and spatial continuities in ICESat data, and provide promising solutions for the continuous study of surface elevation changes in the mountain glaciers of this area.

Recently, altimeter satellites have been employed to study the changes in the material balance of the frozen soil in the Qinghai–Tibet Plateau [5]; however, studies on glacier elevation variations at small watershed scales are relatively lacking. To verify the effectiveness of altimeter radar in monitoring mountain glacier changes and to integrate different altitude sensor data, this study selected

ICESat data and Cryosat-2 data pertaining to the Yigong River Basin obtained during the 2003–2015 period to determine the glacier elevation changes during this period through effective data screening and joint inversion.

2. Data process

2.1. Outliner elimination of elevation footprint

Considering the snow seasons in study area, this paper chose the summer data for elevation anayls. After applying pretreatment to the data from ICESat GLAS and CryoSat 14-2/L2I, which cover the glaciers in the Yigong River Basin, the elevation value at the measurement point was found to contain several errors in the two types of height measuring products. The superposition of the results of altimetry data and existing DEM data could effectively eliminate the outliers. Specifically, the height measurement data were preprocessed to generate point files, the corresponding elevations obtained from the ASTER GDEM and SRTDEM (version 4) were extracted for an overlay analysis, and points with significant differences from the DEM altitude readings were removed. Almost all the glaciers in the Yigong River Basin were accumulated on top of a mountain, and the average elevation was >4700 m; therefore, data with values <0 could be directly removed. In addition, the glaciers exhibited a rapid reduction in the surface elevation from the top of the mountain to the valley; hence, the elevation near the top of the mountain was taken as the upper limit of the elevation distribution of the glaciers.

In the combined use of ASTER GDEM and SRTDEM (version 4) to judge the elevation anomaly, only abnormal values with a significant difference can be removed. The continuous distribution of footprints ensures that the elevation data meet the statistical law for the same satellite altimetry track. Therefore, the method of normalized density of the elevation points was used to eliminate the secondary abnormal elevation points. For each footprint, the data acquisition satisfied the normal distribution; therefore, the normal distribution function was taken as the density distribution function of each data point, and the density curve was superimposed with the calculated mean and standard deviations. Herein, the density integral of the footprint on the same altimeter orbit was 1. The threshold was screened from the distribution images obtained through the normalized density function, where the first trough before a significant increase in the density was the lower limit, and the first trough after a significant decrease in the density was the upper limit.

2.2. Spatial matching analysis of DEM data

The effective elevation interval of the two altimeter products for the Yigong River Basin could be obtained through the above two methods. Precise spatial location matching was based on the accurate calculation of the altitude difference using the DEM. In some regions, the offset of the DEM was typically calculated by arranging GPS measurement points in the field or by extracting typical points through high-precision remote sensing images [6, 7].

Because of the difficulty in achieving field GPS validation for the Yigong River Basin, the offset of the computed ridge lines, rather than the offset of the feature points, was calculated to evaluate the accuracy of the spatial distributions obtained using ASTER GDEM and SRTM DEM [8]. First, two types of DEM ridge lines within the basin were extracted, which ensured not only the integrity of the ridge line (SRTM DEM exhibited good data integrity) but also the accuracy (ASTER GDEM exhibited a high spatial resolution). Based on this result, the location information of 200 ridge lines (in Figure 1) was randomly extracted. Subsequently, the KML files were generated and loaded using Google Earth, and the corresponding ridge position information was also extracted for spatial matching evaluation. Since Google Earth uses high-resolution WorldView-II images, a precise visual interpretation of the ridge lines could be realized. Finally, the position adjustment in the DEM was performed by calculating the offset distance between the two ridge lines.

Because the ridge lines extracted from the Yigong River Basin were approximately along 30° north latitude and 93° east longitude, the relative latitude and longitude values were used to express the location of each ridge line, and the scatter diagrams could clearly reflect the migration. The latitude value was divided by 30, and the longitude value was divided by 93. All the ridge lines in the Yigong River Basin were extracted using the above method. The results in Figure 1 demonstrated a high accuracy of the SRTM DEM in terms of the spatial location matching degree for the basin, except for a few ridges lines in the southeast region. The calculations also showed that the overall deviation in the SRTM satisfied the accuracy requirement. Considering its own spatial resolution, the DEM-based calculation method proved feasible.

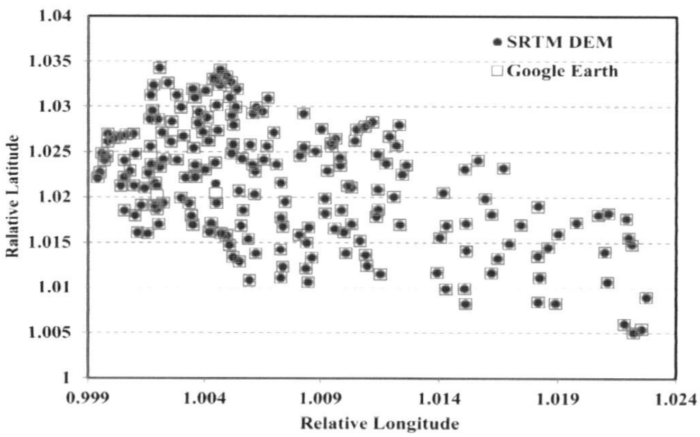

Figure 1. The Geometrical offset correction of SRTM in Yigong Zangbu Basin.

2.3. Elevation trend simulation

By extracting the elevation values of the corresponding positions in SRTM DEM using the elevation points and then calculating the height difference in each sequential track, a set of time-dependent altitude difference sequences could be obtained from the elevation points covering the glaciers. The change trend in the elevation could be obtained by linear fitting of the height difference and time. Robust regression was used to analyze the variation trend in the elevation. After the first fitting of the height difference in this method, the calculation function yielded the same weight to each fitting point, and as the number of iterations increased, the regression function reduced the weights of fitting points far from the trend line after each regression analysis. This reduced the impact of outliers on the trend results, and the iterative process was ended when the regression coefficients tended to converge [9]. The glacier elevation variation in the Yigong River Basin was fitted using the robustfit function in MATLAB with default parameters [10, 11].

3. Result

3.1. Filter valid data

After applying a combination of the above methods to remove the elevation outliers, elevation values beyond 4681.38–5877.34 m and 4586–5944.97 m were considered outliers for the ICESat/GLAS1 and CryoSat-2/L2I products, respectively. These abnormal elevation points were accordingly filtered while retaining points statistically considered to be effective (in Figure 2). As listed in Table 1, the quality of CryoSat-2 data was lower than that of ICESat data. Although the number of ICESat sampling points covering the Yigong River Basin was less than that of Cryostat-2 sampling points, the effective points accounted for 84.42% of the total elevation points in the case of ICESat but only 48.82% in the case of CryoSat-2. Overall, the valid elevation data retained by the two altimetry methods were both objective; hence, they satisfied the measurement coverage of large mountain glaciers in the basin and are suitable for height change analyses.

Table 1. Selected valid elevation points using normaliztion[a].

Data type	Threshold(m) Lower limit	Threshold(m) Upper limit	Points Valid Points	Points Total Points	VPR(%)[a]
ICESat/GLAS14	4681.38	5877.34	748	886	84.42
CryoSat-2/L2I	4586.40	5933.97	1263	2587	48.82

[a] VPR: Valid Points Radio

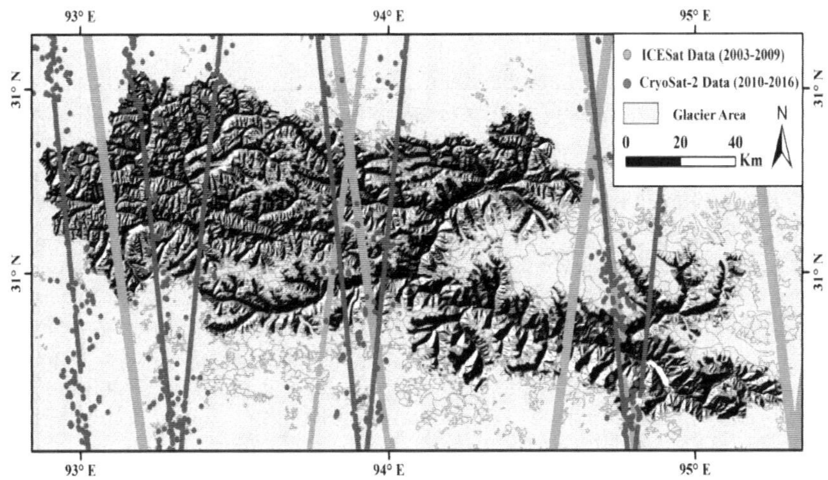

Figure 2. The elevation locations are marked as color dots in Yigong Zangbu Basin.

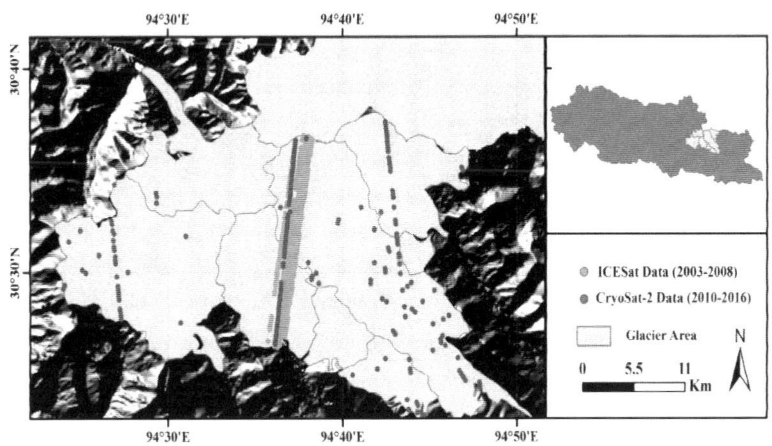

Figure 3. The footprints distribution on Large Double Valley Glacier.

Generally, glaciers are more widely distributed in the eastern and northern parts of the Yigong River Basin than in the west and south. Compared with polar regions that have a continuous form of glacier distribution, such as Greenland, most of the glaciers are in the form of separate glacier

IOP Conf. Series: Earth and Environmental Science **865** (2021) 012036 doi:10.1088/1755-1315/865/1/012036

aggregates and are in peak areas of high-altitude mountains with small areas and large slopes, except for one particular complete glacier block. Despite ICESat and CryoSat-2 data having a high-density data coverage on a global scale and good application in polar regions, including glacier and sea ice areas, neither the spatial distribution nor the time series of the altimeter data was ideal (in Figure 3) in local areas such as the Yigong River Basin. Specifically, when there are non-glacial zones in the orbit, the high point density and time discontinuity in the orbit are low. After screening, continuous and overlapping data bands were found in the large duplex valley glaciers in the basin. These overlapping data bands in the two height measuring products were used to extract and analyze the surface elevation change in the large duplex valley glaciers.

3.2. Analysis of elevation change

The ICESat/GLAS altimeter has demonstrated good application in the glacial regions of the Qinghai–Tibet Plateau, and many useful scientific results have been obtained through different research methods. Currently, the most popular application of CryoSat-2/L2I products is the backward inference of the sea ice distribution and sea ice thickness; however, L2I data have rarely been applied to inland glaciers. For an accurate transmission of the two altimetry data products to reflect the glacier elevation changes in the Yigong River Basin, the height difference was fitted in three segments (e.g., ICESat changes, CryoSat-2 change, and overall change). Any unreasonable value of the height difference (e.g., CryoSat-2 yielded a value as high as 10 m in 2013) was eliminated to ensure the validity of the trend fitting in the two types of height measurement data. The calculations showed dramatic changes in the valley glaciers in the Yigong River Basin. A comparison between Landsat images of less cloud cover in the summer of 2003–2015 and SAR image interference pairs in the corresponding time period showed a significant change in the glacier morphology during this period, the large duplex valley glacier exhibited an increasing tendency of splitting into small glaciers, and the continuous decrease in the glacier surface elevation was an important manifestation of glacier retreat in this basin.

Table 2. Surface elevation change results after data processing.

		Double Valley Glacier in NE Tibet
Data type	Years	Elevation Change(m)
ICESat/GLAS14	2003	3,62
	2004	-
	2005	1.85
	2006	2.67
	2007	-0.07
	2008	3.70
	2009	-
CryoSat-2/L2I	2010	3.87
	2011	-4.22
	2012	5.54
	2013	-
	2014	-11.22
	2015	-6.23
	2016	-9.90

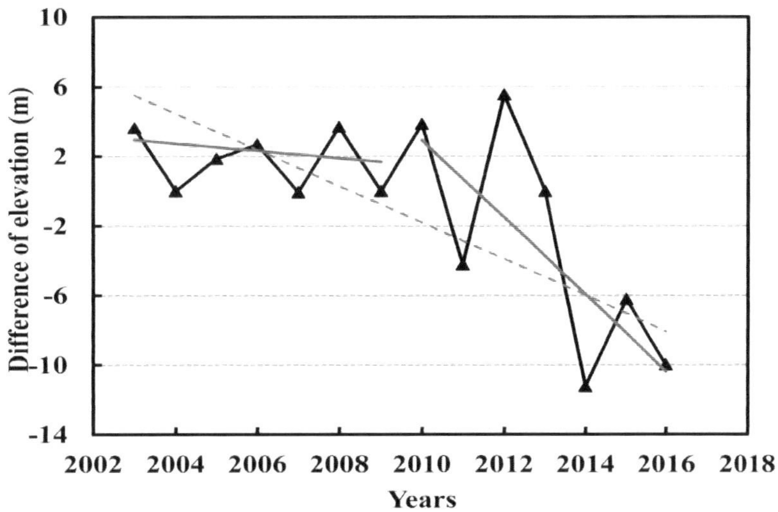

Figure 4. Change in surface elevation of front region of Double Valley Glacier.

For the largest glacier community composed of large compound duplex valley glaciers in the Yigong River Basin, the surface elevation showed a continuous retreat (in Table 2 and Figure 4), with a combined fitting correlation coefficient of 0.72 and a withdrawal rate of $-1.05\pm0.53ma^{-1}$. The fitting results of ICESat and CryoSat-2 segments were also significantly correlated. From the segment fitting results, the overall trend is that a relatively stable glacier form is maintained, except for the small glaciers on the edge. Hence, the elevation changes were stable, and the withdrawal rate was $-0.21\pm0.42ma^{-1}$. During the 2010–2015 period, the glacier community was in a state of accelerated retreat, and the withdrawal rate was $-2.22\pm0.44ma^{-1}$, though the elevation fluctuated significantly.

4. Conclusion and discussion

As height measurement data were used to study the surface elevation changes in the large glaciers near the Yigong River Basin (double the ancient glacier), two types of laser and radar ellipsoidal system conversions were applied to the two-stage products, and the reference to the same plane effectively combined GLAS 14 and L2I data; thus, a more continuous monitoring of the glacier elevation changes in the Yigong River Basin over the 2003–2015 period could be established. The overlapping GLAS 14 and L2I data in the duplex valley glacier were fitted through the iterative weighted least-squares method. The results showed that:

The surface elevation change rate of the large valley glacier community in this basin during 2003–2008 was -0.94m/yr as determined by ICESat/GLAS14. Neckel et al. used ICESat data to extract glacier elevation changes in the entire Qinghai–Tibet Plateau from 2003 to 2009, and found that the change rate of the glacier surface elevation in the Yigong River Basin was -0.81 ± 0.32m/yr [12], which is consistent with the extraction results of this study. Wu Hongbo used ICESat/GLAS and GRACE satellite data to calculate the changes in the mass balance of high Asian glaciers, with the results suggesting that the surface elevation of the glaciers in the Tanggula section and Hengduan mountains decreased 2.06 ± 0.20m and 2.43 ± 0.31m during the 2003–2009 period, respectively [13], which is again consistent with our results.

CryoSat-2/L2I data are mainly algorithm-based designed for polar sea ice and ice sheet areas, and the error threshold for inland areas is stricter, resulting in higher error values when L2I data are applied to inland area [14]. Since the CyoSat-2 data based on SAR interferometry, the fresh snow and bare ice can seriously affect the results due to its penetration capability. The altimeter data showed that the error was significantly higher than that of ICESat, despite the strict removal of abnormal points and the careful calculation of the height difference. The monitored elevation change rate of the glacier in the basin during the 2003–2016 period was -1.36m/yr. While no direct results were available for comparison and verification, the L2I elevation difference based on the baseline B can be utilized as a supplement to the GLAS data and as a reference for studying the continuous elevation changes in the basin.

Notably, however, the proposed DEM spatial superposition analysis and the method of eliminating extreme values using the normalized density function are all based on the product itself, i.e. [15], reducing the range of effective data to optimize the product and improve the accuracy of the result. In fact, there were differences in the spatial distributions of the footprints falling on the glacier, making it difficult to eliminate some of the elevation outliers. In addition, the Yigong River Basin has a variety of glaciers, including large and widely distributed complex valley glaciers as well as ice bucket glaciers and suspended glaciers with small areas and a high partition degree but in aggregation. Therefore, altimeter data can hardly be evenly distributed over the glacier surface. It is impossible to establish an even and accurate plane elevation distribution that includes all the types of glaciers; this has a certain negative effect on the subsequent trend fitting of the glacier surface elevation differences.

References

[1] Chuter S J, Bamber J L 2015 Antarctic Ice Shelf Thickness from Cryosat-2 Radar Altimetry *Geophysical Research Letters* **30** 4
[2] Rémy F, Flament T, Michel A, et al 2015 Envisat and SARAL/AltiKa Observations of the Antarctic Ice Sheet: A Comparison Between the Ku-band and Ka-band *Marine Geodesy* **38** sup1
[3] Helm V, Humbert A, Miller H 2014 Elevation and elevation change of Greenland and Antarctica derived from CryoSat-2 *The Cryosphere* **8** 4
[4] Sebastuan B S, Louise S S 2017 Implications of changing scattering properties on Greenland ice sheet volume change from Cryosat-2 altimetry *Remote Sensing of Environment* **190**
[5] Resti A et al. 1999 The Envisat radar altimeter system (RA-2) *ESA bulletin* **98** 8
[6] Slobbe D C, Ditmar P, Lindenbergh R C 2009 Estimating the rates of mass change, ice volume change and snow volume change in Greenland from ICESat and GRACE data *Geophysical Journal International* **176** 1
[7] King M A, Coleman R, Freemantle A J, et al 2009 A 4-decade record of elevation change of the Amery Ice Shelf, East Antarctica *Journal of Geophysical Research: Earth Surface (2003–2012)* **114** F1
[8] Lee H, Shum C K, Tseng K H, et al 2013 Elevation changes of Bering Glacier System, Alaska, from 1992 to 2010, observed by satellite radar altimetry *Remote Sensing of Environment* **132**
[9] Kääb A, Berthier E, Nuth C et al. 2012 Contrasting patterns of early twenty-first-century glacier mass change in the Himalayas *Nature* **488** 412
[10] Fox J 2014 Robust regression: Appendix to an R and S-PLUS companion to applied regression
[11] O' Leary D P 1990 Robust regression computation using iteratively reweighted least squares *SIAM Journal on Matrix Analysis and Applications* **11** 3
[12] Neckel N, Kropacel J, Bolch T 2014 Glacier mass change on the Tietan Plateau 2003-2009 derived from ICESat laser altimetry measurements *Environment Research Letters* **9** 9
[13] Wu H B 2020 Stuyies on glacier mass balance in High Asia retrived By ICESat-GLAS data and GRACE time-varying gravity field *Acta Geepdaetica et Cartographica Sinica* **49** 4
[14] Fricker H A, Padman L 2012 Thirty years of elevation change on Antarctic Peninsula ice shelves from multimission satellite radar altimetry *Journal of Geophysical Research: Oceans (1978–2012)* **117** C02026
[15] Kropáček J, Neckel N, Bauder A 2013 Estimation of volume changes of mountain glaciers from ICESat data: an example from the Aletsch Glacier, Swiss Alps *The Cryosphere Discussions* **7** 4

Geochemical behavior of Mo and W in the CO_2-rich mineral waters occurring in the East Elbrus area

G A Chelnokov[1, 3], V Y Lavrushin[1], A S Aidarkozhina[1] and Q Guo[2]

[1]Geological Institute Russian Academy of Sciences, Pyzhevsky lane 7, bld. 1. 119017, Moscow, Russia;
[2]State Key Laboratory of Biogeology and Environmental Geology, School of Environmental Studies, China University of Geosciences, 430074 Wuhan, Hubei, PR China

[3]Corresponding author's e-mail: geowater@mail.ru

Abstract. Molybdenum and tungsten concentrations in cold CO_2-rich mineral groundwaters of east Elbrus area lie in two ranges: 0.01-0.43 ppb and 0.01-410 ppb. The temperature of the water varies from 7.9 to 17.6 °C. Mo concentrations have irregular behaviour with temperature and do not demonstrate a significant correlation with any parameters. Tungsten concentrations are strongly and positively correlated only with As, indicating either common sources or common processes controlling their grades. The highest W concentrations are exclusively associated with Na–HCO_3–Cl water types, with a circum-neutral condition ($6<pH<7$) and lower Eh (-60-+45 mV). The dominant species of Mo and W in mineral waters of eastern Elbrus are molybdate ion (MoO_4), and $[W_7O_{24}]^{6-}$ polyanion. Based on the risk assessment results, for two potential contaminants (Mo and W), the concentration level of W exceeded the environmental quality standard, which suggests a severe to moderate ecological risk and, therefore, needs to be considered as the priority pollutant for the studied area.

1. Introduction

The concentrations of Mo and W and processes controlling their chemistry in groundwaters are of environmental interest. From the one side, the enrichment of these elements in aqueous systems is useful to unravel geochemical processes (pollution, water-rock reaction, mixing with geothermal fluids etc.). However, recently, more studies suggest that the exposure to elevated tungsten concentrations may cause various pathological changes or diseases, like rhabdomyosarcoma, lung inflammation, DNA damage in bone marrow, etc. [1].

Several investigations have revealed W and other microelements concentrations in mineral and surface waters within the Baksan river Valley (East Elbrus area), which connected with Tyrnyauz Mo-W deposit [2, 3, 4]. Unfortunately, they contain poorly understood features. Therefore, the adequate geochemical data for tungsten and molybdenum in natural water of the study area are still limited. This article represents new geochemical results on Mo and W in CO_2-rich mineral groundwaters of the East Elbrus area.

2. Sampling and analytical procedures

For the present study, chemical data on ten mineral water manifestations within Baksan river Valley (East Elbrus area, Figure 1), collected by authors during fieldwork in 2016, were interpreted. Sample points were represented by boreholes and spring with water temperatures ranging from 7.9 to 17.6°C.

The water samples were filtered through 0.45 μm mixed cellulose ester filters (Advantec, Japan) and collected in acid-washed, high-density polyethylene sample bottles. Waters for the cation analysis were acidified to pH<2 with ultrapure HNO_3. Water temperature, conductivity, and pH were measured directly in the field using Mettler Toledo probe. Major cations and anions were analyzed by ion chromatography. Carbonate species were titrated in-situ with 0.1 N HCl. Trace elements concentrations in groundwater were determined by ICP-MS (Agilent 7500) analysis. Trace elements, Mo and W concentrations were analyzed by ICP-MS (Agilent 7500 and ELEMENT XR) in the Analytical Department of GIN RAS (Moscow). The chemical results of the water samples are presented in Table 1.

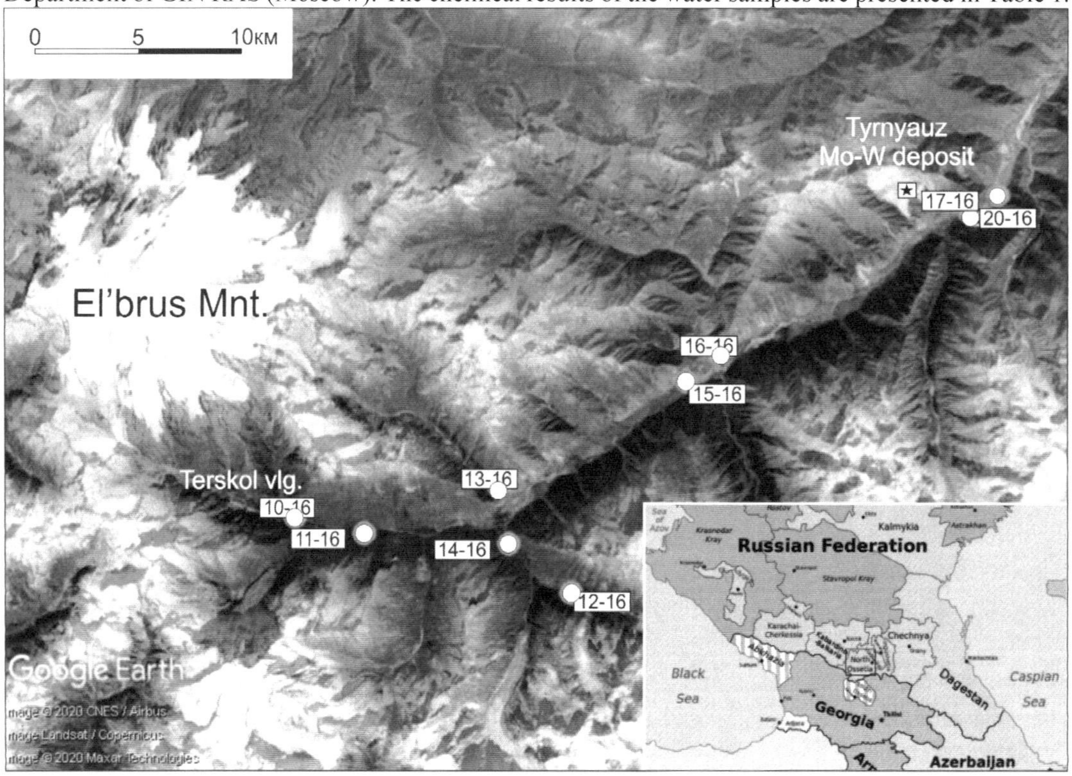

Figure 1. Location of sampling sites, Tyrnyauz Mo-W deposit, and Elbrus volcano.

3. Results and discussion

According to classification diagrams (Figure 2), three chemical types of waters are presented (Table 1). The Ca-HCO_3 type has TDS 0.4-1.0 g/l, temperature less than 10°C, slightly acidic to neutral pH, and slightly reducing conditions (Eh from 0 to 78 mV). Geochemical characteristics of this water show that it is shallow groundwaters with a short period of circulation and their mineralization controlled by CO_2-water-rock interaction processes.

Na-HCO_3 mineral waters with TDS 0.7- 5.0 g/l also display slightly acidic to neutral pH and more oxidizing conditions (Eh from 45 to 112 mV). These waters often have elevated concentrations of SO_4, up to 512 ppm.

Alkali carbonate Na-HCO_3-Cl groundwaters are generally cold (T=16.6–17.6°C), demonstrate highest TDS (5.6-6.1 g/l) and reducing conditions (Eh vary from -60 to 45 mV). Concentrations of SO_4 and Fe vary in a wide range (Table 1).

All studied waters show good Na, Cl and B correlations (r=0.93). These values are well-correlated with the geological history of the region since that the main part of are comprises Late Paleozoic biotitic granites and Middle Paleozoic metamorphic crystalline schists [5]. The highest TDS are recorded in mineral waters far from the Elbrus mnt., on an elevation of about 1300 m, in a sector where is found a contact between biotitic granites and Carbonic limestones (points 17-16, 20-16). Lowest TDS = 0,4 and 0,7 mg/l characterized for springs located on upper parts of the territory (2000 m a.s.l.) (points 10-16-12-16).

The Cl and Br concentrations are highly correlated in all analyzed groundwaters (Figure 3). Mass ratios of Cl to Br range from 358 to 1157 (average 501) for all saline groundwaters, which is an indication of fresh water and NaCl dissolution components.

Table 1. Chemical composition of the studied mineral water [a].

	Unit	Springs							Boreholes	
		10-16	14-16	11-16	12-16	13-16	15-16	16-16	17-16	20-16
		Ca-HCO3		Na-HCO3					Na-HCO3-Cl	
TDS	g/L	0.4	1.0	2.2	0.7	2.0	2.6	5.0	6.1	5.6
T	°C	9.9	7.9	11.6	8.5	10.3	14	16.4	16.6	17.6
pH	Unit	6.1	6.8	6.2	6.5	6.8	6.5	6.5	6.5	6.7
Eh	mV	2.3	78	112	45	103	84	96	-60	45
Na	ppm	5.7	34	304	108	224	283	829	1508	1542
Ca	ppm	74	172	159	75	193	251	406	213	85
Mg	ppm	18	26	65	16	75	124	88	48	15
K	ppm	2.8	3.6	21	14	15	41	84	96	113
Cl	ppm	8.1	81	228	179	195	179	1043	1605	1524
SO_4	ppm	11	6.6	11.5	8	111	512	252	9.6	312
HCO_3	ppm	327	673	1364	313	1242	1177	2279	2526	2006
Fe	ppm	13	11	21	13	4.4	25	6.4	73	1.8
Al	ppm	0.1	0.04	0.05	0.2	0.04	0.05	0.07	0.04	0.03
Mo	ppb	0.07	0.13	0.14	0.17	0.43	0.10	0.18	0.4	0.001
W	ppb	0.08	-	0.02	0.01	-	-	-	0.55	411

[a] Numbers of springs: 10-16– Terskol, 11-16– Badaevka, 12-16– Djan Tugan, 13-16– Irik-Narzan, 14-16 – Adil-Su, 15-16 – Verhni Narzan, 16-16– Nizhni Narzan, 17-16 – borehole, 20-16– Tyrnyauz old borehole. Dash-lower detection limit.

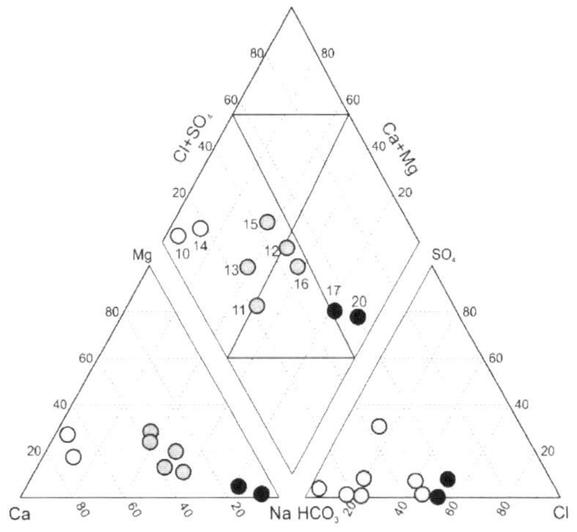

Figure 2. Classification of studied waters using Piper diagram. Numbers according to Table 1.

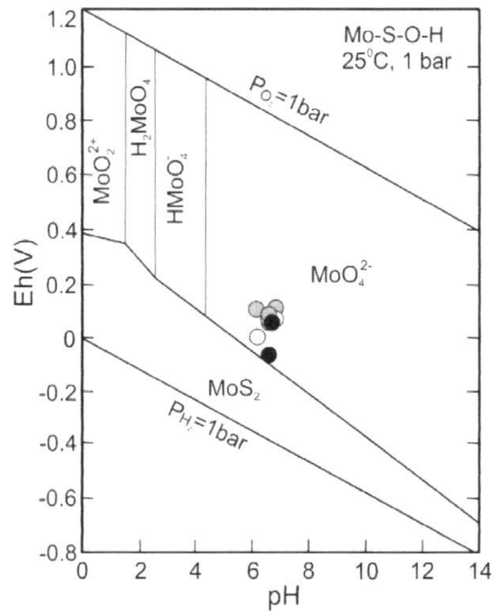

Figure 3. Compositions of molybdenum bearing natural waters in the Eh–pH diagram for the Mo–O–H–S system at 25°C, P = 1 bar. [6]

These correlations are unexpected in the Elbrus geological framework, which consists of granites and schists. But this data is in agreement with isotopic and geochemical data presented by [7]. They demonstrate that the salt and the gas composition of these waters were formed at significant depths and high temperatures. The contribution of connate waters, especially in the Elbrus, is not excluded. Upwelling of deep-seated waters was accompanied by decreasing temperature and dilution by meteoric waters. Mineral waters efficiently interact with rocks through leaching.

The measured concentrations of Mo in all water types are much lower the limits of the Russian drinking water (250 ppb) and the WHO (2011; 70 ppb) standards (Table 1). Maximum Mo concentrations (0.4-0.43 ppb) are associated with Na–HCO_3 and Na-HCO_3-Cl waters. There is no Mo anomaly even in the water influenced by Tyrnyauz W-Mo deposit (point 17-16, 20-16). Statistical analyses show that correlation Mo and main parameters and ions are very weak. Maximum coefficients revealed for pH (r=0,3), Ca (0.27), Fe (0.3) and Al (0.27). Figure 3 demonstrates that molybdate (MoO_4^{2-}) is the dominant species in the studied mineral waters. Molybdenite itself is highly insoluble in water [6], and it is likely to have been oxidized to molybdate and got introduced to the groundwater. Molybdate is the dominant species at a pH > 6. If pH values drop, the molybdate anion becomes protonated. When pH values drop below 4.4 [8], the molybdate anion will fully protonate to $MoO_3(H_2O)_3$. Thus, the migration state of Mo in cold water is highly dependent on Eh-pH.

Russian drinking water standards is limited W concentration as 50 ppb, and most of studied waters demonstrate much lower tungsten concentration (0.01 to 0.55 ppb). Anomalous high W concentrations (411 ppb) were found in one borehole on the vicinities of the Tyrnyauz town (the area of Tyrnyauz W-Mo deposit, point 20-16, Table 1).

Several mineral water samples with high pH (6.8) do not contain W (#13-16, #14-16). The most probable is that these waters apparently are associated with areas lacking any significant source of tungsten. The highest concentrations of W (0.5-411 ppb) are exclusively associated with Na–HCO_3-Cl water. In general, high pH and low redox potential are favorable for the enrichment of tungsten in water. However, water with lowest redox potential (-60 mV) does not demonstrate high W concentrations. On the other hand, this mineral water provides conditions for tungsten minerals deposition requires low redox conditions and available Fe and Mn (or Ca) ions. These ions were not detected in the water sample with high W concentration [9]. Statistical analysis demonstrates that W have good correlation with As (r=0.99), Eh (0.6), Mg (0.5), Na (0.5) and Ca (0.5). The significant and positive correlation between concentrations of W and As may indicate that they have similar sources or that similar processes are controlling their concentrations. Tungsten and $As_{(V)}$ are anions in water and have similar adsorption properties characterized by decreasing adsorption with increasing pH. Whatever the source of W (mineral dissolution, geothermal water, etc.), the high pH, low Fe and Ca of these mineral water samples results in less relative adsorption and greater aqueous concentrations of W.

According to Smedley & Kinniburgh [8], there is a lack of up-to-date thermodynamic and kinetic data for many important reactions for Mo and W in the natural environment and these limits the ability of current geochemical models to predict its fate and transport. Is well known that the WO_4^{2-} anion is the predominant species of natural waters at pH above 4.5 at 25°C. But recent experiments with Raman spectroscopy coupled with capillary technique [10] reveal that $[W_7O_{24}]^{6-}$ polyanion is the predominant species (90%) under circum-neutral condition (6<pH<7) at T 25-100°C. and only 5-10% of the total dissolved W concentration represents by WO_4^{2-}. Even more, there is not W precipitation at temperature interval 25-100°C and pH from 1 to 5.5. According to with Carocci [10] the main species in these conditions are $[W_{10}O_{32}]^{4-}$, $[W_7O_{24}]^{6-}$ and WO_4^{2-}.

4. Conclusion

The conducted research allowed to identify the main features of the distribution and water migration of Mo and W in the CO_2-rich mineral waters of the East Elbrus area. It was found that the concentrations of Mo and W vary significantly, from 0.01-0.43 ppb for Mo and 0.01-410 ppb, for W. Temperature of the water changing from 7.9-17.6 °C. In these conditions, concentrations of Mo have irregular behavior with temperature and do not demonstrate some significant correlations with any parameters. Mo concentrations in waters are not controlled by molybdenite solubility. The dominant species of Mo in mineral waters of eastern Elbrus are molybdate ions (MoO_4^{2-}). Tungsten concentrations are strongly and positively correlated only with As, indicating either common sources or common processes controlling their contents. Tungsten is mobile during water-rock interaction and its concentrations tend to increase with increasing primary rock dissolution and temperature. Highest concentrations of W are exclusively associated with $Na–HCO_3–Cl$ water types, circum-neutral condition (6<pH<7) and lower Eh (-60-+45 mV). Our new data allowed us to suggest that $[W_7O_{24}]^{6-}$ polyanion is a primary species of tungsten, whereas WO_4^{2-} dominates in more alkali environment.

Based on the risk assessment results, for two potential contaminants (Mo and W), the concentration level of W exceeded the environmental quality standard, which suggests a severe to moderate ecological risk and, therefore, needs to be considered as the priority pollutant for the studied area.

Acknowledgements
The reported study was funded by RFBR and NSFC, project number 21-55-53005.

References
[1] Guo Q, Li Y, Luo L, 2019 *Sc. Total Environment* 657 1523
[2] Chirkova E, Baranovskaya E, Kharitonova N, Lavrushin V, Chelnokov G, Bragin I 2019 *E3S Web Conf.*, **98** 01009
[3] Gurbanov A, Bogatikov O, Gazeev V, Leksin A 2015 *Doklady Earth Sc.* **464** 967
[4] Lavrushin Y V, 2012 Subsurface Fluids of the Greater Caucasus and its Surrounding. 348
[5] Gurbanov A, Gazeev V M, Bogatikov O A et al., 2004 Elbrus active Volcano and its geological history Russian J. Earth Sc. ***6*** 257
[6] Ryzhenko B N, 2010 *Geochemistry International.* **48** 407
[7] Dubinina E O, Chernyshev I V, Bortnikov N S, et al., 2005 Geochem. Int., No. 10, 988
[8] Smedley P, Kinniburgh D, 2017 *App. Geo.* **84** 387
[9] Seiler R L, Kenneth G, Stollenwerk J, Garbarino R. 2005 Factors controlling tungsten concentrations in ground water, Carson Desert, Nevada *App.Geo.* **20** 423
[10] Carocci E, 2019 Tungsten transport and deposition in magmatic-hydrothermal environments: the example of Panasqueira (Portugal). Earth Sciences. Université de Lorraine, English.

Characteristics of heavy metal pollution in soils of a typical copper smelting site in China

Jieqiong Zheng[1], Qianling Ma, Li Li and Shangyong Jin

BGRIMM Technology Group, Beijing

[1]Corresponding author's e-mail: 920295159@qq.com

Abstract. During copper smelting, discharge of waste gas, waste water and waste residue has resulted in heavy metals pollution in soils. In a typical copper smelting site in China, arsenic pollution was the most serious, which could be found in soils within 6 m. All the measured heavy metals including Cu, Pb, As, Cd, Zn, Hg, Ni, Cr exceeded background concentrations in the study area. Key contamination areas of the site distributed nearby the uncontrolled emission source and under the leakage area of waste water and waste residue. This study is of great significance to the soil management and control of copper smelting industry.

1. Introduction

Heavy metal pollution in soil is the hot topic in the international community. Heavy metals can be accumulated over a long period. They destroy the soil ecological environment resulting in the loss of soil quality and migrate to the surrounding surface water and groundwater because of the precipitation eluviation effecting water environment quality. More worryingly, heavy metal pollution in soil has brought a serious threat to human health through bioaccumulation [1]. It is well known that smelting non-ferrous metal activity is one of the important sources of heavy metal pollution. In the process of smelting, heavy metals in waste gas, waste water and waste residue are discharged to the environment. Ore mining and processing are responsible for 13% of global mercury emissions [2]. In China, they are 24.39% and 26.36% of lead and znic respectively [3]. And numerous investigations have shown that the soil around the non-ferrous smelter is contaminated with heavy metals [4-8]. Heavy meatals emitted during production deposite from the atmosphere on to the soil surface or seep into the soil along with wastewater. But the majority of previous research has focused on the soil around the plant. Very little attention has been paid to the characteristics and causes of contamination inside the site.

The present study was carried out inside a typical copper smelting site in China.The smelter has been in operation for 13 years and is now disused. The objective of this investigation was to identify characteristic pollutants, key pollution areas and key pollution sources.

2. Materials and methods

2.1. Study area

The copper smelting site in the study covers about 288km^2 and can be divided into office region, production region and living region (shown in Figure 1). The smelter was built in 2006 and shut down in 2019. Its designed annual capacity was 125,000 tons of cathode copper, 547,000 tons of sulfuric acid and 330,000 tons of copper smelting slag recycling.

IOP Conf. Series: Earth and Environmental Science **865** (2021) 012038 doi:10.1088/1755-1315/865/1/012038

Figure 1. The plan of copper smelter.

During the 13 years of operation, it has emitted a large amounts of organized and unorganized soot containing heavy metals. The organized pollution sources included soot from materials transfer, exhaust gas from acid production and soot of slag crushing. The unorganized emission sources come from raw material preparation, copper smelting, and solid waste dumps. The wastewater was treated by the sewage treatment station and returned to production. And all solid waste was safely disposed of. In the site there was a permanent storage yard for sewage neutralizing residue.

2.2. Soil sampling and analysis

A total of 445 soil samples were collected from 198 sites in the bare area of the factory and 4 sites in the workshop region of the factory during Sptember 2019 to April 2020 when the smelter was shut down (shown in Figure 1). Soil samples each of approximately 1.0 kg were taken from different depths (0-0.5m, 0.5-1.5m, 1.5-3.0m, 3.0-6.0m). Then they were air dried, sieved and stored in plastic bags for analysis.

Soil samples were digested with aquaregia by microwave digestion instrument. And then the soil digests were analyzed for Cu, Zn, Ni, Pb, Cd, Cr using inductively coupled plasma-mass spectrometry (ICP-MS) and for Hg and As using atomic fluorescence spectrophotometer [9-10].

2.3. Data analysis

Contaminations of heavy metals in soil samples were analyzed by the enrichment factor (EF) and the pollution index (PI).

EF = Measured concentration in soil / Background concentration in soil

PI = Measured concentration in soil / Standard concentration in soil

Where background concentration in soil references "The element background values in Chinese soils" [11], standard concentration in soil references "Soil environmental quality Risk control standard for soil contamination of development land (GB36600-2018)" [12].

3. Results and discussion

Total concentrations of 8 trace elements in soil samples in the study area are shown in Table 1. The As, Hg, Ni, Pb, Cd, Cu, Zn, and Cr concentrations in the soils were ranging from 2.75 $mg\cdot kg^{-1}$ to 1210 $mg\cdot kg^{-1}$, 0.006 $mg\cdot kg^{-1}$ to 0.905 $mg\cdot kg^{-1}$, 14.6 to 940 $mg\cdot kg^{-1}$, 7.48 to 2108 $mg\cdot kg^{-1}$, <0.07 to 15.1 $mg\cdot kg^{-1}$, 12.2 to 23498 $mg\cdot kg^{-1}$, 24.4 to 1567 $mg\cdot kg^{-1}$, and <2 to 346 $mg\cdot kg^{-1}$. According to the table 1, the concentrations of all the measured elements in the surface soil smaples at 0-0.5m depth were the highest. They decreased with increasing depth. As, Ni, Pb and Cu exceed risk screening standard values and risk intervention standard values of residential land. The results of pollution evaluation are shown in Table 2. The max risk screening pollution indices (PI) of As, Ni, Pb and Cu were 60.5, 6.27, 5.27, 11.75. The exceeding standard rates were 48.37%, 19.19%, 7.60% and 6.40%. As was the typical pollutant factor in the copper smelter. Excessive amounts of arsenic were found in both surface soils and subsurface soils, enven at 3-6m depth. Compared to background concentrations in the control area, all eight elements in the study were elevated. The maximum enrichment factor of Cu was 3507, much higher than those of others elements. The degree of enrichment decreased as follows: Cu>Pb>As>Cd>Zn>Hg>Ni>Cr.

Table 1. Measured concentrations, standard concentration and background concentration.

Element		As	Hg	Ni	Pb	Cd	Cu	Zn	Cr
0-0.5m	Max	1210	0.905	940	2108	15.1	23498	1567	346
	Min	4.83	0.006	9.2	8.94	<0.07	14.6	24.4	27.4
	Median	96.45	0.15	171.28	117.13	2.09	638.52	139.16	138.55
0.5-1.5m	Max	447	0.022	697	749	4.79	4292	470	288
	Min	2.75	0.20	14.6	7.48	<0.07	12.2	52.2	<2
	Median	47.43	0.11	35.99	98.10	1.68	597.59	136.97	128.23
1.5-3m	Max	317	/	14.6	367	/	32.83	/	/
	Min	2.78	/	14.6	367	/	32.83	/	/
	Median	32.83	/	14.6	367	/	32.83	/	/
3-6m	Max	142	/	/	/	/	/	/	/
	Min	2.35	/	/	/	/	/	/	/
	Median	11.94	/	/	/	/	/	/	/
Standard concentration	Risk screening values	20	8	150	400	20	2000	/	/
	Risk intervention values	120	33	600	800	47	8000	/	/
Background concentration		7.5	0.04	19.5	17.2	0.053	6.7	59.1	41.4

All values are in $mg\cdot kg^{-1}$.

Table 2. The largest PI of As, Ni, Pb, Cu in different function divisions of the soil.

Function Division	Risk Screening PI				Risk Intervention PI			
	As	Ni	Pb	Cu	As	Ni	Pb	Cu
Office and living quarter	2.15	<1	<1	<1	<1	<1	<1	<1
Equipment maintenance area(east side of laboratory)	42.8	6.27	1.87	1.11	7.13	1.57	<1	<1
Meltshop	17.3	<1	2.75	2.15	2.88	<1	<1	<1
Acid plant	60.5	<1	5.27	12.75	9.58	<1	2.64	2.94
Raw materials storage area	48.85	<1	2.52	2.90	8.14	<1	<1	<1
Acid tank field	8.16	4.74	1.27	<1	1.36	1.19	<1	<1
Water-quenched slag yard	12.1	5.28	<1	<1	2.02	1.32	<1	<1
Flux crushing station	19.35	<1	1.58	3.27	3.23	<1	<1	<1
Slag beneficiation plant	9.85	4.96	<1	1.10	1.64	1.24	<1	<1
Hazardous wastes yard	2.27	3.31	<1	<1	<1	0.08	<1	<1
Below the sewage disposal	15.70	/	/	/	1.78	/	/	/
Below the meltshop	0.28	<1	/	/	<1	<1		
Below the raw materials storage area	2.68	<1	/	/	<1	<1		
Below the sewage tank of acid plant	59.5	/	/	/	9.08	/	/	/

From the perspective of horizontal space distribution (shown in Figure 2), arsenic contamination was ubiquitous. And the pollution areas in the copper smelter are relatively aggregated distribution. Heavily polluted areas were acid plant, raw materials storage area, equipment maintenance area and meltshop. Flux crushing station was polluted area because of the leeward of raw materials storage area. By analyzing the soil below the cement ground in key areas such as the sewage treatment station, meltshop, the material stock yard, and the acid plant, it can be found that the arsenic in the soil below the cement ground in key areas exceeded risk screening standard values of residential land. Moreover, the arsenic in the soil below the cement ground in the sewage treatment station and the acid plant also exceeded risk intervention standard values.

The field soil contamination was associated with the discharge of exhaust gas, wastewater and solid waste on the ground [13-15].

(i) Uncontrolled emission of soot containing heavy metals on site might contaminate the soil in the green belt of the smelting workshop, the material stock yard, and both sides of material transportation roads, etc. by atmospheric deposition. According to the survey result, the content of heavy metal was less in the surrounding area of the hazardous wastes yard covered with dust screens. Evidently, the controls of ground hardening and uncontrolled waste gas are essential for reducing soil contamination on site.

(ii) Wastewater on site might contaminate the soil under the wastewater treatment station, the sewage tank and the waste acid transportation pipeline of the acid plant. Damaged impermeable membrane of the wastewater collection tank is a major cause for contaminating deep soil. Hence, enterprises should strengthen management to prevent wastewater from leaking and venting.

(iii) Solid waste piles on site might contaminate the soil in the bare ground of the equipment maintenance area and the water-quenched slag yard through leaching after precipitation by means of vertical infiltration. Based on the survey findings, no significant soil contamination is found in the hazardous wastes yard that is strictly anti-seepage. Apparently, solid wastes containing heavy metals must be stored and treated according to certain standards.

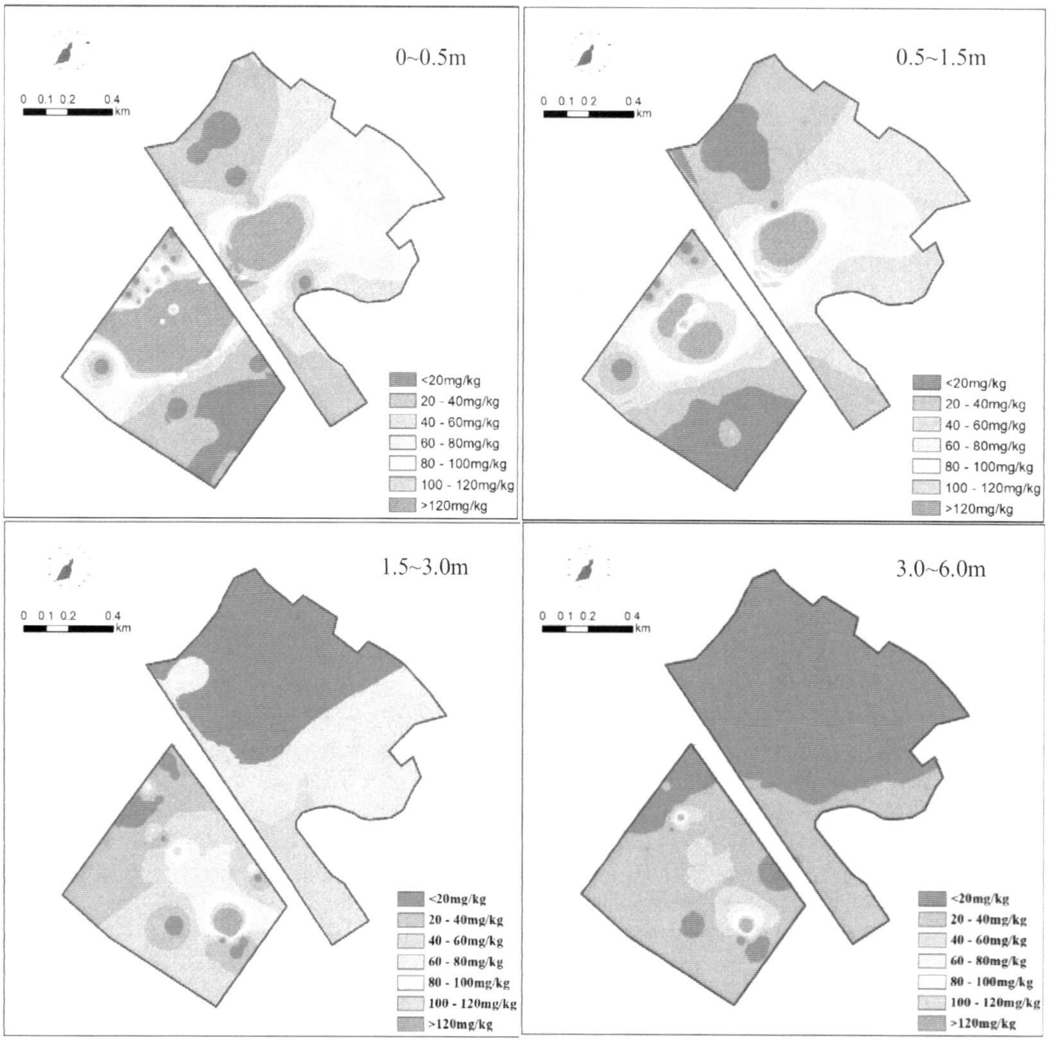

Figure 2. The As concentration distribution at different depths.

4. Conclusion

Arsenic, nickel, lead, and copper were specific pollutants of the soil in copper smelter, of which, arsenic contamination could be found in the most areas. After collecting the soil survey statistics of other copper smelting sites in the north, we found that arsenic contamination was common in the soil of copper smelting sites. Other heavy metals might not exceed risk screening standard values and risk intervention standard values although they are higher than the background values.

Key contamination areas of the site involve the material preparation area, the material stock yard, the smelting area, wastewater collection and treatment, both sides of the material transportation road, the maintenance area, the acid-making area, and pipe area, which are primarily caused by uncontrolled discharges of heavy metals, vertical infiltration of wastewater, and leaching of solid waste.

Acknowledgements

This work was funded by "the National Key Research and Development Program of China (No. 2018YFC1800401)".

References

[1] Duruibe J.O., Ogwuegbu M.O.C., Egwurugwu J.N. 2007 Heavy metal pollution and human biotoxic effects *International Journal of Physical Sciences* **2** p 112-118

[2] Pirrone N., Cinnirella S., Feng, X., et al. 2010 Global mercury emissions to the atmosphere from anthropogenic and natural sources. *Atmospheric Chemistry and Physics* **10** p 5951-5964

[3] Zhang X., Yang L., Li Y., et al. 2011 Estimation of lead and zinc emissions from mineral exploitation based on characteristics of lead/zinc deposits in China *Transactions of Nonferrous Metals Society of China* **21** p 2513-2519

[4] Nilo L, Marcelo A, Manuel B, et al. 2020 Human Health Risk Assessment from the Consumption of Vegetables Grown near a Copper Smelter in Central Chile *Journal of Soil Science and Plant Nutrition* **20** p 1472-1479

[5] Vojtěch Ettler. 2016 Soil contamination near non-ferrous metal smelters: A review *Applied Geochemistry* **64** p 56-74

[6] Behnam K, Farid M, Nasim AE. 2015 Soil Trace Elements Contamination in the Vicinity of Khatoon Abad Copper Smelter, Kerman Province, Iran *Toxicology and Environmental Health Sciences* **7** p 195-204

[7] Kumar A., Tripti, Maleva M., et al. 2020 Toxic metal(loid)s contamination and potential human health risk assessment in the vicinity of century-old copper smelter, Karabash, Russia *Environmental Geochemistry and Health* **42** p 4113-4124

[8] Behnam K., Farid M., Nasim A. E. 2015 Soil trace elements contamination in the vicinity of Khatoon Abad copper smelter, Kerman province, Iran *Environmental geochemistry and health, Toxicology and Environmental Health Sciences* **7** p 195-204.

[9] Han L., Chen R., Liu Z., et al. 2021 Sources of and Control Measures for PTE Pollution in Soil at the Urban Fringe in Weinan, China *Land* **10** p 762-762

[10] Liu X., Chen MB., Li LZ., et al. 2016 Contaminant Characteristics and Health Risk Assessment of Heavy Metals in Soils from Lead-Zincs Melting Plant in Huize County, Yunnan Province, China *Journal of Agricultural Resources and Environment* **33** p 221-229

[11] China National Environmental Monitoring Center 1990 The element background values in Chinese soils *China Environmental Science Press* p 33

[12] Ministry of Ecology and Environment of the People's Republic of China 2018 Soil environmental quality Risk control standard for soil contamination of development land (GB 36600-2018) p 3

[13] Yang W.C., Qi S.H., Xing X.l., et al. 2020 Study on the contribution mechanism of Daye Smelter to the heavy metals in the surrounding soil *Environmental Science & Technology* **43** p 110-115

[14] Yu D., Wang J., Wang Y., et al. 2021 Identifying the Source of Heavy Metal Pollution and Apportionment in Agricultural Soils Impacted by Different Smelters in China by the Positive Matrix Factorization Model and the Pb Isotope Ratio Method *Sustainability* **13** p 6526

[15] Amaya F.U., Cristina L.M., Enrique R.M., et al. 2009 Source identification of heavy metals in pastureland by multivariate analysis in NW Spain *Journal of Hazardous Materials* **165** p 1008–1015

Land suitability and economic value of ecotourism development in Masjid Lake of Kuantan Singingi District, Indonesia

Zulkarnaini[1] **and Nawari**[2, 3]

[1]Professors of Environmental Science Department, Postgraduate Program of Riau University, Pekanbaru, Indonesia, 28127
[2]Master of Environmental Science, Postgraduate Program of Riau University Pekanbaru, Indonesia, 28127

[3]Corresponding author's e-mail: n4wari@gmail.com

Abstract. Masjid Lake is one of the Objects and Attractions of Nature Tourism (ODTWA) in Riau Province, which is located in Koto Kari Village, Central Kuantan Sub-District, Kuantan Singingi District. This area is potential to support the economic in the non-oil and gas sector. However, the ecotourism development in this area is not yet optimal because it is not supported by integrated ecotourism development planning. In order to support integrated ecotourism planning, this research was conducted to determine the suitability and carrying capacity of the environment, conduct economic valuation and formulate sustainable ecotourism management strategies. Data type have been used include secondary and primary data. Secondary data include thematic map, rainfall, number of visitor and current ecotourism management system. Primary data include bio-geophysical aspects of water in the lake and environment surrounding the lake, tourist characteristics and management status. Secondary data compiled from relevant sources and formal institutions, while primary data collected with observation and survey. The land suitability analysis used were weighted linear combination (WLC) method followed by the Douglas method to calculate the carrying capacity of the environment. The economic valuation of ecotourism was assessed by the traveling cost method (TCM). Meanwhile sustainable ecotourism management strategy was prepared based on evaluation of the sustainability of management with multi-criteria analysis (MCA) and SWOT-AHP. Based on the analysis, it can be explained that suitable area (S1+S2+S3) for ecotourism development in Masjid Lake around 21.87 ha or 95.10% of Masjid Lake ecosystem, Meanwhile, 1.1 ha (4.9%) not suitable (N) for ecotourism development. From 21.87 hectares of suitable land, it is known that the physical carrying capacity (PCC) value = 9,574 visitors/ day, real carrying capacity (RCC) = 6,755 visitors / day and effective carrying capacity (ECC) = 3,378 visitors / day. While the actual tourist visit around 54 visitor/day is not exceeded environmental carrying capacity. The willingness to pay (WTP) value of tourists whom visiting to Masjid Lake around IDR 184,701 / visit / year, giving an actual economic value =IDR 3.64 billion / year, it smaller than available potential economic value=IDR 260.95 billion / year. Based on the MCA and SWOT - AHP analysis, it is shows that the sustainability status of ecotourism management in Masjid Lake classified as poor category with sustainability index=40.0%. However, based on the external and internal factors analysis, it shows that the current situation of ecotourism management in Masjid Lake classified in the favourable category because it has more strengths and opportunities than weaknesses and threats. The strategy that must established to improve current ecotourism management in Masjid Lake is a growth-oriented strategy, as follows: (1) Established an integrated sustainable ecotourism management plan (ISEMP) that accommodates all sustainable ecotourism development aspects; (2) Using the ISEMP as the basis to assess the feasibility a community-based ecotourism management; (3) Improve staff capacity and community participation for continuous ecotourism innovations; (4) Implement a digital-based ecotourism product marketing system according to market trends; and (5) Improving the ecotourism development accountability system and built ecotourism management reputation continuously.

1. Introduction

Ecotourism development is one of the alternative development sectors in Riau Province besides mining and, timber extraction in forestry and plantation sectors. This is a necessity considering the economic growth in Riau Province during the 2011 - 2018 period tended to decreases with a growth rate of around 2.74%/year. The poor growth of economic caused by the negative growth of mining sector. On the other hand, the growth of non-oil and gas sector tends to increases. This shows that so far development in Riau Province still relies on natural resources extraction directly. The dominant natural resources extraction activities include; oil, gold and coal mining [1]. Meanwhile, the aggressiveness of non-oil and gas growth triggered by an increase in the pulp and paper forest, palm oil and rubber plantation sectors has not been accompanied by efforts to improve the environment quality. This condition has caused negative impact on environment made several natural disasters in Riau Province. During 2018 recorded that floods that occurred have resulted in 144,211 victims, damaged 13,729 housing units and 88 public facilities [2].

Kuantan Singingi is one of the districts in Riau Province that is impacting by environmental damage with the main activity of gold mining, conversion of forest land to oil palm and rubber plantations. This environmental damage has caused a natural disaster with the worst impact in Riau Province. It has been recorded 49,820 people were recorded as victims, damaging 7,615 housing units and 6 public facilities during 2018 [2]. On the other side, Kuantan Singingi Regency has several potential natural resources that can be developed as tourism objects to improve the local community's economy and at the same time could to reduce environmental damage. One of the natural tourism objects that have the potential to be developed into an ecotourism object is the Masjid Lake. However, the development has yet been optimal, which is caused by the integrated ecotourism management plan hasn't prepared.

The effort to develop ecotourism in Masjid Lake is one form of environmentally friendly use as long as the principles of sustainable tourism are well implemented. The principles of ecotourism development according to the Indonesian Minister of Home Affairs Regulation No.33 of 2009, include: (1) There is compatibility between types of ecotourism and conservation characteristics; (2) Providing economic benefits for local communities; 3). Contains environmental education; (4) Provide satisfaction and experience to visitors; (5) Involving community participation; and (6) accommodating local wisdoms. Based on the description above, the principle of ecotourism development is to balance the three pillars, namely achieving the sustainability of economic, social and ecological functions.

To ensure the sustainability of ecotourism development in Masjid Lake, it is necessary to study the Ecotourism Management in the Masjid Lake area of Kuantan Singingi Regency. This study can be a comprehensive effort to formulate a sustainable ecotourism management model by analyzing several aspects needed in the preparation of a sustainable ecotourism management plan, which includes: (1) The Land suitability and environmental carrying capacity aspect of ecotourism potential; (2) The economic value; (3) Sustainable ecotourism development strategy in Masjid Lake.

2. Material and method

The research was conducted in Masjid lake which located in geographic coordinate 101° 32'00'' - 101° 33'30'' E and 0° 33'30'' - 0° 32'30'' S. The lake is placed in Koto Kari village, Kuantan Tengah sub district, Kuantan Singigi District and Riau Province (Figure 1). Data collection was conducted at June - July 2020.

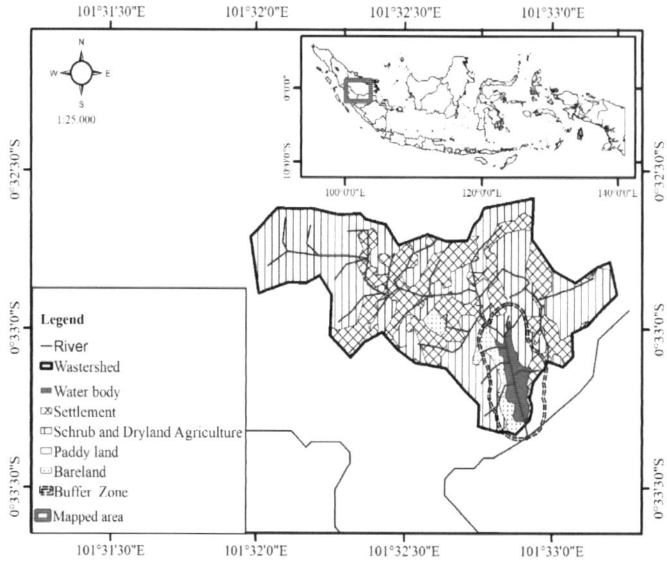

Figure 1. Research location.

Types and data sources used in the study includes the data sources for land suitability analysis, environmental carrying capacity, economic valuation and the status of ecotourism management analysis. The data for land suitability analysis consists: (1) Physical aspects of water in the lake: water depth, current velocity, colours, and water odour; (2) Bio-geophysical aspect surrounding the lake: soil texture, slope, rainfall, land cover, attractiveness; and (3) proximity to roads and proximity to settlements. Tourist characteristics data used for economic valuation of ecotourism include the frequency of tourist visits, travel costs, age, duration of activity, level of income and tourist education. Meanwhile, the data to formulated ecotourism management strategy include criteria and indicators (C&I) for sustainable ecotourism management, weight and rating of each C&I as well as the current sustainability status of ecotourism management.

The data collection of bio-geophysical data is carried out by using primary and secondary data collection methods. Secondary data were obtained through related institutions authorities including thematic map (space regulation, soil, government administration), rainfall, and satellite imagery: DEM SRTM-30 to generate slope and Landsat-8 satellite imagery for landcover map. Primary data collection of bio-geophysical aspects was carried out by field observations including physical condition of the waters, soil sampling, landscape attractiveness and validate the thematic map have been prepared. The data collection on tourist characteristics was carried out by interviewing to tourists who visited Masjid Lake. Likewise, data collection on sustainability status and ecotourism management strategies is carried out by interviewing key stakeholders.

The sampling technique used in the bio-geophysical aspect of primary data collection is systematic sampling with random start. The starting point of the observation is placed randomly, while the next observation point is placed systematically with distance between plots (grid) of 100 x 100 m. The number of samples is calculated based on population characteristics [3]. The sampling technique for observing the characteristics of tourists was carried out using purposive sampling. The number of samples visitors was calculated based on cross sectional sampling [4]. While data collection on the status and management strategy of ecotourism was carried out by using the snowball sampling method [5].

The analysis method of land suitability was carried out using the weighted linear combination (WLC) method using a set of land suitability criteria and indicators (C&I) collected from various sources (Table 1) and supported by the ArcGIS 10.7 software. The determination of land suitability C&I was carried out using the multi-criteria analysis method (MCA) [6]. After the land suitability map expressed, the analysis continued with environmental carrying capacity analysis is carried out using the Douglas method [7]. The economic valuation method for ecotourism development was conducted using the traveling cost method (TCM) [8]. In this analysis, the ecotourism demand curve is analysed using multiple linear regression analysis [9]. Meanwhile, the method of formulating ecotourism management strategies begins with an evaluation of the sustainability of ecotourism development using MCA [10] and and is continued with a SWOT-AHP analysis [11] and [12]. Factor analysis influenced of the status of sustainable management using the generalized boosted regression model (gbm) of R-Statistics software [13].

Table 1. Criteria and indicators (C&I) land suitability for ecotourism development in lake [a].

Criteria	Indicators	Unit of measurement	Land suitability class				References
			Very suitable (S1)	Moderate suitable (S2)	Marginal suitable (S3)	Not Suitable (N)	
Water physical condition in the lake	Water depth	m	0-5	5-7	7-10	>10	[14] modified
	Wave velocity	m/s	< 0.15	0,15-0,30	0,30-0,45	> 0,45	
	Water colours	class	Clear	Fairly turbid	Turbid	Very turbid/brown	
	Water Odour	class	odourless	Fairly odour	odour	very odour	
Bio-geophysical condition surrounding the lake	Soil texture	class	Fine loamy	Coarse loamy	Sandy loamy	Sandy	[15] modified
	Slope	%	0-5%	5-25%	25-35%	≥ 35%	[16]
	Rainfall	mm/y	<1,540	1,540-1,632	1,632-1,729	≥1,729	[17] modified using *quantile* analysis ([18] and [19])
	Landcover	class	Primary of secondary forest	Schrub or dryland agriculture	Paddyland or bare land	Settlement	[20] and [21] modified
Natural attractiveness	Natural attractiveness	%	≥ 95.2	88.5 - 95. 2	81.8 - 88.5	< 81.8	[22] modified using quantile analysis ([18] and [19])
Accessibility	Proximity from roads	m	< 500	500-750	750-1,000	≥ 1,000	[15] modified
	Proximity from settlement	m	<500	500-1000	1,000-5,000	≥ 5,000	[15] modified

[a] Sources: Study of literature

3. Results and discussion

3.1. Masjid Lake condition and tourist characteristics

The interpretation of latest Landsat-8 Satellite Imagery and DEM SRTM-30, shown that the landcover in Masjid Lake consist of watershed area (DAS) covering an area of 155.6 ha with an area of a water body or lake covering an area of 5.7 ha (3.6%), agriculture dry land covering an area of 97.4 ha (62.6%), residential area of 46.1 ha (29.6%) and open land area of 6.4 ha (4.1%). (Figure 1 and Table 2).

Table 2. The distribution of landcover area in Masjid Lake watershed [a].

Landcover	Area (Ha)	Proportion (%)
Water body	5.7	3.6
Settlement	46.1	29.6
Shrub and dryland agriculture	97.4	62.6
Bare land	6.4	4.1
Total watershed area	155.6	100.0%

[a] Sources: Landsat-8 data analysis (2020).

Based on interviews with Masjid Lake managements, it is known that the average number of tourist visits to Masjid Lake is 54 tourists per day or the equivalent of 19,710 tourists / year. The interviews result with 41 tourist respondents explained that all tourists who come to the Masjid lake was from various place in Kuantan Singigi District, namely from Central Kuantan Sub District and its surroundings area with various income, age and education levels (Table 3). The frequency of tourist visits from Kuantan Tengah Sub District is relatively more frequent to visit the to this destination compared to other districts. Its due the area closer to this lake, which results in lower travel costs compared to tourists from other sub-districts. The level of demand for the tourism market is influenced by various travel costs and the socio-economic character of tourists [8]. Increasing travel costs will reduce tourism demand. The farther the origin of tourists from tourist destination area will increase travel costs which include transportation and accommodation costs [23].

Table 3. Tourist characteristics in Masjid Lake [a].

Tourist Origin	N [b]		F[b] (tr/y) [c]	C[b] (IDR/tr) [c]	R[b] (IDR/m) [c]	LT[b] (h/tr) [c]	E[b] (y) [c]	A[b] (y) [c]
	T[b]	%						
Gunung Tidar	9	22.0	5.5	52,316	3,898,861	2.8	11.7	39.2
Kuantan Tengah	18	43.9	6.2	23,639	3,371,712	3.8	12.7	40.8
Sentajo Rayo	11	26.8	5.3	96,424	3,685,384	2.7	12.0	33.0
Singingi	1	2.4	5.3	100,086	2,900,694	2.9	9.0	16.6
Singingi Hilir	2	4.9	4.9	144,366	2,487,197	25	10.5	34.6
Total or average	41	100.0	5.4	83,366	3,268,769	2.9	11.2	32.8

[a] Sources: primary data analysis (2020).
[b] Field note: N=T number of tourist respondent, F= visit frequency, C= travelling cost, R=tourist income, LT= duration of tourist activity, E=tourist education, A=tourist age.
[c] Unit of measurement note: tr= number of trips, h=hours, y=year, m=month.

Based on sex, majority of tourists who came to Masjid Lake were male (70.7%), while female tourists were 29.3%. Most of the male tourists' activities are for picnicking (55.2%) and the remaining 44.8% come to Masjid Lake for fishing. While the majority of female tourists (91.7%) come to Masjid Lake for a having a picnic (Table 4).

Tabel 4. The distribution of tourist in Masjid Lake based on sex and activities [a].

Sex	Fishing		Picnics		Total	
	n[b]	%	n[b]	%	n[b]	%
Male	13	44,8	16	55,2	29	70,7
Female	1	8,3	11	91,7	12	29,3
Total	14	34,1	27	65,9	41	100,0

[a] Sources: Primary data analysis (2020)
[b] n= Number of tourists

3.2. Land suitability analysis for ecotourism development

3.2.1. Determination of weights and ratings of land suitability C&I. Based on multi-criteria analysis (MCA) and AHP involving key stakeholder, namely the main stakeholders in the management of Masjid Lake which consists of the Local Tourism Office, Local Development Planning Agency of Kuansing District and the Koto Kari Village Apparatus have succeeded in compiling weights / priorities and ratings of criteria and indicators (C&I) suitability of land for ecotourism development in Masjid Lake as shown in Table 5.

Table 5. Weight of C&I land suitability for ecotourism development [a].

Criteria (C)	Weight of Criteria (w)	Indicator (I)	Weight of indicator (v)
Water physical condition in the lake	0,30	Water depth	0,08
		Wave velocity	0,08
		Water colours	0,08
		Water Smells	0,08
Bio-geophysical condition surrounding the lake	0,30	Soil texture	0,08
		Slope	0,08
		Rainfall	0,08
		Landcover	0,08
Natural attractiveness	0,30	Natural attractiveness	0,30
Accessibility	0,10	Proximity from roads	0,05
		Proximity from settlement	0,05
Total	1,00		1,00
CR	0.000[b]		

[a] Sources: Primary data analysis (2020)
[b] Consistency ratio (CR) < 0.1 (10%) states that the key informant's opinion is consistent so that it can be used for further analysis

Rating or score on all environmental suitability C&I was carried out by dividing each weight from the AHP analysis by the maximum weight (Table 6).

Table 6. Score or rating of land suitability C&I [a].

Suitability degree	Weight/priority (w_i)	*Score / rating* (r_i)
R1	0,49	1,000
R2	0,27	0,551
R3	0,15	0,306
RN	0,09	0,184
CR	0,011[b]	

[a] Sources: Primary data analysis\ (2020)
[b] Consistency ratio (CR) < 0.1 (10%) states that the key informant's opinion is consistent so that it can be used for further analysis

After each C&I of suitability have been assessed. Then all weight and rating values of C&I were entered in a geographic information system (GIS) based database. The data is arranged in a 30x30m pixel raster data format. The calculation of the final score of land suitability was calculated using the weighted linear combination, sum product of multiplication results between the weights and ratings of each criterion and land suitability indicator for ecotourism development using Arcgis 10.7 software. The maximum score produced is 1 and the lowest score is 0. The land suitability class was determined by quantile analysis, which divides the distribution of the total suitability score distribution into 4 equal parts. In this analysis the maximum total score is 1 and the lowest score is 0. So that the determination of the score using the quantile method as described by Prakash (2003) in [16] is shown in Table 7.

Table 7. Score range of land suitability class [a].

Land suitability class	Score range	Suitability level
S1	0,75-1	Very suitable
S2	0,5-0,75	Moderate suitable
S3	0,25-0,5	Marginal suitable
N	< 0,25	Not suitable

[s] Sources: Secondary data analysis (2020)

3.2.2. Land suitability analysis result. The results of the weighted linear combination (WLC) analysis shown that 95.1% of the existing land area (21.9 ha) is included in the suitable land class for ecotourism development, which consists of 40.8% in the very suitable class (S1), 23.5% of land with medium suitable class (S2) and 30.8% including marginal suitable (S3). Meanwhile, 4.9% classified as not suitable for development into an ecotourism area (Figure 2 and Table 8). The high percentage of suitable land class for tourism development in Masjid Lake is influenced by the physical quality of the waters that support the development of ecotourism that low flow velocity, adequate lake depth and no pollution which maintained the color and odor of the waters naturally. Environmental conditions around the waters of Masjid Lake also support ecotourism activities that soil conditions Inseptisol and Entisol which good soil textures with relatively flat slopes and sufficient rainfall. Land cover which is dominated by shrubs and dry land farming. Likewise, accessibility to the Masjid Lake ecotourism location is relatively close with roads network and settlements making it easier for tourists to access the location and get accommodation. Meanwhile, the thing that decreases the level of land suitability for ecotourism development is that there is a small buffer zone of the Masjid Lake which has been converted into a settlement which can reduce the attractiveness value of ecotourism. The natural landscape factor has a positive influence on the level of tourist visits, where eco-travelers prefer to visit areas that are relatively unspoiled [24].

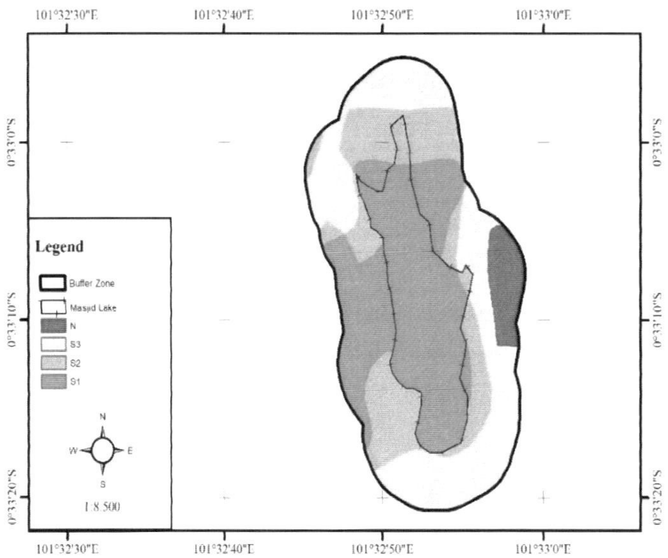

Figure 2. Land suitability map of ecotourism development in Masjid Lake.

Table 8. The distribution of land suitability class in Masjid Lake [a].

Land suitability class	Area(ha)	Proportion (%)
N	1.1	4.9
S3	7.1	30.8
S2	5.4	23.5
S1	9.4	40.8
Total	23,0	100,0

[a] Sources: Primary data analysis (2020)

3.3. *Environmental carrying capacity*

Environmental carrying capacity analysis started by estimation the physical *carrying capacity* (PCC) value. In this process, the suitable land area value for ecotourism development (A) is the total of land in class S1 + S2 + S3, which is 21.8712 ha or equal 218.712 m2. The optimal land area to support picnic tourism (B) is 65 m2/visitor [25]. Based on interviews with tourists, it is known that the average length of tourist activity around 3.2 hours / tourist. The average service time (ST) for Masjid Lake tourism objects based on interviews with the managers around 9 hours/day. So, the value of the rotation factor (RF) = 2.85 visitor/day. Thus, it can be seen that the value of the PCC Masjid Lake is able to accommodate tourist visits around 9,574 visitors/day. Taking into account the correction factor for rainfall erosivity (Cfr), slope (Cft) and soil erodibility (Cfs), it is known that the real carrying capacity (RCC) of Masjid Lake can accommodate about 6.755 visitors/day or 70.6% of the PCC value. The effective carrying capacity (ECC), which is by taking into account the current management factor, the carrying capacity of Masjid Lake is only able to manage 50% of the RCC or is able to serve tourist visits of 3,378 visitors/day (Table 9). Based ECC compared with actual tourist visit explained that current number of visitors come to Masjid Lake is still lower than the available effective carrying capacity. The main factor caused ecotourism management in Masjid Lake yet optimal is due to the lack aspects of ecotourism management include institutions and the inadequate capacity of management staff. Likewise, the level of community participation still low has caused the ecotourism area in Masjid lake was damaged to fires and land conflicts. This situation also supported by lack of financial aspect, especially the manager's motivation to increase investment in the natural tourism facilities and utilities construction to attract tourists. Improved sustainable ecotourism management requires several prerequisites which include community cohesion, equitable benefit sharing, commitment to conservation and management, in a political, social, economic and environmental context [26].

Table 9. Environmental carrying capacity of ecotourism estimation summary [a].

PCC Estimation						
A[b] (m^2)[c]	B[b] (m^2/v)[c]	N[b] (v/d)[c]	T[b] (h/v)[c]	ST[b] (h/d)[c]	RF[b] (v/d)[c]	PCC[b] (v/d)[c]
218,712	65	54	3.16	9.0	2.85	9,574
RCC and ECC Estimation						
PCC[b]	Correction Factors				RCC[b] (v/d)[c]	ECC[b] (v/d)[c]
	Cfr[b]	Cft[b]	Cfs[b]	Cfm[b]		
9,574	0.98	0.97	0.74	0.50	6,755	3,378

[a] Sources: Primary data analysis (2020)

[b] Field note : A= land suitable class area B=optimum area required for picnic , N= actual number visitor/day, T= average tourist activity duration, ST= service time of Masjid Lake, RF= Rotation factor, PCC=physical carrying capacity, RCC= Real carrying capacity, ECC=effective carrying capacity, Cfr = correction factor rainfall erosivity, Cft= correction factor of slope, Cfs= correction factor of soil erodibility and Cfm= correction factor of management.

[c] Unit of measurement note: v=visitors, d=day, h= hours, m=meter.

3.4. *Economic valuation of ecotourism*

The economic valuation analysis of ecotourism development in Masjid Lake, shown that the factors influence the frequency of tourist visits (tourist demand) to the Masjid Lake were travel costs and length of tourist visits, while age, income and tourist education wasn't significant. The relationship between the frequency of tourist visits and the cost of travel is negatives, meaning that the increases the travel costs, follows with decreases frequency of visits. Meanwhile, the relationship between the frequency of visits and the duration of tourist activity is positives which means that the longer the tourist activity causes the increases of frequency of tourist visits. The tourist demand curve model at Masjid Lake is expressed with this following equation:

$$Q = 2.72867455 - 0.00001740 * C + 0.19152741 * T$$

Where Q is the demand for tourism or the frequency of tourist visits (visits / year), C is the travel costs incurred by tourists during a visit to Masjid lake (IDR/ trip) and T is the length of tourist activity in one visit (hours/ trip). By knowing the imminent curve, the tourist surplus value or willingness to pay (WTP) is available to estimated by calculated the integral of demand curve function. The integration process was used lower limit is minimum cost incurred by tourists and above limit value is maximum cost spent by tourists during a visit to Masjid Lake. Based on the results of interviews with visitors, it is known that the minimum cost spent by tourists is IDR 10,839 / trip, while the maximum cost is IDR 147,920 / trip. So that the tourist surplus value or WTP is obtained of IDR. 184,701 / trip (Table 10).

Table 10. WT estimation summary of ecotourism development in Masjid Lake [a].

Minimum Cost (IDR/trip)	Maximum Cost (IDR/trip)	WTP Value (IDR/trip)
10,839	147,920	184,701

[a] Sources: Primary data analysis (2020)

The actual economic value (AEV) ecotourism development in in Masjid Lake was expressed by multiply the WTP value with average tourist visits / year who come to the Masjid Lake = 19,710 visitors/year, so the AEV = IDR 3.64 billion / year. This actual economic value is smaller than the potential economic value (PEV) which multiplication result of WTP and ECC = 1,412,852 tourists / year, PEV= IDR 260.95 billion / year (Table 11). The direct economic value (DEV) received by ecotourism managers at Masjid Lake is still smaller AEC and PEC value, it held because the tourists who came are not charged entrance fees and parking fees. In fact, retribution of parking fees and entrance tickets can be used as financial input to improve current ecotourism destination by constructing more complete facilities & utilities and provide marketing system. It will increase the attractiveness and visit of tourists from various regions, both local and international tourists. Thus, the development of ecotourism will further contribute to improving the economy of the community around Masjid Lake. Optimization of ecotourism levy revenue can not only reduce pressure on ecotourism due to changes in market supply and demand, but can also help ecotourism destination areas to obtain ecological compensation funds [27].

Table 11. Economic valuation summary of ecotourism development in Masjid Lake [a].

Average annual visitor (visitors/y)	ECC (visitors/y)	WTP (IDR/y/tr)	AEV (IDR/y)	PEV (IDR/y)
19,710	1,412,852	184,701	3,640,448,175	260,954,516.671

[a] Sources: Primary data processing (2020)

3.5. Sustainable ecotourism management strategy

3.5.1. Sustainability status of ecotourism management in Masjid Lake. Based on the results of sustainable ecotourism management evaluation using MCA and SWOT-AHP. The criteria and indicators of sustainable ecotourism management produced by cross tabulation between management dimensions (natural resource and cultural heritage management, human resource management and institutions and financial management) with the sustainability dimension (sustainability of economic, social and ecological functions) produced 16 C&I items with equal priority. Based on the analysis, it is known that the sustainability status of ecotourism management in Masjid Lake is classified in the low category with sustainability index = 40.0% of 5 scale or equal 1.8 (CIFOR, 1999). Based on SWOT-AHP analysis of internal and external environment of current ecotourism management shows that, the index value of strength (S) is 65.0% of the maximum scale 5, weakness (W) is 44.0%, opportunity (O) is 60.0% and threat (T) is 45.0%. The status of ecotourism management in Masjid Lake could be described in Cartesians coordinates [(SW), (OT) = [(65.0 - 44.0 = 21.0), (60.0 - 45.0 = 15/0)] = (21.0, 15.0), placed in quadrant I as GROWTH status in the SWOT analysis scatter diagram [12] (Figure 3).

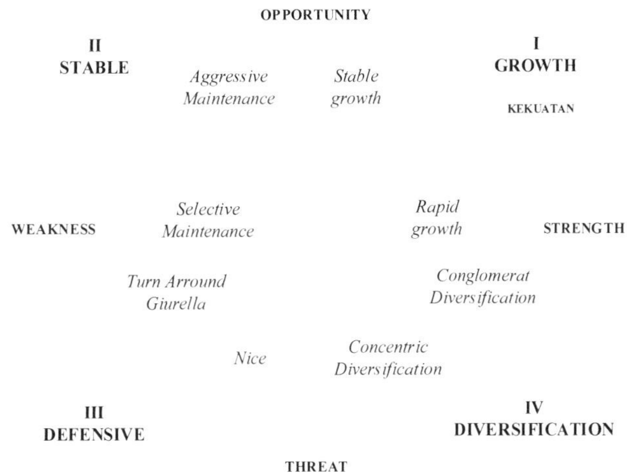

Figure 3. *Scatterplot* of SWOT - AHP describes ecotourism management status in Masjid Lake.

The factor analysis using the generalized boosted regression model (gbm), shown that the factors affect the sustainability status of ecotourism management in Masjid Lake are currently 38.5% influenced by human resource management and institutions, around 28.7% are influenced by financial management, while 19.9% are influenced by natural resources and cultural management. (Figure 4). The factor of human resources and institutional capacity determines a management process, especially in the allocation of financial and natural resources in achieving sustainable management [28].

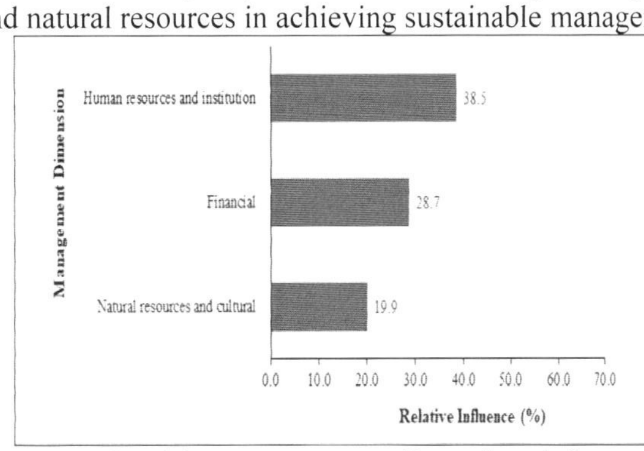

Figure 4. Graphic management dimension influences to sustainability status of ecotourism management in Masjid Lake.

The sustainability status of human resource and institutional management of ecotourism in Masjid Lake categorized low with sustainability index=33.9%. The factors that affect the sustainability of the human resources and institutional management were the level of local community participation (RI=26.8%), the availability of ecotourism management institutions (RI=22.3%), the competency of management staff (RI=21.0%) and compliance with regulations (RI=15.9%) (Figure 5). Currently the management of ecotourism in Masjid Lake still initiated by the Koto Kari village and Kuantan Singngi District government. On the other side, local community participation is still low and the management not supported by settle community-based ecotourism management institutions. The low of community participation and the absence of a management institution have hampered the development of ecotourism in Masjid Lake. Several fires due to the conversion of agricultural land have lowered the attractiveness of Masjid Lake. Increasing the participation of local communities in community-based ecotourism management institutions will produce perspectives and behaviours that support the achievement of conservation of natural and cultural resources [29].

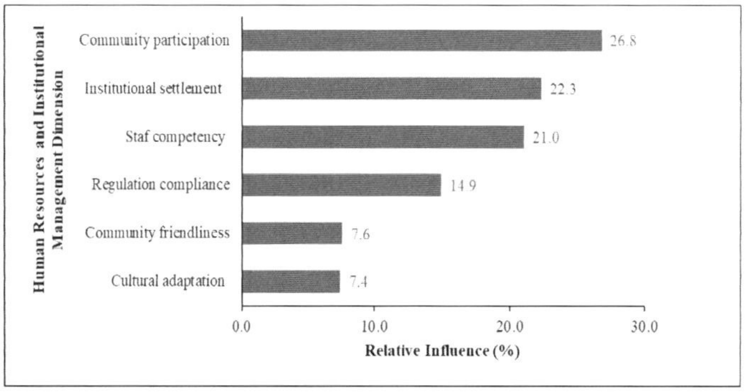

Figure 5. Graphic C&I influences to sustainability status of human resources and institutional management of ecotourism in Masjid Lake.

The sustainability status of the financial management aspects of Masjid Lake categorized low with sustainability index of around 32.2%. The factors determine for the sustainability of the financial management aspects were largely determined by the accountability of financial resource management (RI=60.8%), followed by marketing of ecotourism (RI=25.6%) and increasing the local economy (RI=13.5%) (Figure 6). The Ecotourism management of Masjid Lake was still carried out independently without imposing entrance and parking fees to tourists. Funding sources for management still depend on local government budgets. Without income from ecotourism activities, the sustainability of ecotourism development in Masjid Lake will be difficult to materialize. The ecotourism development in Masjid Lake was also not supported by a good ecotourism marketing system, this is indicated by the number of tourist visits that are still lower than the potential carrying capacity available. An accountable financial management system through management of revenue and expenditure budgets as well as a good marketing system will improve the sustainability of the ecotourism management of Masjid Lake in the future. Ideally, ecotourism development requires an optimum number of investment and marketing funds for the development of ecotourism based on natural resource conservation. Ecotourism activities will have a positive economic impact on managers and local communities. However, the management of this financial aspect requires preventive action against misuse of funding through accountable financial management [30].

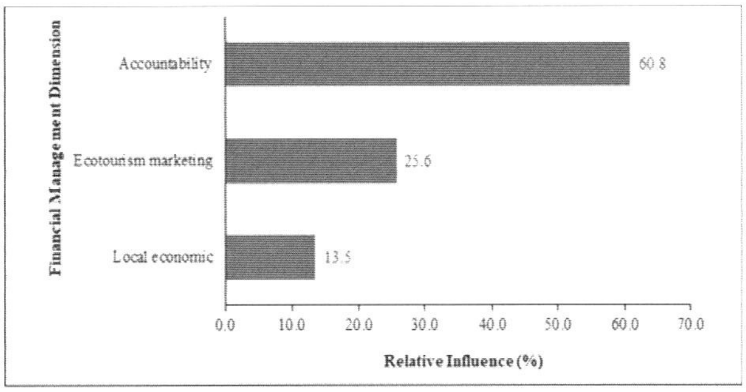

Figure 6. Graphic C&I influences to sustainability status of financial management of ecotourism in Masjid Lake.

Based on the results of the analysis, it is shown that the sustainability status natural resources and cultural management in Masjid Lake classified moderate with sustainability index = 53.9%. Strong factors affecting the sustainability status of the dimensions of natural resources and cultural management were land suitability planning (RI=20.9%), environmental carrying capacity planning (17.8%) environmental interpretation programs (17.2%), local knowledge and wisdom (14,8%) and natural resource protection (13.2%) (Figure 7). Currently, ecotourism management in Masjid Lake has not been supported by an integrated ecotourism management plan that considers suitability, environmental carrying capacity and the economic valuation of ecotourism. This causes the managers not have sufficient information to establish an ecotourism development programs that is able to balance the interest of tourists with the land suitability - carrying capacity of the environment. A further implication due to this lack of integrity causes the managers to be unable to increase the number of tourist visits. Meanwhile, on the one hand, social activities around Masjid Lake very intensive that required prevention against damage to natural resources. Integrated sustainable planning using information on land suitability, carrying capacity and economic valuation can become guidance to create a roadmap for achieving a more sustainable ecotourism management [31]. The implementation of an ecotourism management plan linked to national development can be done by integrating it with development in the other sectors (cross-sectoral) and also integrating it in a cross-structural development hierarchy (district-provincial-central government). It aims to achieve synergy in the development of ecotourism in the concept of sustainable national development [32]. Inline with the concept of village development, ecotourism development can be carried out by developing a tourism village business through the establishment of village-owned enterprises that involve active community participation [33].

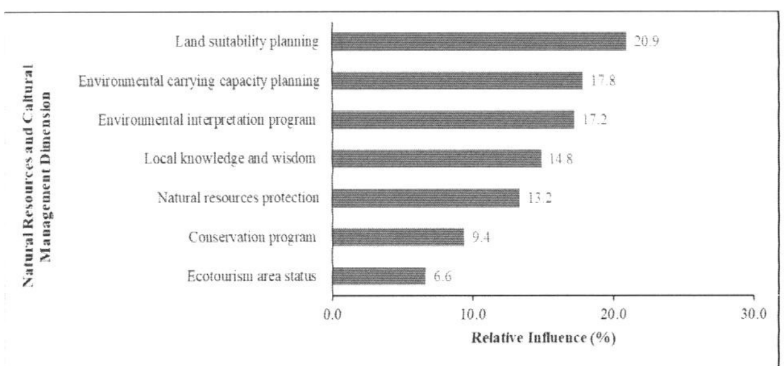

Figure 7. Graphic C&I influences to sustainability status natural resources and cultural management of ecotourism in Masjid Lake.

3.5.2. Sustainable ecotourism management strategy in Masjid Lake. Based on the sustainability evaluation and the results of the SWOT-AHP analysis, shown that the current situation of ecotourism

IOP Conf. Series: Earth and Environmental Science **865** (2021) 012039 doi:10.1088/1755-1315/865/1/012039

management in Masjid Lake yet optimal. However, management potential in the future is in a favourable situation. Existing ecotourism management has great potential strengths and opportunities so that it can be utilized to reduce weaknesses and overcome threats. So, the strategy that must be applied in this condition is to support a rapid growth strategy (Figure 2). By considering the situation of strengths and opportunities along with the factors that affect the management dimension (Figure 3-7) in a more profitable direction, rapid growth-oriented strategies that can be developed to reduce weaknesses and overcome threats in improving ecotourism management in Masjid Lake were as follows:

1. Established an integrated sustainable ecotourism management planning (ISEMP) of Masjid Lake by taking into account the aspects of suitability, environmental carrying capacity and integrated economic valuation across sectors and structures (village, district, provincial and national governments) as well as the application of collaborative management in community-based ecotourism management accompanied by strengthening of the Masjid Lake ecotourism management institution. This can be integrated with village development programs in the form of the establishment of village-owned ecotourism enterprises
2. Using the ISMEP as the basis for the feasibility of developing an community based ecotourism management in Masjid Lake to obtain an investment allocation for enhances curreent ecotourism facilities and utilities and increasing the attractiveness and competitiveness of ecotourism products
3. Increase staff capacity and community participation for innovation in ecotourism development continuously at Masjid Lake by collaboration with universities, research institutions and private companies in Riau Province
4. Implement a digital-based ecotourism product marketing system by considering the ecotourism demand or market trend
5. Improve the ecotourism management accountability system which is integrated to village development program and build a reputation continuously for ecotourism market increment.

4. Conclusion

Based on the analysis, the conclusions that can be drawn from this study are as follows: (1) The area of the Masjid Lake classified as suitable for ecotourism development is 21.87 ha or 95.1% of the total area, meanwhile, 1.1 ha (4.9%) is land that is not suitable land. From 21.87 ha of land suitable known that the PCC value = 9,574 visitors/day, RCC = 6,755 visitors/ day and the ECC value = 3,378 visitors / day. The actual tourist visits around 54 visitors/day was not exceed the environmental carrying capacity; (2) The willingness to pay (WTP) value of tourists visiting Masjid Lake is IDR 184,701/ visit/year, while the current actual economic value of ecotourism development was around IDR 3.64 billion / year, which still smaller than the available potential economic value around IDR 260.95 billion/year; (3) Based on management evaluation using AMK and SWOT - AHP analysis, it is known that the sustainability status of ecotourism management in Masjid Lake classified as low category with a sustainability index of 40%. However, the analysis of internal and external factors shows that the management of ecotourism in Masjid Lake in the future will be in a favorable situation because it has more strengths and opportunities than weaknesses and challenges. The strategy that must be used to improve ecotourism management in the Majid Lake is a rapid growth-oriented strategy, namely using the existing strengths and opportunities to eliminate weaknesses and challenges. This strategy can be implemented through: (1). Established an integrated sustainable ecotourism management planning (ISEMP) that accommodates all aspects of sustainable ecotourism development; (2) Using ISEMP as the basis to assess the feasibility of community based ecotourism management in Masjid Lake; (3) Improve staffs capacity and community participation for continuous ecotourism development innovations; (4) Implementing a digital-based ecotourism product marketing system by considering the trend of ecotourism demand; and (5) Increasing the accountability management system that is integrated to village development program and building reputation to tourists continuously.

Greeting note
The authors would like to thank the various parties who have provided support for this research. Especially to Postgraduate Program of Riau University and the Local Government of Kuantan Singingi Regency who have fully supported the completion of this research. Hopefully this research can be used as information and literature to build integrated sustainable ecotourism management at the research location and on a wider scale in the future. This research methodology potentially become reference to implement in the other site. Of course, there space for improvements especially establish specific C&I of land suitability of each tourist activities, continued with assessing it carrying capacity, economic valuation and formulate management strategy.

References
[1] Bappenas. (2019). Laporan perekonomian Provinsi Riau. (2018). Kantor Kementerian Pembangunan Nasional. Jakarta.
[2] BNPB. (2020). Data dan informasi bencana alam Provinsi Riau 2019. http://bnpb.cloud/dibi/tabel2a. Accessed at 21 February 2020.
[3] Sadono, R, Murdawa B., Soeprijadi, D., & Nawari. (2011). Biometrika hutan volume-i. Metode Statistika. Interlude. Yogyakarta.
[4] Lemeshow S, Hosmer, Jr.D.W., Klar J, & Lwanga S.K. (1990). Adequacy of sample size in health studies. John Wiley & Sons. New York.
[5] Sugiyono. (2014). Metode penelitian pendidikan pendekatan kuantitatif, kualitatif, dan R&D. Alfabeta. Bandung.
[6] Mendoza, G.A. & Prabhu, R. (2000). "Development of a methodology for selecting criteria and indicators of sustainable forest management: a case study on participatory assessment". Environmental Management, Volume 26(6), 659-673.
[7] Zacarias, D.A., Williams, A.T., & Newton, A. (2011). Recreation carrying capacity estimations to support beach management at Praia de Faro Portugal. Applied Geography, Volume 31 (3), 1075-1081.
[8] Fauzi, A. (2004). Ekonomi sumber daya alam dan lingkungan: Teori dan Aplikasi. PT Gramedia Pustaka Utama. Jakarta.
[9] Nawari. (2010). Analisis regresi dengan Microsof Excel dan SPSS 17. PT Elexmedia Komputindo. Jakarta.
[10] CIFOR. (1999). Panduan untuk menerapkan analisis multi kriteria dalam menilai kriteria dan indikator. SMK Grafika Mardi Yuana. Bogor.
[11] Argyris, C. (1985). Strategy, change and defensive routines. Pitman, Boston.
[12] Rangkuti & Freddy. (2004). Manajemen persediaan. Jakarta: PT Raja Grafindo Persada. Jakarta
[13] Friedman, H.J. (2002). Stochastic gradient boosting. Computational Statistics & Data Analysis, Volume 38(4), 367-378.
[14] Yulianda, F. (2007). Ekowisata bahari sebagai alternatif pemanfaatan sumberdaya pesisir berbasis konservasi. Makalah Seminar Sains pada Departemen Manajemen Sumberdaya Perairan. Fakultas Perikanan dan Ilmu Kelautan IPB. Bogor.
[15] Bali, A., Seyed, M.M, Borhan, R., Nematollah, K., & Mir M.K.Z. (2015). A spatial decision support system for ecotourism development in Caspian Hyrcanian Mixed Forests Ecoregion. Bol. Ciênc. Geod., sec. Artigos, Volume 1 (2), 340-353.
[16] Bunruamkaew, K & Murayama, Y. (2012). Land use and natural resources planning for sustainable ecotourism using gis in Surat Thani, Thailand. Sustainability Journal, Volume 4, 412-429.
[17] Asmamaw, D., & Gidey, E. (2018). Identification of potential eco-tourism site suitability using ahp and gis, a case of Hugumburda Forest and Its Surrounding Areas, Ethiopia. Adv Envi Was Mana Volume 1, 1-4.
[18] Parzen, E. (2004). Quantile probability and statistical data modeling. Statist. Sci., Volume 19(4), 652-662.

[19] Lopez, J.M.R., Heiderb, K., & Scheffran, J. (2016). Frontiers of urbanization: identifying and explaining urbanization hotspots in The South of Mexico City using human and remote sensing. Applied Geography, Volume 79, 1-10.

[20] Boyd, S.W. & Butler, R.W. 1995. Review of the development of ecotourism with respect to identifying criteria for ecotourism for Northern Ontario. Report for Department of Natural Resources/ Forestry, Ministry of Natural Resources. Sault Ste. Marie, Ontario.

[21] Makhdoum, M. (2006). Principles of land-use planning. 7th edn. Tehran University Press. Tehran.

[22] Dirjen PHKA. (2003). Pedoman analisis daerah operasi obyek dan daya tarik wisata alam (ADO-ODTWA). Kantor Direktorat Jenderal PHKA. Bogor.

[23] Nicolau, J.L., & Más, F.J. (2006). The influence of distance and prices on the choice of tourist destinations: the moderating role of motivations. Journal of Tourism Management, Volume 27(5), 982-996.

[24] Chen Y., Sun B., Liao S.B., Chen L., & Luo S.X. (2016). Landscape perception based on personal attributes in determining the scenic beauty of in-stand natural secondary forests. Ann. For. Res., Volume 59(1), 91-103.

[25] Fandeli, C., & Muhammad. (2009). Prinsip-prinsip dasar mengkonservasi lanskap. Gadjah Mada University Press. Yogyakarta.

[26] Foucat, S.A. (2002). Community-based ecotourism management moving towards sustainability, in Ventanilla, Oaxaca, Mexico. Journal Ocean & Coastal Management Volume 45(8), 511-529.

[27] Jing, Y & Funcai, H. (2011). Research on management of ecotourism based on economic models. Energy Procedia Volume 5, 1563–1567.

[28] Benn, S., Edwards, M., & Williams, T. (2014). Organizational change for corporate sustainability, third edition. Routledge, London.

[29] Stem, C.J, Lassoie, J.P., Lee, D.R., & Deshler, D.D. (2003). Community participation in ecotourism benefit: the link to conservation practices and perspectives. Society and Natural Resources, Volume 16, 387–413.

[30] Tisdell, C.A. (2003). Economic aspects of ecotourism: wildlife-based tourism and its contribution to nature. Sri Lankan Journal of Agricultural Economics, Volume 5(1), 83-95.

[31] Kastarlak, B.I. & Barber, B. (2014). Fundamentals of planning and developing tourism. Pearson Education Limited. Essex, UK.

[32] Bhuiyan, A.H., Siwar C., Ismail., S.M., & Islam R. (2012). The role of ecotourısm for sustaınable development in East Coast Economıc Regıon (Ecer) Malaysıa. OIDA International Journal of Sustainable Development, 3(9), 53-60.

[33] Nugroho, I., Pramukanto, F.H, Negara, P.D., Purnomowati, W., & Wulandari, W. (2016). Promoting the rural development through the ecotourism activities in Indonesia. American Journal of Tourism Management, Volume 5(1), 9-18.

Efficient removal of lead ions from aqueous solution using C-TiO_2 adsorbent

Qi Luo[1], Guiqin Liu[2], Li Wang[3, 4], Wukang Wang[1], Ying Wang[1], Hao Wang[1] and Wen Xu[1]

[1]Zhejiang Light Industrial Products Inspection and Research Institute, Hangzhou 310018, China;
[2]Zhejiang Institute of Product Quality and Safety Science, Hangzhou 310013, China;
[3]Zhejiang A&F University, Hangzhou 311300, China

[4]Corresponding author's e-mail: 14791240@qq.com

Abstract. Carbon-doped TiO_2 (C-TiO_2) was fabricated using a facile hydrothermal treatment. It was applied as an adsorbent to remove Pb(II) from the aqueous solution. The morphology and properties of the adsorbent were characterized with TEM, XRD, SAED and XPS. Furthermore, the applicability of Langmuir isothermal model for Pb(II) onto C-TiO_2 was examined. The adsorption capacity of the adsorbent and the determination of Pb(II) were evaluated with ICP-MS. The maximum adsorption capacity is 28.7 mg g^{-1}, and this method has broad application prospects in areas of environmental purification.

1. Introduction

The presence of heavy metal ions in the environment is of serious concerns due to their toxicity. Pb, in particular, is a natural element predominantly used in industrial applications, and wastewater contains a wide range of Pb(II) concentrations [1,2]. It may affect children's intelligence or cause negative effects on the metabolic process even at low dosages [3]. Unlike organic pollutants that are usually susceptible to biological degradation, Pb(II) does not degrade into harmless end products [4]. Adsorbents with good adsorption capacities are applied to interact with metal ions to reduce metal contamination [5-7]. Therefore, an effective adsorbent for removing Pb(II) is suitable for water treatment.

The use of nano-adsorbent in environmental remediation has attracted considerable interest owing to their activity [8-10]. Remarkably, TiO_2 which has properties such as non-toxicity, relative cheapness, and high specific surface area, is becoming a new functional material [11-13]. Carbon-doped TiO_2 (C-TiO_2) has attracted much interest owing to their potential in the removal of organic pollutants. But the adsorption capacity of modifications of TiO_2 using carbon for heavy metal ions has received less attention. The aim of this paper is to assess the ability of C-TiO_2 to adsorb Pb(II) from an aqueous solution.

Here, C-TiO_2 were fabricated using a hydrothermal process and post-calcination. The morphology and chemical composition of products were characterized by transmission electron microscopy (TEM), X-ray diffraction (XRD), selected area electron diffraction (SAED) and X-ray photoelectron spectroscopy (XPS). An approach for the effective removal of Pb(II) via C-TiO_2 at room temperature was reported.

2. Experimental section

2.1. Instruments and reagents
The main instruments included a Talos F200X high-resolution transmission electron microscope (Thermo Scientific, USA), an EscaLab 250Xi X-ray photoelectron spectrometer (Thermo Scientific), a Grant LMC-3000 centrifuge (Cambridge, UK), a NexION 350 inductively coupled plasma mass spectrometer (PerkinElmer, USA), a SevenExcellence pH meter (Mettler Toledo, USA).

Titanium butoxide (TBT), low-density polyethylene (LDPE), nitric acid and ethanol were of at least analytical grade, and purchased from Macklin (China). The 1000 mg L^{-1} Pb(II) standard solution was purchased from the Nonferrous Metals and Electronic Materials Analysis Center (China). 18.2 MΩ cm water was prepared with a Dura pro 12 water purifier (China).

2.2. Preparation of C-TiO_2 and characterizations
200 mL 98% ethanol (as solvent), 8.0 g LDPE (as carbon source) and 13.6 mL TBT (as a titanium source) were successively added to a beaker under stirring. The resulting liquid was transferred to a PPL-lined hydrothermal reactor when stirring at 170 °C for 5 h. After cooling to room temperature, the precursors were washed with ethanol and water, and then dried in an oven at 80 °C for 30 min; they were calcined in a muffle furnace at 490 °C for 3 h to obtain C-TiO_2 powder.

C-TiO_2 particles were scattered in ethanol by ultrasonic dispersion for 5 min, and then made of 3 mm diameter wafers into TEM. The microstructure of the adsorbent was observed under an accelerating voltage of 200 kV. XRD analysis was performed using Cu-Kα radiation (λ = 1.541 Å) with a Cu X-ray tube at 30 kV. XPS spectra were collected at a power of 150 W using monochromatized Al-Kα radiation, and the binding energy was scanned from 100 to 700 eV. The SAED analysis of the adsorbent was also performed.

2.3. Adsorption test
Similar amounts of C-TiO_2 were added to colorimetric tubes, and the pH value of the solution was adjusted to 6.5 using diluted nitric acid. The standard solutions of Pb(II) were transferred to microcentrifuge tubes after 2 min of shaking. Followed by centrifugation at 1500 rpm for 10 min and filtration, the Pb(II) concentrations were determined in 5 mL supernatant.

In kinetic energy discrimination mode, the mass spectrometer was attuned by 1.0 μg L^{-1} mass spectrometric fluid, and Pb(II) was then determined by inductively coupled plasma mass spectrometry (ICP-MS) using high purity helium as the collision gas and high purity argon as the carrier gas.

3. Results and discussion

3.1. Morphology and properties of C-TiO_2
Nanosized adsorbents were three-dimensional crystallines with partly aggregation (Figure 1*A*). The average grain size of C-TiO_2 was counted as 27 nm (Figure 1*B*). The lattice spacings were 0.36 and 0.18 nm corresponding to (101) and (200) crystal planes of high-resolution TEM observation (Figure 1*C*), which were regarded as the crystal plane of the anatase phase. In general, both XRD and high-resolution TEM analysis can measure the phase of TiO_2 (including anatase, rutile and brookite). For XRD analysis (Figure 1*D*), the diffraction peaks at 25.3°, 37.9° and 48.1° were assigned to the diffraction from (101), (004) and (200) crystal planes of the anatase phase, respectively.

The phase of adsorbent was further identified using SAED pattern (Figure 1*E*), and regarded as polycrystalline TiO_2, indicating the crystalline shape of the as-synthesized C-TiO_2. Brunauer-Emmett-Teller (BET) method was used to investigate the specific surface and pore volume of C-TiO_2. The specific surface area was determined to be 371 m^2 g^{-1}, and the pore volume was 0.65 cm^3 g^{-1}.

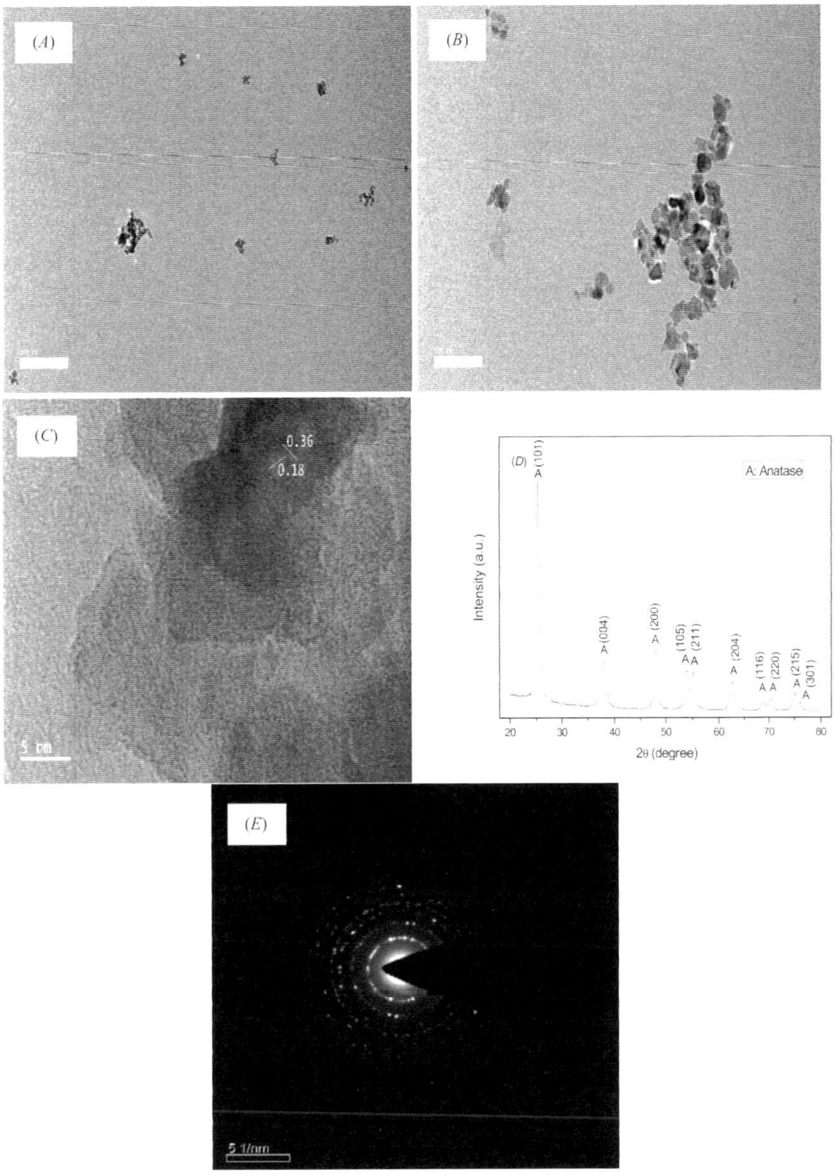

Figure 1. TEM images (*A*,*B*), high-resolution TEM image (*C*), XRD pattern (*D*) and SAED pattern (*E*) of C-TiO_2.

The C 1s, Ti 2p, O 1s and Ti 2s core levels were observed over an XPS survey spectrum (Figure 2*A*). The high-resolution XPS spectra (Figure 2*B–D*) had C element confirmed by two peaks at 284.7 and 288.8 eV, which were assigned to C–C and C–O bonds. The Ti 2p spectrum displayed two typical peaks of Ti–O at 464.4 and 458.5 eV. These were assigned to the binding energy of Ti $2p_{1/2}$ and Ti $2p_{3/2}$. The O 1s spectrum could be deconvoluted into two individual component peaks located at 529.7 and 531.7 eV. They were attributed to surface O–Ti and O–H groups. As shown in Table 1, the C content was about 9.4% as a result of XPS, revealing the elemental generation of C-TiO_2.

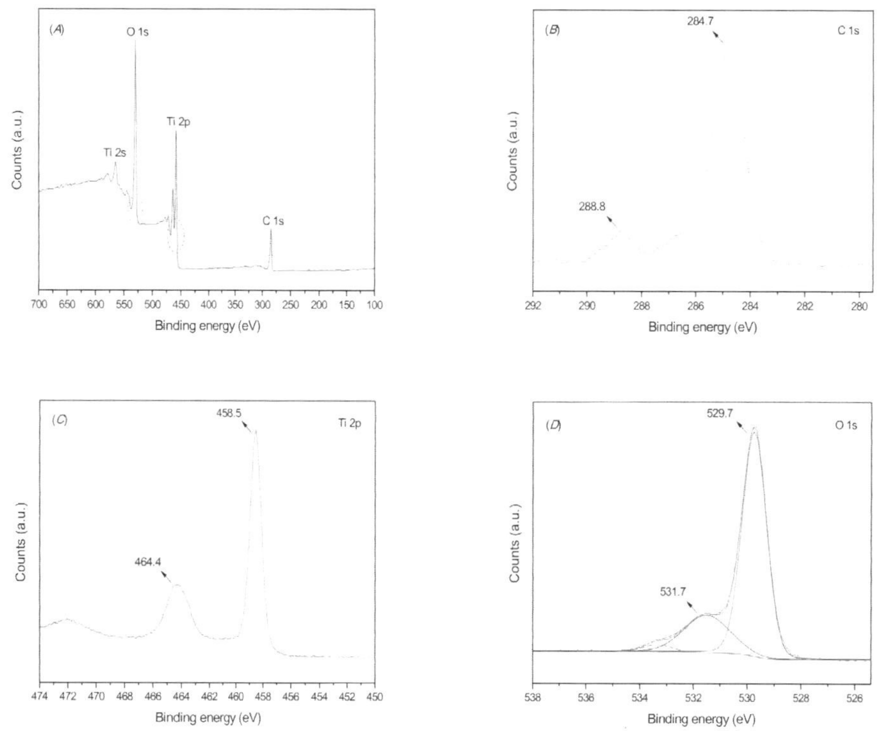

Figure 2. XPS survey spectrum (*A*) and high-resolution XPS spectra of C 1s (*B*), Ti 2p (*C*) and O 1s (*D*) for C-TiO_2 as a function of binding energy.

Table 1. C, Ti and O contents of as-prepared adsorbent.

Element	XPS (%)
C	9.37
Ti	29.89
O	60.74

3.2. Removal of Pb(II) from water using C-TiO_2 adsorbent

The metal ion concentrations of the collection supernatant via C-TiO_2 were performed using ICP-MS. The net intensities of Pb-208 were in good linear relation to known concentrations. The instrument detection limit (IDL) of Pb(II) was defined as three standard deviations (SD) of eleven measurements of the blank reagent (IDL = 0.006 μg L^{-1}). Adsorption studies were performed at room temperature starting at different concentrations of Pb(II), and the adsorbent dosage was selected as 0.1 g/25 mL. Each test was repeated in triplicate, and the error bars represent SD. The metal ion retained onto the adsorbent is calculated using the following equation:

$$q_e = \frac{(C_0 - C_e) \cdot V}{m} \quad (1)$$

where C_0 and C_e are the initial and final concentrations of Pb(II) (mg L^{-1}), term q_e is the amount adsorbed per specified mass of adsorbent (mg g^{-1}), V is the volume of solution (mL), and m is the mass of adsorbent (mg). The adsorption percentage (A_p, %) of the Pb(II) from solution can be calculated by Equation (2):

IOP Conf. Series: Earth and Environmental Science **865** (2021) 012040 doi:10.1088/1755-1315/865/1/012040

$$A_p = \frac{C_0 - C_e}{C_0} \times 100\% \qquad (2)$$

The A_p of Pb(II) increased with the contact time, and equilibrium was reached within 100 min (Figure 3). The 95% removal efficiency occurred with small fluctuations when the content of Pb(II)/C-TiO_2 was 19.3 mg g^{-1} (Figure 4). The adsorption curves suggested that Pb(II) exhibited affinity to C-TiO_2 in weakly acidic solutions. The speciation of the adsorbent was an important factor in the adsorption process.

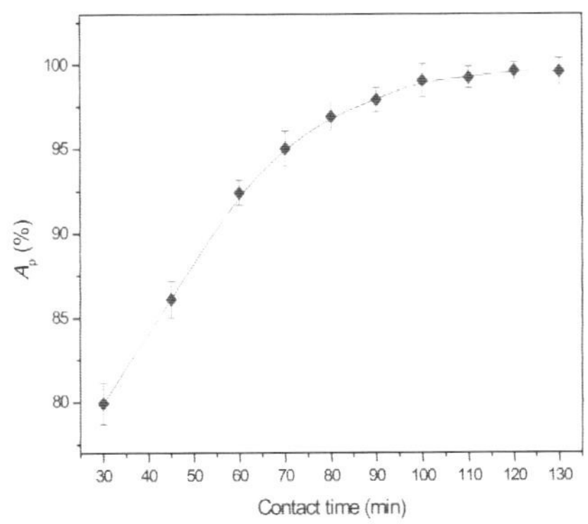

Figure 3. Effect of contact time on the adsorption of Pb(II) (initial concentration = 30 mg L^{-1} and pH = 6.5).

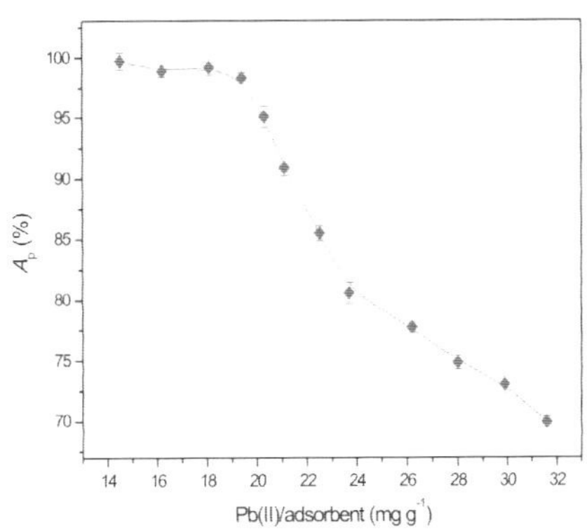

Figure 4. Adsorption percentage plot of Pb(II) (contact time = 100 min and pH = 6.5).

Isothermal studies show how metal ion distributes between adsorbent surfaces and metal ion solutions at equilibrium. The Langmuir isotherm is represented by the linear Equation (3), and the maximum adsorption capacity (K, mg g^{-1}) is determined from a plot of C_e/q_e and C_e:

$$\frac{C_e}{q_e} = \frac{C_e}{K} + \frac{1}{K \cdot b} \qquad (3)$$

where b is the Langmuir constant (L mg^{-1}). K of C-TiO_2 for the removal of Pb(II) were determined to be 28.7 mg g^{-1}. The removal ability of bare TiO_2 for Pb(II) was 16.4 mg g^{-1}.

The proposed adsorption mechanism is as follows: Carbon and O–H groups on the adsorbent surface play a part for Pb(II) adsorption. Lead ions were adsorbed onto active sites, while these sites changed with increasing content of metal/C-TiO_2: A_p decreased, and the active sites were filled.

According to the calculation of K, the adsorbent amount required is 100/28.7 mg L^{-1} in the case of Pb(II). 3.4 mg adsorbent can be mixed with 1.0 L water to obtain suspensions at pH = 6.5. The quality of adsorbent consumed is small (<3.5 mg L^{-1}). This reliable method presents a competitive capacity versus other adsorbents previously reported (Table 2). The adsorption capacity of C-TiO_2 is higher than other materials, indicating practical significance for Pb(II) removal and future use of nonmetal-doped TiO_2.

Table 2. Comparison of adsorption capacity of adsorbent materials for Pb(II).

Adsorbent material	Pb(II) (mg g^{-1})	Reference
Cassava stem pith	3.7	[14]
Activated carbon	8.7	[15]
Cashew-nut shell	19.5	[16]
Jackfruit wood sawdust	4.4	[17]
Oryza sativa husk	6.1	[18]
C-TiO_2	28.7	This work

4. Conclusion

In summary, C-TiO_2 was prepared through hydrothermal method and post-calcination. The adsorption capacity and its application for the removal of Pb(II) were fully evaluated. Pb(II) was adsorbed onto active sites, and the sites were filled with increasing content of metal/C-TiO_2. In terms of removing heavy metal ions from contaminated water, the adsorbents can be mixed with water in weakly acidic condition and effectively utilized for 100 min. These integrated features will make the C-TiO_2 have attractive applications in areas of environmental purification.

References

[1] Metwally S S, El-Sherief E A, Mekhamer H S 2020 *Separation Science* **55** 635
[2] Khani R, Sobhani S, Beyki M H 2016 *Journal of Colloid and Interface Science* **466** 198
[3] Green A, Saia M 2019 *Pediatrics* **143** e20183567
[4] Verma R, Gupta B D 2015 *Food Chemistry* **166** 568
[5] Yuan R H, Liu W H, Teng Y J, Nie J, Ma S Z 2015 *Spectroscopy and Spectral Analysis* **35** 1276
[6] Ale A, Bacchetta C, Rossi A S, Galdopórpora J, Desimone M F, Gervasio S, Cazenave J 2018 *Ecotoxicology and Environmental Safety* **148** 976
[7] Rahman P M, Mujeeb V M A, Muraleedharan K 2017 *International Journal of Biological Macromolecules* **97** 382
[8] Zhu H Q, Lou C Y, Liu F R, Jiang G J, Ye X Y, Yuan R H 2021 *Key Engineering Materials* **880** 139
[9] Chen D, Tu N, Si C, Yin M, Wang X 2019 *Water Science & Technology* **80** 884
[10] Liu W H, Yuan R H, Teng Y J 2014 *Chinese Journal of Luminescence* **35** 1124
[11] Putri K, Annisa A, Husain S, Rodiansono R 2020 *Bulletin of Chemical Reaction Engineering & Catalysis* **15** 35
[12] Lin Y T, Weng C H, Lin Y H, Chen F Y 2013 *Separation and Purification Technology* **116** 114
[13] Kanan S, Moyet M A, Arthur R B, Patterson H H 2020 *Catalysis Reviews* **62** 1
[14] Tintin M, Fauzan M, Ayatullah A, Achmad C, Pratikno H 2020 *Materials Science Forum* **981** 331
[15] Lakshmikandhan K, Ramadevi A 2019 *Water SA* **45** 374
[16] Kamchai N, Rapeeporn P, Wikanda S 2020 *Scientific Reports* **10** 3343
[17] Mutiara T, Setyaningsih L, Chafidz A, Panandita B, Raharjo R 2019 *IOP Conference Series: Materials Science and Engineering* **543** 012094
[18] Kaur M, Kumari S, Sharma P 2020 *Biotechnology Reports* **25** e00410

Environmental and economic sustainability in pricing strategy of carbon tax: application of environmental Kuznets curve and real options approach

Chuan-Chuan Ko[1, 3], Chien-Yu Liu[2], Chunmei Liu[1], Xiangyong Yuan[1] and Yuanshan Li[1]

[1]Department of School of Economics and Management, Huizhou University, Huizhou, Guangdong
[2] Department of Food and Beverage Management, Jinwen University of Science & Technology, New Taipei City, Taiwan

[3]Corresponding author's e-mail: kochuan@hzu.edu.cn

Abstract. The issue of global warming has become one of the serious factors affecting sustainable economic development with the dynamics of economic development interact with carbon emissions. The pricing strategy for carbon tax is one of the most feasible methods to achieve the goal of "net zero" in carbon emissions. This paper mainly discusses how the government prices carbon tax without affecting the target profit of the firms. This paper combines the dynamic real options approach (ROA) and the environmental Kuznets curve (EKC) to construct an economic model for the government to levy carbon tax. This paper goals the government how to levy carbon tax on industry to accelerate industrial transformation. Assume that the uncertain changes in the growth of the industry are dependent on the carbon tax, and the changes follow the geometric Brownian motion when carbon emissions change, the government imposes different carbon taxes on industry in order to control carbon emissions within a reasonable range. If industry reduces carbon emissions, the government will also reduce carbon tax. The threshold is the upper limit of the government's carbon tax on industry. In order to accelerate the industry investment in innovative clean energy technology and equipment, the threshold provides a reference basis for the government to impose carbon tax on industry and achieve a win-win strategy for economic development and environmental protection.

1. Introduction

Tcvetkov mentioned that global warming is an existential threat to mankind, and a rapid energy transition is essential. This will be the decisive social, political and technological challenge facing the 21st century [1]. The carbon emissions generated in the industrial production process have significant negative externalities to the environment and there is a high correlation with climate change. The Intergovernmental Panel on Climate Change (IPCC) report proposes that net carbon emissions must reach "net zero" in 2050. This paper mainly discusses how the government can impose a carbon tax on firms without affecting their target profits. Therefore, this paper will focus on the issue of environmental pollution taxes. Using the hypothesis of the environmental Kuznets curve (EKC), a country's national income and the deterioration of its environmental quality will show an inverted U-shaped relationship. This mainly refers to the deterioration of environmental quality in the early stage of economic development as the national income increases, and the environmental quality gradually deteriorates.

However, when the economic development reaches a turning point and the industrial structure is gradually transformed into a technology-intensive industry, the environmental quality can be improved. This paper will discuss how the government sets a reasonable carbon tax. The carbon tax can accelerate industry to promote industrial investment in innovative clean energy technology and equipment for energy saving and carbon reduction in order to control the continued deterioration of environmental quality, and then improve environmental quality.

This paper studies carbon emissions from an industrial perspective, and discusses the high degree of between carbon emissions and economic growth. Guo et al. pointed out that some main industries not only consume a lot of fossil energy but also release a lot of carbon emissions. It also affects energy consumption and carbon emissions in other industries. Therefore, the government should attach great importance to these industries [2]. Zhang and Zhang studied the environmental Kuznets curve (EKC) assumption is valid for China, and the curve between carbon emissions and economic growth is in an inverted U shape [3]. Song et al. recommend that reducing carbon emissions through present technology is more difficult than reducing other pollutants. Improving energy efficiency and developing clean energy are two economically viable options in the future. However, without low-carbon technology, pursuing green growth can have a negative impact on production activities [4]. How to accelerate industrial investment in innovative clean energy technology and equipment have become important research topics. Generally, the traditional methods used to evaluate the feasibility of research and development of innovative production technology investment projects include: profitability index (PI), internal rate of return (IRR), and payback period (PP), accounting rate of return (AAR) and net present value methods (NPV). But these methods are more suitable for static investment environment. In a complex and uncertain investment environment, managers' investment strategies should adopt a dynamic decision analysis model. Among them, the real options approach (ROA) is more able to deal with the complicated investment environment than traditional evaluation methods (Myers [5]; Dixit and Pindyck [6]). Saługa et al. proposed that real option valuation is a good assessment of the value of this flexibility. It is proposed that for coal gasification technology, real option valuation is more suitable than traditional discounted cash flow technology [7]. Balibrea-Iniesta had developed real options approach to evaluate photovoltaic energy projects with a capacity greater than 100 kW in France [8]. Deeney et al. proposed a practical-related decision support tool based on real options to help the actual evaluation of technology R&D investment, and demonstrated the decision support tool. And use it to evaluate the R&D investment potential of CO2 recycling technology for the production of energy commodities [9]. Ko et al. used a dynamic evaluation method of real options approach (ROA) to construct a decision-making model. This model explored the government's policy of levying carbon tax and subsidies to reduce carbon emissions, and rewarding companies for investment in upgrading green energy equipment. This model provided a reference basis for firms to invest in green energy equipment and government policies to control carbon emissions [10].

The important task of controlling pollution emissions is the innovation of low-carbon technology (Balsalobre-Lorente et al. [11]). And it is necessary to formulate regulations and policies for the environment of energy consumption and carbon emissions (Chen et al. [12]). From the financial point of view, this paper intends to use the real options approach (ROA) combined with the environmental Kuznets curve (EKC) model to evaluate the threshold for the government to impose carbon tax. This is also the upper limit for the government to impose carbon tax on the industry. This threshold carbon tax promotes the feasibility of industrial investment in innovative clean energy technology and equipment and carbon reduction. This threshold provides the government's most appropriate reference basis for levying carbon tax. While achieving economic development, environmental quality can also be improved.

2. Decision model

This paper combines the real options approach (ROA) and the environmental Kuznets curve (EKC) to construct an economic model for the government to levy carbon tax. Suppose that a country's economic growth and environmental quality degradation will show an inverted U-shaped relationship (EKC).

2.1. The assumption

The paper aims to control the carbon emissions of each industry in the world within a reasonable range. Due to industrial demand shocks, or due to individual firm product demand shocks, unpredictable changes occur in the output of different firms. Therefore, the government imposes different carbon taxes on changes in carbon emissions during the production process of firms. The carbon tax changes follow a random process, and the changes constitute firm-specific shocks. In addition, carbon emissions are also impacted by another overall random process. Specifically, we assume that the government's changes in the carbon tax $P_x(t)$ imposed on each firm-specific in the market are as follows:

$$P_x(t) = S \times W(t) \times Y(Q_v) \tag{1}$$

In the Eq. (1), the S is firm-specific shocks. Assuming in a competitive market, $W(t)$ is the industry-wide shock. Q_v is the industrial output. Q_e is the carbon emissions in the industrial production process. Among them, carbon emissions $Q_e = Y(Q_v)$ is a function of Q_v in the industrial production process. Therefore $Q_e = Y(Q_v)$ is a concave function of Q_v. That is, if the output value increases at the beginning, carbon emissions will also increase. However, as the output value increases and the economy reaches a certain turning point, the industrial structure must be gradually transformed. The industry must develop innovative technologies for energy saving, carbon emission reduction and purchase new equipment. In this way, even if firms increase production capacity, they can also reduce carbon emissions. At the same time, in order to accelerate industrial transformation, the government has controlled carbon emissions within a reasonable range. The carbon emission tax levied by the government must be sufficient to induce firms to invest in energy-saving, carbon emission-reducing innovative technologies and purchase new equipment to reduce carbon emissions.

Now consider industry-wide uncertainty. If the carbon tax imposed by the government to control carbon emissions does not affect the target profit of the firm, the new firms will enter the market, and carbon emissions will increase accordingly. If the government levies a higher carbon tax, the firm must choose to invest in innovative clean energy technology and equipment, or withdraw from the market. If the firms continue to produce, it must choose to invest in sunk costs and reduce carbon emissions. Where industry-wide shock $W(t)$ and the government levies carbon tax $P_x(t)$ will follow the geometric Brownian motion process:

$$dW(t) = \alpha W(t)dt + \sigma W(t)dZ(t);\, dP_x(t) = \alpha P_x(t)dt + P_x W(t)dZ(t) \tag{2}$$

Where Eq. (2), α is the expected growth rate of $W(t)$ and $P_x(t)$. σ is the standard deviation of $W(t)$ and $P_x(t)$. $dZ(t)$ is the increments of the standard Wiener process. Here, $E[dZ(t)]=0$, $E[dZ(t)]^2=dt$. In addition to considering the carbon emissions generated by changes in output, firms have an impact on the government's carbon tax. At the same time, firms need to respond to the increase in the amount of carbon emissions in the environment when new firms entering the market are uncertain in the competitive market. In this way, the government will impose a higher carbon tax, and firms will also increase production costs. In order to sustainably operate, increase target profits and corporate social responsibility, firms should choose to invest in innovative clean energy technology and equipment. Therefore, when the carbon emissions $Q_e = Y(Q_v)$ changes with Q_v, the government levies carbon tax $P_x(t)$ and industry-wide uncertainty $W(t)$ to form an equal rate change.

2.2. Decision model

The government levies carbon tax for firms, which affects the investment in innovative clean energy technology and equipment. The path of levying carbon tax depends on the carbon emissions $Q_e = Y(Q_v)$. Therefore, the cost of corporate carbon emissions $C(P_x(t), Y(Q_v))$ is a function of carbon emissions $Q_e = Y(Q_v)$ and carbon tax $P_x(t)$. This study denotes by $C(P_x(t), Y(Q_v))$, the expected future emission cost discount value is shown as Eq. (4):

$$E\left[\int_0^{\infty} C(P_x(t),Y(Q_v))\times e^{-rt}dt\right]=\frac{P_x(t)\times Y(Q_v)}{r-\alpha} \tag{3}$$

Where, r is the risk-free discount rate. It is expected that the carbon tax that firms will pay in the future is the discount value of carbon emission $Q_e = Y(Q_v)$ multiplied by the government's carbon tax per ton of carbon emissions $P_x(t)$.

Assuming that firms invest in innovative clean energy technology and equipment, they need to invest in sunk cost I. When the government's carbon tax $\overline{P_x(t)}$ is raised to the upper limit, firms will face the sunk cost of choosing to invest. Otherwise, firms will withdraw from the market due to the excessively high production costs caused by the government's carbon tax. When the government levies the present value of the carbon tax is greater than or equal to the depreciation expense of the sunk cost $P_x(t)\times Y(Q_v)/(r-\alpha)\geq g\times I$. Then the firm will consider the sunk cost of investment I. In this way, the firms can reduce carbon emissions and at the same time reduce the cost of carbon emissions during production. Wherever g is the depreciation rate of the new equipment. If $P_x(t)\times Y(Q_v)/(r-\alpha)< g\times I$, the firms will not change the production model. At the same time, the industry continues to have new firms entering the market. Because the carbon tax levied by the government is not enough to threaten the target profits of enterprises, it can't induce firms to invest in clean energy to renew production equipment motivation.

Under the assumption that changes in the government's carbon tax follow the geometric Brownian motion process. The government levies different carbon taxes in accordance with the uncertain changes in industrial carbon emissions. Then the government's levy of carbon tax affects firms' investment in innovative clean energy technology and equipment. The management flexibility value is $V\left[C(P_x(t),Y(Q_v))\right]$. Using Itô's Lemma theorem (Itô [13]), its government-policy value of management flexibility is as follows Eq. (4):

$$V\left[C(P_x(t),Y(Q_v))\right]=V_{P_x(t)}\left[C(P_x(t),Y(Q_v))\right]dP_x(t)+\frac{1}{2}V_{P_x(t)P_x(t)}\left[C(P_x(t),Y(Q_v))\right]d(P_x(t))^2 \tag{4}$$

Whereby $V_{P_x(t)}\left[C(P_x(t),Y(Q_v))\right]$ and $V_{P_x(t)P_x(t)}\left[C(P_x(t),Y(Q_v))\right]$ are the first and the second-order differential equations derived from $V\left[C(P_x(t),Y(Q_v))\right]$ for $P_x(t)$. Using the value matching condition (VMC): At the optimal threshold $P_x(t)^*=\overline{P_x(t)}$, the management flexibility strategy value $V\left[C(P_x(t),Y(Q_v))\right]=A_1P_x(t)^{\beta_1}$ is equal to the current value of the expected government's carbon tax (the production cost of carbon emissions in the production process of the firms) $E\left[\int_0^{\infty} C(P_x(t),Y(Q_v))\times e^{-rt}dt\right]$ minus the depreciation expense $g\times I$. This value matching condition (VCM) satisfies the value-uniqueness condition, such as in Eq. (5). Then, the marginal value is equal in the first-order derivative function, that is, it satisfies the conditions of equal marginal value. This is a smooth-pasting condition (SPC), such as in Eq. (6) (Dixit and Pindyck [6]):

$$V\left[C(P_x(t),Y(Q_v))\right]=E\left[\int_0^{\infty} C(P_x(t),Y(Q_v))\times e^{-rt}dt\right]-g\times I \tag{5}$$

$$\frac{\partial V\left[C(P_x(t),Y(Q_v))\right]}{\partial P_x(t)}=\frac{\partial E\left[\int_0^{\infty} C(P_x(t),Y(Q_v))\times e^{-rt}dt\right]}{\partial P_x(t)} \tag{6}$$

$P_x(t)^*=\overline{P_x(t)}$ is the optimal threshold of carbon tax imposed by the government in the production process of firms, and its management flexibility value $V\left[C(P_x(t),Y(Q_v))\right]$ is showed as in Eq. (7):

$$V\left[C(P_x(t),Y(Q_v))\right]=\begin{cases} A_1P_x(t)^{\beta_1}, P_x(t)<P_x(t)^* \\ \dfrac{P_x(t)\times Y(Q_v)}{r-\alpha}-g\times I, P_x(t)<P_x(t)^* \end{cases} \tag{7}$$

When $P_x(t)<P_x(t)^*=\overline{P_x(t)}$, the carbon tax imposed by the government is less than the optimal carbon tax. Therefore, the government's levy of carbon tax will increase the production cost of the firm, which will not affect the target profit of the firm. Hence, there are still new firms entering the market in this

industry. Carbon emissions will continue to increase. At this time, the government will adopt a strategy of continuously increasing the carbon tax to curb the industry's carbon emissions. Until $P_x(t) \geq P_x(t)^* = \overline{P_x(t)}$, when the government's carbon tax is greater than or equal to the optimal threshold, the firm will choose to invest in innovative clean energy technology and equipment, and invest in sunk cost I. Or they cannot reach the target profits of the firm and withdraw from the market because they need to pay a too high carbon tax. From Eq. (15), use value matching condition (VMC) and smooth-pasting condition (SPC) to solve the optimal threshold, as in Eq. (8) (Dixit and Pindyck [6]):

$$\begin{cases} A_1 P_x(t)^{\beta_1} = \dfrac{P_x(t)\times Y(Q_v)}{r-\alpha} - g\times I \\ \beta_1 A_1 P_x(t)^{\beta_1 - 1} = \dfrac{Y(Q_v)}{r-\alpha} \end{cases} \tag{8}$$

Reorganize Eq. (8), firms invest in innovative clean energy technology and equipment, invest in sunk costs, and reduce carbon emissions. The optimal threshold $P_x(t)^* = \overline{P_x(t)}$ for the government to levy carbon tax, such as Eq. (9):

$$P_x(t)^* = \overline{P_x(t)} = \frac{\beta_1}{\beta_1 - 1}\times\frac{g\times I\times(r-\alpha)}{Y(Q_v)} \tag{9}$$

In Eq. (9), $Y(Q_v)$ is a function of carbon emissions in the production process of the firms $Y(Q_v) = Q_e$, and its change is affected by the output Q_v. Therefore $Y(Q_v)$ is a concave function of Q_v. This paper assumes that it is an environmental Kuznets curve (EKC), and the output and carbon emissions present an inverted U-shaped relationship, and its function is as Eq. (10):

$$Y(Q_v) = a\times Q_v^2 + b\times Q_v + c \tag{10}$$

Then Eq. (10) is a function of carbon emissions, assuming that the output Q_v and carbon emissions $Y(Q_v)$ present an inverted U-shaped relationship. Now, $a<0$, $a>0$, $c\geq 0$. After sorting out Eq. (9) and Eq. (10), the optimal threshold for the government to levy carbon tax, such as Eq. (11):

$$P_x(t)^* = \overline{P_x(t)} = \frac{\beta_1}{\beta_1 - 1}\times\frac{g\times I\times(r-\alpha)}{a\times Q_v^2 + b\times Q_v + c} \tag{11}$$

Solve the optimal threshold of the government's carbon tax $P_x(t)^* = \overline{P_x(t)}$ from the above construction model. When the government levies carbon tax $P_x(t)^* = \overline{P_x(t)}$, firms will choose to invest in sunk costs to innovate clean energy technology and update production equipment. When the government formulates carbon tax to reach the optimal threshold $P_x(t)^* = \overline{P_x(t)}$. In consideration of the firm's target profit, the firm must choose to invest in sunk cost innovative clean energy technology and equipment. Firms can reduce carbon emissions, reduce carbon tax costs, and increase profits. This optimal threshold $P_x(t)^* = \overline{P_x(t)}$ provides a reference for the government to formulate environmental protection policies, levy carbon taxes, and force firms to invest in innovative clean energy technology and equipment.

3. Numerical examples

The paper first assumes that the industrial output and carbon emissions appear an inverted U-shaped relationship, which is the hypothesis of the environmental Kuznets curve (EKC). The numerical example uses the carbon emissions of industrial products from 1990 to 2017 [14] are dependent variables $Q_e = Y(Q_v)$. In addition, the data of the total domestic industrial production from 1990 to 2017 [15] are independent variable Q_v. Using Polynomial Regression to derive the output value Q_v and carbon emission function $Y(Q_v)$ in the production process of the firms, the parameter value is as Eq. (12):

$$Y(Q_v) = a\times Q_v^2 + b\times Q_v + c = -0.58\times Q_v^2 + 1.80\times Q_v + 4.97 \tag{12}$$

The parameter values are $a=-0.58$, $b=1.80$ and $c=4.97$. Then the maximum value is obtained by the first-order derivative function Eq. (12), and the maximum value is as Eq. (13):

$$Y_{Q_v}(Q_v) = 2\times(-0.58)\times Q_v + 1.80 = 0,\ Q_v = 1.55 \tag{13}$$

Then, the change in the growth rate of domestic industrial production from 1990 to 2017 [15] will affect the government's taxation of carbon emissions $P_x(t)$. Based on this data, the expected growth rate and standard deviation of the government's carbon tax $P_x(t)$ are evaluated. The related exogenous variable assumptions are shown in Table 1:

Table 1. Exogenous variables.

Exogenous variables	Significance	Value
α	α is expected growth rate of $W(t)$ and $P_x(t)$.	0.02
σ	σ is standard deviation of $W(t)$ and $P_x(t)$.	0.20
r	Risk discount rate.	0.15
Q_v	Industrial output (trillion).	1.55
a	This parameter is a function of output and carbon emissions in the production process of the firm ($Y(Q_v)$)	-0.58
b	This parameter is a function of output and carbon emissions in the production process of the firm ($Y(Q_v)$)	1.80
c	This parameter is a function of output and carbon emissions in the production process of the firm ($Y(Q_v)$)	4.97
I	Innovative clean energy technology and equipment to invest in fixed costs (Unit: million dollars).	7.75
g	Depreciation rate of new production equipment.	0.20

Derived from the external variables and numerical data in Table 1, the numerical analysis results: The optimal threshold for the government to impose carbon tax is $P_x(t)^* = \overline{P_x(t)} = 498.54$ (dollars). This threshold is when the government imposes carbon tax of 498.54 (dollars) per ton of carbon emissions, firms will choose to invest in innovative clean energy technology and equipment in Taiwan. If the government levies carbon tax $P_x(t) < P_x(t)^* = 498.54$ (dollars), the carbon tax imposed by the government will not affect the target profit of the firm, and the firm will choose to delay investment. When the government levies carbon tax to reach the optimal threshold $P_x(t)^* = \overline{P_x(t)} = 498.54$ (dollars), the cost of the tax affects the target profit of the firms. At this time, firms will choose to invest in innovative clean energy technology and equipment to reduce carbon emissions. At the same time, the variable cost of carbon tax can be reduced. It also cooperates with the government's environmental protection policy and can fulfill corporate social responsibilities. The optimal threshold can provide a reference basis for the government to formulate a carbon tax policy and for firms to invest in innovative clean energy technology and equipment.

4. Conclusion

From a financial point of view and a flexible management strategy perspective, this paper combine the real options approach (ROA) with the Environmental Kuznets Curve (EKC) model to construct a carbon taxation strategy model for controlling carbon emissions. Discuss the environmental protection issues of how the government can effectively control carbon emissions. In the model, it is assumed that the industrial development model conforms to the Environmental Kuznets Curve (EKC) model. When the industry reaches a certain turning point, the industrial structure must be transformed to innovate clean energy technology and update production equipment to improve carbon emissions. Under the assumption of a perfectly competitive market, industry-wide shock $W(t)$ and carbon tax $P_x(t)$ will follow the geometric Brownian motion process. Use the real options approach to construct the government's decision-making model for levying carbon tax and derive the optimal threshold for the government to levy carbon tax $P_x(t)^*$. The threshold is for the government to impose an optimal carbon tax on firms. If the government levies carbon tax on firms to reach this threshold, the firms will choose to invest in the development of innovative clean energy technology and equipment to improve carbon emissions; otherwise, they will choose to withdraw from the market. The analysis results of the above numerical examples are shown, when the government imposes carbon tax of 498.54 (dollars) per ton of carbon emissions, the cost of the carbon tax will affect the firm's target profit. Firms will choose to innovate clean energy technology and update production equipment, otherwise they must exit the market.

The results of this study mainly provide the government how to pricing carbon tax without affecting the target profit of the firms and the GDP. Accelerate firms to innovate clean energy technology and update production equipment to improve carbon emissions. This method can effectively control carbon emissions and achieve a win-win strategy for economic development and environmental protection.

Acknowledgments
The authors would like to thank the Supported by The Professorial and Doctoral Scientific Research Foundation of Huizhou University, for financially supporting this research under Contract No. 2020JB072.

References

[1] Tcvetkov P 2021 Climate policy imbalance in the energy sector: Time to focus on the value of CO2 utilization. *Energies* 14(2) 411 https://doi.org/10.3390/en14020411
[2] Guo J, Zhang Y J, Zhang, K B 2018 The key sectors for energy conservation and carbon emissions reduction in China: evidence from the input-output method *Journal of Cleaner Production* **179** 180–190
[3] Zhang Y, Zhang S 2018 The impacts of GDP, trade structure, exchange rate and FDI inflows on China's carbon emissions *Energy Policy* **120** 347-353
[4] Song M, Zhu S, Wang J, Zhao J 2020 Share green growth: Regional evaluation of green output performance in China *International Journal of Production Economics* **219** 152-163
[5] Myers S C 1977 Determinants of corporate borrowing *Journal of Financial Economics* **5** 146-175.
[6] Dixit A K, Pindyck R S 1994 Investment under Uncertainty. New Jersey: Princeton University Press.
[7] Saługa P W, Grzesiak P, Kamiński J 2020 Valuation of decision flexibility and strategic value in coal gasification projects with the option-to-switch between different outputs *Energies* 13(11) 2826 https://doi.org/10.3390/en13112826
[8] Balibrea-Iniesta J 2020 Economic analysis of renewable energy regulation in France: A case study for photovoltaic plants based on real options *Energies* **13(11)** 2760 https://doi.org/10.3390/en13112760
[9] Deeney P, Cummins M, Heintz K, Pryce M T 2021 A real options based decision support tool for R&D investment: Application to CO2 recycling technology *European Journal of Operational Research* **289(2)** 696-711
[10] Ko C C, Liu C Y, Chen Z Y, Zhou J 2019 Sustainable development economic strategy model for reducing carbon emission by using real options approach *Sustainability* **11(19)** 5498 https://doi.org/10.3390/su11195498
[11] Balsalobre-Lorente D, Shahbaz M, Roubaud D, Farhani S 2018 How economic growth, renewable electricity and natural resources contribute to CO2 emissions? *Energy Policy* **113** 356-367
[12] Chen X, Huang B, Lin C T 2019 Environmental awareness and environmental Kuznets curve *Economic Modelling* **77** 2-11
[13] Itô K 1951 On stochastic differential equation memories *American Mathematical Society* **4** 1-51
[14] 2019 Republic of China national greenhouse gas emissions inventory report Taiwan Environmental Protection Administration http://www.epa.gov.tw
[15] National Statistics, R. O. C. https://www.stat.gov.tw/

IOP Conf. Series: Earth and Environmental Science **865** (2021) 012042 doi:10.1088/1755-1315/865/1/012042

Engineering performance's evaluation of integrated backfill materials

Xiang Fan[1, 4], Jinyang Du[2], Facheng Yi[3], and Yangjie Feng[1]

[1]Tian Fu College of Southwestern University of Finance and Economics, Sichuan, China
[2]Shenghong Construction Co., Ltd, Chendu, Sichuan, China
[3]Southwest University of Science & Technology, Mianyang, Sichuan, China

[4]Corresponding author's e-mail: 920391362@qq.com

Abstract. In the process of deep geological disposal, bentonite, illite, zeolite, kaolinite, quartz sand and other materials were used as filling materials for filling in waste tin and rock. This paper proposed integrated backfilling materials (ZBP_8, ZBP_7, ABP_8, ABP_7 and Xinjiang bentonite Na^+) etc and five groups of formula, with a series of geotechnical experiments measured the engineering properties, such as glial price, expansion solution, liquid limit, plastic limit of compacted expansive, unconfined compressive strength, elastic modulus, Poisson's ratio and coefficient of permeability. Compared with the GMZ-1 bentonite, the engineering performance of various proportions meeted the backfill material disposal requirements and could be used as backfill material candidate for the integrated materials, but proportion of zeolite and palygorskite should not be too large, generally better from 20% to 30%. The specific engineering properties were as follows: Under the condition of the same proportion, ABP mixture was higher than the ZBP mixture of liquid limit. In the ABP mixture and the ZBP mixture, dry density at $1.7g/cm^3$-$1.8g/cm^3$ and moisture content between 15%-20% of the compaction performance were the best. The compaction performance of ZBP mixture was better than that of ABP mixture under the same condition. Different ratio of ZBP and ABP composite minerals under no load condition, it found that ultimately expansion and expansion force with the increase of bentonite. The expansion performance of ABP mixture was better than that of ZBP mixture. Under the same condition, the unconfined compressive strength, elastic modulus and Poisson's ratio increased with the increase of the content of bentonite in the integrated backfill material. Under the condition of 100 MPa compaction, the permeability coefficient of ZBP and ABP mixture was 2 orders of magnitude higher than that of GMZ-1. The permeability coefficient of Xinjiang bentonite under 50MPa was very close to that of GMZ-1 bentonite under 100MPa.

1. Introduction

In deep geological disposal, the buried high-level waste underground projects known as geological disposal repository. The disposal repository was generally designed using a conceptual model of "multiple barrier systems". But the backfill material, as the material filled between the waste container and the geological body, should have strong adsorption, water insulation, thermal conductivity and mechanical buffering performance. The backfill materials studied abroad have been studied for more than 30 years [1, 2, 3]. The main backfill materials studied are bentonite, illite, zeolite, kaolinite, quartz sand, attapulgite and so on. According to the requirements of high radioactive waste for backfilling materials, Pusch (1977, 1980), through analysis, comparison and relevant experiments, believed that bentonite with montmorillonite as the main component was the most suitable material [4, 5]. Due to the differences in structure and composition, Weststic et al. (1982) believed that sodium bentonite was more suitable for backfilling materials than calcium bentonite [6]. Sweden, the United States, Canada and other countries had focused on a large number of laboratory tests and field tests on bentonite. Previous studies had shown that bentonite was an ideal backfilling material, and adding quartz sand to bentonite was conducive to increasing the thermal conductivity [7]. The research in this field in China began in 1986. Through the investigation and screening of bentonite deposits nationwide [8], it was determined that gaomiaozi bentonite deposit in xinghe county, Inner Mongolia was the first choice for the backfilling material supply base of geological disposal repository of high-level radioactive waste in China. However, most of the backfilling materials studied at home and abroad were single minerals, and there were few studies on compound minerals [9, 10]. Cui suli, liu jisheng, zhang huyuan et al. [11] proposed a new concept of bentonite -- sand mixture as high buffer backfilling materials, and achieved certain results.

2. Selection and integration of backfill materials

In previous research, it was found that the mixture of zeolite bentonite (ZB) and attapulgite bentonite (AB) with the ratio of 3:7 And 2:8 had better engineering properties [12, 13]. In order to improve the thermal conductivity of the mixture, 10% pyrite was added to the mixture, and the engineering properties of ZBP_7, ZBP_8, ABP_7 and ABP_8 were measured as follows in Table 1. Remarks: Z was zeolite, B was Xinjiang Altay sodium bentonite (the base material of this experiment), P was pyrite, A was attapulgite, $XJNa^+$ was Xinjiang Altay sodium bentonite. The subscript Arabic numeral was the percentage of bentonite in the total mass of the mixture. The basic physical properties of these four groups of mixed soils were compared with those of Gaomiaozi bentonite and Altai bentonite in Xinjiang, as shown in Table 1.

Table 1. Comparison of physical properties between backfill material and Inner Mongolia bentonite.

Mineral \ Properties	Colloid value (ml/15g)	Swelling capacity (ml/g)	Swelling value (ml/3g)	Liquid limit (%)	Plastic limit (%)	Proportion
GMZB-1 [14]	53-73	9.3-20	8.6-11.8	90-120	30-44	2.4-2.6
ZBP_7	92	11.0	21.0	73.8	37.1	2.32
ZBP_8	92	12.0	25.0	89.9	39.5	2.51
ABP_7	94	10.5	23.5	89.5	35.6	2.12
ABP_8	97	13.0	26.0	90.5	38.5	2.23
$XJNa^+$	>100	28.0	42.0	82.3	27.0	2.52

It could be seen from Table 1 that the colloidal value of the ZBP mixtures and ABP mixtures were higher than that of the bentonite of gaomiaozi in Inner Mongolia. Because the substrate of the mixtures in this paper was the bentonite of altai in xinjiang, whose colloidal value content was >100[12]. At the same time, the swelling capacity and swelling value of altai bentonite (XJNa+), ZBP mixtures and ABP mixtures were larger than those of gaomiaozi bentonite in Inner Mongolia. The result was closely related to the attribute of bentonite and the content of montmorillonite. The reason was that the content of montmorillonite was higher, the colloid value and the expansibility of the mixture were higher, which could effectively block the migration of nuclides from the waste water pool to the external environment. The results showed that the plastic limit, liquid limit and void ratio of ZBP and ABP mixtures were also close to those of the Gaomiaozi bentonite (GMZB), which could meet the requirements of backfill materials.

3. Comprehensive evaluation of engineering performance of integrated backfill materials

3.1. Comparison of compaction performance

In this paper, the sample of the outer radius was 70 mm, the inner radius was 30 mm, and the height 20 mm was a fan-shaped ring [15], which through self-made mold, using hydraulic universal testing machine, the loading at 0.5 mm of rate. Under the same moisture content and dry density, the compaction load required for compacting backfill material was measured. The results were shown in Table 2 and Figure 1.

Table 2. Compaction performance of integraed material comparised with bentonite of Inner Mongol.

Samples	Water content (%)	Dry density (g/cm3)	Wet density (g/cm3)	Void ratio	Pressing load (MPa)
GMZB-1 [14]	15.00	1.75	2.08	22.45	35.45
GMZB-2 [16]	15.21	1.72	2.02	24.04	32.65
ZBP_8	15.66	1.68	1.94	38.10	27.36
ZBP_7	15.96	1.71	2.02	33.33	26.43
ABP_8	15.58	1.76	2.03	31.82	27.78
ABP_7	15.44	1.74	1.99	34.88	25.60
$XJNa^+$	15.27	1.72	2.05	28.21	30.24

It can be seen from Figure 1 and Table 2 that the compaction load of Gaomiaozi bentonite was slightly larger at the same moisture content and dry density, which may be caused by the different content of montmorillonite. The compaction characteristics of bentonite were determined by the structure of montmorillonite with water swelling. The higher the content of montmorillonite, the more difficult it was to compact, which was consistent with the result obtained by Liu Yuemiao (2001) [17]. In order to get the same dry density backfill material, a certain amount of zeolite or attapulgite was added to improve the compaction performance of bentonite. Through the research on the compaction molding performance of integrated backfill material, when the compaction load was greater than 50MPa, the compaction dry density changed little with the increase of compaction load [13]. Therefore, when the moisture content was not less than 10% or 25%, the dry density of backfill material was less than 1.7g /cm^3, which could be compacted within 50 MPa. In addition, the relationship between the maximum dry density and the optimal moisture content of ZBP moisture or ABP moisture were also obtained by light compaction and heavy compaction. The experimental results showed that adding a certain amount of zeolite or attapulgite could significantly reduce the plasticity index of backfill material, but in the compaction test, it could obtain a larger maximum dry density and a smaller optimal moisture content, so the properties of pure bentonite were improved. Secondly, the experimental results showed that the maximum dry density obtained by heavy compaction was larger than that obtained by light compaction, that was the compaction energy was greater, the sample density was greater, and the water content was smaller. The experimental results were consistent with the research of many foreign scholars [17, 18].

3.2. Comparison of expansion performance

In this paper, the integrated backfill material samples were prepared according to the optimal moisture content, the prefabricated dry density of samples were about 1.7g/cm^3, and the compaction load was 20MPa. Five groups of mixtures, such as ZBP, ABP and xjna +, were put into the dilatometer to adjust the initial reading of the meter. Then, distilled water was slowly injected to measure the expansion rate and the expansion force of the integrated backfill material under load condition without load. The results were shown in Figure 2 and Table 3.

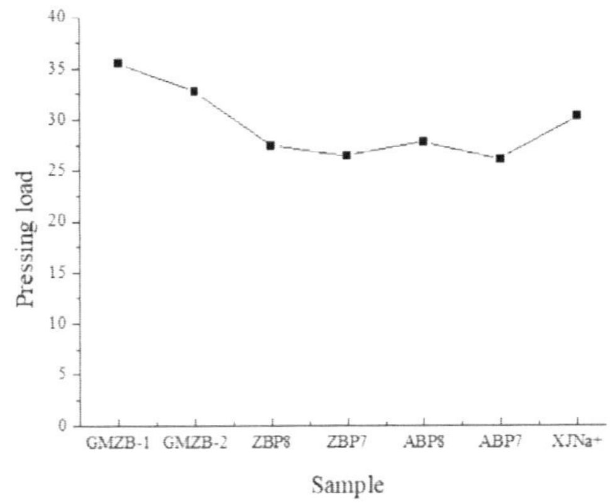

Figure 1. Compaction performance of the backfill material with Inner Mongolia bentonite.

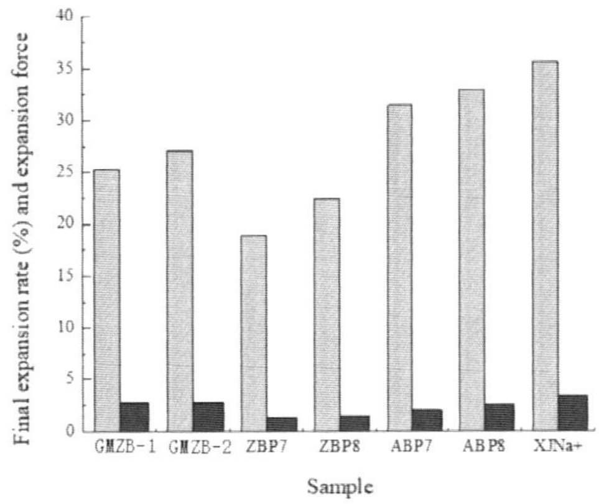

Figure 2. Backfill materials and swelling properties of bentonite in Inner Mongolia.

Table 3. Compaction and swelling performance of integrated material.

Samples	Pressing load (MPa)	Moisture content (%)	Wet density (g/cm^3)	Compaction density (g/cm^3)	Final expansion rate (%)	Expansion force (MPa)
GMZB-1[14]	20	22.04	1.69	1.39	25.3	2.7
GMZB-2[16]	30	22.04	1.73	1.46	27.1	2.8
ZBP_7	20	15.37	2.01	1.74	18.9	1.3
ZBP_8	20	15.78	1.99	1.72	22.3	1.4
ABP_7	20	15.35	1.95	1.69	31.4	2.0
ABP_8	20	15.26	1.97	1.71	32.9	2.5
$XJNa^+$	20	15.42	1.97	1.71	35.6	3.4

It can be seen from Figure 2 and Table 3 that the compacted density of Inner Mongolia bentonite increased with the increase of compressive pressure under the condition of the same water content. Different samples had different densities under the same pressure, which may be due to the different content of montmorillonite in the samples. The higher the content of montmorillonite, the smaller the dry density. In the same ratio, the expansion force of ABP mixture was larger than that of ZBP mixture, mainly because zeolite was non-viscous, while attapulgite contains some montmorillonite. The results also showed that the swelling force of the mixture was not as great as that of Inner Mongolia, which was mainly related to the low content of bentonite in the mixture.

3.3. Compression performance comparison

The samples were prepared according to the volume and dry density calculation required by the quality of the wet earth, and poured the pressure within the sample device, through the electro-hydraulic servo universal testing machine pressure to the required density. The instrument used for the compressive strength test was a hydraulic universal testing machine with the diameter of 61.8 mm and the height of 50 mm. The unconfined compression and deformation tests of compacted bentonite were carried out in accordance with the industrial standard of the People's Republic of China "rules for rock test of water conservancy and hydropower projects (SL264-2001)". The loading rate of unconfined compression and deformation test was 0.1~0.5 mm/min. Strain gauge method was used for strain measurement. The unconfined compressive strength, elastic modulus and poisson's ratio of ZBPmixture, ABP mixture and $XJNa^+$ mixture to water content and dry density were measured. The results were shown in Table 4.

Table 4. The unconfined compression properties of the backfill material and Inner Mongolia bentonite.

Samples	Moisture content (%)	Dry density (g/cm3)	Unconfined compressive strength (MPa)	Elastic modulus (GPa)	Poisson's ratio
GMZB-1[14]	15.0	1.76	4.85	2.92	0.48
GMZB-2[16]	15.0	1.58	2.17	0.99	0.42
ZBP_7	15.51	1.72	2.92	2.52	0.47
ZBP_8	15.24	1.74	2.71	2.59	0.50
ABP_7	15.31	1.71	2.53	2.07	0.55
ABP_8	15.06	1.67	2.34	1.98	0.58
$XJNa^+$	15.46	1.68	2.44	1.81	0.60

It can be seen from Table 4 that under the same moisture content of the gaomiaozi bentonite, the greater the dry density, the greater the unconfined compressive strength and elastic modulus. Under the same condition, the unconfined compressive strength, elastic modulus and poisson ratio of integrated backfill materials increased with the increase of bentonite content in the total mass of the mixture. The differences in the Table 4 may be caused by the bentonite content in the mixture, the particle size and strength of zeolite, bentonite and attapulgite in the mixture. In bentonite content was 70%, the same conditions the unconfined compressive strength and elastic modulus of ABP mixture were higher than ZBP mixture, but the poisson's ratio of ABP mixture were lower than ZBP mixture. This showed that adding attapulgite maked deformationc and low intensity easily. It had to do with attapulgite containing a certain amount of montmorillonite, so adding a mixture of zeolite could get greater compressive strength. In addition, at the same dry density, the unconfined compressive strength and elastic modulus decreased with the increase of moisture content, indicating that dry density and moisture content were important factors for the compaction and strength of backfill materials.

It could be seen from Table 5 that under the same dry density, with the increase of the pressure, the permeability coefficient decreased. The difference of pressure between 100MPa and 10MPa was 10 times, and the difference of permeability coefficient was 20~50 times. However, for the same proportion of ABP and ZBP mixture, when the compression load was 50MPa and 100MPa, the permeability difference was not very obvious, which indicated that when the compression load was greater than 50MPa, the mixture could achieve compaction, which was consistent with the compaction performance of the overall backfill material [15]. Secondly, the measured permeability coefficient of integrated ZBP_7 mixture and ABP_7 mixture was high 2 orders of magnitude than zhen-yao shen [19] Inner Mongolia in 100MPa. However, the measured permeability coefficient of single xinjiang altay bentonite under the pressing load of 50 MPa near zhen-yao shen high temple of bentonite under 100 MPa. The results showed that the permeability of backfill material was related to both dry density and pressure. The higher the pressure and dry density of the mixture in the same proportion, the worse the permeability. It was also related to the mineral composition. Suppose the mixture of 10% pyrite mineral powder could also lead to porosity increased, accelerated the permeability of mixture. Thirdly, under the same conditions of compression load and bentonite proportion of the mineral, permeability coefficient was smaller than ABP mixture. The reason was that the zeolite was non-viscous minerals, did not have water retention, the zeolite increased the porosity, the permeability increased rapidly.

Table 5. The infiltration parameters of the integrated backfill materials with different compaction. forces and bentonite in Inner Mongolia.

Samples	Pressload (MPa)	Moisture content (%)	Wet density (g/cm^3)	Compaction density (g/cm^3)	Porosity (%)	Permeability coefficient (10-11cm/s)
GMZB-1[19]	100	20.38	2.08	1.79	28.7	0.01
XJNa+	100	15.4	2.08	1.68	20.7	0.016
ZBP7	100	15.80	2.11	1.82	27.6	2.11
ABP7	100	15.36	2.07	1.69	20.23	1.15
ZBP7	50	15.9	2.01	1.70	20.21	3.67
ABP7	50	15.4	2.13	1.69	19.57	1.841
ZBP8	50	15.5	2.02	1.66	20.06	0.964
ABP8	50	15.2	2.03	1.72	20.09	0.365
ZBP7[16]	10	9.44	2.27	2.07	22.24	42.68

It could be seen from Table 5 that under the same dry density, with the increase of the pressure, the permeability coefficient decreased. The difference of pressure between 100MPa and 10MPa was 10 times, and the difference of permeability coefficient was 20~50 times. However, for the same proportion of ABP and ZBP mixture, when the compression load was 50MPa and 100MPa, the permeability difference was not very obvious, which indicated that when the compression load was greater than 50MPa, the mixture could achieve compaction, which was consistent with the compaction performance of the overall backfill material [15]. Secondly, the measured permeability coefficient of integrated ZBP_7 mixture and ABP_7 mixture was high 2 orders of magnitude than zhen-yao shen [19] Inner Mongolia in 100MPa. However, the measured permeability coefficient of single xinjiang altay bentonite under the pressing load of 50 MPa near zhen-yao shen high temple of bentonite under 100 MPa. The results showed that the permeability of backfill material was related to both dry density and pressure. The higher the pressure and dry density of the mixture in the same proportion, the worse the permeability. It was also related to the mineral composition. Suppose the mixture of 10% pyrite mineral powder could also lead to porosity increased, accelerated the permeability of mixture. Thirdly, under the same conditions of compression load and bentonite proportion of the mineral, permeability coefficient was smaller than ABP mixture. The reason was that the zeolite was non-viscous minerals, did not have water retention, the zeolite increased the porosity, the permeability increased rapidly.

4. Conclusion and outlook

Through the author's research in recent years, a certain proportion of integrated backfill material was proposed, and each index could meet the requirements of domestic deep geological disposal of high-level radioactive waste. This paper mainly studied the engineering performance comparison of ZBP_8, ZBP_7, ABP_8, ABP_7, Xinjiang Altay bentonite Na^+ and with Gaomiaozi bentonite Na^+ in Inner Mongolia.The specific conclusions were as follows:

First, taking Xinjiang Na^+ bentonite as base material, zeolite, attapulgite, pyrite and other materials could be added into the mixture according to a certain proportion. In the laboratory, zeolite or attapulgite with 20% - 30% content and bentonite with 70% - 80% content were 2:8 and 3:7, and pyrite with 10% content as additive were selected as candidate materials. The engineering performance of each ratio meeted the disposal requirements of backfill materials.

Secondly, in ZBP and ABP mixtures, the swelling capacity and gum value increased with the increase of bentonite content. The specific gravity index of the samples was greater than 2, and the plasticity index was greater than 10. The liquid limit of ABP mixture was higher than that of ZBP mixture at the same ratio.

Thirdly, under the conditions of dynamic compaction and static compaction, the best compaction performance could be obtained when the dry density of ZBP and ABP mixture was 1.7 g/cm^3-1.8 g/cm^3 and the moisture content was 15%-20%. The experimental results showed that the compaction performance of ZBP mixture was better than that of ABP mixture. In practical engineering, it was suggested that the proportion of zeolite mixture or attapulgite mixture should not be too large, generally 20%-30% was better.

Fourthly, by measuring the final expansion of ZBP and ABP mixture with different proportions under no load condition and the expansion force under load condition, it was found that the expansion performance of bentonite was better with the increase of bentonite, and the expansion performance of ABP mixture was better than that of ZBP mixture. In addition, the experiments showed that the initial moisture content, compaction load and dry density had great influence on the expansion amount and expansion force.

Fifthly, the unconfined compressive strength, elastic modulus and Poisson's ratio of different proportions of ZBP and ABP mixture were related to water content and dry density. Under the same conditions, the unconfined compressive strength and elastic modulus of ZBP mixture were much larger than those of ABP mixture, but the Poisson's ratio decreased.

Finally, the results showed that the permeability coefficients of ZBP, ABP and xjna$^+$ decreased with the increase of bentonite content at 100MPa or 50MPa, and the change of permeability coefficient with time basically conforms to Darcy's law.

References

[1] Borje T, Bert A 1986 Migration of fission products and actinides in compacted bentonitet. *Sweden: SKB* **13** 86
[2] Borje T 1983 Radionuclide diffusion and mobilities in compacted bentonite *Sweden: SKB* **8** 83
[3] Radhokrishra H S, Chan H T 1989 *Can. Geotech. J* **26** 629
[4] Pusch R 1977 *Sweden: SKB*. **33**
[5] Pusch R 1980 *Teknisk Rapport*, Stockholm, Sweden. **16** 80
[6] Westsik J H 1982 *Material Research* **01** 329
[7] Ivan D, Jana F, Ivan J 2006 *Applied Clay Science*. **28** 223
[8] Liu Y M, Chen Z R 2001 *Acta mineralogica Sinica*. **21** 541
[9] Hou W B, Liu X D, et al. 1999 Beijing: Atomic Energy Press. **12** 116
[10] Zhu X 1992 *Beijing: Science Press*. **3** 662
[11] Cui S L, Liu J S, Zhang H Y 2012 *Symposium on underground waste disposal*. **39** 124
[12] Hou L 2008 *Southwest University of science and technology*. **07** 56
[13] Fan X, Yi F C, *China nonmetallic mineral industry guide*. **02** 15
[14] Liu Y M, Xu G Q 1998 *Radiation protection* **290**
[15] Fan X, Du J Y, Yi F C 2011 *China non metallic mineral industry guide*. **05** 28
[16] Fan X 2010 *Southwest University of science and technology*. **21** 24
[17] Liu Y M, Xu G Q, Liu S F, Chen Z R 2001 *Uranium geology*. **1** 44
[18] Japan Nuclear Cycle Development Institute 2000 *Desing and Engineering Technology*. **1** 225
[19] Shen Z Y, Li G D, et al. 1998 *Dam observation and geotechnical test*. **01** 26

Research on the influencing factors of reclaimed wastewater reuse and countermeasure in China

Leixiang Wu[1], Panwei Zhang[1], Xingchen Liu[1, 2], Ang Li[1], Zhuowei Wang[1] and Ronghao Guan[1]

[1]State Key Laboratory of Simulation and Regulation of Water Cycle in River Basin, China Institute of Water Resources and Hydropower Research, Beijing, 100038, China

[2]Corresponding author's e-mail: 841656911@qq.com

Abstract. Reclaimed water is regarded as the "Second water source" of the city, which plays an important role in alleviating the shortage of urban water resources and water pollution. By summarizing the current situation of reclaimed wastewater reuse in China, probing into the problems existing in the utilization of reclaimed water, some restrictive factors such as laws and regulations in reclaimed wastewater, water price system, water quality problems, and public participation were analyzed in this paper. Some suggestions in reclaimed wastewater reuse were put forward in this article.

1. Introduction

With the global urbanization expansion and the high-speed development of the relevant industry, the human consumption of water is increasing day by day. Since the beginning of the 20th century, the consumption of water has increased nearly 8 times, and water shortage has become a common phenomenon in most countries. Due to the uneven spatial and temporal distribution of water resources, special national conditions, water conditions and economic development requirements, the problems of water resources and water environment in China are becoming more and more complicated, and the national water safety situation is getting more serious. In the process of implementing the most strict water resources management system, with the water consumption approaching the total water consumption control index in some areas, it has become an inevitable choice for these areas to open up sources and reduce consumption [1]. Reclaimed water development and utilization can replace conventional water resources, improve the recycling efficiency of water resources, and which plays a positive role in alleviating the contradiction between supply and demand of water resources. Promoting the utilization of reclaimed water not only opens up the "Second water source" of urban water use, but also is one of the important contents of water-saving social construction. Meanwhile, it can effectively reduce the pressure caused by pollutant emissions on the ecological environment.

In April 2015, "Focusing on cities in water-scarce and water-polluted areas, improve the reuse facilities of reclaimed water" was proposed by Water Pollition Prevention and Control Action Plan in China [2]. By 2020, "Incorporating reclaimed water into the unified allocation system of water resources" has become an important measure for the Ministry of Water Resources to promote national water conservation and implement the "water conservation first policy" in the new era. It can be seen that the utilization of unconventional water resources such as reclaimed water plays an important role. However, the proportion of reclaimed water utilization in China accounts for less than 1.5% of the total water supply, which is still far behind the planned target [3]. Therefore, it is necessary to study the status quo of reclaimed water utilization in China, analyze the influencing factors and problems of reclaimed water utilization, and put forward suggestions to improve the utilization rate of reclaimed water, so as to improve the scale and efficiency of reclaimed water utilization and break the bottleneck that restricts the development and utilization of reclaimed water.

2. Status of reclaimed water utilization

According to the China Water Resources Bulletin from 2009 to 2019, the reclaimed water utilization in China has shown a rapid growth trend in the past 11 years, from 2.15 billion m^3 in 2009 to 11.61 billion m^3 in 2019 (Table 1 and Figure 1) [4], an increase of nearly 6 times. Regarding the construction of reclaimed water treatment facilities, there has been rapid growth in recent years. As presented in Figure 2, through the investment in fixed assets for urban sewage treatment and reclaimed water utilization in the past 11 years, it can be seen that since 2009, fixed asset investment in this area has grown rapidly. Fixed asset investment in urban sewage treatment and reclaimed water utilization in China has increased from 41.86 billion yuan in 2009 to 80.37 billion yuan by 2019. The daily production capacity of reclaimed water increased nearly 3.5 times from 10.34 million m^3 in 2009 to 44.29 million m^3 in 2019. The length of the reclaimed water pipeline has increased from 2250 km in 2009 to 12140 km in 2019, an increase of nearly 5.4 times [5]. In this paper, reclaimed water investment facilities construction data are from the 2019 China Urban and Rural Construction Statistical Yearbook [5].

Table 1. Urban sewage treatment and reclaimed water utilization in China from 2009 to 2019 [5].

Year	Number of Sewage Treatment	Treatment capacity of municipal sewage treatment plant / million m^3/d	Urban sewage treatment rate /%	Production capacity of reclaimed water /m^3/d	Reclaimed water utilization / billion m^3
2009	1214	90.52	75.25	1034	2.15
2010	1444	104.36	82.31	1082	2.83
2011	1588	113.03	83.63	1389	3.32
2012	1670	117.33	87.3	1453	4.43
2013	1736	124.54	89.34	1761	4.82
2014	1807	130.87	90.18	2065	4.65
2015	1943	140.28	91.9	2317	5.26
2016	2039	149.10	93.44	2762	5.92
2017	2209	157.43	94.54	3588	7.13
2018	2321	168.80	95.49	3578	8.55
2019	2471	178.63	96.81	4429	11.61

At present, investment in the utilization of reclaimed water in various parts of China is extremely unbalanced, of which Beijing has the most investment. Taking 2019 as an example, Beijing invested nearly 1.42 billion yuan in reclaimed water utilization, accounting for 29.5% of the investment in reclaimed water utilization in China. Other provinces and cities invested less, among them, 11 provinces such as Hubei and Liaoning have no investment in fixed assets of reclaimed water [4]. From this point of view, although the construction of reclaimed water facilities in China has developed rapidly, the investment in various regions is extremely uneven, and some provinces and cities are not enthusiastic about investment and construction.

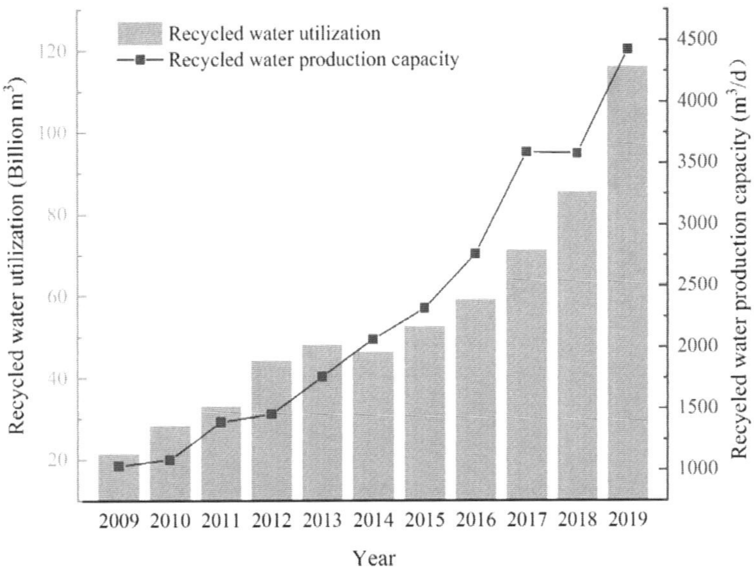

Figure 1. Consumption and production capacity of recycled water in China from 2009 to 2019.

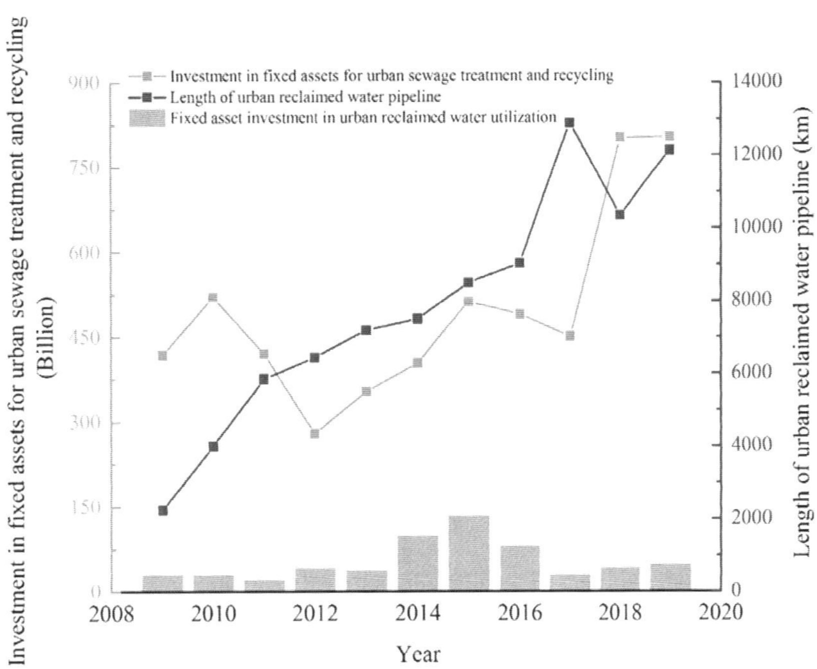

Figure 2. Investment in fixed assets for urban sewage treatment and reclaimed water conservancy and length of reclaimed water pipeline network in China from 2009 to 2019.

3. Analysis of influencing factors of reclaimed water

This study investigated the development, utilization and investment of reclaimed water, and construction of sewage treatment facilities in China in the recent 11 years, and conducted correlation analysis. According to the correlation analysis of factors affecting the development and utilization of reclaimed water (Figure 3), the annual reclaimed water utilization has an obvious positive correlation with the fixed asset investment of the whole society ($r=0.882$, $p<0.01$), and is positively correlated with the production capacity of reclaimed water and the total amount of sewage treatment ($r>0.86$, $P<0.05$). The production capacity of reclaimed water has an obvious positive correlation with Gross Domestic Product (GDP), fixed asset investment of the whole society, financial revenue, pipeline length and the construction of municipal public facilities ($r>0.91$, $p<0.01$). The total amount of sewage treatment has positive correlation with the production capacity of reclaimed water, pipelines length, the construction of municipal public facilities, financial revenue and GDP ($r>0.94$, $p<0.01$).

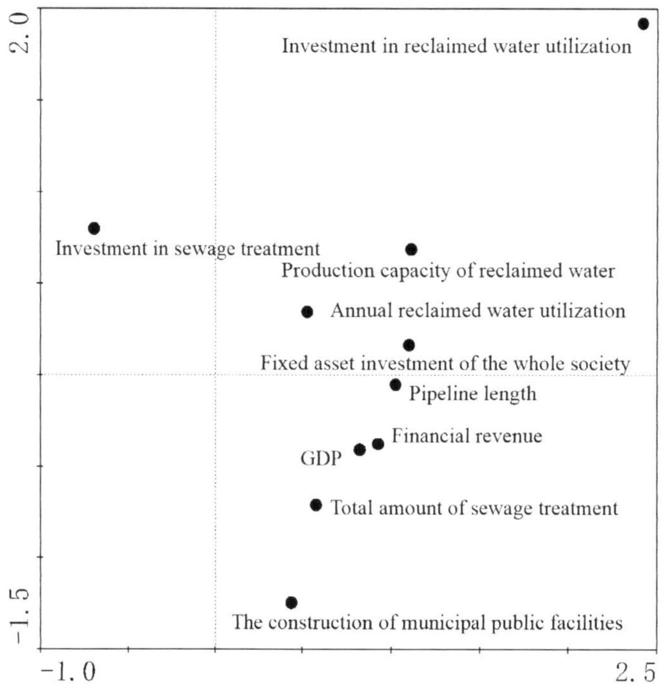

Figure 3. RDA on influencing factors of reclaimed water development and utilization.

The pipeline length is obviously positively correlated with the construction of municipal public facilities, the fixed assets investment of the whole society, financial revenue and GDP ($r>0.94$, $p<0.01$). The construction of municipal public facilities has an obvious positive correlation with the fixed asset investment, financial revenue and GDP ($r>0.94$, $p<0.01$). The fixed assets investment of the whole society has an obvious positive correlation with financial revenue and GDP ($r>0.99$, $p<0.01$), and there is an significant positive correlation between financial revenue and GDP ($r=0.997$, $p<0.01$). Among them, the annual reclaimed water utilization is less relevant to investment in reclaimed water utilization and investment in sewage treatment, mainly due to the lack of investment incentive policies for the sewage recycling industry, which results in the lack of investment sources. The construction of reclaimed water reuse projects requires a lot of funds. According to data, the cost of laying the trunk lines of the reclaimed water pipeline network alone is as high as 3 million yuan/km [4].

Due to lagging pipeline network construction and imperfect supporting facilities, the reclaimed water cannot be sent out, and the actual utilization is far less than the designed production capacity, and the long-term low-load operation affects the benefits. At present, the construction of supporting facilities for reclaimed water projects in many regions is still dominated by government investment, and the investment and financing system for reclaimed water projects has not been established [6].

There is a lack of diversified investment and financing channels, private capital and foreign investment have not been fully attracted. In addition to the above factors, the price of tap water has little impact on the development and utilization of reclaimed water, mainly because the price of traditional water resources in China has always been at a relatively low level, and water price system cannot stimulate consumers to actively adopt reclaimed water. Besides the water resource fees, consumers have basically not paid other fees for self-provided water sources (including surface water and groundwater), and their direct water costs are lower than the price of reclaimed water, causing some companies to refuse to use reclaimed water. Moreover, the charging mechanism of reclaimed water has not been established in China. There is no clear and reasonable standard to follow for the pricing of reclaimed water. The pricing process shows greater arbitrariness and subjectivity. This unreasonable price restricts the development of reclaimed water utilization [7]. There is no clear and reasonable standard to follow in the pricing of reclaimed water, and the pricing process is more arbitrary and subjective, which limits the development of reclaimed water utilization [8].

4. Problems existed in the utilization of reclaimed water

In recent years, although the reuse of reclaimed water has been paid more attention and developed rapidly in China, compared with developed countries, there is still a big gap and deficiency in the reuse of urban sewage. Mainly manifested in the following areas.

4.1. Policies and regulations on the utilization of reclaimed water

At the national level, apart from the exemption of value-added tax for the reclaimed water enterprises included in the comprehensive utilization of resources, there is no institutionalized financial subsidy and price incentive policies, and there is no comprehensive reclaimed water law that can regulate the interests of all parties [9]. From a local perspective, according to incomplete statistics, a total of 98 local regulations, government regulations and normative documents related to reclaimed water have been promulgated in China, including 8 provincial regulations, 6 provincial government regulations, 52 provincial normative documents, 11 municipal regulations, 11 municipal government regulations, and 10 municipal normative documents [10]. Formally, about 70% of the policies and regulations related to recycled water are scattered among laws, regulations, and normative documents such as sewage treatment, environmental protection, and resource utilization. Only 12 cities have issued specific regulations and norms for recycled water. Most of the documented countries are based on macroscopic "encourage" and "vigorous development" terms, and lack specific and clear incentive policies, supporting measures, and reward and punishment systems [10]. For example, the government can give priority to certain environmental protection project loans or reduce related taxes for enterprises that self-finance to build reclaimed water facilities. The promulgation of favorable policies will greatly increase the enthusiasm of enterprises in the development and utilization of reclaimed water, and promote the development of reclaimed water utilization.

4.2. Water price system of reclaimed water

The price of reclaimed water is the core element of the market. Under the condition that the quality and quantity of reclaimed water can satisfy the security and stability, the reasonable price mechanism can generate economic incentives for the demand for reclaimed water [11]. Only when the prices of reclaimed water, surface water, and groundwater have a certain degree of price comparison, so that the public feels that the use of reclaimed water is “profitable”, can reasonable water consumption be guided and promotion of reclaimed water be promoted. The current water price structure in most areas of China is unreasonable, and a complete water price system has not been formed [12]. The price of tap water is obviously low, and in some areas, the cost of water production and water supply cannot even be compensated. The low price of tap water makes it difficult for reclaimed water to reflect the price advantage [13]. In addition to water resources fees, the users basically pay no other fees. And their direct water cost is lower than the price of reclaimed water, resulting in some enterprises refused to use reclaimed water.

4.3. Water quality of reclaimed water
Water quality is a key indicator of reclaimed water utilization. The quality of reclaimed water is greatly affected by the treatment capacity of upstream sewage treatment plant. If the quality of the upstream water is unstable, the treatment capacity of the reclaimed water plant may be exceeded [14]. By comparing the water quality standards for different purposes such as reclaimed water for agricultural irrigation, industrial production, landscape water, and municipal miscellaneous use, it can be found that except for the water quality standards for reclaimed water for agricultural irrigation, which basically meets Class IV standards, all other indicators only meet Class V standards [15]. Some users need to perform secondary treatment on the reclaimed water again to reach the utilization standards. Different industries have different requirements for the quality of reclaimed water, which makes the transportation and large-scale use of reclaimed water more difficult [16].

4.4. Public perception
At present, reclaimed water utilization projects in some areas are small in scale, limited in scope of reuse, and the government's insufficient publicity has made the public's understanding of reclaimed water limited. Affected by traditional concepts, the public has concerns about the quality of reclaimed water, and the acceptance level is still relatively low [17]. In addition, the lack of public information on water scarcity, principles and procedures for setting water prices makes it difficult to participate in and supervise the management of water resources, which affects the promotion of wastewater recycling to some extent.

5. Suggestion on the utilization of reclaimed water
Solidly advance the work of "incorporating unconventional water sources such as recycled water, rainwater, and brackish water into the unified allocation of water resources" in the "Ten-point Water Plan". Accelerate the introduction of national or industry guidelines for the promotion of the use of recycled water, and guide local governments to incorporate water resources and water environment management into the integration of water services. It is recommended to strengthen the long-term planning of reclaimed water utilization in cities with water shortages, and conduct detailed analysis and research on the feasibility and overall planning of reclaimed water utilization. On this basis, the scientific and rational medium-and long-term planning and implementation program of reclaimed water utilization are formulated. The planning should be forward-looking, to avoid the development of the city to a certain stage, and the reclaimed water pipeline network becomes a restrictive factor of reclaimed water utilization. The relevant national ministries and commissions should issue "Urban Reclaimed Water Utilization Regulations" and other industry-leading policies as soon as possible to guide water-deficient cities to promote the utilization of reclaimed water in accordance with local conditions. Clear legal guarantee should be provided from the aspects of industrial policies, departmental responsibilities and legal applicability to solve the problems existing in the supervision and management, planning, construction investment, operating expenses, layout of reclaimed water plants and price of reclaimed water.

6. Conclusion
With the acceleration of urbanization, the form of water shortage has become increasingly prominent. Promoting advanced sewage treatment and reclaimed water utilization has become the primary way to alleviate the contradiction between water supply and demand, improve the efficiency of comprehensive utilization of water resources, and reduce water pollution. Meanwhile, it is also an inevitable choice for building a resource-saving and environment-friendly society. To formulate laws and regulations on the use of reclaimed water, establish and improve legal guarantees, improve technologies for the use of reclaimed water, and increase awareness of the use of reclaimed water will be important approaches to solve water shortages. Therefore, the use of reclaimed water will play a very important role and far-reaching significance in solving the problem of water shortage in China and even the world.

Acknowledgments
This work was supported by the National Key Research and Development Program of China (No. 2017YFC0403503-3) and the National Natural Science Foundation of China (No. 51879279).

References

[1] He P, Phan L, Gu G, Hervouet G 2001 Reclaimed municipal wastewater--a potential water resource in China *Water Science & Technology A Journal of the International Association on Water Pollution Research* **43(10)**: 51-8
[2] Rajashree Hajare, Pawan 2021 Labhasetwar, Pranav Nagarnaik. A critical review of applications of QMRA for healthy and safe reclaimed water management *Environmental Modeling & Assessment*
[3] Bao C, He D M 2019 Scenario modeling of urbanization development and water scarcity based on system dynamics: A case study of beijing–Tianjin–Hebei urban agglomeration, China *International Journal of Environmental Research and Public Health* **16(20)**
[4] Ministry of Water Resources of the People's Republic of China 2020 2009-2019 China Water Resources Bulletin *Beijing: China Water Resources and Hydropower Press*
[5] Ministry of Housing and Urban-Rural Development of the People's Republic of China 2021 2019 Urban and Rural Construction Statistical Yearbook *Beijing: China Planning Press*
[6] Zhai X L, Zang Y Q 2019 Application of reclaimed water to the field of social life. *Shandong Water Resources* **(11)**:20-21
[7] Li N 2021 Research on optimization of urban wastewater reclamation and reuse *Modern Industrial Economy and Informationization* **11(03)**:65-66
[8] Fan Y P 2014 Research on comprehensive benefit evaluation and policy impact of reclaimed water utilization *University of Chinese Academy of Sciences*
[9] Li C, Sun Y Y, Shen H Y, Xue Xiao 2010 Foreign reclaimed water reuse policy and its enlightenment to China *Environmental Science & Technology* **33(S2)**: 626-627
[10] Li M, Jin Y C 2005 Developing mechanism of reuse water price *Journal of Hebei University of Engineering (Natural Science Edition)* **(02)**: 90-92
[11] Jing Guo 2020 The study on the effect of government subsidy considering ecological environmental benefits of recycled water reuse water *Xi'an University of Architecture and Technology*
[12] Duan T, Liu X J 2007 Study on demand for urban reused water *Ecological Economy* **(04)**: 151-153
[13] Ma Z, Zhou F 2012 The current situation of my country's water price policy and its countermeasures *Environmental Protection* **(19)**: 54-57
[14] Yuan Y 2019 Investigation of reclaimed water quality in wastewater treatment plant and study on its resue in thermal power plants *Xi'an University of Architecture and Technology*
[15] Zhang C M, Liang J, Liu W Y 2021 Comparative study on the bacterial diversity and antibiotic resistance genes of urban landscape waters replenished by reclaimed water and surface water in Xi'an, China *Environmental science and pollution research international*
[16] Wei F,Yang X P, Wang Y, Huo M X 2020 Loopholes in the current reclaimed water quality standards for clogging control during aquifer storage and recovery in China *Water Cycle* **1**
[17] Li L, Liu X J, Zhang X Y 2021 Public attention and sentiment of recycled water: Evidence from social media text mining in China *Journal of Cleaner Production* **303**

Mesoporous TiO_2 sphere with high adsorbability: Adsorption behavior study of Pb(II) from water

Hao Ding[1], Xiangyu Ye[2, 3], Xiaodong Wu[1], Quan Liang[1], Tianyi Hu[2] and Difan Wang[2]

[1]Eco-environmental Science Research and Design Institute of Zhejiang, Hangzhou 310007, China;
[2]Zhejiang Light Industrial Products Inspection and Research Institute, Hangzhou 310018, China

[3]Corresponding author's e-mail: yexiangyu1987@163.com

Abstract. Mesoporous TiO_2 sphere was prepared using a template-free solvothermal treatment. The crystalline size, morphology and property of the adsorbent were characterized with transmission electron microscopy and electron probe microanalysis/wavelength dispersive spectroscopy. Moreover, the applicability of three isothermal models for Pb(II) onto mesostructured TiO_2 was investigated. The Pb(II) adsorbability of adsorbent was evaluated using inductively coupled plasma mass spectrometry. The maximum adsorbability of mesoporous TiO_2 sphere for Pb(II) is 35.2 mg g^{-1}, and this material has bright application prospects for the removal of toxic metal ions.

1. Introduction

Removal of toxic metal ions has become a focus in the field of public health and ecological systems [1,2]. Pb is natural element predominantly used in industrial applications, and some river contains a wide range of Pb(II) concentrations [3,4]. It may affect children's intelligence or cause negative effects on metabolic process even at low dosages [5]. Hence, an effective adsorbent for removing Pb(II) is suitable for water treatment. Adsorbents with good adsorption capacities are applied to interact with metal ions to reduce the metal contamination [6]. Currently, the widely employed methods for water purification are based on physicochemical processes [7-9].

Mesoporous adsorbent in environmental remediation has attracted considerable interest due to its specific surface area and activity [10]. TiO_2 has properties such as high corrosion resistance, relative cheapness and high specific surface area, becoming a new adsorbent material [11,12]. The aim of this paper is to assess the ability of mesoporous TiO_2 spheres (MTs) to adsorb Pb(II) from water. In this paper, MTs were prepared using a template-free method. The morphology and composition of products were characterized by transmission electron microscopy (TEM) and electron probe microanalysis/wavelength dispersive spectroscopy (EPMA/WDS). An application in efficient removal of Pb(II) via MTs at room temperature was reported.

2. Experimental section

2.1. Instruments and reagents
The main instruments used included a G2 F30 transmission electron microscope (FEI, USA), a JXA-8530F PLUS electron probe microanalyser (JEOL, Japan), a Grant PCV-2400 centrifuge (Cambridge, UK), a NexION 350X inductively coupled plasma mass spectrometer (PerkinElmer, USA), a S210-K pH meter (Mettler Toledo, USA) and a magnetic hotplate stirrer (Talboys, USA).

99% titanium butoxide, 99% glycol, carbamide, acetone and ethanol were of at least analytical grade, and purchased from Aladdin (China). The 1000 mg L^{-1} Pb(II) standard stock solution was purchased from Zhongke Quality Inspection Biotechnology (China). Water (resistivity of 18.2 MΩ cm) was prepared with a Smart-DUV water purifier (China).

2.2. Synthesis of MTs and characterizations
The synthesis of MTs was conducted by treating the following steps. First, titanium glycolate was prepared as precursor. 0.5 mL titanium butoxide was added to 30 mL glycol with stirring, and then it was added to 110 mL 99.7% acetone by staying for 5 min. After centrifuging, the precipitate was washed with ethanol, and then dried in an oven at 70 °C to obtain the precursor powder. Next, 0.9 g titanium glycolate was dispersed into the mixed liquid composed of 150 mL ethanol and 3.8 g carbamide. The well-stirred liquid was transferred to a PTFE-lined reactor when stirred at 160 °C for 10 h. After cooling to room temperature, the product was collected by centrifugation. Based on solvothermal reaction, the mesoporous materials were obtained via hydrolysis.

MTs were scattered in ethanol by ultrasonic treatment for 10 min, and then made into wafers of 3 mm in diameter. The TEM image of the adsorbent was observed under an accelerating voltage of 300 kV. Qualitative analysis of the C/Ti/O composition was carried out using EPMA/WDS operated at 15 kV.

2.3. Adsorption measurement
MTs were added to aliquots of colorimetric tubes, and the pH value of the solution was adjusted by diluted nitric acid. The standard solutions of Pb(II) were pipetted into colorimetric tubes, and brought to volume with water. They were transferred to microcentrifuge tubes after 1 min of shaking, and then centrifuged at 2000 rpm for 5 min to obtain supernatant.

The mass spectrometer was attuned by 1.0 μg L^{-1} mass spectrometric fluid in KED mode to obtain stable signals and high sensitivities. The Pb(II) concentrations of the collection supernatant via MTs were performed by inductively coupled plasma mass spectrometry (ICP-MS).

3. Results and discussion

3.1. Crystalline size, morphology and properties
TiO_2 fabricated by solvothermal treatment appeared to be mesoporous spheres (Figure 1). MTs were well-shaped, whereas the average grain size was counted as 91 nm, which possessed advantages of uniform particle size distribution and obvious porous structure. High-resolution TEM observation combined with BET method were used to investigate the specific surface area and pore volume of MTs. The specific surface area was determined to be 523 $m^2 g^{-1}$, and the pore volume was 0.86 $cm^3 g^{-1}$. The mesostructured material can interact with the contaminant to adsorb more toxic metal ions.

Due to the fine resolution of the EPMA plane scan (Figure 2), it was possible to visualise the element distribution on the surface of as-prepared material. For micro-area elemental analysis, the contents of Ti, O and C in MTs were revealed by EPMA/WDS results (Table 1), indicating the composition of MTs.

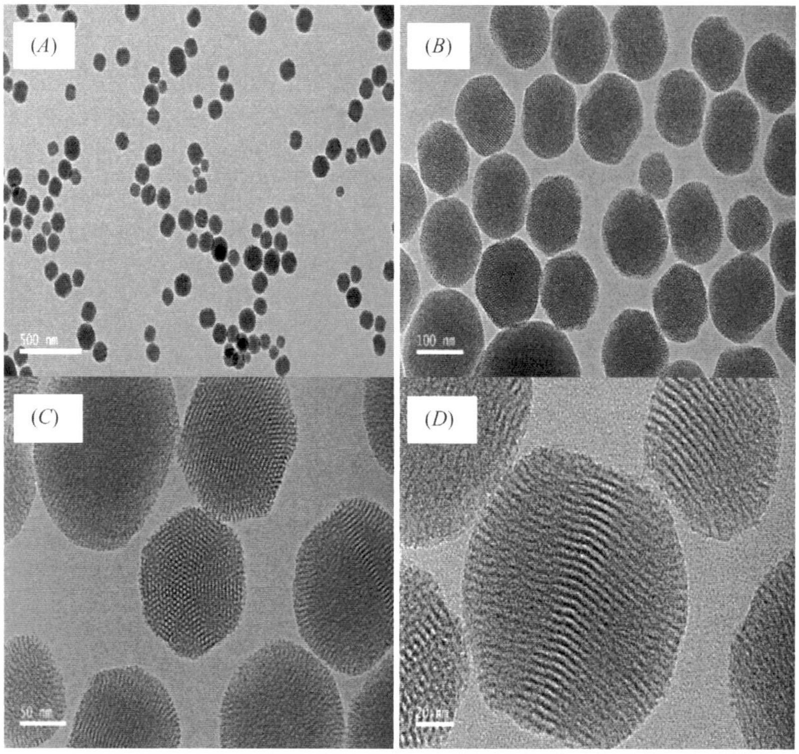

Figure 1. TEM (*A*,*B*,*C*) and high-resolution TEM (*D*) images of MTs.

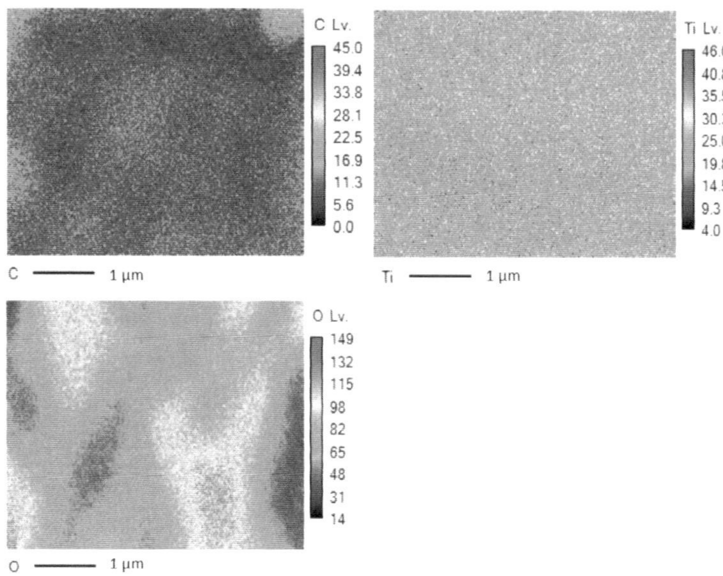

Figure 2. EPMA/WDS mapping of as-prepared MTs.

IOP Conf. Series: Earth and Environmental Science **865** (2021) 012044 doi:10.1088/1755-1315/865/1/012044

Table 1. Ti, O and C contents of MTs.

Element	EPMA/WDS (%)
Ti	31.04
O	66.81
C	2.15

3.2. Removal of Pb(II) from aqueous solution using MTs

The net intensities of Pb-208 were in good linear relation to known concentrations, and the correlation coefficient (R) was 0.9996. Adsorption tests were performed at 23±1 °C starting at different concentrations of Pb(II). Each test was repeated in triplicate, and the error bars represent standard deviation. The metal ion retained onto MTs is calculated:

$$q_e = \frac{(C_0 - C_e)\cdot V}{m} \quad (1)$$

where C_0 and C_e are the initial and final concentrations of metal ion (mg L^{-1}), q_e is the amount adsorbed per specified mass of adsorbent (mg g^{-1}), V is the volume of solution (mL), and m is the mass of adsorbent (mg). The adsorption percentage (A_p, %) of the metal ion from solution is calculated by Equation (2):

$$A_p = \frac{C_0 - C_e}{C_0} \times 100\% \quad (2)$$

A_p increased with contact time, and equilibrium was reached within 90 min (Figure 3). The A_p increased with rising pH value, and a maximum A_p was obtained at pH 5.6–7.8 (Figure 4). Since the precipitation of metal hydroxide relies on pH value, its pH-dependence is applied to removal procedures of metal ions. In the adsorption process at pH >6.0 for Pb(II) might exist different behaviors, and the precipitation of Pb(II) contributed to uptake of the metal ions at high pH values of solutions. Therefore, the adsorption behavior of Pb(II) onto MTs was studied at pH = 6.0 for subsequent experiments. Pb(II) exhibited a high affinity onto MTs in faintly acidic solutions. Consequently, the pH value of the solution affected the surface charge and speciation of the adsorbents.

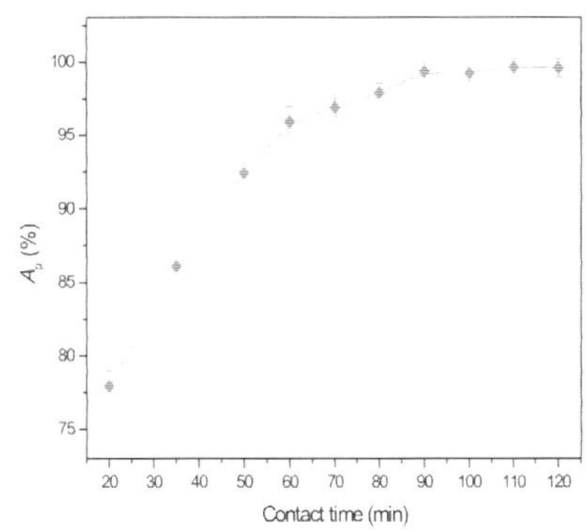

Figure 3. Effect of contact time on Pb(II) adsorption using MTs (initial concentration = 30 mg L^{-1}, pH = 6.0, and adsorbent dosage = 4.0 g L^{-1}).

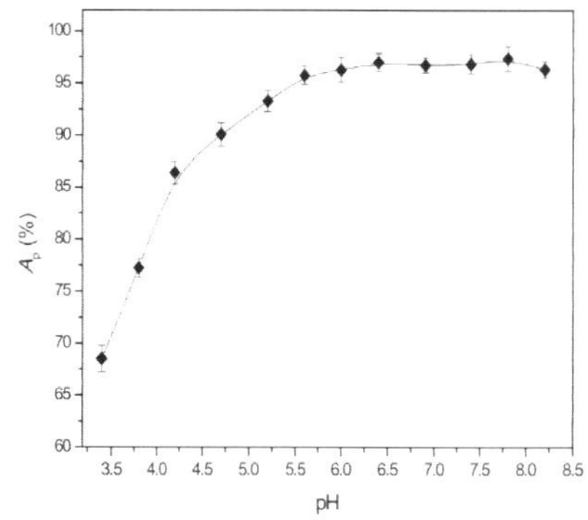

Figure 4. Effect of pH value on Pb(II) adsorption using MTs (initial concentration = 30 mg L^{-1}, contact time = 90 min, and adsorbent dosage = 4.0 g L^{-1}).

3.3. Three isothermal models for Pb(II) adsorbed onto MTs

The model parameters were estimated with R values. The Langmuir isotherm is presented by the linear Equation (3), and the maximum adsorbability (K, mg g^{-1}) is determined:

$$\frac{C_e}{q_e} = \frac{C_e}{K} + \frac{1}{K \cdot b} \quad (3)$$

where b is the Langmuir constant (L mg^{-1}). K and b for the removal of Pb(II) were determined from the slope and intercept of lines.

The Freundlich isotherm equation is:

$$\log(q_e) = \log(k_f) + \frac{1}{n} \cdot \log(C_e) \quad (4)$$

where k_f and n are the Freundlich constants.

The Temkin isotherm model is described by Equation (5):

$$q_e = B \cdot \ln(k_t) + B \cdot \ln(C_e) \quad (5)$$

where k_t and B are is the Temkin constants.

Table 2 displays the characteristics of adsorption isotherms expressed by the parameters of plots. The R^2 value of the Langmuir isotherm for Pb(II) was higher than those of Freundlich and Temkin isotherms. The three models closely followed this order: Langmuir > Temkin > Freundlich.

Proposed mechanism in the adsorption process using mesoporous TiO_2 is as follows: polar groups e.g., hydroxyl can interact with metal ions to form complexation reactions. In fact, the adsorption behavior involves ion exchange as the interaction force between Pb(II) and MTs. Metal ions were adsorbed on specific sites, growing into more porous structure. These indicated that the Pb(II) was adsorbed onto the surface of MTs through binding with polar groups.

Table 2. Estimated isothermal models and their constants for Pb(II) onto MTs.

Cation	Langmuir isotherm			Freundlich isotherm			Temkin isotherm		
	b	K	R^2	$1/n$	k_f	R^2	B	k_t	R^2
Pb(II)	0.011	35.2	0.971	0.280	4.26	0.959	0.176	1.79	0.963

2.5 mg adsorbent was mixed with 1.0 L water to obtain suspensions at pH = 6.0. There was only minor amount of adsorbent consumed, which presented a competitive adsorbability versus other adsorbents previously reported (Table 3). The Pb(II) adsorbability of MTs was higher than other materials, which could be useful in treatment of water containing toxic metal ions. Finally, the results conveyed a potential for removing Pb(II), providing practical application of mesostructured materials.

Table 3. Comparison of adsorbability of adsorbent materials for Pb(II).

Adsorbent material	Pb(II) (mg g^{-1})	Reference
Torrefied poplar-biomass	30.0	[13]
PVC-acetylacetone composite	33.1	[14]
Chicken egg shell powder	33.7	[15]
Poly-o-toluidine/triethanolamine	31.3	[16]
Chitosan-N-lauroyl	31.5	[17]
Eggshell powder-sericite	33.9	[18]
Mesoporous TiO_2 sphere	35.2	This work

4. Conclusion

MTs were synthesized through a solvothermal method using titanium glycolate as precursor. Series of characterizations suggested that the MTs exhibited high crystallinity, uniform morphology and high specific surface area. The adsorbability of MTs and the mechanism of adsorbent interaction were discussed. This study had evidenced that the contact time and pH value of the solution were important factors for Pb(II) adsorption. For water purification, the adsorbents were mixed with water at pH = 6.0 with high adsorption efficiency, and used to treat for 90 min. This approach with suitable conditions can be utilized to prevent Pb(II) contaminant in wastewater from industries.

References

[1] Mercante L A, Andre R S, Schneider R, Mattoso L H, Correa D S 2020 *New Journal of Chemistry* **44** 13030
[2] Che X H, Yuan R H, Tang L C, Xiong Y D, Wu M F, Zhu M H 2021 *Journal of Physics: Conference Series* **1978** 012009
[3] Wei J, Duan L, Wei J, Hoffmann E, Meng X 2021 *Journal of Environmental Sciences* **101** 135
[4] Akbari-Adergani B, Rahnama S, Shirkhan F 2017 *J Mazandaran Univ Med Sci* **27** 162
[5] Zhang P, Ouyang S, Li P, Sun Z, Ding N, Huang Y 2020 *Journal of Cleaner Production* **246** 118728
[6] Qu J, Li Y, Song T, Huang S, Jin Y 2020 *Ecological Engineering* **142** 105639
[7] Yuan R H, Liu W H, Teng Y J, Nie J, Ma S Z 2015 *Spectroscopy and Spectral Analysis* **35** 1276
[8] Liu W H, Yuan R H, Teng Y J 2014 *Chinese Journal of Luminescence* **35** 1124
[9] Sandouqa A, Al-Shannag M, Al-Hamamre Z 2020 *Renewable Energy* **151** 103
[10] Vu D, Li Z Y, Zhang H N, Wang C 2012 *Journal of Colloid & Interface Science* **367** 429
[11] Yuan R H, Wu T H, Hu T Y, Shi L, Shi J 2020 *IOP Conference Series: Materials Science and Engineering* **892** 012014
[12] Chen D, Tu N, Si C, Yin M, Wang X 2019 *Water Science & Technology* **80** 884
[13] Demey H, Melkior T, Chatroux A, Attar K, Sébastien T, Miller H, Muriel M 2019 *Chemical Engineering Journal* **361** 839
[14] Elzbieta R, Katarzyna W 2019 *Polymers* **11** 513
[15] Kim D, Hwang S, Kim Y, Jeong C, Pyo H, Sang R 2019 *Bulletin of the Korean Chemical Society* **40** 1156
[16] Nafisur R, Fazeel K, Mohd N 2020 *Environmental Technology & Innovation* **17** 100634
[17] Vieira C L, Neto F O S, Carvalho-Silva V H, Signini R 2019 *Journal of Environmental Chemical Engineering* **7** 103070
[18] Choi H 2019 *Water Science and Technology* **79** 1922

Object-based multi-features Wetland classification method of GF-2 PMS imagery

Leishi Chen[1, 2, 5], Qinzhe Han[1, 2], Yilin Lin[3, 4], Dongsheng Du[1, 2] and Jianbo Deng[1, 2]

[1]Hunan Key Laboratory of Meteorological Disaster Prevention and Reduction, Changsha 410118, China
[2]Hunan Meteorological Research Institute, Changsha 410118, China
[3]Faculty of Land Resource Engineering, Kunming University of Science and Technology, Kunming 650093, China
[4]Key Laboratory of Geospatial Information Integration Innovation for Smart Mines, Kunming 650093, China

[5]Corresponding author's e-mail: chenleishi0917@gmail.com

Abstract. Wetland plays an important role in the earth's ecosystem. However there are still a lot of problems to classify and extract wetlands through high-resolution satellite imagery. We propose an object-based multi-features wetland classification method of GF-2 PMS imagery, taking Yuanjiang, Hunan Province as the study area, selecting multi-dimensional object features including spectral, geometric, terrain and texture features to train the CART classifier and evaluate the accuracy of the classification results. The results show that the method can obtain accurate wetland classification results, which provides technical reference for wetland classification based on GF-2 PMS imagery.

1. Introduction

Wetland is an important part of the earth natural ecosystem and has the reputation of being the kidney of the earth. It is a key task for mankind to protect wetland resources, which requires meticulous and accurate wetland classification. Traditional remote sensing wetland classification is mainly based on medium-resolution multispectral images [1]. Studies have shown that Landsat (TM) 2–4 bands are helpful for detecting understory vegetation, and relatively bands 5 and 7 are helpful for identifying water and freshwater swamps. The Landsat 8 satellite launched in 2013 has greatly improved the accuracy of wetland classification, but this type of data is difficult to distinguish small and medium-sized wetland plots. Meter-level high spatial resolution multispectral images represented by GeoEye, WorldView and GF series can accurately identify broken wetland blocks [2, 3]. With high spatial accuracy and low cost, they provide a reliable source of wetland classification data. However, due to the finer texture and geometric information of high-resolution multi-spectral imagery, the corresponding classification work often has the problem of "salt and pepper" noise and "same objects with different spectra, and different objects with the same spectra" problem [4]. Therefore, we propose an object-based wetland classification method of GF-2 PMS imagery to solve the problem.

2. Study area and data sources

2.1. Study area
Yuanjiang City is adjacent to the West Dongting Lake, which connects the four rivers of Xiang, Zi, Yuan and Li, and is named after the place where the Yuan River flows into Dongting Lake, as is shown in Figure 1. The city has a resident population of 697,700. It has a subtropical humid monsoon climate with an average annual temperature of 16.9°C and an average rainfall of 1322mm. The city is surrounded by lakes on three sides, the river network is densely covered with broken terrain, and the tidal flats and lakes are densely covered. It is a typical Dongting Lake wetland city with moderate scales, which can be used as a typical research area for wetland classification.

2.2. Data sources
Select the GF-2 PMS imagery taken on March 27, 2018 was selected as the source data, including panchromatic band and multi-spectral band, complete geometric correction and radiometric correction to ensure the accuracy of the imagery (Figure 1). The auxiliary data includes the boundary of the study area and DEM data with a precision of 30 meters.

In addition, 300 sampling points were manually located based on field survey data and GOOGLE EARTH historical images. Randomly select 200 training samples, and the remaining 100 are used as accuracy evaluation samples. There is no overlap between the two types of samples.

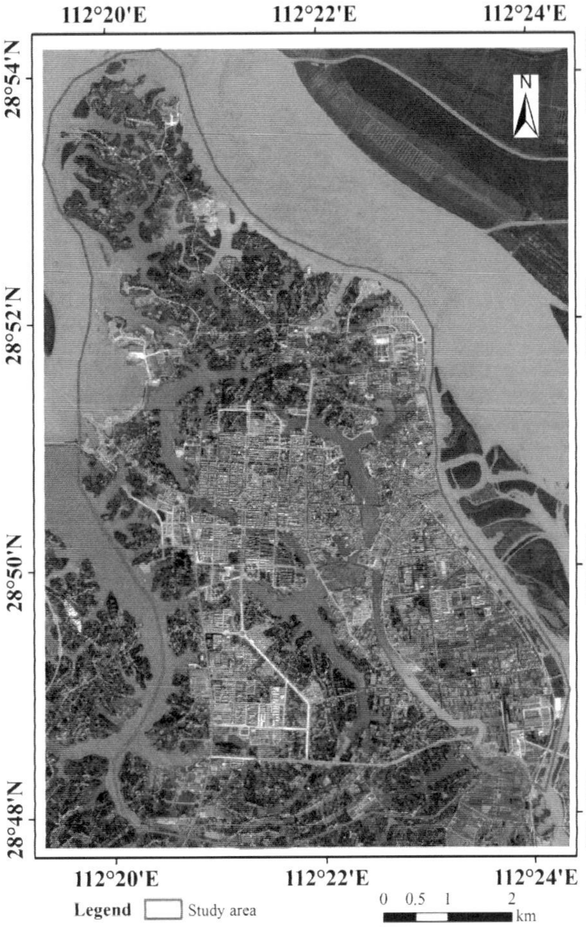

Figure 1. Overview of the study area.

2.3. Preprocessing
The GF-2 PMS image contains 1 panchromatic band and 3 multispectral bands. By ENVI software, geometric corrections were performed on panchromatic and multispectral data respectively, and the cubic convolution interpolation method was used in the orthorectification experiment. Image registration ensures that the displacement deviation of panchromatic and multi-spectral images is smaller than a single multi-spectral pixel by manually selecting 30 control points with the same name.

The fusion of panchromatic and multispectral images uses the Gram-Schmdit method, which combines the radiation information of the multispectral band with the high resolution of the panchromatic band to obtain a multispectral image with a spatial resolution of 1 meter. Then perform absolute calibration and FLAASH atmospheric correction on the fused remote sensing image. The calibration coefficient and spectral response function are provided by the China Resources Satellite Application Center (http://www.cresda.com/CN/Downloads/dbcs/index.shtml). The image is cropped to obtain the preprocessed image of the study area.

Table 1. Wetland classification system in the study area.

Categories	Description
Lake wetland	Permanent lakes.
Paddy wetland	Fields where aquatic crops such as rice are grown.
Pond wetland	Aquaculture ponds, artificial water landscape.
Dry farming	Fields where xerophytic crops are grown.
Green space	Orchards, pastures, plantations and urban green spaces.
Building	Land and internal roads covered by buildings such as urban and rural settlements, independent settlements and factories.
Bare land	Bare land or land with less than 5% vegetation coverage.
Road	Urban paved roads covered with concrete or asphalt.

3. Methodology

3.1. Classification system
The establishment of a classification system is the prerequisite for wetland classification [5]. Based on the distribution of wetlands in Yuanjiang City and the spectral characteristics of the source data imagery, a wetland classification system has been developed (Table 1).

Table 2. Features attributes for classification.

Features	Description
Spectral	
Value of each band	Band1, band2, band3 and band4
Customized	NDWI, NDVI, DVI and RVI
Geometric	
Shape	Border index and shape index (unit: pxl)
Range	Area, perimeter, width and length/width (unit: pxl)
Location	The distance from the center of object to 4 each boundary (unit: pxl)
Texture	GLCM homogeneity, contrast, dissimilarity, entropy, mean and StdDev
Terrain	DEM (unit: m)

3.2. Classification features
Use the multi-scale segmentation method to segment the imagery to be classified [6], and a large number of patches are the objects that need to be classified. The spectral, geometric, texture, and terrain features of the objects are extracted for high-precision classification of land [7]. The feature attributes for classification are shown in Table 2.

According to the topography, climate and hydrological characteristics of Dongting Lake wetland, one water body index and three vegetation indexes are selected to characterize the characteristics of the object [8]. Spectral features include 4 band values of the source data after preprocessing, and 4 custom spectral indices, namely: Normalized Water Index (NDWI), Normalized Difference Vegetation Index (NDVI), Difference Vegetation Index (DVI) and Ratio Vegetation Index (RVI), defined as follows:

$$NDWI = (GREEN - NIR) / (GREEN + NIR) \tag{1}$$

$$NDVI = (NIR - RED) / (NIR + RED) \tag{2}$$

$$DVI = NIR - RED \tag{3}$$

$$RVI = NIR / RED \tag{4}$$

where RED, $GREEN$ and NIR are band value of each object.

The geometric features include the shape feature, range feature, and location feature of the object, and 30-meter accuracy DEM elevation data of each object is extracted and used as a topographic feature for classification. A total of 6 texture features based on gray level co-occurrence matrix (GLCM), which are commonly used in object-based classification and have high correlation, are selected for texture features, which are selected based on the principal component analysis method. In order to reduce noise and avoid the salt and pepper effect, a 3×3 scale moving window is selected through multiple experiments with a step size of 1. The gray-level co-occurrence matrix is used to calculate the texture features of the imagery in 6 omnidirectional directions, and a total of 24 in 4 bands are obtained. Experiment shows that after adding customized spectral features and texture features, the optimal separation distance of the classification samples increases from 2.04 to 2.12.

3.3. Classification method
Based on the object-based imagery classification principle, after completing the multi-scale segmentation of remote sensing imagery, the extracted classification features are used as the training data set, and the CART (Classification and Regression Tree) classifier based on the GINI index is trained to conduct multi-scale and multi-feature experiments [9, 10]. Its expression is as follows:

$$Gini(D,A) = \frac{|D_1|}{|D|}Gini(D_1) + \frac{|D_2|}{|D|}Gini(D_2) \tag{5}$$

where $Gini(D,A)$ is the GINI index of set D under the condition of known characteristics A. If the value of $Gini(D,A)$ is larger, the uncertainty of the sample is also greater, so the characteristic that satisfies the minimum value of $Gini(D,A)$ is going to be selected.

The CART decision tree algorithm performs Boolean operations on the branch nodes. If the judgment condition is true, the left branch of the node, otherwise, the right branch, and the operation obtains the binary decision tree. When the number of layers of the decision tree reaches the preset maximum value, or the samples in all leaf nodes belong to the same category or the number of samples is 1, the CART decision tree algorithm stops growing, and the training of the classifier is completed.

4. Results and discussions

4.1. Results of classification
The classification result of the study area is shown in Figure 2. Various large and small lake wetlands in the study area have been effectively identified; paddy field wetlands are widely distributed on the north and southeast sides of the construction area, which is the largest wetland category; pond wetlands are

mainly distributed in the east of the study area. In particular, the results of small wetlands by this classification method conform to the actual distribution.

4.2. Precision analysis

The accuracy evaluation of the classification results uses the Khat method based on the confusion matrix, and uses the sample points constructed in 2.2 of the paper to select 100 as the accuracy verification samples [11]. The segmentation pattern obtained by the imagery segmentation is assigned based on the verification sample to construct the TTA Mask inspection pattern and the classification accuracy is evaluated with pixels as the statistical unit.

The results show that the overall accuracy of the classification results obtained by this method is 85.01%; the Kappa coefficient is 0.8219. Among all wetland categories, the maximum user accuracy is 91.23% of lake wetlands, and the minimum is 80.07% of pond wetlands; the maximum mapping accuracy is 89.22% of lake wetlands, and the minimum is 83.65% of paddy wetlands. Among all non-wetland categories, the maximum user accuracy is 84.50% of bare land and the minimum is 78.68% of green space; the maximum mapping accuracy is 90.19% of green space, and the minimum is 79.64% of bare land. Because this method is more sensitive to water, the extraction accuracy of wet land is significantly higher than that of other classifications of land.

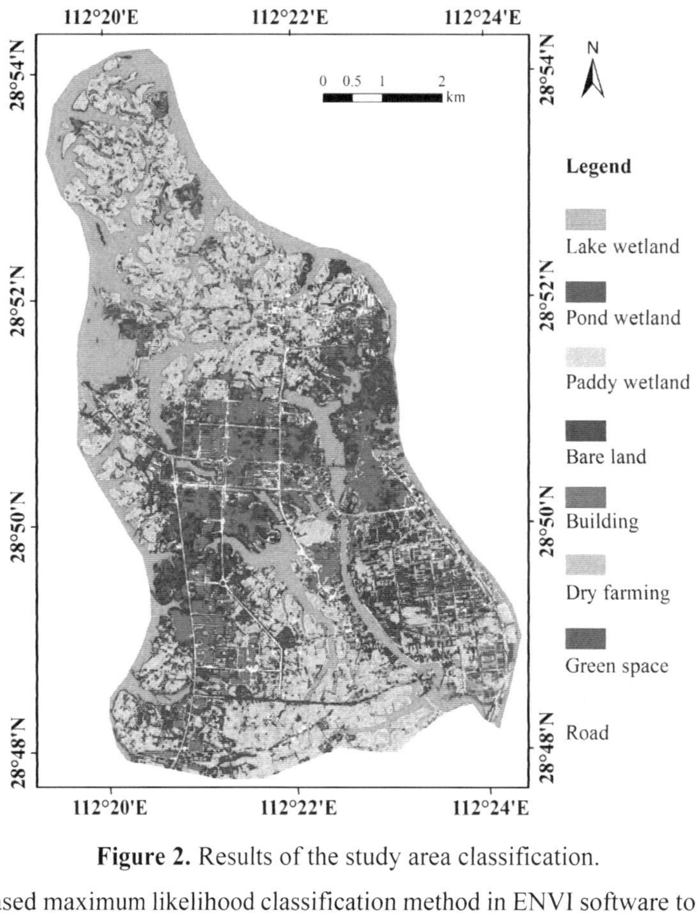

Figure 2. Results of the study area classification.

Using pixel-based maximum likelihood classification method in ENVI software to conduct a control experiment, use the 200 sample points (mentioned in section 2.2) as the training sample, and the classification features are the spectral features in Table 2, including value of 4 bands and 4 customized spectral features, the comparison of the local classification results of the two is shown in Figure 3 [12]. The overall total score accuracy of the control experiment group is 74.26%; the Kappa coefficient is

0.7092. The overall total score accuracy of the control experiment group was 74.26%; the Kappa coefficient was 0.7092. Therefore, the classification accuracy of our proposed method is much higher than that of the experimental group.

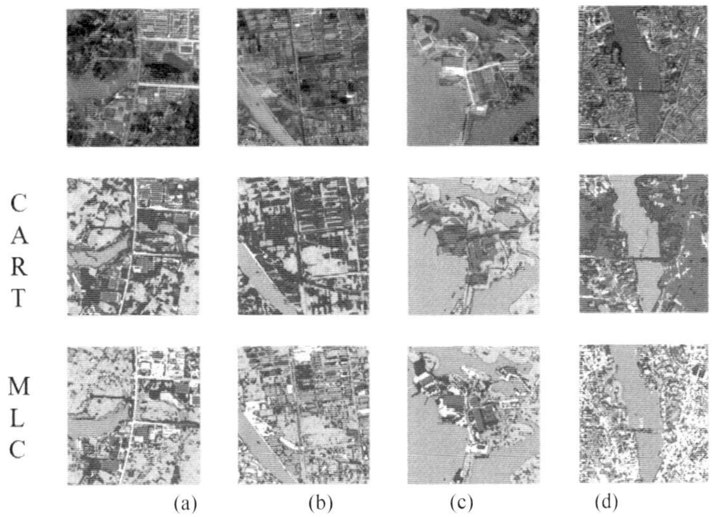

Figure 3. Comparison of partial results of the two classification methods.

5. Conclusion

From a partial point of view, the classification accuracy of the CART method in roads and building land types is significantly higher than that of the MLC method. From an overall point of view, our classification method combines the advantages of multi-features and CART two theories, and the overall classification accuracy of the classification in the experimental area reaches 85.01%; the Kappa coefficient is 0.8219, and the constructed wetland classification method is based on spatial consistency and in terms of overall classification accuracy. By adding texture features and custom spectral features, the separation of the classified samples is improved, compared with other object-oriented wetland classification methods, the proposed methods have obtained more accurate and credible classification results.

The method still has the following limitations: First, the selection process of training samples is too complicated, and certain prior knowledge is required to select the classification samples, which requires professionals to operate and spend a lot of time; Secondly, the calculation time of this method is too relatively long to be suitable for large classification areas. It can be considered to use machine learning and neural network methods to improve the selection process of training samples, increase the degree of automation of the method, and the classification process to raise computational efficiency based on the principle of parallelism.

Acknowledgements

This work was supported by the key project of Hunan Meteorological Bureau (Grant No. XQKJ18A003).

References

[1] Radeva K, Nedkov R, Dancheva A 2018 Application of remote sensing data for a wetland ecosystem services assessment in the area of Negovan village *C. Remote Sensing for Agriculture, Ecosystems, and Hydrology XX. International Society for Optics and Photonics* **10783** 107830Y

[2] Mao D, Wang Z, Du B, et al 2020 National wetland mapping in China: A new product resulting

from object-based and hierarchical classification of Landsat 8 OLI images *J. ISPRS Journal of Photogrammetry and Remote Sensing* **164** 11-25

[3] Yan F, Liu X, Chen J, et al 2017 China's wetland databases based on remote sensing technology *J. Chinese Geographical Science* **27** 374-388

[4] Kaplan G, Avdan U 2017 MAPPING AND MONITORING WETLANDS USING SENTINEL-2 SATELLITE IMAGERY *J. ISPRS Annals of Photogrammetry, Remote Sensing & Spatial Information Sciences* **2017** 4

[5] Shen G, Yang X, Jin Y, et al 2019 Remote sensing and evaluation of the wetland ecological degradation process of the Zoige Plateau Wetland in China *J. Ecological Indicators* **104** 48-58

[6] Guigues L, Cocquerez J P, Le Men H 2006 Scale-sets image analysis *J. International Journal of Computer Vision* **68** 289-317

[7] Guo M, Li J, Sheng C, et al 2017 A review of wetland remote sensing *J. Sensors* **17** 777

[8] Anaya-Acevedo J A, Escobar-Martínez J F, Massone H, et al 2017 Identification of wetland areas in the context of agricultural development using Remote Sensing and GIS *J. Dyna* **84** 186-194

[9] Hu Z, Zhang Q, Zou Q, et al 2018 Stepwise Evolution Analysis of the Region-Merging Segmentation for Scale Parameterization *J. IEEE Journal of Selected Topics in Applied Earth Observations and Remote Sensing* **11** 2461-2472

[10] Drăguţ L, Tiede D, Levick S R 2010 ESP: a tool to estimate scale parameter for multiresolution image segmentation of remotely sensed data *J. International Journal of Geographical Information Science* **24** 859-871

[11] Wang X, Wang C, Wang B, et al 2020 Spatio-temporal characteristics of land cover changes of Liangjiang new destrict in China during 2010-2018 *C. IOP Conference Series: Earth and Environmental Science. IOP Publishing* **569** 012047

[12] Berhane T M, Lane C R, Wu Q, et al 2018 Comparing pixel-and object-based approaches in effectively classifying wetland-dominated landscapes *J. Remote sensing* **10** 46

IOP Conf. Series: Earth and Environmental Science **865** (2021) 012046 doi:10.1088/1755-1315/865/1/012046

Simulation on plateau power recovery of turbocharged diesel engine matched with electronic unit pump

Yupo Ma[1], Yu Feng, Zengbin Wu, Changcheng Wang, Tongzhou Liao, Zhanyu Zhan, Zhenyu Zuo, Guoning Shi

China North Vehicle Research Institute, Beijing, 100072, China

[1]Corresponding author's e-mail: mayupo2015@163.com

Abstract. Aiming at the problem of power reduction of supercharged diesel engine equipped with electronic unit pump in plateau operation, a simulation model is established by using one-dimensional software BOOST. Under the condition that the maximum combustion pressure and exhaust temperature are basically unchanged, the variation of maximum power, optimal injection advance angle and excessive air coefficient with inlet absolute pressure at typical speed is analyzed, which has important guiding significance for the control strategy of diesel engine in plateau operation.

1. Introduction

Diesel engines are widely used in industries, agriculture and transportation. In plain areas, turbocharged diesel engines have excellent performance on the aspects of power, efficiency and fuel consumption [1, 2]. However, when it works in plateau areas, many serious problems will arise. These problems can cause deterioration of performance and increase of harmful emissions [3, 4].

The main characteristics of plateau climate environment are low atmospheric pressure, air density, air oxygen concentration and average air temperature, as shown in Table 1. When the supercharged diesel engine works in plateau area, the intake pressure decreases, and the air quality in the cylinder decreases. When the speed and fuel supply are constant, the excess air coefficient decreases, so the exhaust temperature and specific fuel consumption will increase significantly, while the maximum power and maximum combustion pressure will decrease greatly. The more obvious the deterioration of diesel engine performance occurs with the higher altitude [5].

Table 1. Parameters of atmosphere at different altitudes.

Height (m)	Atmosphere Pressure (kPa)	Atmosphere Temperature (K)	Air density ($kg.m^{-3}$)	Oxygen density ($kg.m^{-3}$)
0	101.3	288	1.1981	0.251
1000	90	282	1.1117	0.226
2000	81	275	0.9733	0.204
3000	70.1	268	0.8919	0.187
4000	61.6	261.5	0.8063	0.169

For highly intensified supercharged diesel engines, the maximum explosion pressure and exhaust temperature are the two most important limiting parameters [6]. Excessive maximum explosion pressure will increase the mechanical load of the cylinder head, while excessive exhaust temperature

IOP Conf. Series: Earth and Environmental Science **865** (2021) 012046 doi:10.1088/1755-1315/865/1/012046

will increase the thermal load of the exhaust system. In the working process of the engine, they need to be strictly controlled and cannot exceed the prescribed limits for a long time. In this study, under the condition that the maximum combustion pressure and exhaust temperature are basically unchanged, the variation laws of parameters such as maximum power, optimal injection advance angle and optimal excess air coefficient with supercharging pressure at the maximum engine speed are analyzed by one-dimensional BOOST simulation software, which has important guiding significance for the control strategy of diesel engines working at plateau areas.

2. Establishment and calibration of simulation model

2.1. Main parameters of original diesel engine

In this study, a V-type eight-cylinder supercharged diesel engine is taken as the research object, as shown in Table 2, the maximum displacement is 16L, the compression ratio is 14:1, the maximum power is 650kW (engine speed is 2500r/min), and the maximum torque is 2780N.m (engine speed is 1800r/min). The combustion chamber of the engine adopts the double swirl combustion chamber developed by Beijing Institute of Technology [7, 8], which improves the combustion speed and shortens the combustion duration to a certain extent. Delphi electronic unit pump oil supply system is adopted in the oil supply system, in which the plunger diameter is 11mm, and the oil supply cam adopts the deceleration cam profile designed by Beijing Institute of Technology, which improves the injection pressure and basically realizes the ideal constant pressure oil supply characteristics [9-11].

Table 2. The original diesel engine main parameters.

Bore diameter & Stroke (mm)	132&145
Compression ratio	14:1
Combustion chamber	double swirl
Ignition sequence	1-8-4-5-7-3-6-2
Start injection pressure (MPa)	20
Nozzle number	8
Nozzle diameter (mm)	0.27
Max torque (N.m)	2780@(1800r/min)
Max power (kW)	650@(2500r/min)

2.2. Simulation model

There is no supercharger model in the model, but the inlet pressure to be simulated is directly input at the inlet of the intake manifold, and a 200mm pipe is added to the exhaust pipe to simulate different exhaust backpressure by changing its diameter. As shown in Figure 1:

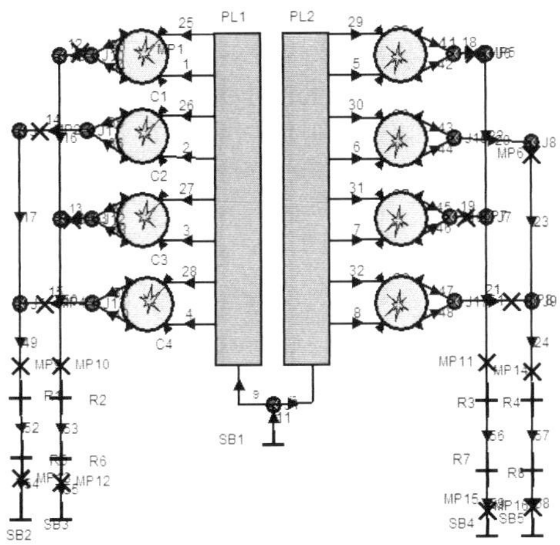

Figure 1. Simulation model of diesel engine.

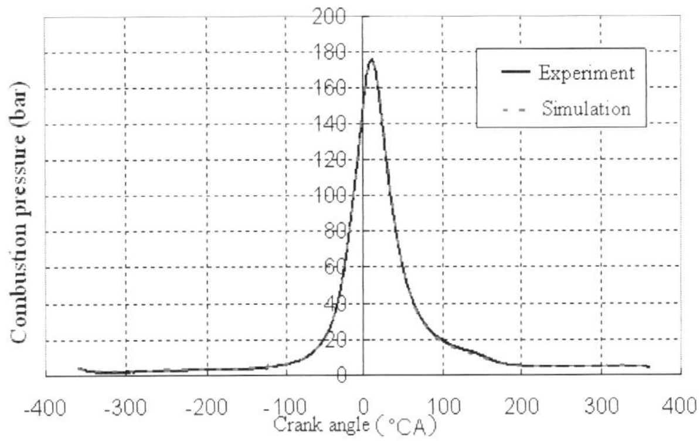

Figure 2. Comparison of experimental and simulation results of cylinder pressure curve.

2.3. *The calibration of simulation model*

When the ambient pressure is 101kPa, the above simulation model is calibrated according to the experimental data of the typical working point of the engine, which mainly includes parameters such as cylinder pressure curve, power, fuel consumption, excessive air coefficient, injection advance angle and exhaust temperature, as shown in Figure 2 and Table 3: It can be found that the simulation model has been calibrated very well and can be used to conduct the simulation research for this diesel engine.

Table 3. Comparison between experimental and simulation results of different performance parameters.

Performance parameters	Experiment results	Simulation results	Performance parameters	Experiment results	Simulation results
Power (kW)	651.4	651.7	Specific fuel consumption(g/kW.h)	232.2	232.0
Excessive air coefficient	1.53	1.53	Inlet temperature(℃)	82.5	82.7
Maximum combustion pressure(bar)	174	173.7	Inlet pressure(bar)	3.37	3.37
Exhaust temperature (℃)	748.5	748	Exhaust pressure(bar)	3.42	3.45

IOP Conf. Series: Earth and Environmental Science **865** (2021) 012046 doi:10.1088/1755-1315/865/1/012046

3. Simulation analysis of diesel engine plateau performance

By changing the intake pressure, simulate on the plateau working performance of diesel engine have been conducted. Firstly, the simulation was carried out without optimization matching, and the main problems to be solved were found out. Then, according to the limitation of maximum combustion pressure and exhaust temperature, the system optimization matching simulation analysis of all injection parameters is carried out, and the optimal injection advance angle and injection quantity under different intake pressures are determined.

3.1. Simulation analysis on plateau performance of diesel before systematical optimization

In order to directly reflect the influence of altitude on the performance of diesel engine, based on the experimental data when the engine is in plain, the diesel engine is simulated by changing the intake pressure independently while keeping the intake manifold temperature, injection advance angle and cycle fuel supply unchanged. Figures 3-7 showed the simulation results of power, specific fuel consumption, excessive air coefficient, maximum combustion pressure and exhaust temperature before turbocharger.

From the above figures, it can be seen that when the intake pressure decreases from 3.4bar to 2.4bar, the power decreases by 55kW, the specific fuel consumption increases by 22 g/(kW.h), the excessive air coefficient decreases by 0.48, the maximum combustion pressure decreases by 37bar, and the exhaust temperature before turbocharger increases by 160℃. This is because when the injection advance angle and circulating fuel supply are unchanged, the decrease of intake pressure will cause the decrease of intake air volume. As a consequence, longer combustion duration, lower combustion efficiency and more post combustion happens.

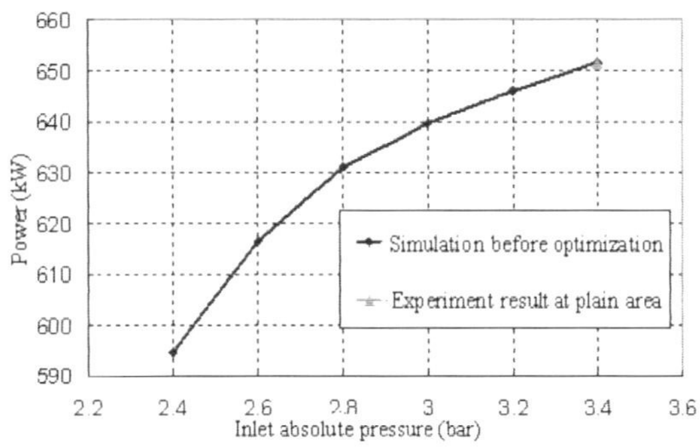

Figure 3. Simulation of power before optimization.

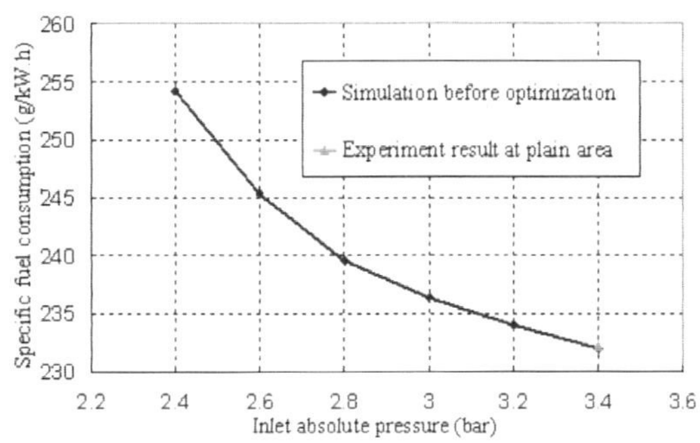

Figure 4. Simulation of SFC before optimization.

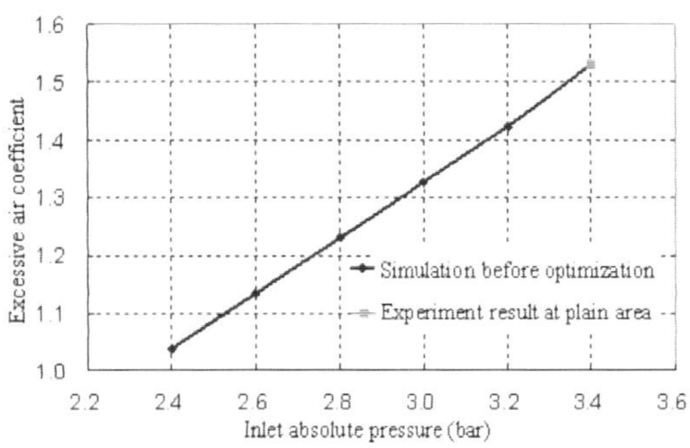

Figure 5. Simulation of EAC before optimization.

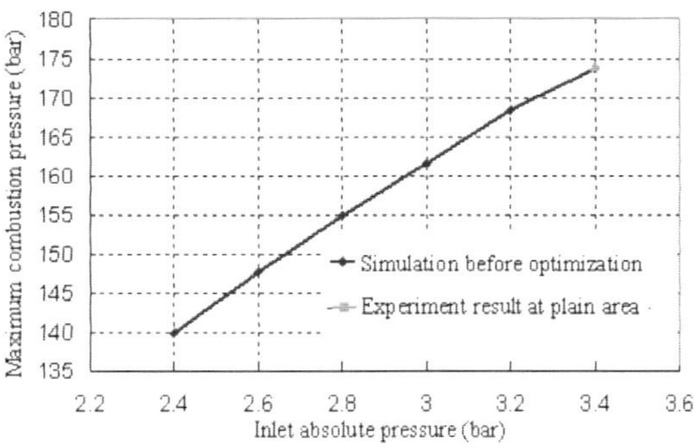

Figure 6. Simulation of MCP before optimization.

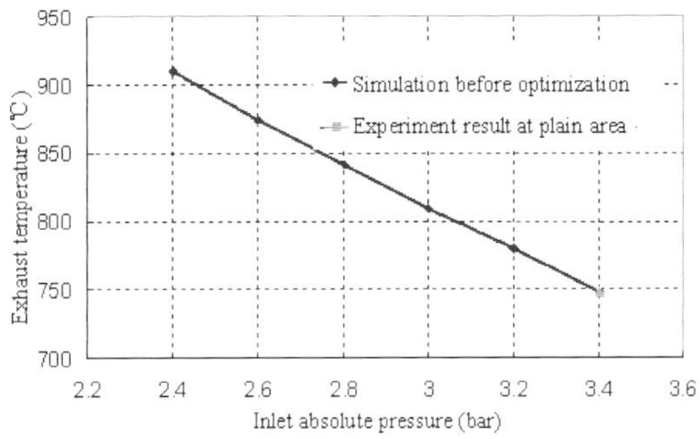

Figure 7. Simulation of ET before optimization.

It should be pointed out that with the decrease of intake pressure, the exhaust temperature before vortex rises obviously, which will greatly reduce the mechanical strength of exhaust pipe and turbocharger. In some severe cases, it will directly damage the exhaust system, resulting in huge faults and losses. Therefore, when the intake pressure decreases, the control parameters of the engine need to be adjusted accordingly to ensure its normal and safe operation. In addition, the highest combustion pressure also has a great influence on the reliability of the engine, which needs to be strictly controlled during the adjustment process of the engine.

3.2. Simulation analysis on plateau performance of diesel with systematical optimization

To sum up, when the intake pressure decreases, the performance and reliability of the engine are seriously deteriorated, so it is necessary to take corresponding measures to optimize the control of the engine. The principle of optimization in this study is to obtain the maximum power under the condition that the maximum combustion pressure and exhaust temperature before vortex do not exceed the results of plain operation. According to the experimental results in Table 3, the constraint conditions are given: maximum combustion pressure less than 175bar, exhaust temperature before turbocharger less than 750℃.

For diesel engine with electronic unit pump oil supply system, injection advance angle and fuel supply per cycle are two flexible control parameters, which can be adjusted in a wide range by adjusting the position of start-up current and the length of injection pulse width. Theoretically, the greater the injection advance angle, the greater the proportion of premixed combustion, so the higher the maximum combustion pressure and the lower the exhaust temperature. When the injection advance angle is constant, the larger the fuel supply per cycle is, the longer the combustion duration is, so the exhaust temperature increases correspondingly, but the maximum combustion pressure is basically unchanged. This is because the maximum combustion pressure is basically only related to the premixed combustion process, so the combustion pressure quickly reaches the maximum at the initial stage of the combustion process, which has nothing to do with the length of injection pulse width.

In the process of optimization, for a certain intake pressure, the relationship between the highest combustion pressure and the injection advance angle can be obtained by changing the injection advance angle, and the best injection advance angle under the intake pressure can be determined according to the constraint condition of the highest combustion pressure. Then, by changing the fuel supply per cycle, the relationship between exhaust temperature and it can be obtained.

IOP Conf. Series: Earth and Environmental Science **865** (2021) 012046 doi:10.1088/1755-1315/865/1/012046

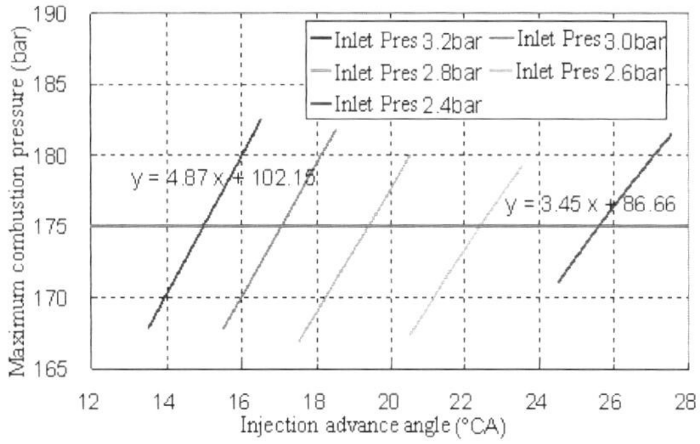

Figure 8. The variation of MCP with IAA.

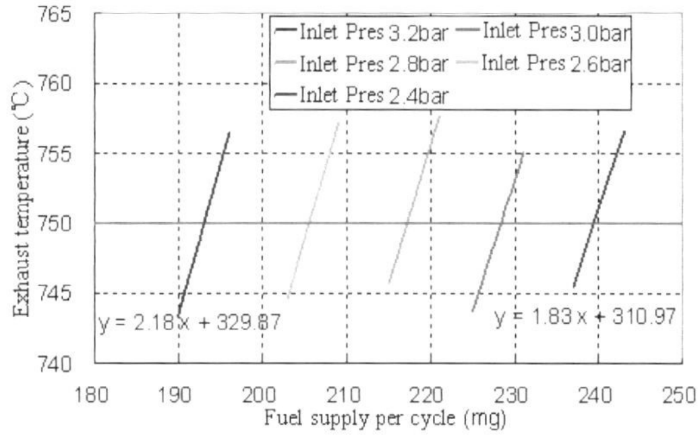

Figure 9. The variation of ET with FSPC.

Figure 8 shows the variation of the maximum combustion pressure with the injection advance angle under different intake pressures. According to the constraint value of the highest combustion pressure, the best injection advance angle under different intake air pressures can be determined. In addition, it can be found that under different intake pressures, the maximum combustion pressure basically increases linearly with the injection advance angle, but its slope changes. When the intake pressure is 3.2 bar and 2.4 bar respectively, the maximum combustion pressure increases by about 5 bar and 3.5 bar for every 1°CA increase of the injection advance angle. Therefore, the higher intake pressure, the faster the maximum combustion pressure increases with the injection advance angle.

After the optimal injection advance angle being determined, the exhaust temperature before turbocharger and its variation law can be obtained by adjusting the fuel supply per cycle, as shown in Figure 9. According to the constraint value of exhaust temperature before turbocharger, the best fuel supply per cycle can be obtained. In addition, it can be found that under different intake pressures, the exhaust temperature before turbocharger basically increases linearly with the increase of fuel supply per cycle, and the slope is basically unchanged, that is, for every 1mg increase of fuel supply per cycle, the exhaust temperature before turbocharger increases by about 2 ℃.

Figures 10-12 show the relationship between the optimal injection advance angle, the optimal fuel supply per cycle and the optimal excessive air coefficient with the intake pressure. In the process of optimization, the variation intervals of fuel supply per cycle and injection advance angle are 1mg and 0.5°CA, respectively. As shown in Figure 10, with the decrease of intake pressure, the optimal injection advance angle increases parabolically, that is, the lower the intake pressure, the faster the optimal injection advance angle increases. As shown in Figure 11 and Figure 12, with the decrease of intake pressure, the optimal fuel supply per cycle and the optimal excessive air coefficient decreases linearly. When the intake pressure decreases by 0.1bar, the fuel supply per cycle and the optimal excessive air coefficient decreases by 5.8mg and 0.015 respectively.

Figure 13 shows the relationship between optimized diesel engine performance and intake pressure. It can be found that the optimized maximum power decreases linearly with the decrease of intake pressure, and it decreases about 17kW for every 0.1bar decrease of intake pressure. The minimum specific fuel consumption is basically unchanged at first, and then increases significantly. In addition, when the intake pressure increases, the minimum specific fuel consumption tends to increase, because the injection advance angle is small at this time, which deviates from the most economical injection position.

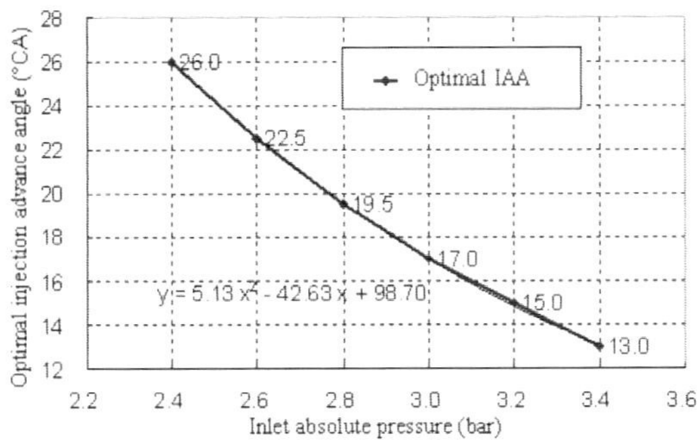

Figure 10. The variation of optimal IAA with IAP.

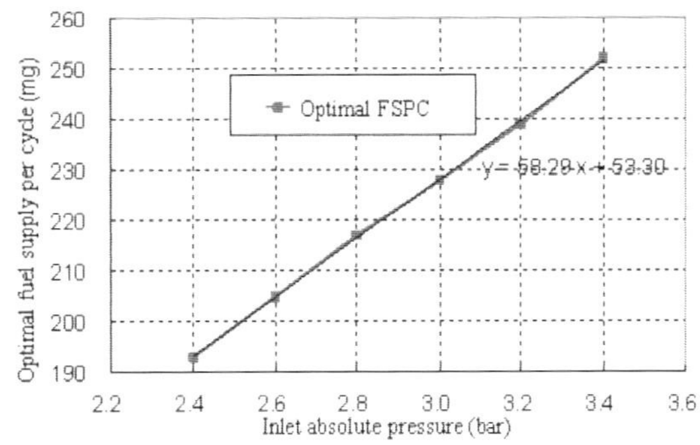

Figure 11. The variation of optimal FSCP with IAP.

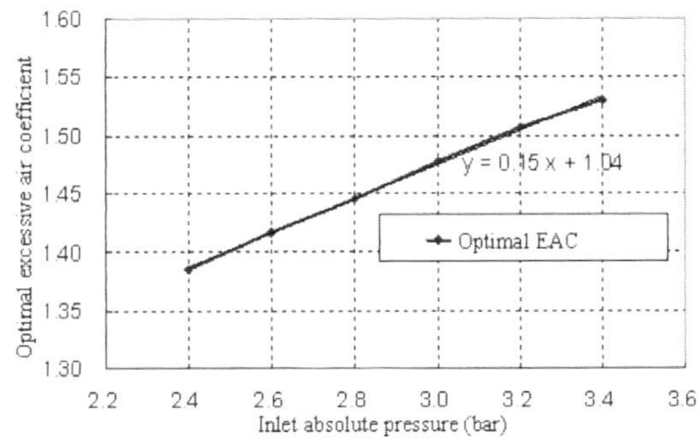

Figure 12. The variation of optimal EAC with IAP.

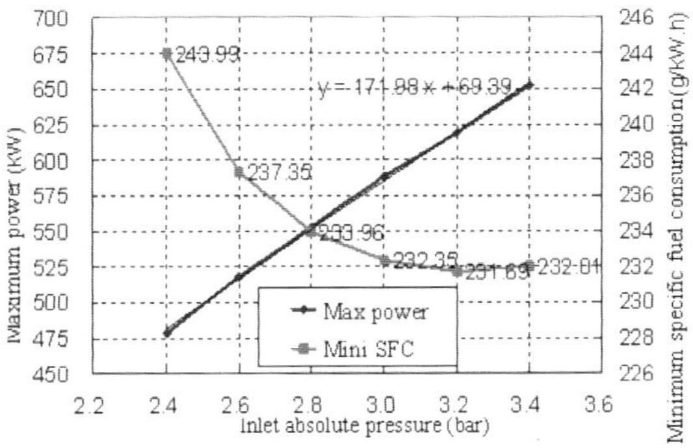

Figure 13. The maximum power and minimum SFC after optimization.

4. Conclusions

Aiming at the problem that the performance of diesel engine deteriorates significantly due to the drop of intake pressure when working in plateau environment, a simulation model of a V-type eight-cylinder supercharged diesel engine is established by using one-dimensional calculation software BOOST. The simulation model is calibrated according to the experimental data of the typical point of the engine in plain area. Firstly, the simulation calculation is carried out by changing the intake pressure alone, and then the influence of injection advance angle on the maximum combustion pressure and the influence of fuel supply per cycle on the exhaust temperature before turbocharger are analyzed respectively. Finally, under the condition that the maximum combustion pressure and exhaust temperature before turbocharger do not exceed the plain experimental results, the engine is optimally controlled, and the optimal injection advance angle, fuel supply per cycle, excessive air coefficient, the maximum power and the lowest specific fuel consumption under different intake pressures are obtained. The main conclusions are as follows:

(1) Before optimization, when the intake pressure is reduced alone, the engine performance such as power and specific fuel consumption obviously deteriorates, among which the exhaust temperature before turbocharger deteriorates most seriously, and it increases by about 16 ℃ for every 0.1bar decrease of intake pressure, which has a serious impact on the reliability of the engine.

(2) Under different intake pressures, the maximum combustion pressure increases linearly with the increase of injection advance angle, and the higher the intake pressure, the faster the maximum combustion pressure increases with injection advance angle.

(3) Under different intake pressures, with the increase of fuel supply per cycle, the exhaust temperature before turbocharger increases linearly, and its slope is basically independent of intake pressure.

(4) With the decrease of intake pressure, the optimal fuel supply per cycle, excessive air coefficient and the maximum power all decrease linearly. The optimum injection advance angle basically increases parabolically. When the intake pressure is greater than 3bar, the minimum specific fuel consumption remains basically unchanged.

References

[1] Yan J, Chou S K, Desideri U, Xia X 2014 Innovative and sustainable solutions of clean energy technologies and policies (Part I). *Appl Energy* **130**: 447-79

[2] Chou S K, Chua K J, Ho J C, Ooi C L 2004 On the study of an energy - efficient greenhouse for heating, cooling and dehumidification applications. *Appl Energy* **77**: 355-73

[3] Shi X, Li W X 2011 Simulation on Plateau Performance of Diesel Engine Matched with Two-stage Sequential Turbocharging System. *Acta Armamentarii* **32(4)**: 397-399

[4] Zhang L F, Liu S, Zhang H Y et al 2017 Study on Fuel-Air Synergistic Control of Regulable Two-Stage TurbochargedDiesel Engine Working at High Altitude. *Diesel Engine*, **39(3)**: 1-3

[5] Li H L, Shi L, Deng K Y, et al. 2015 Calculation Research on Variable Altitude Adaptability of Regulated Two-Stage Turbocharging System. *Chinese Internal Combustion Engine Engineering*, **36(3)**:1-5

[6] Shang H K, Dong C L, He J F, et al 2017 Research on Compression Ratio and Fuel Injection Advance Angle Optimization of Two-stage Turbocharged Diesel Engine. *Acta Armamentarii* **(36)**: 20-25

[7] Shang Y, Liu F S, Li X R, Wu J 2012 Research on Parametric Design Method of Combustion Chamber on Diesel Engine. *Advanced Material Research*. p.383-390

[8] Li X R, Sun Z Y, Du W 2010 Research and Development of Double Swirl Combustion System for a DI Diesel Engine. *Combustion Science and Technology*: 1029-1049

[9] Qiu T, Liu X H, Liu F S, Yu L 2008 The Study for Cam Profile Design of EUP. *Transactions of CSICE*, **26(5)**: 476-478

[10] Zhang Z 2017 Methodology to parametric design of cam profile for electronic unit pump. *Energy*, **139**: 170-172

[11] Wang P, Liu F S, Li X R 2012 Driving characteristics for electronic control unit pump fuel injection system. In: 2012 International conference on computer distributed control and intelligent environmental monitoring. *IEEE*. p.264-7

Research on ecological design of Junhe Wetland Park in Pingyi County, Linyi, Shandong Province

Laichen Zhang[1], Jianwu Ma[2, 3], Yalin Wang[2]

[1]Guanlu Construction Co., Ltd. Linyi, Shandong Province
[2]Golden Mantis School of Architecture, Soochow University, Jiangsu, China

[3]Corresponding author's e-mail: 478646560@qq.com

Abstract. With the development of cities, wetlands are gradually being eroded. How to restore its natural state and make rational use of wetland resources is the top priority for the construction of wetland parks. Riverine wetland is the most common type of wetland to build an ecological structure. Only by restoring the wet ecology at the intersection between the river and the city can human and nature live in harmony. This article takes the Junhe Wetland Park as an example to study how to design the river wetland park with ecological methods, focusing on the ecological functions of the revetment and vegetation design, and create a wetland ecological landscape.

1. Introduction

Wetland refers to a kind of geographic landform characterized by wet environment. Its main characteristics are long-term infiltration of the ground surface and perennial water-filled land. As the "kidney of the earth", wetlands have a rich variety of terrestrial and aquatic organisms. The high biodiversity has great ecological value. The classification of wetlands generally includes natural wetlands (such as swamps, lakes, rivers, etc.), created wetlands (such as fish ponds, reservoirs, paddy fields, etc.), and sea areas whose water depth does not exceed 6m at low tide [1].

Riparian wetlands are the most common type of wetland. It is a shoal that exists between rivers and land. It is affected by river hydrology and is submerged by water throughout the year or intermittently. It exchanges materials and energy between upstream and downstream. Rivers are closely related to human production and life. It also plays an important role in the development of the city. With regional construction and development, the process of urbanization continues to oppress the living space of urban river wetlands. Water conservancy facilities such as dams and sluices have split the flow of the river, and the geometry of the river channel and the rigidity of the revetment have changed the method of runoff of the river [2], the reclamation of river farmland and the discharge of domestic sewage have also greatly destroyed the original river wetland ecological environment, and the large-scale development of riverside residential areas and commercial areas also affected the water storage and flood control capacity of river wetlands, so the country has strengthened In the context of wetland construction, it is particularly important to promote the naturalization of river wetland restoration and the construction of river wetland parks through ecological design methods.

2. Current problems in river wetland design

The development of rivers and cities is symbiotic, so the habitats of river wetlands are extremely susceptible to human activities. In addition, due to the lack of understanding of wetlands in the process of urban development, there are often many problems in the early planning.

One is that over-development destroys the ecology, and the traces of over-artificialization have seriously affected the natural habitat of the wetland. At present, both banks of urban rivers are over-developed. The original natural revetments are covered by hardened cement revetments. A large number of artificial design elements have destroyed the original plant habitat, the vegetation structure is single, and the ecological environment is affected; second, it is not developed or utilized. Separate the river from the surrounding functions of the city. Although meeting ecological safety is a basic requirement in river planning and design, it is also necessary to take into account related factors such as recreation, culture, leisure, and popular science education [3], and rational development and utilization will maximize the value of wetlands; third, wetlands functional zoning of the park is unreasonable, it is necessary to delineate the boundaries of protected areas, development areas and semi-development areas. For example, the London Wetland Park is open to visitors, but for protected areas, the building of the bird watching house is integrated into nature, and all observation activities are carried out inside the building to minimize the interference to wetland organisms [4]; fourth is the construction of municipal pipe networks affect the runoff of the wetland and destroys the continuity of the river wetland water cycle. The vertical rigid revetment also impairs the water storage and flood control capacity of the wetland. Fifth, the planning and design are simple, without regional cultural characteristics. Many national wetland parks are planned and constructed as scenic spots, City parks, water parks and other modes [5]. Sixth, the impervious surface affects the normal progress of rainwater circulation. In order to deal with the increased surface runoff, drainage pipes are often used, ignoring the role of evaporation and infiltration in the wetland system [6], blocking the wetland water circulation process.

Facing the common problems in the construction of wetland parks, the following will take the Junhe Wetland Park as an example to discuss how to use ecological design to better develop and utilize wetlands, and to make education and entertainment in one. Provide visitors with a wetland activity space that combines ecology, leisure, aesthetics and popular science.

3. Current status and problems of Junhe Wetland Park

Junhe Wetland Park is located in Pingyi, Linyi City, Shandong Province. It is close to the city's main roads and has convenient transportation. The surrounding land types are mainly residential areas and hotels. It is located in the suburbs of the city and has a small flow of people. The current river ecological protection is good in portions of rural areas (Figure 1).

Main current issues: (1) River bank space is occupied by farmland. Chemical fertilizers and pesticides used in agricultural production will flow into the river along with surface runoff and rainwater to pollute the river water. At the same time, the use of plastic film cloth affects the landscape on both sides of the riverine wetland. The ecological continuity of the current wetland is broken in areas of development (Figure 2). (2) Insufficient ecological diversity. The site itself has rich elevation differences, but the current situation gives people an open visual experience. It fails to make full use of topographical changes and lacks interest. The current vegetation type is single, mainly planting crops, and there are few ground cover plants that cause the land to be exposed. The seasonal changes of plants are not obvious, the color tone is single, the wetland space is not rich enough, and the landscape effect is not obvious; the current revetment is mostly in a natural and undeveloped state, lacking hydrophilic space, poor interaction with water, and planting of the revetment is sloppy, lacking a sense of rhythm. (3) The existing buildings are brick houses, the style is not outstanding, the regional culture is not much integrated, and the local characteristics and humanistic care are lacking (Figure 3).

Figure 1. The original wetland is ecological.

Figure 2. The landscape on both sides of the river is poor.

Figure 3. The current state of the building is poor.

4. Ecological design of Junhe Wetland Park

The current ecological foundation of the river is good. The shoal islands in the center of the river have become habitats for wild birds such as egrets and wild ducks. The poplar forests planted on the gentle slope of the north bank for decades have been preserved, and the riverside landscape is designed as a natural ecological wetland. The organic combination of the ecological blue lung of the Junhe wetland and the ecological green lung of the riverside green space are integrated (Figure 4).

In terms of ecology, the emphasis is on the design of revetment ecological landscapes and plant landscapes, and makes full use of topographical changes to create different viewing effects of changing scenes, increase waterside viewing functions, and meet people's needs for waterside activities.

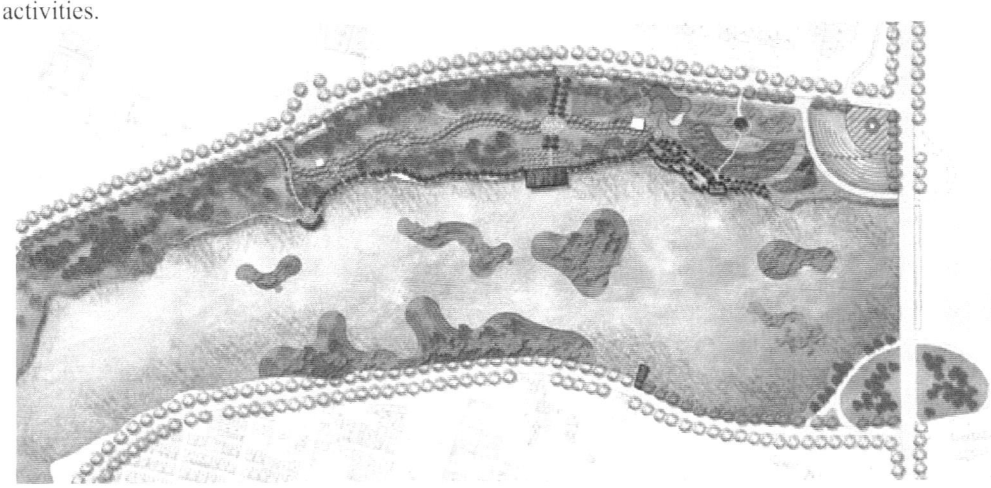

Figure 4. Design master plan of Junhe Wetland Park.

4.1. Ecological design of revetment

Revetment is a transitional zone between land and water. It plays a very important role in wetland landscape design and ecological protection. Revetment construction has developed from natural materials such as branches and rocks to hardened forms such as concrete and mortar blocks [7], river ecology has been severely damaged, the space for biological reproduction in the water-land transition zone has shrunk, and the biodiversity has been drastically reduced. Therefore, the design of the shore must be returned to the ecology to reduce the impact of excessive artificial traces. Aiming at the surrounding scenery sources of Junhe River, the designed revetments are divided into the following two types:

Natural ecological revetment: the main form of the current revetment, the gentle slope enters the water, and the main increase of rocks is used for stabilization. Natural native tree species are planted on the bank, and different aquatic, wet, and xerophytic plants are planted at different locations according to the changes in the water level of the river. Plants and revetment materials mostly use natural materials such as stones, wooden stakes, sand and stones to protect the growth and reproduction space of animals and plants (Figure 5).

Figure 5. Natural ecological revetment.

Recreation-type ecological revetment: the form of revetment is a combination of nature and man-made. It is mainly in the areas close to the waterside activities (Figure 6). There are various hydrophilic sites such as water trestle bridges, hydrophilic platforms, and river walks. Elevated wooden platforms can not only reduce the impact on wetlands, but also increase the recreational and ornamental functions, while being close to the natural scenery, they can also consolidate the soil and protect the embankment.

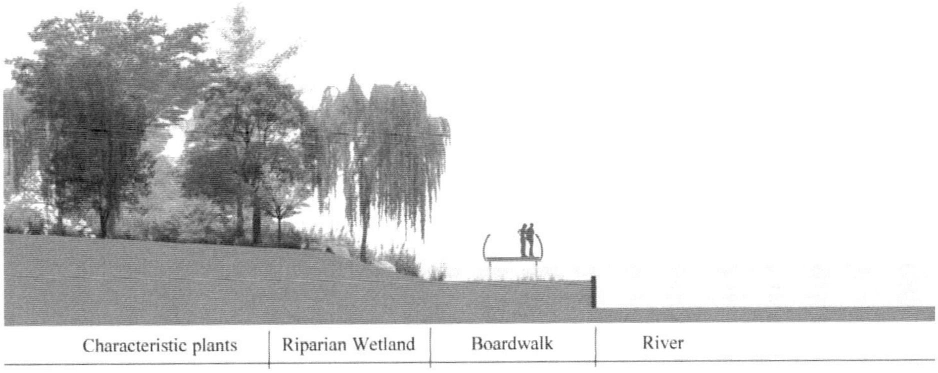

Figure 6. Recreation-type ecological revetment.

4.2. Plant ecological design

Wetland plant configuration must first be based on the preconditions of protecting and restoring the wetland ecosystem and protecting the biological characteristics of the wetland [8]. In view of the problem of the exposed surface in the Junhe Wetland, extensive planting of ground cover is a better solution. Generally referring to low plant groups, perennial herbs and low bushes, dense shrubs, etc. They are laid in open fields or suitable for various environments such as shady and humid forests and forest gaps to cover the ground [9]. Ground cover plants can cover the bare ground, maintain soil and water, and enrich the landscape.

In terrestrial areas, flowering ground cover plants can be planted to create colorful road and greenway landscapes. Woody plants are mostly flowering plants. In spring, you can plant magnolia, cherry blossoms, and begonias. In summer, there are roses, beautiful cherry blossoms, hydrangea, etc. In autumn, there are ginkgo, red maple, cosmos, etc. Winter has plum blossom, privet, boxwood, holly, etc. Highlight the seasonal changes, so that the four seasons are different and the seasons can be different. Figure 7 and Figure 8 show the colorful ground cover plants.

Figure 7. The combination of sculpture and ground cover plants.

Figure 8. The combination of architecture and ground cover plants.

A metasequoia forest can be planted in the wet zone by the river to create a water forest landscape, highlighting the light and shadow landscape effect of forest reflections on the water surface (Figure 9). There should be a variety of aquatic plants, and the planting of emerging, floating, and submerged plants reflects a sense of rhythm.

IOP Conf. Series: Earth and Environmental Science **865** (2021) 012047 doi:10.1088/1755-1315/865/1/012047

Figure 9. Effect picture of light and shadow of water forest.

5. Conclusion

The construction of a wetland park must first meet the ecological conditions. Based on ecological restoration, the wetland habitat will be expanded through the re-natural transformation of the revetment and the enrichment of plant communities, attracting the habitat of amphibians and birds, and the combination of natural revetment and vegetation to form a rich and seasonal space. The distinct plant populations ensure the natural habitat of the wetland to the greatest extent. The technical and economic indicators of Junhe Wetland Park imply that green space and water surface account for a relatively large amount, 84.17% (Table 1). Hardened pavements such as walking trails are all permeable pavement, and the hardened area is small. Wetland parks are different from ordinary parks. Without disturbing the wetland resources, its functional design tends to popularize science and protection rather than viewing, resting, and touring, and to achieve a characteristic wetland landscape.

Table 1. A table of technical and economic indicators.

		Data	
Name	Unit	Quantity	Proportion
Total land area	m^2	488850.0	100%
Building (visitor service center, public toilet)	m^2	255.0	0.05%
Permeable asphalt car road	m^2	54102.0	11.07%
Permeable concrete road	m^2	5498.0	1.13%
Permeable paving	m^2	11020.0	2.25%
Wooden walkway		2538.0	0.52%
Ecological parking lot		3975.0	0.81%
Green space	m^2	148722.0	30.42%
Water surface	m^2	262740.0	53.75%

Acknowledgements

This article was funded by the Jiangsu Forestry Science and Technology Innovation and Promotion Project "UAV High-Broad Spectrum Wetland Ecological Environment Monitoring and Wetland Vegetation Structure Optimization" (No.: LYKJ [2019]32).

References

[1] Deng X M 2015 Analysis on the planning method of the functional area of the river-type national wetland park Wetland *Science and Management* **11** 4-7
[2] Shao S W 2019 Research on ecological restoration design of urban river-type wetland park *Beijing Forestry University*
[3] Pan Y, Lv F, Lei Y 2015 Analysis on the urban river landscape planning and design method based on landscape urbanism——Taking the Changliyong Wetland Park in Zhaoqing new district as an example *China Famous City* **11** 57-63
[4] Bu J H, Wang Y 2005 The operation mode and design concept of London Wetland Park *Huazhong Architecture* **02** 103-105
[5] Wu H J, Dan X Q, Shu Y, Liu S H, Huang Y 2015 China National Wetland Park: current status, challenges and countermeasures *Wetland Science* **13** 306-314
[6] Ma J W, Acos S 2011 The art of rainwater management in American landscape design *Chinese Landscape Architecture* **27** 93-96
[7] Song T T 2020 Research on the design method of wetland ecological revetment landscape——Taking Changdang Lake National Wetland Park as an example *Popular Literature* **06** 118-119
[8] Zhang X B,Wang R, He J, Liang Z S 2016 A preliminary study on the plant landscape design of a river wetland park——Taking the Qianhu National Wetland Park in Shanxi as an example *Wetland Science and Management* **12** 8-11
[9] Flower Teaching and Research Group, Department of Landscape Architecture, Beijing Forestry University 1990 Flower Science *China Forestry Publishing House*

IOP Conf. Series: Earth and Environmental Science **865** (2021) 012048 doi:10.1088/1755-1315/865/1/012048

Analysis of soil nematode community of alpine meadow in the middle of the Tibetan Plateau

Lei Hou[1, 2, 3], **Huiying Xue**[1, 4]

[1]College of Resources & Environment, Tibet Agricultural and Animal Husbandry University, Nyingchi 860000, Tibet, China
[2]Key Laboratory of Forest Ecology in Tibet of Ministry of Education, Nyingchi 860000, Tibet, China
[3]Center for Ecological Research, Northeast Forestry University, Harbin 150040, China.

[4]Corresponding author's e-mail: 472425717@qq.com

Abstract. The alpine meadow of Tibet Plateau is extensively distributed, and the soil nematode community may exhibit certain regional differences. To understand whether there were differences among soil nematode communities, soil samples were collected in the Bangjietang alpine meadow which is in the middle of the Tibet Plateau, and the composition of soil nematode communities was determined by high-throughput sequencing technology. It was found that from 0-10 cm to 10-20 cm, the composition of soil nematode community changed significantly, the relative abundance of Chromadorea at the class level decreased while that of Enpolea increased, the relative abundance of Rhabdtida at the order level decreased while that of Dorylaimida increased. There were only eight genera of soil nematodes at the genus level, which was significantly different from the Northern Tibet region. The soil nematode communities were dominated by a bacterivorous trophic group. The soil chemical properties such as soil organic matter (SOM), available nitrogen (AN), total nitrogen (TN), total phosphorus (TP), Zn, Cu, Pb, and Cd affected the class and order level of soil nematodes, specifically, they both had negative effects on Chromadorea and Rhabiditida, and positive effects on Enoplea and Dorylaimida. In general, our results indicate that the composition of soil nematode communities in the alpine meadow which was in the middle of the Tibet Plateau was similar but different from that in Northern Tibet Plateau, specifically, they were mainly bacterivorous trophic group, but Bangjietang alpine meadow contained few genera, and the dominant nematode communities were different at class and order levels.

1. Introduction

Soil organisms have received much attention in recent years due to their ecosystem functioning, which has central roles in global biogeochemistry, influencing the fertility of soils and the exchange of gases with the atmosphere [1]. As a dominant component of soil organisms, the nematode is the most abundant and accounts for an estimated four-fifths of all animals on land [2, 3]. They feature in all major trophic levels of the soil food web [4], play crucial parts in regulating carbon and nutrient dynamics, and are a good indicator of biological activity in soils [5-7]. However, despite our understanding of soil nematode at a global pattern [8], we still lack a basic understanding of biogeographical patterns in the Tibet Plateau, where is a crucial area but rarely studied. As we knew, the alpine meadow is one of the most important and extensively distributed ecosystems in the Tibet Plateau where is very fragile and sensitive to anthropogenic disturbance [9]. Previous studies on alpine meadow soil nematodes were concentrated in the northern Tibet Plateau [10-12], but no studies were undertaken on other alpine meadows of the Tibet Plateau. Therefore, the information lacking the soil nematodes in other regions of the Tibet Plateau alpine meadow is impeding the understanding of biogeography in nematodes in the Tibet Plateau.

The traditional morphological identification method is used to study the diversity of nematodes, which is time-consuming and laborious [13]. The development of high-throughput sequencing technology provides a convenient tool for soil biological community research [14] and results are considered to have better consistency in nematodes [15]. Here, we used high-throughput sequencing methods to describe the soil nematode diversity and abundance in Bangjietang alpine meadow which is located in the middle of the Tibet Plateau to examine the difference of alpine meadow soil nematode communities in areas with different latitude in the Tibet Plateau. It provided a data foundation for the biogeographical pattern of the soil nematode communities in the Tibet Plateau alpine meadow.

2. Materials and Methods

2.1. Study site

In August 2019, we sampled soils in Bangjietang alpine meadow with a mean elevation of about 4450 m (92°26′19.9″E, 29°54′28.27″N), in the middle of the Tibet Plateau, belonged to Tibet Autonomous Region, China. The location has mean annual precipitation of 550 mm, a mean annual temperature of -3.8 ℃, and a mean annual wind speed of 3 m/s. The main plants include *Kobresia pygmaea*, *Stipa purpurea*, *Androsace gmelinii*, *Potentilla ansrina*, *Pedicularis reaupinanta*, *Lamiophlomis rotate* and *Lancea tibetica*, etc [16].

2.2. Soil samples and determination of soil chemical properties

The soil samples were collected with a cylindrical collector at the depth of 0-10 cm and 10-20 cm from the Bangjietang alpine meadow which covered an area of 15 km^2. Soil was randomly selected from 3 different locations and mixed into 3 samples for each soil layer. They were packed in polyethylene bags labeled with the sample information and were kept on ice during transportation to the laboratory. The samples were air-dried and gravel, roots, and various residues removed. We crushed part of each soil sample and passed it through a 1-mm square-aperture sieve. Soil fertility determination was undertaken as follows: total nitrogen (TN) was assayed by use of the semi-trace Kelvin method, total phosphorus (TP) was assayed by use of the sodium carbonate melting method, available phosphorus (AP) was assayed by use of the hydrochloric acid-ammonium fluoride method, available nitrogen (AN) was assayed by use of the alkali hydrolysis diffusion method, and soil organic matter (SOM) was determined using the high-temperature external hot dichromate oxidation-capacity method. Soil heavy metal determination was described as: the total amounts of Cu, Zn, Pb, and Cd were determined by flame atomic absorption spectrometry (F-AAS) with a nitric acid-hydrofluoric acid-perchloric acid digestion system, pH was measured using the potential method [17].

2.3. High-throughput sequencing

2.3.1. DNA extraction and PCR amplification. Soil DNA was extracted according to the instructions of the E.Z.N.A. SOIL kit (Omega Bio-Tek, Norcross, GA, USA). The DNA concentration and purity were determined by NanoDrop2000, and DNA extract was checked by 1% agarose gel. PCR amplification was performed using NF15' -GGTGGTGCATGGCCTTCTTAGTTAGTTTTAGTTt-3' and 18Sr2bR 5' -TACAAAGGGCAGGGACGTAAT-3' primers [14, 18].

2.3.2. Illumina Miseq sequencing and analysis. The data were analyzed on the free online platform of Majorbio I-Sanger Cloud Platform (www.i-sanger.com). Specifically, the NCBI (NT) was used as a sequencing data comparison database. The vegan package in R was used for redundancy analysis (RDA)

IOP Conf. Series: Earth and Environmental Science **865** (2021) 012048 doi:10.1088/1755-1315/865/1/012048

to ascertain the correlation between soil nematode communities and soil chemical properties. The trophic group composition map of soil nematodes was prepared by GraphPad Prism 5.

3. Results

3.1. Analysis of sequencing data and nematode community diversity

Random sampling of optimized sequences and construction of rarefaction curves by the number of sampled sequences and the number of OTUs they can represent. OTUs (Operational Taxonomic Units) are the same markers artificially assigned to a taxonomic unit (strain, species, genus, group, etc.) in phylogenetic or population genetics studies to facilitate analysis. In rarefaction curves, both the 6 soil samples attended saturation plateau (Figure 1a). The results of rarefaction curves showed that OUTs were similar in different soil layer samples. The different soil layers shared 13 OTUs of the total 30 OTUs (Figure 1b).

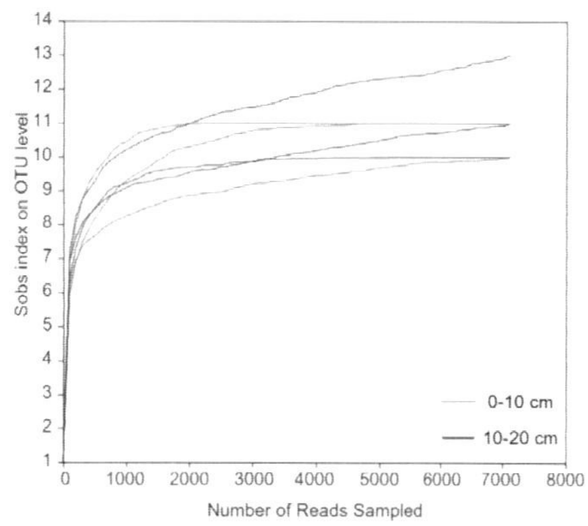

a

b

Figure 1. Diversity analysis of Bangjietang alpine meadow soil nematode sequence reads. (a) Rarefaction curves for nematode OTUs in the different soil layers, (b) OTU Venn diagram of the soil nematode communities.

3.2. Nematode community composition analysis at the class and order levels

The classification of sequences demonstrated nematode community differences in different soil layers at the class level and order levels. In this study, the sequences were clustered into 3 nematode classes (2 identified classes) and the abundance of each class varied in two soil layers. It was obvious that Chromadorea was the most dominant class in the depths of 0-10 cm, accounting for 68.31%, followed by unclassified_p_Nematoda (31.32%) and Enpolea (0.37%). In the depths of 10-20 cm, the above 3 classes all accounted for more than 25% of the total population in descending order of unclassified_p_Nematoda, Chromadorea and Enpolea, accounted for 39.65%, 32.26%, and 28.09%, respectively (Figure 2a). These results demonstrated that Chromadorea which was identified as a dominant class and was decreasing by 36.05%, whereas Enpolea was increasing by 27.72% from the surface to deep soil.

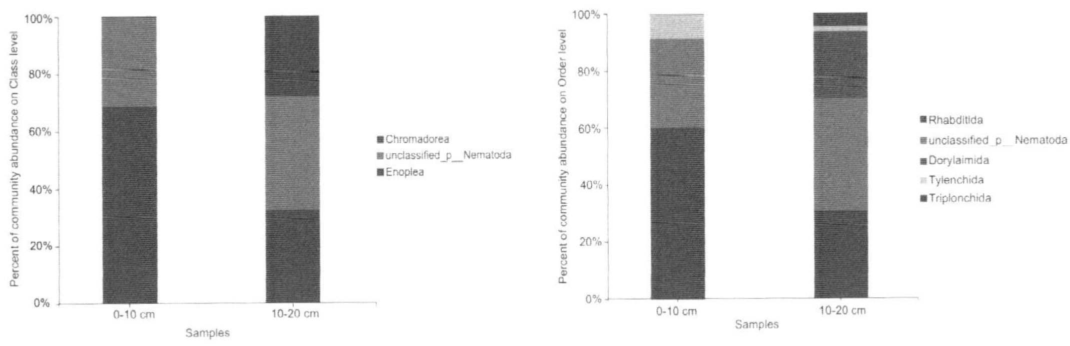

a b

Figure 2. The relative abundances of soil nematode communities at (a) the class level and (b) the order level in the different soil layers.

At the order level, the sequences were clustered into 5 nematode classes (4 identified classes), and the abundance of each order varied in the different soil layers. It was obvious that Rhabdtida was the most dominant nematode order in the depths of 0-10 cm, accounting for 60.00%, followed by unclassified_p_Nematoda and Tylenchida, accounting for 31.32% and 8.31%, respectively. In the depths of 10-20 cm, unclassified_p_Nematoda predominated for 39.65%, followed by Rhabdtida, Dorylaimida, Triplonchida and Tylenchida, accounting for 30.44%, 23.48%, 4.61% and 1.82%, respectively (Figure 2b). The results showed that Rhabdtida which was identified as a dominant order and was decreasing by 29.56%, whereas Dorylaimida was increasing by 23.48% from the surface to deep soil.

3.3. Nematode community composition analysis at the family level and genus level

The heatmap analysis at the family level revealed that nematode communities were classified into 8 families (7 identified families). The overall nematode composition of the families differed in the different soil layers. It was obvious that Cephalobidae and unclassified_p_Nematoda were the most dominant family in the depths of 0-10 cm. In the deeper soil of 10-20 cm, the dominant families were unclassified_p_Nematoda and Cephalobidae (Figure 3a). At the genus level, the heatmap analysis revealed that nematode communities were classified into 8 different genera (7 identified genera). It was obvious that Acrobeloides and unclassified_p_Nematoda were the most dominant genus in the depths of 0-10 cm and 10-20 cm (Figure 3b). The results showed that Cephalobidae and Acrobeloide which were identified were the most dominant family and genus from the surface to deep soil.

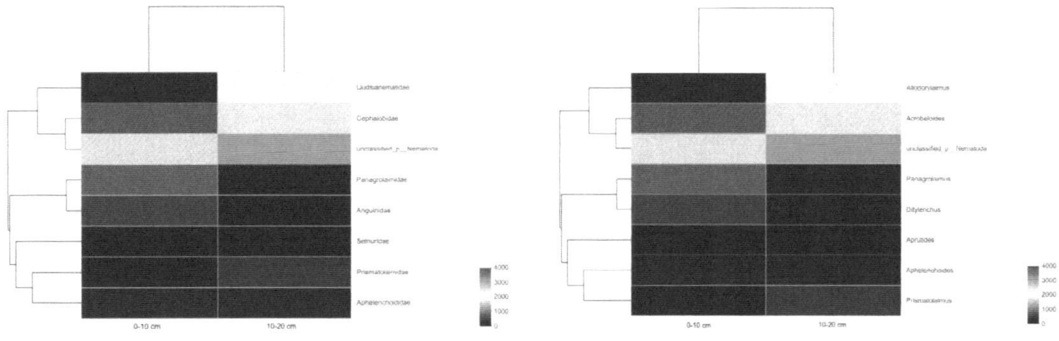

a b

Figure 3. Heatmap analysis of the soil nematode community at (a) the family and (b) the genus level in the different soil layers. The clustering indicates the similarity of certain species among different soil layers. The heat map colors from blue to red represent the relative abundance from low to high.

3.4. Trophic groups composition of soil nematode communities

To display the differences in trophic group composition in the different soil layers, soil nematodes were classified into the trophic group at the genus level. The bacterivorous (Ba) group predominated among trophic groups among the soil nematodes that were identified. Since unclassified_p_Nematoda (unknown) was not classified into a specific trophic group, the soil nematode communities were dominated by bacterivorous, 0-10 cm and 10-20 cm accounted for 60.18% and 35.11%, respectively (Figure 4). It should be noted that omnivores-predator nematodes (Op) were only accounted for 0.07% from 0-10 cm and suddenly increased to 23.39% from 10-20 cm which presented that the deeper soil had more nutrients available for omnivores-predator to use. These results indicated that the bacterivorous were the dominant trophic group in both soil layers.

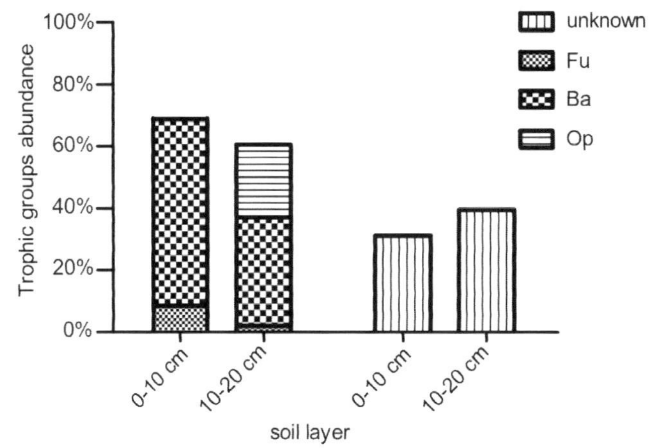

Figure 4. Composition of trophic groups of soil nematode communities in different soil layers (Fu, fungivores; Ba, bacterivores; Op, omnivores-predators; unknown, undetermined).

3.5. Relationships between nematode communities and soil chemical properties

The observed soil chemical properties, including soil fertility and heavy metals, affected soil nematode community composition. In soil fertility, the RDA diagram revealed that SOM, AN, TN, and TP were relatively more important variables than AP in explaining variations in the nematode communities composition in different soil layers at Bangjietang alpine meadow. At the class level, the 1st axis explained 86.56% and the 2nd axis for 13.44% of the variations in soil nematode communities compositions. TP was the most important influence factor to the 1st axis with a correlation coefficient of -0.90, and AP was the most important influence factor to the 2nd axis with a correlation coefficient of -0.83. For the identified nematode class, Enoplea was positively correlated with SOM, AN, TN, and TP, and no correlation with AP while Chromadorea was negatively correlated with SOM, AN, TN, and TP, and positively correlated with AP (Figure 5a). At the order level, the 1st axis explained 83.79% and the 2nd axis for 14.54% of the variations in soil nematode compositions. TP was the most important influence factor to the 1st axis with a correlation coefficient of 0.91, and AP was the most important influence factor to the 2nd axis with a correlation coefficient of - 0.80. For the identified nematode order, Dorylaimida was positively correlated with SOM, AN, TN, and TP, and no correlation with AP while Rhabditida was negatively correlated with SOM, AN, TN, and TP, and positively correlated with AP (Figure 5b).

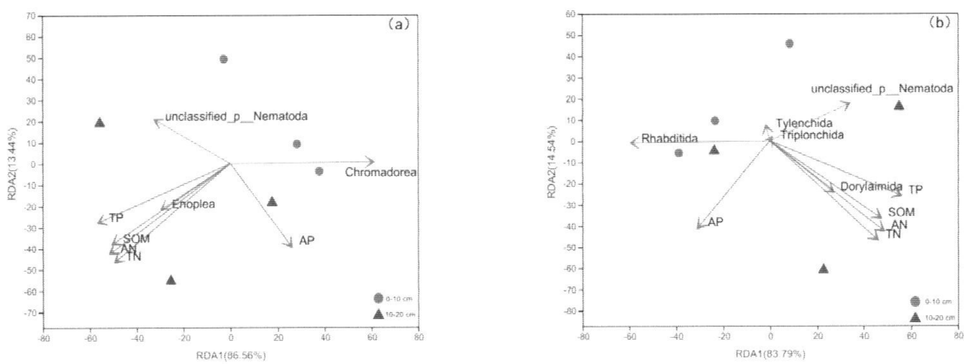

a b

Figure 5. Redundancy analysis (RDA) diagram of soil nematode communities and soil fertility properties at (a) class level and (b) order level.

In soil heavy metal, the RDA diagram revealed that Cu, Pb, Cd, and pH were relatively more important variables than Zn in explaining variations in the nematode communities composition in different soil layers at Bangjietang alpine meadow. At the class level, the 1st axis explained 86.56% and the 2nd axis for 13.44% of the variations in soil nematode compositions. Zn was the most important influence factor to the 1st axis with a correlation coefficient of - 0.89, and Cd was the most important influence factor to the 2nd axis with a correlation coefficient of - 0.99. For the identified nematode class, Enoplea was positively correlated with Zn, Cu, Pb, Cd, and no correlation with pH, while Chromadorea was negatively correlated with Zn, Cu, Pb, Cd, and positively correlated with pH (Figure 6a). At the order level, the 1st axis explained 83.79% and the 2nd axis for 14.54% of the variations in soil nematode communities compositions. Zn was the most important influence factor to the 1st axis with a correlation coefficient of 0.98, and Cd was the most important influence factor to the 2nd axis with a correlation coefficient of - 0.99. For the identified nematode order, Dorylaimida was positively correlated with Zn, Cu, Pb, Cd, and negatively correlated with pH, while Rhabditida was negatively correlated with Zn, Cu, and Pb, and positively correlated with pH (Figure 6b).

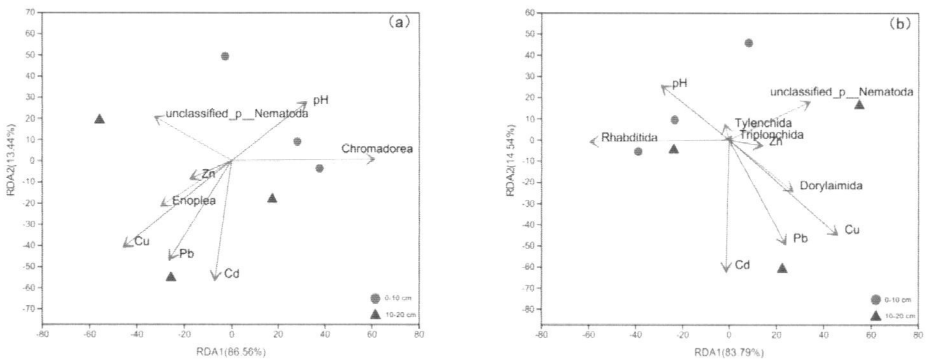

a b

Figure 6. Redundancy analysis (RDA) diagram of soil nematode communities and soil heavy metal contents at (a) class level and (b) order level.

4. Discussion

Our study mainly focused on soil nematode assemblages in the Bangjietang alpine meadow. With the different soil layers, the soil properties affected nematode fauna.

4.1. Soil nematode communities composition

Soil nematode is a type of soil organism that is sensitive to environmental factors [19] and its community distribution is affected by different ecological types [13]. In the same ecological type, due to its wide range, differences may arise in soil nematode communities within the same ecological type. Among the identified soil nematode communities at class level in the Bangjietang alpine meadow in the Tibet Plateau, Chromadorea and Enpolea accounted for 68.31% and 0.37% respectively in soils at 0-10 cm depth, and Chromadorea and Enpolea accounted for more than 25% (32.26% and 28.09%, respectively) at depths of 10-20 cm. Chromadorea decreased and Enpolea increased from depths of 0-10 cm to 10-20 cm. Previous studies in the alpine meadow in northern Tibet Plateau demonstrated that Enpolea and Chromadorea were the two classes with the highest relative abundance, and Enpolea was the dominant nematode class in the 0-25 cm soil layer [20, 21], and studies on soil nematodes in alpine meadows in other parts of the world have been conducted in areas lower than ours and have mostly used traditional identification methods [22, 23]. The difference between the two regions in Tibet Plateau indicated that there were some environmental differences, and it affected the dominant community of soil nematode. For the identified soil nematode communities at order level in the Bangjietang alpine meadow of the Tibet Plateau, Rhabdtida (60.00%) was the most dominant nematode order in the 0-10 cm soil layer, and Tylenchida only accounted for 8.31%; in the 10-20 cm soil layer, Rhabdtida was the most dominant nematode order, accounting for 30.44%, Dorylaimida accounted for 23.48%. With increasing depth from 0-10 cm to 10-20 cm, Rhabdtida decreased while Dorylaimida increased. Studies on alpine meadow soils in the northern Tibet Plateau showed that Dorylaimida was the most dominant nematode order in the 0-25 cm soil layer, followed by Tylenchida [20]. This differed from the results of the present study: the main difference lay in the dominant nematode order in the 0-10 cm soil layer. Comparing the difference between the communities of nematode in Bangjietang and that in northern Tibet Plateau, we found that the alpine meadow in Bangjietang was based on *Kobresia pygmaea*, *Stipa purpurea*, *Androsace gmelinii*, *Potentilla ansrina*, *Pedicularis reaupinanta*, *Lamiophlomis rotate*, *Lancea tibetica*, and so on [16], whereas that in the northern Tibet Plateau was based on *Kobresia Pygmaea*, *Leontopodium nanum*, *K. umilis*, *Potentilla bifurca*, *P. saundersiana*, *Anaphallis lactea*, and so on [24]. This indicated that the distribution of nematodes in different plant communities in the alpine meadow was not only different [12] but also different plant species within the similar plant community might lead to differences in dominant soil nematode communities.

Compared to the northern Tibet Plateau alpine meadow where there were 30 genera in the northern Tibet Plateau [20], there were only 8 nematode genera in the Bangjietang alpine meadow. This phenomenon may be related to the distribution of soil organic matter caused by latitude [8]. To our knowledge, the alpine meadow in northern Tibet lay at a higher latitude, so it was speculated that it had the greater soil nematode genera.

Among the identified nematodes, bacterivorous nematodes accounted for 60.18% in the 0-10 cm soil layer and 35.11% in the 10-20 cm soil layer. Therefore, the nematode trophic group in the 0-20 cm soil layer were mainly bacterivorous, and there were relatively few nematodes in other trophic groups. Our results were consistent with those reported by Xue et al in *Kobresia littledalei* community which in the northern Tibet Plateau alpine meadow [12]. Meanwhile, an increase of bacterivorous nematodes could reflect the increase of soil bacteria and soil organic matter [25], which were consistent with the previous inference.

4.2. Effects of soil chemical properties on soil nematode communities

Nematode communities have been found to be related to environmental variables such as rainfall and temperature [26-28], and our results showed that the soil nematode communities in Bangjietang alpine meadow are related to soil chemical properties. Previous studies have shown that phosphorus or nitrogen, and heavy metal contamination such as Cu, Cd, Pb, Zn have negative effects on nematode communities [29-32]. Combined with this study, phosphorus or nitrogen, and Zn, Cu, Pb, Cd may have negative effects on nematode communities of Chromadorea and Rhabditida, specifically. Besides, as the nematode can obtain resources from soil organic matter, the abundance of nematodes depends on the amount of soil organic matter. In our study, soil organic matter has a positive effect on Enoplea and Dorylaimida.

5. Conclusions

Our results suggest that the Chromadorea and Rhabdtida are the most dominant nematode fauna in Bangjietang alpine meadow where is situated in the middle of the Tibet Plateau, and the bacterivorous nematode is the dominant trophic group. The SOM, AN, TN, TP, Cu, Zn, Pb, Cd are important factors that affect the Enopla, Chromadorea, Rhabditida, and Dorylaimida in the present study, and the SOM may be the most important factor contributing to soil nematode community species in alpine meadows of the northern and central Tibetan Plateau.

Acknowledgements

All authors have read and agreed to the published version of the manuscript.
Funding: This research was funded by Key Laboratory of Forest Ecology in Tibet of Ministry of Education, grant number XZA-JYBSYS-2020-07.

References

[1] Paul E A 2015 *Soil microbiology, Ecology, and Biochemistry 4th edn Elsevier*
[2] Yeates G W 2003 Nematodes as soil indicators: functional and biodiversity aspects *Biol Fert Soils* **37** 199-210
[3] Bardgett R D, Putten W H V D 2014 Belowground biodiversity and ecosystem functioning Nature **515** 505-511
[4] Song M, Liu Y Z, Jing S S 2015 Response of soil nematodes to climate change: a review *Acta Ecologica Sinica* **35** 6857-6867
[5] Freckman D W, Caswell E P 1985 The ecology of nematodes in agroecosystems *Annu Rev Phytopathol* **23** 275-296
[6] Ferris H 2010 Contribution of Nematodes to the Structure and Function of the Soil Food Web *Journal of Nematology* **42** 63-67

[7] Neher D A 2002 Role of Nematodes in Soil Health and Their Use as Indicators *Journal of nematology* **33** 161-168
[8] Hoogen J V D, Geisen S, Routh D, Ferris H, Crowther T W 2019 Soil nematode abundance and functional group composition at a global scale *Nature* **572** 194-198
[9] Xu M H, Xue X 2013 A research on summer vegetation characteristics & short-time responses to experimental warming of alpine meadow in the Qinghai-Tibetan Plateau *Acta Ecologica Sinica* **33** 2071-2083
[10] Xue H Y, Luo D Q, Wang H Y, Qu X L 2017 Effects of free grazing or enclosure on soil nematodes in alpine meadows in North Tibet, China *Acta Pedologica Sinica* **54** 480-492
[11] Xue H Y, Luo D Q, Hu F, Li H X, Wang J S, Qu X L, Wang H Y, Yu B Z, Sun Q 2016 Effect of short-term enclosure on soil nematode communities in an alpine meadow in Northern Tibet *Acta Ecologica Sinica* **36** 6139-6148
[12] Xue H Y, Hu F, Luo D Q 2013 Effects of alpine meadow plant communities on soil nematode functional structure in Northern Tibet, China *Acta Ecologica Sinica* **33** 1482-1494
[13] Geisen S, Snoek L B, Ten Hooven F C, Duyts H, Kostenko O, Bloem J, Martens H, Quist C W, Helder J A, Van der Putten W H 2018 Integrating quantitative morphological and qualitative molecular methods to analyse soil nematode community responses to plant range expansion *Methods Ecol Evol*
[14] Xia W W, Jia Z J 2014 Comparative analysis of soil microbial communities by pyrosequencing and DGGE *Acta Microbiologica Sinica* **54** 1489-1499
[15] Du X F, Li Y B, Han X, Ahmad W, Li Q 2020 Using high-throughput sequencing quantitatively to investigate soil nematode community composition in a steppe-forest ecotone *Appl Soil Ecol* **152** 103562
[16] Wang X T, Gao Y, Wei X H, Sun L, Zhao Y H, Chen D D, Yixi C M, Miao Y J 2014 Effects of different grazing intensities on the soil seed bank of alpine meadow in the bangjietang research station of Tibet *Acta Agrestia Sinica* **22** 750-756
[17] Bao S D 2000 *Soil agrochemical analysis* China Agriculture Press
[18] Zhu X Z, Li Q, Li Y P, Han H B, Ma K P 2015 Eupatorium adenophorum invasion alters soil bacterial community and diversity *Biodiversity Science* **23** 665-672
[19] Fiscus D A, Neher D A 2002 Distinguishing sensitivity of free-living soil nematode genera to physical and chemical disturbances *Ecol Appl* **12** 565-575
[20] Xue B, Hou L, H Y Xue 2019 Research on the characteristics of soil nematode communities in alpine meadow in northern Tibet by using high-throughput sequencing *Acta Ecologica Sinica* **39** 4088-4095
[21] Hou L, Xue B, Xue H Y 2019 Study on soil nematode community of northern Tibetan alpine meadow under simulated warming condition by high-throughput sequencing method *Acta Agrestia Sinica* **27** 443-451
[22] Háněl L 2017 Soil nematodes in alpine meadows of the Tatra National Park (Slovak Republic) *Helminthologia* 54
[23] Reno M, Gmryová E, Erevková A 2020 The Effect of Soil Type and Ecosystems on the Soil Nematode and Microbial Communities *Helminthologia* **57** 129-144
[24] Wei X H, Yang P, Xie Z K, Wang Y J 2003 Distribution and utilization of Kobresia pygmaea type pastures in Naqu prefecture of Tibet *Acta Agrestia Sinica* 67-74
[25] Grffths B S, Ritz K, Wheatley R E 1994 Nematodes as indicators of enhanced microbiological activity in a Scottish organic farming system *Soil Use and Management* **10** 20-24
[26] Nielsen U N, Ayres E, Wall D H, Li G, Bardgett R D, Wu T, Garey J R 2015 Global - scale patterns of assemblage structure of soil nematodes in relation to climate and ecosystem properties *Global Ecology & Biogeography* **23** 968-978
[27] Song D, Pan K, Tariq A, Sun F, Li Z, Sun X, Zhang L, Olusanya O A, Wu X 2017 Large-scale patterns of distribution and diversity of terrestrial nematodes *Applied Soil Ecology* 161-169
[28] Thakur M P, Real I D, Cesarz S, Steinauer K, Reich P B, Hobbie S, Ciobanu M, Rich R, Worm K,

Eisenhauer N 2019 Soil microbial, nematode, and enzymatic responses to elevated CO2, N fertilization, warming, and reduced precipitation *Soil Biology and Biochemistry*

[29] Zhao J, Wang F, Li J, Zou B, Wang X, Li Z, Fu S 2014 Effects of experimental nitrogen and/or phosphorus additions on soil nematode communities in a secondary tropical forest *Soil Biology and Biochemistry*

[30] Park B Y, Lee J K, Ro H M, Kim Y H 2011 Effects of heavy metal contamination from an abandoned mine on nematode community structure as an indicator of soil ecosystem health *Applied Soil Ecology* **51** 17-24

[31] Navas A, Flores-Romero P, Sánchez-Moreno S, Camargo J A, Mcgawley E C 2010 Effects of heavy metal soil pollution on nematode communities after the aznalcollar mining spill *Nematropica* **40** 13-29

[32] Li Q, Zhong S, Li F, Lou Y 2011 Nematode community structure as bioindicator of soil heavy metal pollution along an urban?rural gradient in southern Shenyang, China *International Journal of Environment & Pollution* **45** 297-309

IOP Conf. Series: Earth and Environmental Science **865** (2021) 012049 doi:10.1088/1755-1315/865/1/012049

Influence of underground engineering on surrounding water environment in sedimentary rock distribution area and its treatment principle

Maozhong Tian[1,2,3], Lihua Zhang[1,2], Zheng Hu[1,2] and Guo Guo[1,2]

[1]China Hydropower Consulting Group Guiyang Survey and Design Institute, Geotechnical Engineering Co., Ltd, Guizhou, 550081
[2]Guiyang Engineering Corporation, Guizhou, 550081

[3]Corresponding anthor's e-mail: 1228941658@qq.com

Abstract. Sedimentary rocks are normally characterized by a stratification structure with high connectivity. According to the combinations of rock formations, they can be divided into permeable strata and impermeable strata. Due to the influence of post-stage tectonic events and the Karst phenomenon, the hydrogeological conditions are complicated. The groundwater level varies significantly in sedimentary rocks under the control of topographic variations. For underground engineering construction where the foundation is buried below groundwater level, changes may occur in the replenishment relationship between surface water bodies and groundwater runoff resulting from a large-scale excavation, which consequently makes an adverse effect on the water environment of the surrounding cities. This paper aims to highlight the impact of underground engineering projects on the water environment in Karst areas. Based on the study on characteristics of different types of sedimentary rocks in southern China, such as Karst development characteristics, water-rich characteristics of rock masses, groundwater environmental conditions in cities with a mountainous topography, and types of underground engineering, this paper analyzes the impact modes and treatment principles of projects distributed in spots, large expanse, and line under different geological conditions. The research results have been widely used in the construction of urban subways in a highly developed Karst stratum in southern China, which fully satisfies the requirements of projects.

1. Introduction

In recent years, with the rapid development of the national economy and urban construction, to improve the ground environment and make full use of land resources, large-scale urban underground projects have been continuously carried out in small and medium-sized cities across the country, consequently, the urban development has had a great impact on the surface water and groundwater environment around the projects areas, where various issues have been reported, such as surface water reserves reduction, water level decrease, or even disappear, In some cases, the decline of the groundwater level has changed the original groundwater environment, causing a series of engineering geological problems and social issues [1-3]. Thus, the risk assessment of construction projects has become a priority during the planning of new urban areas [4].

For soft rock and clastic sedimentary rocks, the rock mass itself is relatively impervious. Underground engineering construction in soft rock or clastic sedimentary rocks has little influence on the surrounding water environment. Whereas the impact is more serious in chemical sedimentary rocks, due to the development of internal bedding structure, the presence of faults and joints cutting, and the development of local karst pipelines and cavities, underground engineering construction may have a greater impact on the surrounding water environment, due to the relatively high permeability of such areas. In the practice of underground engineering construction in sedimentary rock areas, problems such as land subsidence, water logging in depression, anti-floating failure, groundwater pollution or water level decline, lake water level decline, and spring disappearance often occur in the surrounding area of underground engineering construction due to the change of surrounding water environment [5-7]. Yan Xuexin et al. [8] researched on ground subsidence caused by decompression and precipitation of foundation pits from the aspects of engineering hydrogeological characteristics, influencing factors, comprehensive zoning, and settlement estimation. Zhu Bangyan et al. [9] used InSAR and geological data to comprehensively analyze the evolution characteristics and causes of land subsidence in Hexi, Nanjing. Di Shengping [10] took Boxing County of Binzhou as the engineering background, based on the detailed hydrogeological and engineering geological data of the study area and the ground subsidence monitoring data over the years systematically analyzed the dynamic distribution of groundwater and the evolution characteristics of land subsidence in the area. Based on existing data and actual investigation and research, Su Guifen et al. [11] comparatively analyzed the changes in the cyclic evolution conditions of a karst water system in the middle section of Huaying Mountain under natural conditions and artificial activities.

At present, it is relatively rare to systematically study the impact of spotted, patchy, and linear projects underground engineering on the surrounding water environment in the sedimentary rock distribution areas, and the research on the mode and treatment principle. The change of groundwater environment may induce new engineering problems of the built or to be built projects. Zhang Yang et al. [12], Hou Guibao et al. [13], Gao Yong et al. [14] Based on the hydrogeological conditions of the linear engineering area, analyzed the main reasons for the change of groundwater aquifer structure and flow field caused by the impact of the project on the groundwater environment during the construction and operation period, and put forward the treatment measures; Zhuang Xufeng et al. [15], He Su et al. [16] analyzed the interaction between tunnel engineering and underground karst water system and the engineering problems caused by them with engineering examples.

The current research focuses mainly on understanding the impact the construction projects on groundwater dynamic in karst terrains, in order to determine practical treatment measures.

Based on summarizing the characteristics of underground engineering, geological structure, and hydrological characteristics, as well as the characteristics of water environment around the project in the sedimentary rock area, this paper analyzes the geological structure, groundwater level, runoff conditions and the relationship between the site and its surrounding areas, studies the influence degree and influence mode of the project construction on the surrounding water environment, and puts forward the treatment principles. It has a certain guiding significance to the sustainable development of the city and the safety of underground engineering construction.

2. Underground engineering characteristics

The underground projects are classified based on their characteristics; nature, scale and size, and the surrounding groundwater environment conditions; thus, the impact on underground environment varies according to the differences on the characteristics of different projects.

Under different water environment conditions, even the nature and scale of the project are the same, the impact of the construction on the surrounding water environment is also different. At present, domestic underground projects are classified according to their uses: underground traffic engineering, underground civil air defense engineering, underground national defense engineering, underground storage engineering, underground industrial engineering, underground commercial engineering, underground municipal pipeline engineering, etc. This article analyzes the relationship between underground engineering and the water environment, mainly classified and explained according to the scale of the project, the plane distribution, and the buried depth.

2.1. Project scale and plane distribution
According to the scale of the project and the plane distribution mode, it is classified as spotted project, patchy project and linear project.

Spotted underground projects: generally small in scale, mainly single underground buildings, mostly underground garages and underground air defenses.

Patchy underground projects: The scale of flaky underground engineering is relatively large, and it is mainly distributed in flaky underground buildings, such as underground garages in large residential areas, large underground shopping malls, and underground workshops. Linear underground engineering: Linear underground engineering is generally linear distribution with large extension length, mainly including urban subway engineering, urban tunnel engineering, urban comprehensive pipe gallery and underground municipal pipeline.

2.2. Buried depth of underground engineering
According to the characteristics of mountainous cities in Gui Zhou province, combined with the buried depth of underground engineering, underground engineering can be divided into three types: shallow underground engineering, middle underground engineering and deep underground engineering.

Shallow underground engineering generally refers to underground engineering buried depth less than 10m, mainly including underground garages, underground pedestrian passages, urban water, electricity, gas, communications, and other public facilities.

Middle underground engineering refers to the underground engineering buried depth between 10m and 30m, mainly including urban integrated pipe corridors, underground shopping malls, part of the subway, groundwater treatment plants, etc.

3. Geological structure and hydrological characteristics
As per definition, deep underground engineering mainly refers to underground projects with the foundation's depth is greater than 30m, such as urban underground rail transit, municipal highway tunnels, and other important underground engineering facilities.

Underground engineering construction is closely related to the surrounding rock-soil combination characteristics and geological structural development. The degree and scale of karst development also have a greater impact on the engineering geological environment in the chemical sedimentary rock distribution area. Therefore, to analyze the impact of underground engineering on the surrounding water environment, firstly, the rock and soil texture characteristics and permeability performance in the engineering area should be analyzed and demonstrated.

3.1. Lithohydrological feature
In Guizhou province, the sedimentary rock strata can be roughly divided into three types, clastic sedimentary rocks, claystone, and chemical sedimentary rocks. According to the different lithological characteristics, the water yield property and water permeability performance of various rock mass combinations are described as follows:

3.1.1. Clastic sedimentary rocks. There are mainly sandstone, quartz sandstone, siltstone, argillaceous siltstone, sandy mudstone and other formations. It is more exposed in southern and northern Guizhou, with a distribution area of about 10%-20%. It is a thin-bedded, medium to thick-bedded, and thick-bedded structure, with developed internal bedding structure, argillaceous infilling, tight binding, and good connectivity. The rock mass is relatively impervious and weakly water yield property. Rock mass permeability is mainly controlled by the degree of weathering, interlayer structural planes, faults, and joint development. Generally, the highly weathered rock masses, faults zones, and highly fractured rock masses have high water permeability. The water permeability of the moderately weathered to slightly weathered rock mass is relatively low. In water resources and hydropower engineering, the relatively impervious stratum is often used as the bottom of an impervious curtain.

IOP Conf. Series: Earth and Environmental Science **865** (2021) 012049 doi:10.1088/1755-1315/865/1/012049

3.1.2. Clay rocks. Mainly claystone, shale, mudstone, calcareous mudstone, exposed everywhere in Guizhou, with a distribution area of about 5~10%. Extremely thin-bedded, thin-bedded, and medium-thick bedded structure, with well-developed bedding and layers, infilling with the gouge, and good interlayer cementation. Since the grain diameter of the rock texture is less than 0.005mm, the rock mass is considered impermeable. Even if there are faults and joints, most of them are infilling gouge and cemented, and there are no strong water permeability problem. It can be used as a better waterproof supporting stratum in engineering construction.

3.1.3. Chemical sedimentary rocks. There are mainly carbonate rock strata, which can be divided into limestone and dolomite. The limestone category includes limestone, biological limestone, siliceous limestone, flint limestone, micrite limestone, and carbonaceous limestone, dolomite limestone, etc. Dolomite categories include dolomite, lime dolomite, and micrite dolomite. The chemical sedimentary rocks have a large distribution in Guizhou, accounting for about 70-80% of the area of Guizhou Province. The rock structure is thin-bedded, medium-thick bedded, thick-bedded, and massive-bedded, with well-developed bedding, good connectivity, and poor interlayer bonding relative to clastic rocks and clay rocks. It has a certain water permeability. It is a water-bearing formation with relatively good water yield properties. The water permeability is mainly controlled by the development of bedding, faults, fractures, and karst. The seepage and water gushing phenomena in underground engineering construction mainly occur in such stratum, which is the main object of research on the impact of the water environment around the project.

3.2. Tectonohydrogeological feature

The geological structure of foundation rock mass mainly refers to the fault and the fracture structure produced in the process of crustal movement. The scale of the geological structure, the width of the fracture zone and fault affected zone, and characteristics of the filling cement have a greater impact on the groundwater environment. So it is one of the main objects of underground engineering construction research.

3.2.1. Tectonohydrogeological feature of regional fracture. Generally, the regional fracture refers to a large-scale fault structure with an extension length more than 10km, which has the characteristics of large extension length, large stratum fault distance, large fracture zone and large range of fault affected zone, and it is easy to become a collection and runoff channel of groundwater. According the mechanical properties of the fault, it can be further divided into compressive fracture and tensile fracture. Fractures formed under general compressive stress, the width of the fracture zone and affected zone is relatively small compared to the tensile fault, and most of them are filling cemented type, its water conductivity is relatively poor. The faults formed under the action of tensile stress have relatively large fracture zone and influence zone width, poor filling and cementation properties, and relatively good water storage and hydraulic conductivity.

3.2.2. Tectonohydrogeological feature of associated (secondary) fault structures. Mainly affected by regional faults, a series of small-scale secondary faults are formed around the main fault. The secondary fault generally extension length of 1~5.0km, and is distributed in cross and parallel with the main fault. The fault distance, fracture zone and affected zone are relatively small compared to regional faults. They are also the collection and runoff channels of groundwater, but their scale is limited.

3.2.3. Hydrogeological characteristics of joints (fissures). The water yield and conductivity property of the joints (cracks) in the rock mass are mainly controlled by joints density, aperture, and degree of rock weathering. Generally, highly weathered rock masses and rock masses near fracture zones have high joint (crack) density, poor filling and cementation, large aperture, and relatively good watery and conductivity. As the weathering degree of the rock mass decreases and further away from the faulted

IOP Conf. Series: Earth and Environmental Science **865** (2021) 012049 doi:10.1088/1755-1315/865/1/012049

structure, the joints (cracks) are less developed, and most of them are filled and closed, and the watery and hydraulic conductivity property is gradually weakened.

3.3. Karst hydrogeological characteristics

In the carbonate strata distribution area, due to the interaction of soluble rock and groundwater, inside the structurally developed rock masses, karst fissure crevices, caves, karst pipelines, and karst underground rivers of various sizes are often formed, which become groundwater collection, storage, and runoff channels. The types of karst water include fissure water, karst cave water, karst pipeline water, and karst underground river. They have the characteristics of dispersion, good permeability, and large variation of water volume. It is mainly distributed and exposed in linear flow, pulse flow, and streamflow.

4. Characteristics of water environment around the project

With the continuous development of urban construction in China, people's demand for material and cultural living standards of the city is increasing. Most of the living environments should be chosen in the beautiful areas with mountains and waters, and the corresponding municipal supporting projects and ancillary facilities will be built accordingly. Therefore, there are different surface water and groundwater environments around the projects. Its environmental characteristics are the key objects of underground engineering to research the surrounding water environment.

4.1. Surface water environment characteristics

4.1.1. River system. In generally large and medium-sized cities, most of them have one or more rivers and water systems, some passing through the city, and some passing around the city. In addition to the main rivers in mountainous cities, there are 1 to 2 or more tributary water systems distributed in a branch shape, as shown in Figure 1. Some constructions, urban roads, and subways are mainly distributed along both banks of the river, ranging from 50 to 200m away from the river bank. Some urban subways even cross the riverbed to connect the two banks. The river system has the characteristics of strong fluidity, large variation of water level, and water volume. The morphological characteristics, water surface width, river bed gradient, and the variation of river water level during dry seasons and flood seaons have a greater influence on the foundation depth, design schemes, and engineering treatment measures of the surrounding underground projects, and specific analysis and demonstration are required.

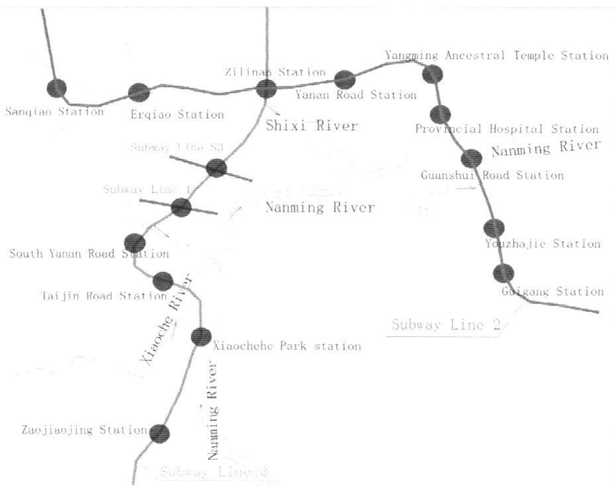

Figure 1. Subway and river distribution in the main urban area.

4.1.2. Lake landscape. In addition to the river system, there are also some lake landscapes in the city or suburban. According to the formation, the lakes can be divided into natural lakes and artificial lakes, with a distribution area ranging from 0.5 to 10.0 km^2, as shown in Figure 2. In general, natural lakes are relatively large while artificial lakes are small. Both of them have the characteristics of a certain surface water (or groundwater) recharge source, gentle water surface, and relatively stable water level. At present, in newly developed urban construction, artificial lakes are mostly used for landscape engineering. Around the lake, most of the urban infrastructures such as housing construction projects, urban road projects, underground rail transit, etc, which have been built or to be built, and the later underground engineering construction has a greater impact on the water environment.

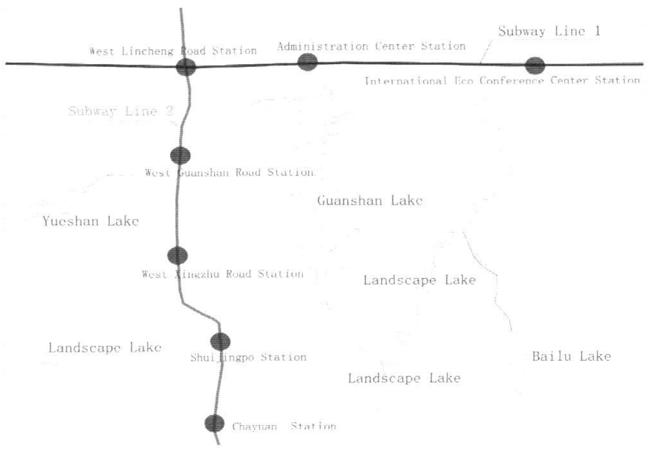

Figure 2. Distribution of subway, lake and artificial lake in an urban area.

4.1.3. Ancient Wells and Springs. Ancient wells and spring groups are generally distributed in groundwater runoff, discharge datum nearby rivers, and lakes. They are exposed individually or in groups. General plains with shallow depth, hilly cities buried deep, for a long time, it has been one of the water sources for urban residents' domestic water, with a long history, and most of them as a focus on the protection of urban landscape. In recent years, due to the construction of large-scale urban buildings, deep foundation pits and a large number of pile foundation projects, it may change the environmental conditions of groundwater runoff and cause well springs to flow interruption. This is the work content that should be studied in the process of underground engineering construction.

4.1.4. Pumping Well Point. The pumping well point mainly refers to the location of the well hole where the groundwater source is collected. At present, most cities have canceled the groundwater supply method and switched to tap water supply. However, some factories and enterprises still use groundwater for cooling cycles, geothermal central air-conditioning, environmental greening, and other water use methods. Most of the wells are used to pump water, and the depths of the wells are mostly between 80 and 150m, mainly shallow groundwater. In urban planning and construction, especially the construction of linear underground works (such as subway), if the pumping well is close to the site, it may lead to the water level decline of the pumping well, or even the depletion of the pumping well, which should be investigated and studied.

4.2. Groundwater environment characteristics
In engineering construction, different groundwater environments have a greater impact on underground engineering construction, and there are many types and ways of impact. This paper mainly analyzes the free phreatic environment, confined water environment, karst underground river pipeline and structured

IOP Conf. Series: Earth and Environmental Science **865** (2021) 012049 doi:10.1088/1755-1315/865/1/012049

water system, which have a great influence on the construction and operation of underground engineering.

4.2.1. Free phreatic environment. Free phreatic environment mainly refers to the phreatic groundwater environment with the free water surface, relatively stable groundwater level, or relatively small hydraulic gradient in the water-bearing rock and soil body with certain permeability. Overall analysis, the groundwater with the characteristics of the free phreatic surface, as shown in Figure 3, its groundwater level and hydraulic gradient are mainly controlled by topography, rainfall, rock, and soil permeability. The water level is relatively stable in the sand and gravel layer, the permeable soil layer, and the weathered rock mass with a range of 3.0~5.0m and a hydraulic gradient of 1~3% in the dry flood season. In mountainous terrain and carbonate rock distribution area, the variation of groundwater level is relatively large, during the dry season and flood season, the variation of water level is 3.0~8.0m, and the hydraulic gradient is generally 3~7%, local area can reach 10~12%. Underground engineering construction is carried out in the free phreatic environment, the first thing to ascertain the groundwater confluence area, groundwater recharge, runoff path, water level, and permeability characteristics of rock mass and soil in the investigation and design stage.

4.2.2. Confined water environment. From the perspective of spatial analysis, the confined water environment is mainly distributed in the aquifer between the upper and lower aquiclude. The aquifer has a certain thickness, high at both ends, low in the middle, groundwater in a permeable body with certain pressure-bearing properties, free recharge and drainage areas at both ends, and a pressure-bearing area in the middle. Its artesian head is mainly controlled by the water level elevation of the free recharge and drainage area and the buried depth of the aquifer, the water volume is determined by the thickness of the aquifer and the water volume in the recharge area, as shown in Figure 3. The confined water environment is mainly distributed in the aquifer between the upper and lower water-resistant strataIn the mountainous sedimentary rock distribution area, such groundwater environment is easily formed in the basin formed by synclinal structure. For the urban tunnel engineering, subway engineering, when passing through such groundwater environment, it is easy to appear water gushing, water permeation phenomenon, the treatment engineering is very difficult, so it should be avoided as far as possible.

Figure 3. Location relationship between underground engineering and phreatic and confined groundwater.

4.2.3. Karst underground river, Karst pipeline and structural water environment. (1) Karst underground river and Karst pipeline

In the distribution area of soluble rock strata, due to the dissolution and erosion of groundwater, underground karst channels of various sizes are often formed in the soluble rock, and underground karst rivers are formed near the base level of groundwater drainage, with a length of several hundred to several thousand meters, the burial depth ranges from 10m to 100 m, and is a drainage channel for groundwater runoff. In underground tunnel engineering, foundation pit engineering, and bored pile foundation construction, the phenomenon of interception and blockage may occur, which will change groundwater runoff path and result in the reduction or disappearance of groundwater users and spring lake landscape downstream.

(2) Tectonic water environment

The tectonic water environment mainly refers to the groundwater system in the fracture zone. Most of the regional fault structures that formed under the action of tensile stress often become groundwater storage and runoff channels due to the relatively large length and width of the fracture zone. The main features of the tectonic water environment are the large confluence area, sufficient recharge, and abundant groundwater reserves. Large structural springs are often formed in the drainage area. Underground engineering should focus on its construction and ecological environmental impact during the planning and design stage.

5. The impact of underground engineering on the surrounding water environment

5.1. Impact on surface water

When the underground project is located near or below the surface water, the possible impacts during the construction process are mainly manifested in two aspects: one is the surface water backflow to pit; second, the water level of surface water decline and the water volume decreases. The details are as follows:

5.1.1. Surface water backflow. For urban deep foundation pit projects and subway line projects located near-surface water and below the water surface elevation, the buried depth of the foundations is mostly between 10 and 60m. Due to the loose structure of the subsurface soil and the influence of weathering, faults, fissures, and karst development on the underlying shallow rock mass, most of the rock mass within the depth of 0~40m has the characteristics of strong water permeability. In the process of foundation pit and tunnel excavation, surface water is easy to be irrigated backwards, resulting in water gushing and water permeation, which often happens in the process of engineering construction. For example, subway lines 1, 2, and 3 in a city of southwestern China all cross the Nanming River and tributary system when passing through the old city. Water and mud gushing occurred many times during the construction phase. A large amount of surface water backflow into the pit may cause the water level of the surface water to descend or dry up. At the same time, it has a great impact on the construction safety of surface landscape engineering and underground engineering. So it is necessary to carry out special waterproof curtains, anti-seepage treatment before underground engineering construction. Under normal circumstances, as long as the proper treatment measures are carried out during the construction period, the later period will have little impact on the surface water, landscape projects with small surface water bodies may cause lake water depletion.

5.1.2. Water level descend, water volume decreases. When the underground engineering and the rock and soil near the surface water are medium or weakly permeable, there may not be obvious surface water inversion during the construction process, but due to the impact of foundation pit and tunnel construction drainage, the hydraulic gradient increases and the seepage volume increases. When the local surface water supply is less than the seepage volume, the water storage capacity of surface lakes and artificial landscapes will decrease, and the water level will descend.

5.1.3. Wells and spring setting off. The influence of underground engineering on the ancient Wells, springs, and Wells (holes) will lead to water quantity reduction and water level descend, and even cause the phenomenon of well and spring setting off.

For example, in a county in western Guizhou, there are 108 springs, known as the "city of hundred springs" reputation. Among them, Huilong Pool is the first of the hundred springs, the "century-old lake". According to the survey, the daily normal water inflow is 20,000 to 30,000m3. After 2008, due to the basement excavation and pile foundation construction of the surrounding and upstream high-rise buildings, the water volume in 2009 gradually decreased. After 2011, the water flow was completely setting off, and the water volume of other springs water flow also reduced or setting off. After investigation, it is considered that it can not be recovered, which indicates that underground engineering has a great impact on the surrounding water environment.

5.2. Impact on the groundwater environment
In urban architecture, underground engineering and groundwater are an interactive community. The runoff conditions and type characteristics of groundwater are different from the scale and location of underground engineering, so there are great differences in the impact on groundwater environment, which need to be analyzed and studied one by one.

5.2.1. Impact of spotted engineering. Spotted engineering mainly refers to small-scale single-type underground structures, such as underground shopping malls, underground garages, underground civil air defense projects, etc, generally buried depth is 10-20m. From the perspective of the geological structure, most of them are located in the soil or shallow rock mass. According to the different characteristics of the groundwater environment, the influence methods are as follows:

(1) Impact on the phreatic environment

Underground and foundation pit excavations are carried out in permeable soils and weathered rock mass. Affected by the drainage of the foundation pit, there is a problem that the groundwater level of the surrounding foundation may descend during the construction process. For landscape projects that use groundwater runoff recharge sources, the recharge may be reduced or cut off during the construction period. However, due to the small scale of the project and the limited impact, most of them can be gradually restored after the completion of the construction, and the overall impact is not significant. As shown in Figure 4.

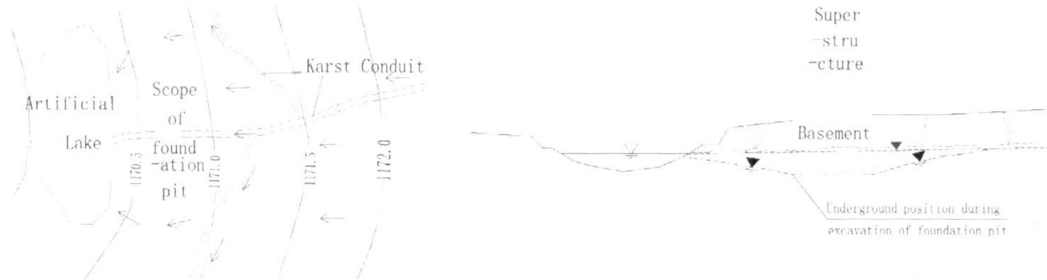

Figure 4. Schematic diagram of influence relationship between underground engineering and groundwater.

(2) Impact on the confined water environment

When the underground project debunks the upper water-impermeable formation into the confined water-bearing formation, it creates artificial gaps in the enclosed water environment. A large amount of water gushing may happen during the excavation of the foundation pit, and there will be a decrease in water volume or a cut-off phenomenon in the pressured drainage area, which will have a relatively large impact. If the anti-seepage project is handled properly during the construction period, the confined water environment may be recovered or partially recovered in the later period, but the treatment project is quite difficult, as shown in Figure 5.

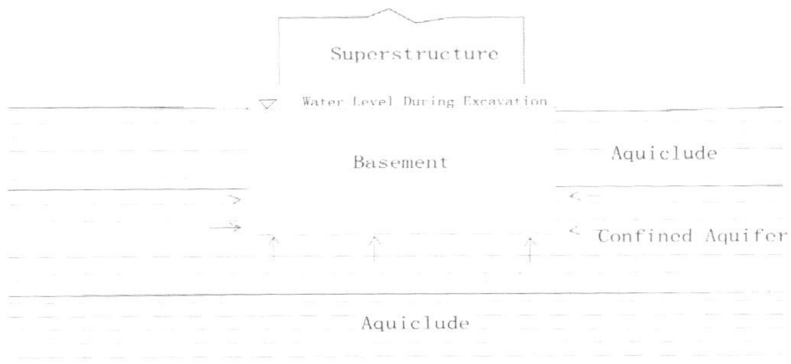

Figure 5. Schematic diagram of influence relation of foundation pit engineering to confined groundwater.

(3) Impact on Karst pipeline water and structural zone water

In solvable strata areas, for shallow-buried water-passing karst pipelines and fractured structures near the water drainage datum, the water-passing sections are mostly small. Even in small-scale single projects, karst pipelines may be cut off and blocked during the construction of foundation pits and pile foundations. Due to the multi-directional characteristics of the karst pipeline, the cut-off karst pipeline water will have runoff diversion, resulting in changes in the karst pipeline water runoff conditions, which have a greater impact on the downstream landscape engineering that requires such groundwater recharge. Generally, it is difficult to recover, so it can only be effectively avoided by taking avoidance measures in the early stage or conducting special design treatment during the construction period, as shown in Figure 4.

5.2.2. The impact of patchy engineering. In large-scale construction sites and patchy underground projects such as large residential areas, underground shopping malls, and underground workshops where large-scale are excavated, when the buried depth of underground structures is below the groundwater level, large-scale excavation of foundation pits has a great impact on the surrounding water environment (as shown in Figure 6), which is mainly reflected in the changes of groundwater level and changes in groundwater runoff conditions. The degree of impact is analyzed as follows:

Figure 6. Water gushing in large foundation pits.

(1) Changes in groundwater level

In sedimentary rock areas, when the buried depth of the underground project is below the free water table, the large area of foundation pit excavation and drainage treatment may cause the surrounding groundwater level to descend in a large area, which has a greater impact.

After the completion of the project, the groundwater level on the upstream surface (or the waterfront surface) can be recovered to its original state or slightly increased, the groundwater level in some areas can rise to 2~3.0m, and the downstream surface cannot be restored basically, and there is a significant water head. For the confined groundwater environment, after the local project has debunked the relative aquifer, the water environment has been damaged or greatly changed, and the water level in the drainage area has generally descended, and the conditions for recovery are generally not available. For shallow-buried karst pipelines, it cannot be recovered after the water is cut off.

(2) Changing groundwater runoff conditions

Large-scale underground projects have greatly changed the underground environment. In addition to the large changes in the groundwater level, the runoff conditions of the groundwater will also change significantly. In sedimentary rock formations of the aquifer, for phreatic groundwater, karst pipeline water, and structural zone water will flow radially along both sides of the surface of the underground structure, while confined groundwater may change from subsurface to surface flow, it is discharged in the form of a rising spring on the construction site.

5.2.3. The impact of linear engineering. For linear projects such as urban subways and tunnels, due to their large extension length and good penetration, they may become groundwater collection and drainage channels during the construction and operation stages, which have a greater impact on the surrounding water environment of the project. This section mainly analyzes and explains the impact of changes in groundwater environmental conditions on the surrounding water environment through the relationship between groundwater runoff conditions and the location of underground projects.

(1) Influence of vertical runoff arrangement

When the extension direction of the underground line project is perpendicular or at a large angle to the direction of the groundwater runoff, and it is located on the upstream surface of the surface water, if the groundwater level is lower than the bottom elevation of the underground project, it will generally not affect the groundwater runoff conditions and make up for the surface water, it does not affect the drainage relationship of surface water and the groundwater flow.

For example, when the buried depth of the project is below the groundwater level or surface water level, the greater impact may be occurred, mainly in two aspects, temporary influence and permanent influence, as shown in Figures 7, 8 and 9.

Temporary impact mainly refers to the temporary change in the groundwater conditions during the construction period, the phenomenon of groundwater level decline, adjacent surface water intrusion, and draining, which occurred during the construction period, will gradually or partially recover during the operation period. It has little impact on the groundwater runoff conditions and the relationship between surface water recharge and drainage.

Permanent impact: Because the runoff channel is cut off, the recharge relationship of groundwater and surface water is changed. Even in the operation period after the completion of the project, the groundwater level is difficult to restore, which has a greater impact on the surface water landscape engineering that relies on the groundwater recharge source. From the time effect analysis, most of the impacts are permanent.

IOP Conf. Series: Earth and Environmental Science **865** (2021) 012049 doi:10.1088/1755-1315/865/1/012049

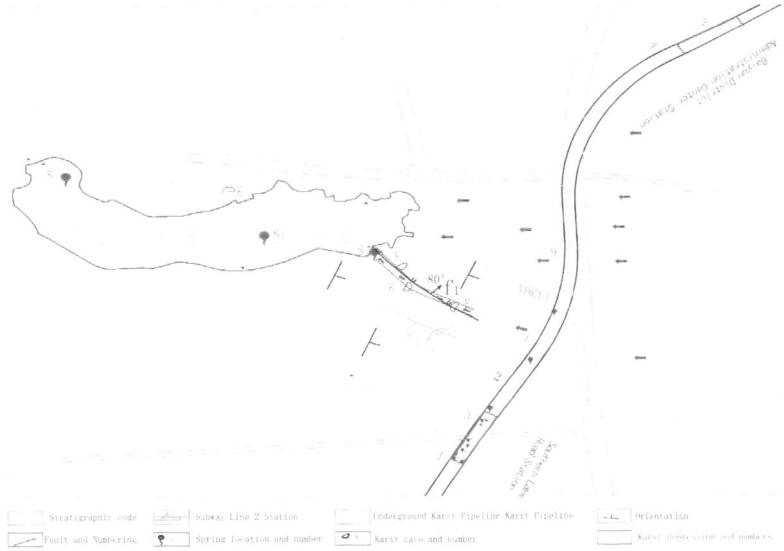

Figure 7. Groundwater runoff relationship map of subway Line 2 in a city.

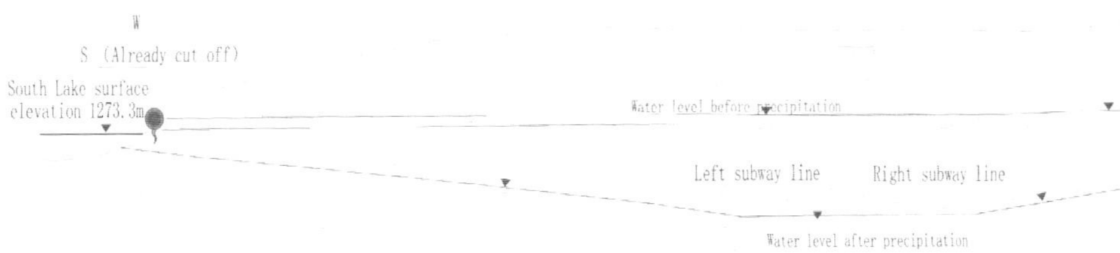

Figure 8. Profile of groundwater runoff variation during subway construction.

Figure 9. Water gushing occurs during underground excavation of linear subway project.

(2) Impact of parallel runoff arrangement

When the linear underground project is arranged in the same direction as the underground runoff and located on both sides of the surface water or below the groundwater level, its impact is mainly in the construction and excavation stage. Underground line projects tend to become groundwater runoff drainage channels, causing the groundwater level around the project to descend, and surface water may appear the surface water flow backward. Generally, since the underground runoff conditions have not been changed, the groundwater level can be restored after the completion of the project, and the impact on the surface water is also small. However, it is important to pay attention to the consolidation and grouting of the surrounding rock of the tunnel during the construction phase. Otherwise, the loose surrounding rock may become an underground seepage channel, affecting the recovery of groundwater l and surface water.

5.2.4. The impact of the Impervious grouting project. For the groundwater environment, seepage control is an engineering treatment method that artificially changes the conditions of groundwater runoff. The main purpose is to prevent the surface water from flow backward and groundwater leakage. The impervious curtain guides the discharge of groundwater in the planned direction to protect the safety of underground projects during the construction period and the impact of water seepage during the operation period. It is generally used in water conservancy and hydropower projects and some underground projects that require high waterproofing.

Although the installation of impervious curtain project effectively protects the seepage problem of underground projects, natural landscapes such as artificial lakes, ancient wells, and springs that need to be recharged by groundwater downstream may cause permanent cut-off and dry up, and have a greater social impact. Generally, the impervious curtain should be used with caution in urban scenic areas and groundwater environmental protection areas.

6. Principle of engineering treatment

In the natural water environment, surface water and groundwater are a dynamic balance of mutual circulation. The construction of underground projects will change this balance, which will adversely affect the supply and demand balance of the surrounding water environment, especially for urban water environment landscape projects. Some impacts are temporary, and some impacts may be permanent. How to reduce the adverse impact on the surrounding water environment during the construction of the project is a problem that the project builders have been discussing and thinking about. The first principle of project management is to protect the safety of the underground project during construction and operation. The second is to protect the surrounding water environment and landscape of the project from being greatly affected by the construction of the project. The above chapters have analyzed and demonstrated the geological structure, groundwater types, environmental characteristics, and impact relationships of the sedimentary area. This section mainly explains the principles of surface water and groundwater environmental treatment and protection.

6.1. Treatment of surface water flow backward

6.1.1. Treatment of adjacent surface water. (1) Underground linear project

Underground municipal road tunnels and subway projects are generally divided into left and right sections from the cross-section, with a distance ranging from 5.0 to 20 m. The diameter of a single tunnel is mostly between 5.0 and 12.0m, there are also a few slightly larger sections. When the line is located near the surface water and is below the normal water level, the surface water flow backward problem may occur during the tunnel construction process, and special engineering treatment measures need to be carried out.

a. Treatment principles

The principles of treatment are as follows: ① Prevent surface water (such as lakes, artificial water landscapes) from dry up due to a large amount of surface water flow backward to the underground tunnel;

② Protect the safety of underground engineering construction operations; ③ Does not affect the recharge and discharge of groundwater in the later period.

b. Treatment measure

① Set up impermeable curtain: The impermeable curtain is always set between the project and the surface water, within the range of 5.0~10m from the surface water boundary. According to the water permeability of the rock mass, 2~3 rows of curtain grouting holes can be arranged in the form of a plum-flower pattern. The upper limit of the curtain is not higher than the construction surface water elevation. The lower limit can be determined according to the water permeability of the rock mass, generally, it is less than 3Lu or the permeability coefficient is less than 3×10^{-5}cm/s, or 4.5~6.0m below the tunnel floor. The length range is limited to the possible influence boundary and can extend along the water boundary, as shown in Figure 10. For the tunnel facing the surface of the groundwater, the circular grouting method can be used to prevent seepage according to the seepage amount.

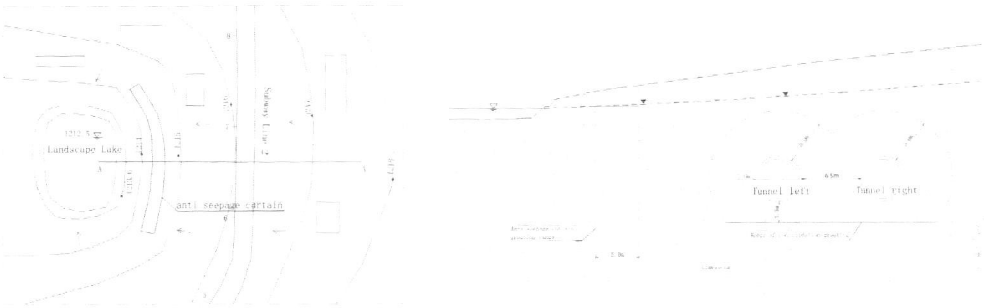

Figure 10. Section of groundwater runoff variation during the subway construction process.

② Tunnel anti-seepage consolidation grouting: This method is carried out in the tunnel. As the construction progresses gradually, it can be divided into two steps according to the strength of the rock mass permeability: (a) For the surrounding rock of tunnels of type IV and V with a permeability coefficient of 1.0～10m/d, In the process of construction, 3~5 pilot water exploration holes of 30~40m depth can be arranged along the axis of the hole or on the side near-surface water at an Angle of 5~10°. Find out the possible water gushing problem in the forward direction. If the amount of water is large, advanced water plugging and consolidation grouting should be carried out. The consolidation grouting holes are arranged radially along the axis on the support surface and the hole depth is 35~40m. As shown in Figure 11, the grouting is segmented from the inside to the outside. If there is a large amount of water inflow, the karst pipeline can cut off by a mold bag or a cable bag and then grouting and consolidated. After the preliminary consolidation is completed, the water-stop grouting is visualized during the tunneling process. The size of the seepage volume determines further treatment measures. If the seepage volume is large, the surrounding rock of the annular section of the tunnel can be grouted twice every 10m. The depth is generally controlled in the range of 6~8m. The number of holes can be appropriately dense on the side close to the surface water. The grouting depth can be a large value. (b) For Type II and III surrounding rocks with permeability coefficient of 0.01~0.1m/d, if the seepage volume is small, the general equipment can be drained without treatment. For the class III and IV rock mass with slightly larger water seepage, annular consolidation grouting can be carried out for every 10m section of the vertical tunnel axis during the construction period, and the grouting holes can be properly encrypted in the part with large seepage. After the grouting treatment, the permeability coefficient of the consolidated body through the water pressure test is generally 0.02~0.05 m/d.

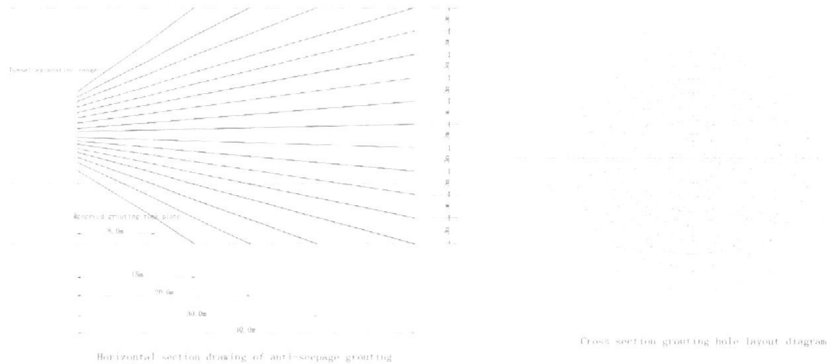

Figure 11. Schematic diagram of advanced anti seepage grouting for whole tunnel section in strong permeable stratum.

(2) Deep foundation pit project

a. For deep foundation pit projects, the main consideration is the surface water flows backward during the construction period and the balance of groundwater recharge in the later period. Treatment measures can be comprehensively considered based on the surface water and groundwater runoff, recharge relationship, and the location of the foundation pit project. In general, appropriate treatment measures can be taken according to the upstream and downstream locations.

b. Engineering treatment measures

When the foundation pit project is located on the upstream side of the surface water recharge, the treatment principle should consider the problem of surface water flow backward and later period recharge sources. From the perspective of inversion treatment, it can be carried out according to the "anti-seepage curtain" scheme in the above-mentioned underground line engineering treatment. Considering that it will not affect the recharge of groundwater to the surface water in the later stage, blind ditch interception and drainage should be set up on the water surface and both sides of the foundation pit. The elevation of the bottom of the blind ditch is 1.0~2.0m below the base level of underground runoff, and the ditch is filled with permeable materials to lead the upstream groundwater to the surface water along the drainage ditch beside the foundation pit, without affecting the recharge and drainage relationship of groundwater and surface water. When the foundation pit project is located on the downstream side of the surface water, the impact of groundwater can be ignored. In terms of treatment measures, if the rock mass is broken and the water permeability is strong, and "anti-seepage curtain" can be set. If the permeability coefficient of the rock mass is less than 0.05m/d, anti-seepage treatment is not necessary, and the groundwater level will gradually recover after the construction and backfilling treatment, and it will not have a large impact on the surface water and landscape.

6.1.2. Treatment of projects cross the surface water. Generally, underground projects that cross the surface water are mainly linear highway tunnels, subways, and city pipe systems. This paper mainly in view of the mountain city subway, municipal tunnel crosses small lakes and rivers, commonly used open excavation and underground excavation method may be produced by surface water flow backward some have effectiveness stronger engineering treatment measures.

(1) Treatment of open-cut excavation method

For shallow-buried underground linear underground projects, the open-cut method is generally used when crossing the surface water. The advantage is that it can effectively avoid the collapse caused by the thin and poor integrity rock mass of the tunnel crown. However, it is necessary to solve the impact of surface water flow backward. In terms of engineering treatment measures, water-proof cofferdams can be installed in the range of 10-20m outside the excavation boundary of the project to divert surface water, and the waterproof cofferdam and the lower rock mass can be grouted together with anti-seepage.

In general, it can be set up 2~3 rows of grouting holes, the row spacing of grouting holes 1.5 m, the grouting curtain is 5~8m deep into the bottom of the tunnel, and the top is slightly higher than the water level of the surface water.

(2) Treatment of underground excavation method

For linear underground projects with a buried depth of more than 20m, the excavation quantity of open excavation is relatively large and uneconomical, and the underground excavation method is generally used for the construction. The treatment measures can be divided into two categories.

a. Ground sealing + consolidation grouting anti-seepage method: Cofferdams can be set on the surface of the tunnel excavation boundary within a range of 8~10m to clear alluvial and silt layers at the bottom of river beds or lakes, then grouting holes are arranged in a plum blossom pattern with a row spacing of 1.5m, and the bottom grouting is 5~6m below the bottom elevation of the tunnel. After the grouting curtain is completed, 0.5~1.0m thick C20 concrete is poured on the surface for sealing treatment. After all anti-seepage work is completed, tunnel construction can be carried out, as shown in Figure 12.

b. Tunnel consolidation grouting anti-seepage method: The specific measures are the same as the above-mentioned linear underground engineering near-surface water treatment measures, and will not be described in detail.

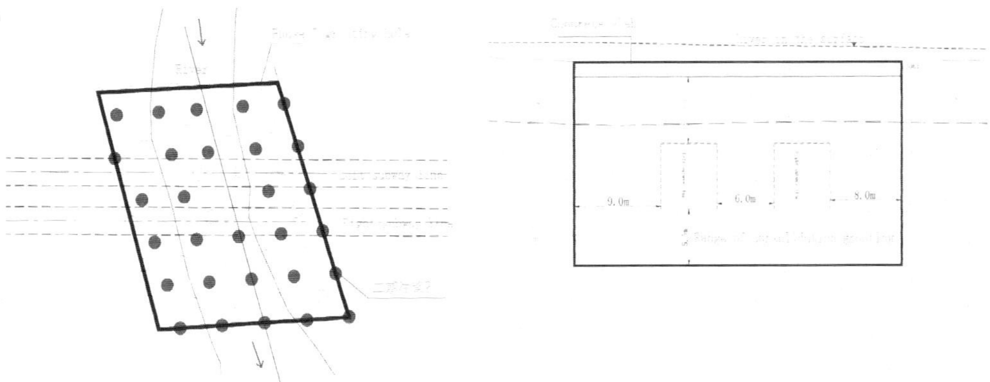

Figure 12. Schematic diagram of the treatment to prevent the river flow backward during the underground excavation tunnel cross the river.

6.2. Treatment of groundwater

Before the groundwater treatment is carried out, it is necessary to find out the groundwater's impact protection object, location and runoff recharge and discharge relationship according to the aforementioned requirements, and make a targeted engineering treatment plan.

6.2.1. Treatment of phreatic groundwater. (1) For linear underground projects:

When the local underground water depth is shallow and the project buried depth is large, the general situation is mainly to consider the treatment of tunnel water gushing during the construction period. The impact on the relationship of underground runoff recharge and drainage is mainly during the construction period. After the completion of the project, the original state can be restored. In principle, the treatment can be carried out in accordance with the tunnel anti-seepage consolidation grouting method. When the groundwater and the buried depth of the project are both shallow, the treatment method of water cut-off and draining should be considered. Set up blind drains on the upstream side of the water surface and within 5~10m away from the project boundary. The elevation of the bottom of the ditch is lower than 1.5~2.0m of underground water level in the dry season. Set up a collecting well every 10~20m, and each collecting well is arranged with a drainage pipe that spans the underground line project to divert and drain the groundwater to the downstream to facilitate the later recharge and recovery of the groundwater. As for the cut-off, it is mainly the anti-seepage treatment during the construction of the tunnel. The anti-seepage consolidation grouting method can be used, but the range is best controlled within 8.0m, which has little effect on the recovery of groundwater in the later period.

(2) Foundation pit projects whose basement is below the groundwater level:

a. For pulsating, linear, and stream-type groundwater, according to the direction of underground runoff, blind drains should be set up under the groundwater level on the upstream side and on both sides to lead the groundwater to the rock mass on the downstream side. In order not to affect the underground runoff, during the excavation of the foundation pit, anti-seepage consolidation grouting should be avoided as far as possible on the surrounding rock mass. A water collection well can be set in the base to pump groundwater into the rock mass and soil mass on the downstream side.

b. For the groundwater and karst pipeline water in the fault fracture zone, the water mainly flows along the fault and karst pipeline, and the water volume is relatively large. Drain and drainage pipes or inverted siphon pipes should be installed on the upstream side of the water outlet to guide the fault zone water and karst pipeline water along the periphery or bottom of the foundation pit to the downstream fault zone and karst pipeline. Other seepage and gushing water can be drained through a collection well, and after the completion of the project, the groundwater environment will not be greatly affected.

(3) Pile foundation engineering:

In the water-conductive fault zone and strong karst development area, there are often holes passing through the water-conductive fault and underground karst pipelines during the construction of the pile foundation project. The water flow varies from 5.0 to 100L/s. Sometimes the water flow can be seen or heard at the top of the hole. Under normal circumstances, most of the construction parties use steel sleeves to protect the walls and then carry out underwater concrete pouring. After the construction of many bored piles, water diversion faults and underground karst pipeline water are cutoffs, leading to downstream wells and springs dry up. The correct method is that first conduct a connection test to find out the relationship between the flow direction of groundwater and wells and springs in the water source protection area. Pipe holes in karst areas should be backfilled with water-permeable blocks, and the upper part should be poured with C15 or C20 concrete. The pile foundation needs to be crossed or avoided, as shown in Figure 13. When the buried depth is relatively shallow, open trench excavation and burying of water diversion pipes can also be used. In short, the treatment measures are based on the principle of not cut off water channels and karst pipes.

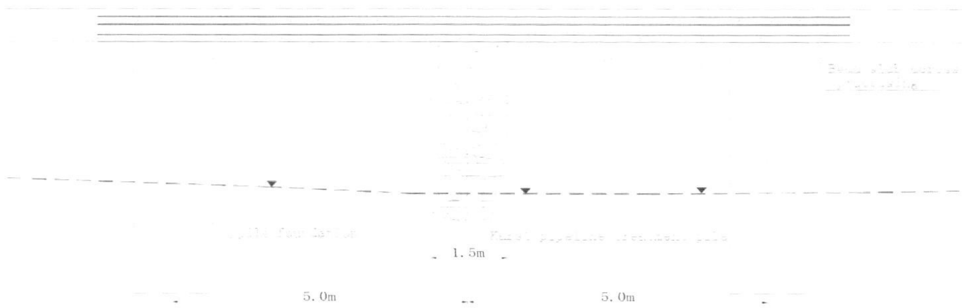

Figure 13. Schematic diagram of pile foundation karst pipeline treatment.

6.2.2. Treatment of confined groundwater. The confined groundwater located between the two water-impermeable layers has the characteristics of a high water head and large water volume. When the underground project passes through or cuts through the upper impermeable stratum, a large number water gushing may occur of the tunnels and foundation pit. Engineering treatment should be based on the principle of not cut off the confined water runoff path, effectively preventing water gushing, and ensuring safe construction.

(1) Linear underground engineering

When a tunnel project passes through a confined aquifer, the main engineering hazard is high-pressure water gushing. In the case of ascertaining the aquiclude and confined aquifer thickness, water head elevation, and rock permeability, high-pressure consolidation grouting can be used to prevent seepage treatment on the surrounding rock of the tunnel. According to the different locations of engineering treatment, it can be divided into ground consolidation grouting and tunnel consolidation grouting, which are described as follows:

a. Ground high-pressure anti-seepage consolidation grouting: According to the plane position of the linear underground project, high-pressure consolidation grouting is carried out in the tunnel and surrounding rocks mass along the axis of the tunnel and vertically downwards from the ground. In general, the range of grouting should be extended to 6~8m outside the boundary of the tunnel, and the grouting boreholes should be arranged in a row spacing of 1.5~2.0m in a plum blossom pattern on the plane. The grouting sequence can be divided into first stage hole and second stage hole interval construction grouting. As shown in Figure 10, the grouting pressure, water-cement mix ratio, admixture, etc. can be determined according to the grouting test. After the completion of grouting, the core pulling test and water pressure test should be carried out. If the grouting does not meet the requirements, reinforcement grouting should be carried out. Tunnel construction can be carried out after the completion of all anti-seepage work.

b. Tunnel anti-seepage consolidation grouting: The procedure of anti-seepage consolidation grouting the same as the above-mentioned grouting procedure of tunnels near-surface water, and will not be described in detail.

(2) Deep foundation pit project

During the excavation of the foundation pit, there may be a large amount of water gushing through the upper water-proof layer of the confined aquifer, which may even endanger the stability of the foundation pit slope. Therefore, anti-seepage treatment is a very important project. The focus is on the bottom of the foundation pit and the surrounding walls. In the case of ascertaining the thickness of the aquiclude, the confined aquifer, the elevation of the water head, and the water permeability of the rock mass, the high-pressure grouting method can be used to consolidate the aquifer around and at the bottom of the foundation pit. The anti-seepage curtain can be divided into ground consolidation grouting and foundation pit consolidation grouting according to different engineering treatment positions, which are described as follows:

a. Ground grouting treatment: Before the excavation of the foundation pit, grouting boreholes shall be arranged in a plum blossom pattern with a row spacing of 1.5~2.0m on the ground. The grouting shall be 4.0~6.0m away from the outer boundary of the foundation pit on the plane. The periphery of the foundation pit on the facade should be connected with the upper impermeable stratum and enter the impermeable layer 2.0~3.0m. The bottom of the foundation pit should be 5.0~6.0m deep below the foundation. As shown in Figure 14. To ensure the quality of grouting during the grouting construction process, the grouting holes should be divided into phase one and phase two to construct the grouting. The grouting sequence is from bottom to top. After the grouting work is completed, the drilling core-pulling inspection and water pressure test should be carried out, when the water permeability is less than 0.03m/d, foundation pit excavation can be carried out.

IOP Conf. Series: Earth and Environmental Science **865** (2021) 012049 doi:10.1088/1755-1315/865/1/012049

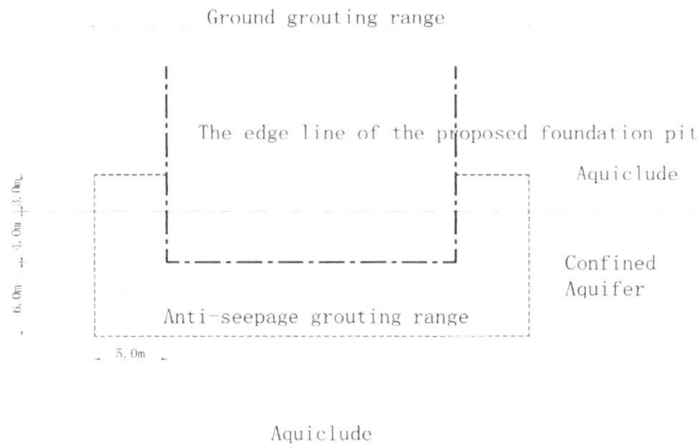

Figure 14. Schematic diagram of ground grouting treatment for foundation pit in confined aquifer.

b. Grouting treatment in the foundation pit: When the local grouting is restricted by the construction conditions, the grouting method in the foundation pit can be used for treatment. In this method, it is advisable to keep 5m away from the pit boundary when the foundation pit is excavated at a depth of 2.0~4.0 m. The grouting boreholes are arranged radially downwards and the diagonal sidewalls, as shown in Figure 13. The scope of grouting is the same as the above-mentioned ground grouting method. The facade shape is shown in Figure 15. The grouting sequence can be carried out from bottom to top and from deep to shallow. After the grouting work is completed, drill core-pulling inspection and water pressure test should be carried out, foundation pit excavation can be carried out after meeting the requirement of water permeability less than 0.03 m/d.

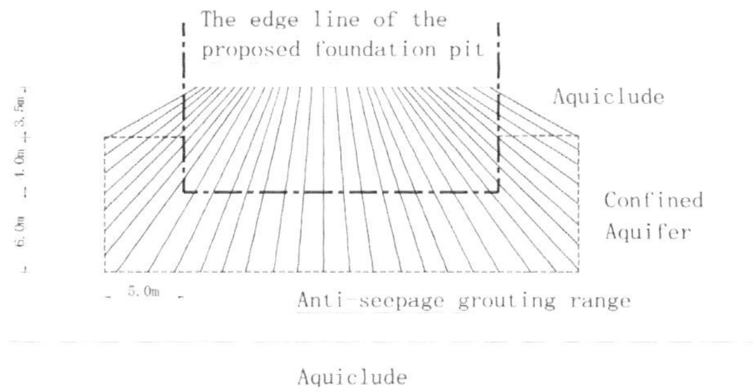

Figure 15. Schematic diagram of grouting treatment in confined aquifer foundation pit.

7. Conclusion

According to the analysis, in the southern sedimentary rock distribution area, especially where the Karst phenomenon was highly developed, impacts on the surrounding water environment caused by different types of projects are concluded. For projects distributed in spots, the impacts are relatively minor, which is much easier to be treated. While, the impacts caused by projects distributed in a large expanse is considerable, where the water environment conditions can generally be recovered by targeted treatment measures. In terms of projects distributed in line, the design and treatment measures for groundwater restoration should be determined first as groundwater runoff conditions have been changed. The current treatment methods generally affect the underground environment by 50 to 70%. Since underground engineering has a sensitive impact on the urban water environment, with a further study on the improvement of treatment measures, the ultimate goal is probably to reach 70 to 90% of the original state of the water environment.

References

[1] Huang Q B, Peng J B, Wang F Y, et al 2019 Issues and challenges in the development of urban underground space in adverse geological environment[J]. *Earth Science Frontiers*, **(3)**: 85-94.
[2] Huang Q B, Yan C C Zhang X N, et al 2020 The impact of urbanization on groundwater quantity, quality, hydrothermal changes and Its countermeasures[J]. *Advances in Earth Science*, **35(5)**: 497-512.
[3] Liu S Z, Tian C, Zhu Z Y, et al 2019 Mechanism research and prediction warning of land subsidence in typical sections of the Yellow River delta[J]. *Acta Geologica Sinica*, **(S01)**: 251-260.
[4] Kang H Z, Chen L, Guo Q Z, et al 2019 An overview of quantification of groundwater recharge in sponge city construction[J]. *Earth Science Frontiers*, **26(6)**: 58-65.
[5] Gao X B, Wang W Z, Hou B J, et al. 2020 Analysis of karst groundwater pollution in northern China[J]. *Carsologica Sinica*, **39(3)**: 287-298.
[6] Huang J M, Deng X W, Hu R Q 2015 The relationship between groundwater and ground collapse and land subsidence in Jinshazhou, Guangzhou city[J]. *Chinese Geology*, **42(1)**: 300-307.
[7] Chen Z Q, Hu Z. 2018 Major geotechnical engineering problems of rail transit project in Guiyang city[J]. *Water Power*, **44(7)**: 20-24.
[8] Yan X X, Yang T L, Lin J X, et al 2019 Influence factors analysis and calculation of land subsidence caused by dewatering of ultra-deep foundation pit[J]. *Journal of Nanjing University (Natural Science)*, **55(3)**: 401-408.
[9] Zhu B Y, Yao F Y, Sun J W, et al 2020 Attribution analysis on land subsidence feature in Hexi Area of Nanjing by InSAR and geological data[J]. *Geomatics and Information Science of Wuhan University*, **(3)**: 442-450.
[10] Di S T, Jia C, Zhang S P, et al 2020 Regional characteristics and evolutionary trend prediction of land subsidence caused by groundwater over exploitation in North Shandong of the North China Plain[J]. *Acta Geologica Sinica*, **94(5)**: 1638-1654.
[11] Su G F, Xu M 2019 Analysis of cyclic evolution condition of a karst water system in central Huaying mountain[J]. *Carsologica Sinica*, **38(2)**: 193-201.
[12] Zhang Y, Li S T, Jin X L, et al 2012 An analysis of the impact of linear engineering construction on groundwater flow field[J]. *Journal of Glaciology and Geocryology*, **34(5)**: 1200-1205.
[13] Hou G B, Kang J W, Feng X, et al 2016 Aanlysis on the effect of the groundwater environment of "the Taiyuan metro line NO.2 project"[J]. *Environmental Engineering*, **(S1)**: 1108-1110.
[14] Gao Y, Liang C, Che Z G 2020 The influence of typical subway project in Nanjing on underground flow dynamics[J]. *Science Technology and Engineering,* **20(28)**: 11711-11717.
[15] Zhuang X F, Sun D 2016 Influence of tunnel construction on karst water: Case analyses[J]. *Carsologica Sinica*, **35(6)**: 681-687
[16] He S, Jiang L W, Zhang G T 2019 Influence of drainage on karst groundwater at an existing tunnel in the east-spring mountains, Chongqing[J]. *Carsologica Sinica*, **38(4)**: 496-501.

Extraction of open-pit mining area based on optimal scales selection and hierarchical classification: A case of Nonoc lateritic nickel

Xian Zhang[1, 2, 3], Wei Li[1, 2], Li Chen[1] and Yaqin Sun[1]

[1]China Aero Geophysical Survey and Remote Sensing Center for Natural Resources, Beijing 100083, China;
[2]Key Laboratory of Airborne Geophysics and Remote Sensing Geology, Ministry of Natural Resources, Beijing 100083, China;

[3]Corresponding author's e-mail: zhangxrs@163.com

Abstract. Traditional remote sensing extraction of open-pit mining area usually has low efficiency. To reduce time of extraction, taking laterite nickel as an example, we proposed a remote sensing extraction strategy of ground objects information in open-pit mining area. Firstly, the optimal segmentation scale parameter (SP) of each category was selected based on the local variance change rate. Secondly, the spectral feature, spatial feature and inter-object relationship feature of high spatial resolution remote sensing images were taken full advantage of feature space construction. Finally, based on the selected SPs and feature space, a top-down hierarchical classification framework was established. The overall accuracy and the Kappa coefficient of the proposed method were 95.49% and 0.9387 by the confusion matrix test, respectively. Compared with the multiple experiments to determine segmentation scales, the proposed method improves the efficiency, meanwhile guarantees the accuracy.

1. Introduction

Extraction of open-pit mining area is of great significance to monitor the mining activities and the mining situation accurately [1]. According to the content of extraction, the relevant researches can be divided into single element extraction and multi-elements extraction. The former is recognition of specific feature in the mining area, such as tailings pond, reservoir, mining area and so on. The latter generally classifies all ground objects in the mining area, which can comprehensively understand their distribution. Combined with multi-period remote sensing images, it can also establish a ground object type transfer matrix to understand the changes of various ground objects.

Traditional extraction methods are mostly manual visual interpretation or machine learning based on pixels. Gao et al extracted tailings ponds in Heilongjiang Province based on ArcGIS software [2]. Ding et al established interpretation marks of various ground objects in the mining area, and extracted various land use types based on ERDAS and ArcInfo platforms [3]. Most of machine learning methods base on pixels use Landsat or other medium spatial resolution images, and commonly use classifiers such as support vector machine and nearest neighbor algorithm [4-5]. However, the former method is usually labor-intensive and inefficient, while the latter method easily leads to salt and pepper noise, and cannot be applied to smaller mining areas due to a lot of mixed pixels in the image. With the popularization of high spatial resolution remote sensing image and the development of image processing technology, object-based image analysis (OBIA) has been applied to the extraction in mining area. In this method, the image is first divided into relatively homogenous objects, and then the objects are used as the minimum unit for classification and extraction. This can not only effectively avoid the impact of noise, but also make full use of the shape, texture and other spatial features of high-resolution images. At present, it is more and more widely used in extraction of multi-elements in open-pit mining area [6-9]. In recent years, deep learning methods have been gradually introduced into the extraction of ground object in mining areas. However, the current relevant studies only involve single element extraction [10-11]. For multi-element extraction, deep learning methods need to be further developed.

Nickel ore is an important strategic resource and a typical open-pit mining area. Multi-element extraction of nickel ore area is helpful to understand the mining situation and environmental changes in the mining area. Therefore, in this paper, the OBIA method was used to extract the laterite nickel ore from Nonoc Island, Philippines. The determination of segmentation scale in OBIA is one of the key factors affecting the extraction accuracy. However, in the current research, the selection of segmentation scale parameter (SP) is all empirical method or multiple experimental method, which not only easily leads to local optimization, but also consumes a lot of time and manpower. This paper introduced the idea of the rate of change (ROC) of local variance (LV) proposed by Drăguţ et al [12-13]. to determine the SP, combined with the feature options, established a top-down hierarchical classification framework, looking forward to improving the efficiency of extraction as well as guaranteeing accuracy.

2. Study area and data

The Nonoc nickel deposit is located in Nonoc Island, which is one of the largest laterites nickel ore in the Philippines [14]. It is estimated there are 112 million tons of nickel ore which is mainly laterite oxide ore [15]. The laterite regolith is thicker and the thickness is generally 3~10 m [16]. This study selected the eastern region of Nonoc nickel deposit (about 10 km^2) as the study area. There are seven types of ground object, including water, vegetation, tailing ponds, road, bare land, mining area and pending mining area.

The PLEIADES image was employed as data source to extract the mineral elements from the study area. It has multispectral imagery with 4 bands (B, G, R, and NIR) of 2 m spatial resolution and panchromatic image of 0.5 m spatial resolution. The pre-processing was carried out by ENVI5.3, including radiometric correction, geometric correction, clip and image fusion.

3. Methods and result

3.1. Selection of optimal segmentation scales

The optimal SP selection method proposed by Drăguţ is based on the rate of change(ROC) of local variance(LV). It is widely recognized that the LV of an image can reveal the spatial structure of an image. Therefore, the rate of change of local variance(ROC_{LV}) between different segmentation scale levels can measure the change of segmentation results at different levels. At the peaks of ROC_{LV} graph, the segmentation objects characterized by relative homogeneity. ROC_{LV} is defined as:

$$ROC_{LV} = \frac{LV_i}{LV_{i-1}} \times 100\% \quad (1)$$

where LV_i is the LV at target level and LV_{i-1} is the LV at next lower level.

Draw a line chart of the LV and ROC_{LV} with the change of SP. When the ROC_{LV} is located at the local maximum value, it can be considered as a more appropriate SP.

Taking 3 as the step size, the ROC_{LV} with the SP of 80~530 was calculated and the line chart was drawn (Figure 1). As can be seen from the Figure 2, when the SP was 95,113,155,278,329,524, ROC_{LV} was the maximum value of a certain range around it, and it was also the higher value of ROC_{LV} in the whole range of segmentation scale. Therefore, the image was segmented at these scales. The results are shown in the Figure 2. As shown in Figure 2, when SP=95, 113 and 155, the segmentation results were more fragmented, which were suitable for bare land scattered among vegetation areas. By contrast, the SP of 95 was smaller, and there was over-segmentation in the extraction of bare land, while the SP of 155 had under-segmentation areas (Figure 2 (a) - (c)). Therefore, bare soil was extracted at the scale of 113. SP of 278 and 329 were in the middle, which were suitable for the extraction of road and vegetation in mining area. The results of 278 scale had some over-segmentation areas, so 329 was adopted (Figure 2 (d), (e)). The result of 524 scale segmentation was large patch, which was suitable for large and homogenous categories, such as mining area, water body and tailing ponds(Figure 2 (f)).

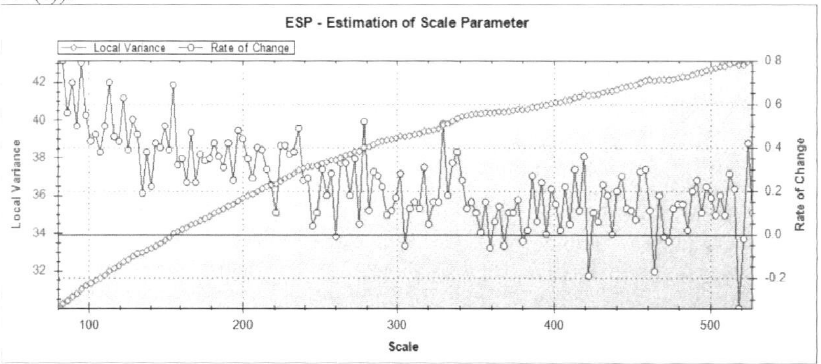

Figure 1. Line chart of the LV and ROC_{LV} with segmentation scale.

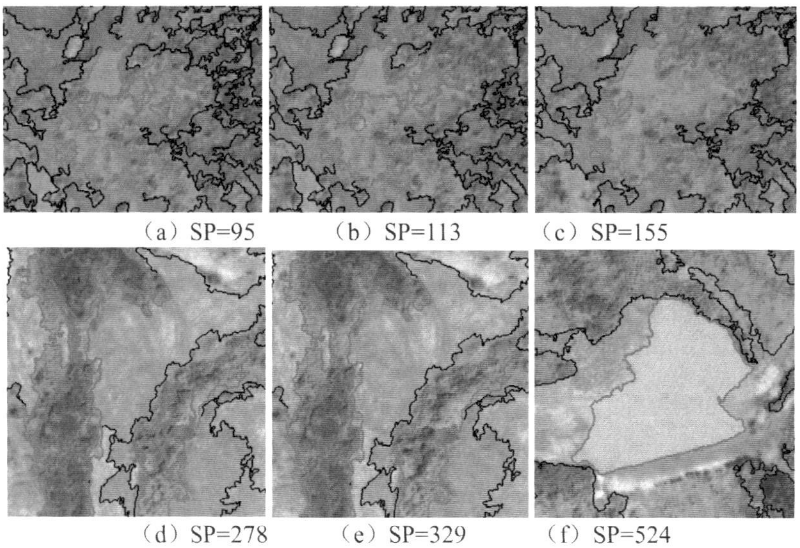

Figure 2. Segmentation results of various SPs.

3.2. Hierarchical classification

Considering that the ground objects in the study area are relatively complex and there is no unified optimal SPs for all of them, adopting the selected SPs above, a top-down hierarchical classification framework was established to extract all types of ground objects, so as to improve the overall extraction accuracy.

3.2.1. Feature space construction. The spectral feature, shape feature and inter-object relationship feature of high spatial resolution remote sensing images were fully utilized to construct feature space. First, samples for each category were selected, and then the separabilities of various features were analysed. Finally, the following features were selected from more than 100 features: mean spectral value of all objects in band 2 (B2), mean spectral value of all objects in band 3(B3), mean spectral value of 4 bands (Brightness), normalized difference vegetation index(NDVI), normalized difference water index (NDWI), length-width ratio (LWR) and adjacency relationship between two categories(Border to). The formulas of NDVI, NDWI and LWR are as follows:

$$\mathrm{NDVI} = (NIR - R)/(NIR + R) \quad (2)$$

$$\mathrm{NDWI} = (G - NIR)/(G + NIR) \quad (3)$$

$$LWR = eig_1/eig_2\ ,\ eig_1(S) > eig_2(S) \quad (4)$$

where *NIR*, *R* and *G* are spectral value of band 4, band 3 and band 2, respectively; *eig* is the eigenvalue of the covariance matrix of an object.

It should be noted that although the texture features of high spatial resolution images are relatively obvious, the calculation of texture features is very complex and time-consuming. Especially for large area like this study area, the efficiency is very low if the texture features are adopted. Therefore, based on the three optimal segmentation scales selected above, the author proposed a top-down hierarchical classification strategy. The bare land and pending mining areas differentiated by texture features in past studies were extracted at different scale levels, combined with inter-object relationship feature, thereby avoiding the utilization of texture features and further improving extraction efficiency.

3.2.2. Classification processing. According to the 3 selected SPs in 2.1, the classification was performed with 3 levels. The flowchart of multi-scale segmentation and hierarchical classification processing is shown in Figure 3.

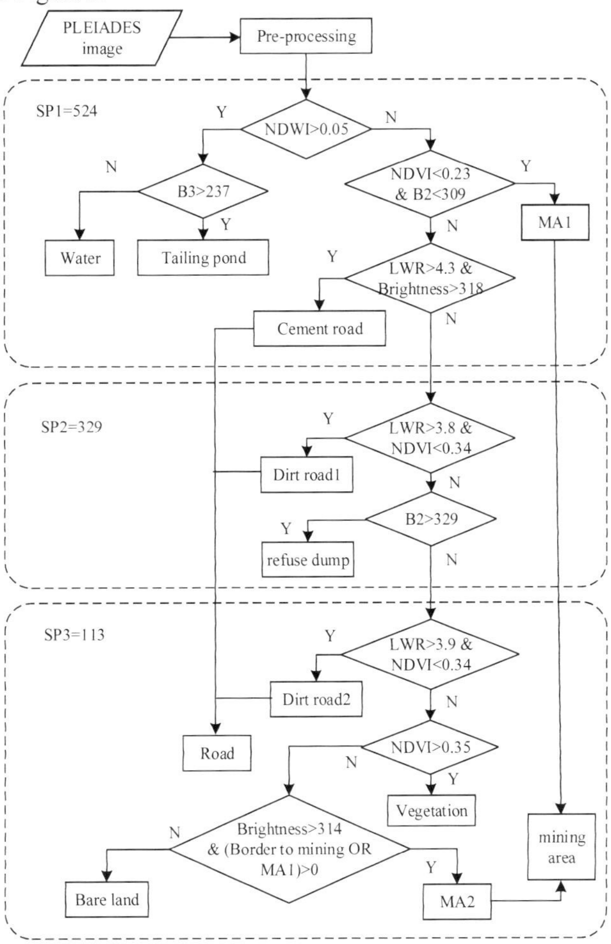

Figure 3. Flowchart of multi-scale segmentation and hierarchical classification.

IOP Conf. Series: Earth and Environmental Science **865** (2021) 012050 doi:10.1088/1755-1315/865/1/012050

(I) SP1=524.

① The objects whose NDWI>0.05 were classified as the water and tailing ponds, where the objects whose B3>237 were water, while the objects whose B3<237 were tailing pond.

② When the NDWI<0.05, if NDVI<0.23 and B2<309, the objects would be a part of mining area (MA1). In the rest of the objects, if LWR>4.3 and Brightness>318, they would be classified as cement road.

(II) SP2=329

In the rest of the unclassified objects, the objects whose LWR>3.8 and NDVI<0.34 were classified as dirt road1, while the (LWR<3.8 or NDVI>0.34) and B2>329 were classified as refuse dump.

(III) SP3=113

① In the rest of the unclassified objects, the objects whose LWR>3.8 and NDVI<0.34 would be the dirt road2, while the (LWR<3.9 or NDVI>0.34) and NDVI>0.35 were classified as vegetation.

② The rest of the unclassified objects were the mixtures of bare land and a part of mining area (MA2), which could be distinguished by the Brightness and Border to. If Brightness>314 and (Border to mining area or MA1), the objects would be MA2, the rest of the unclassified objects were classified as bare land.

③ Finally, the MA1 and MA2 composed mining area, and the cement road, dirt road1 and dirt road2 composed road.

3.2.3. Classification result and accuracy evaluation. In order to evaluate the accuracy of classification result comprehensively and objectively, 13×18 test sample areas were selected from the original image at equal interval. The classification result and distribution of test samples are shown in Figure 4. Indicators such as producer accuracy, user accuracy, Kappa coefficient and overall accuracy of various categories are shown in Table 1.

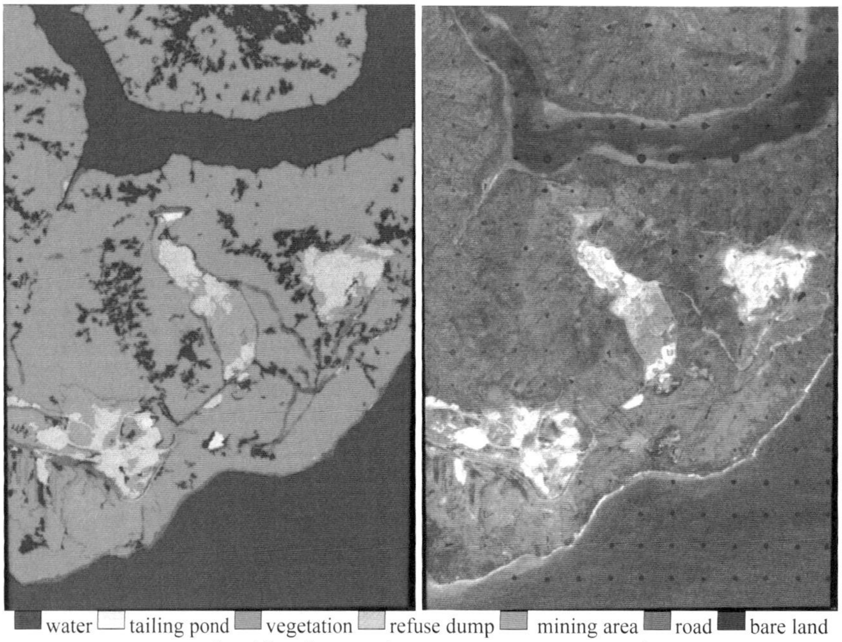

(a) classification result (b) test samples

Figure 4. Classification result and distribution of test samples.

Table 1. Evaluation indicators.

Ground objects	producer accuracy	user accuracy	Kappa per class	overall accuracy	kappa
water	100%	100%	1		
tailing pond	100%	100%	1		
vegetation	99.26%	98.10%	0.989		
refuse dump	90.35%	84.72%	0.8973	95.49%	0.9385
mining area	95.90%	86.64%	0.9544		
road	74.57%	98.78%	0.7368		
bare land	78.90%	90.77%	0.7679		

It can be seen from Figure 4 and Table 1 that the extraction accuracy of each category is relatively high on the whole, with overall accuracy and Kappa coefficient of 95.49% and 0.9385 respectively. For categories of water, vegetation and tailing pond, the producer accuracy and user accuracy are higher than 98%. Among the other four categories, the producer accuracy of refuse dump and mining area is higher (90.35% and 95.90% respectively), while the user accuracy is relatively low (around 85%). On the contrary, the user accuracy of road and bare land is 98.78% and 90.77% respectively, while the producer accuracy is lower (around 75%). The Kappa coefficient of these two categories is also significantly lower than that of other categories. The reason is the confusion of the road and bare land categories with mining area and refuse dump.

4. Conclusions

Based on optimal scale selection, a hierarchical extraction strategy for ground objects in open-pit mining area is proposed in this paper. By selecting the optimal scale in advance, the inefficiency caused by multiple experiments is avoided. In addition, through the design of a top-down three-layer extraction framework, not only avoid over-segmentation and under-segmentation effectively, but also other features were selected based on the feature optimization method that can replace the texture features, thus refraining from complex calculation and further improving the extraction efficiency. The extraction accuracy is also guaranteed, the overall accuracy is 95.49%, and the Kappa coefficient is 0.9385.

Acknowledgments

This work is supported by Youth Innovation Fund Project of Ministry of Land and Resources Key Laboratory of Airborne Geophysics and Remote Sensing Geology (2020YFL32).

References

[1] Sonter L J, Moran C J, Barret D J, et al 2014 Processes of land use change in mining regions *Journal of Cleaner Production* **84** 494

[2] Yongzhi Gao, Jianguo Hou, Yu Chu, et al 2019 Remote sensing monitoring of tailings ponds based on the latest domestic satellite data *Journal of Heilongjiang Institute of Technology* **3** 26

[3] Juan Ding, Yuetong Xu, Yanjie Yang, et al 2008 Information extraction technique of remote sensing images of mining area based on Erdas and Arc/Info *Mining R&D* **3** 43

[4] Capolupo A, Monterisi C, Tarantino E 2020 Landsat images classification algorithm (LICA) to automatically extract land cover information in Google Earth Engine environment *Remote Sensing* **12** 1201

[5] Wenbo Wu, Zhengpeng Zhang 2007 Building of land resources classification system and information extraction based on TM image in mining area *Journal of China Coal Society* **12** 1282

[6] Run Huangfu, Ao Li 2020 Contrastive analysis of extraction methods for ground object information in mining area based on remote sensing image *Journal of Tangshan University* **3**

42
[7] Lanjie Cao, Bing Wu Jinghua Wang, et al 2019 Object-oriented information extraction and analysis of the iron Tailings with GF-1 remote sensing image *Geomatics & Spatial Information Technology* **4** 98
[8] Yating Song, Dong Jiang, Yaohuan Huang, et al 2016 Research on the priority of the land use types extraction of opencast mine area based on object-oriented classification *Remote Sensing Technology and Application* **3** 572
[9] Dingbo Yuan, Chenglin Liu, Guobin Wang 2013 The application and the research of object-oriented method for extraction of mining area information *Remote Sesning Information* **2** 110
[10] Kai Yan, Ting Shen, Zhengchao Chen, et al 2020 Automatic extraction of tailing pond based on SSD of deep learning *Journal of University of Chinese Academy of Sciences* **3** 360
[11] Gallwey J, Robiati C, Coggan J, et al 2020 A Sentinel-2 based multispectral convolutional neural network for detecting artisanal small-scale mining in Ghana: Applying deep learning to shallow mining Remote Sensing of Environment **248** 111970
[12] Drăguţ L, Tiede D, R. Levick S R 2010 ESP: A tool to estimate scale parameter for multiresolution image segmentation of remotely sensed data *International Journal of Geographical Information Science* **6** 859
[13] Drăguţ L, Csillik O, Eisank C, et al 2014 Automated parameterisation for multi-scale image segmentation on multiple layers *ISPRS Journal of Photogrammetry and Remote Sensing* **88** 119
[14] Xueshan Yang, Yuansheng Guo, Baiyou Chen, et al 2013 The distribution and the exploration, development and utilization situation of the lateritic nickel ore resources in the world *Acta Geoscientia Sinica* **z1** 193
[15] Yinliang Cui, Xueshan Yang, Yongguo Jiang, et al 2013 Metallogenic condition prospecting indicator mineralized type lateritic nickel deposit *Acta Mineralogica Sinica* **4** 449
[16] Woodcock, C. and Strahler, A., 1987. The factor of scale in remote sensing. Remote Sensing of Environment, 21, 311-332.

IOP Conf. Series: Earth and Environmental Science **865** (2021) 012051 doi:10.1088/1755-1315/865/1/012051

Artificial intelligence prediction of air traffic flow rate at the Hong Kong International Airport

Kai-kwong Hon[1, 2]

[1]Hong Kong Observatory, Tsim Sha Tsui, Hong Kong, China

[2]Corresponding author's e-mail: kkhon@hko.gov.hk

Abstract. The Hong Kong Flight Information Region (HKFIR) is surrounded by the airspaces of Guangzhou, Sanya, Taipei and Manila. International/regional flight routes to and from the Hong Kong International Airport (HKIA) rank among the busiest in the world. Air traffic flow rate is a complex parameter influenced by a variety of factors such as weather, business operation and regulations, etc. Accurate prediction of flow rate allows airport management to optimise tactical planning and decision making, potentially benefitting airport operations and air traffic efficiency. Leveraging the concept of Big Data, this paper establishes a preliminary algorithm for artificial intelligence prediction of air traffic flow rate at HKIA through supervised machine learning. Results based on 2017 data showed positive short-term prediction skills as well as superior performance over persistence based on the previous day.

1. Introduction

The Hong Kong International Airport (HKIA) is one of the busiest international airports in the world, with total aircraft movements exceeding 400,000 in 2016 [1]. The Hong Kong Flight Information Region (HKFIR), which borders the airspaces of Guangzhou, Sanya, Taipiei and Manila, contains many international and regional flight routes to and from HKIA which rank among the busiest in the world [2]. Accurate prediction of flow rate allows airport management to optimise tactical planning and decision making, potentially benefitting airport operations and air traffic efficiency.

Air traffic flow rate is a complex parameter influenced by a variety of factors such as weather, business operation and regulations, etc. Such a complex, inter-disciplinary problem naturally lends itself to Big Data and machine learning methods [3]. In fact, artificial intelligence techniques have been successfully applied in a number of aviation-related topics, ranging from prediction of turbulence [4] to fuel consumption modelling [5]. In this study, we tackle the problem of short-term prediction of air traffic flow rate, including both arrival and departure rate, at HKIA.

This paper is organised as follows. Section 2 describes the data and method used for establishing the machine learning prediction model. Section 3 describes the verification results. Brief discussions and conclusions are given in Section 4.

2. Data and method

In this section, we describe the data and method used in the establishment of the machine learning model for prediction of air traffic flow rate at HKIA.

IOP Conf. Series: Earth and Environmental Science **865** (2021) 012051 doi:10.1088/1755-1315/865/1/012051

2.1. Aircraft position data

HKO has installed an Automatic Dependent Surveillance – Broadcast (ADS-B) reception system since 2016 [6] for real-time reception of aircraft position data around Hong Kong. The reception antenna is located at the Tai Mo Shan Weather Radar Station (114.12°E, 22.41°N) which stands at over 900 m above sea level and provides unobstructed line-of-sight over its neighbouring airspace, with an effective reception range up to approximately 600 km (Figure 1).

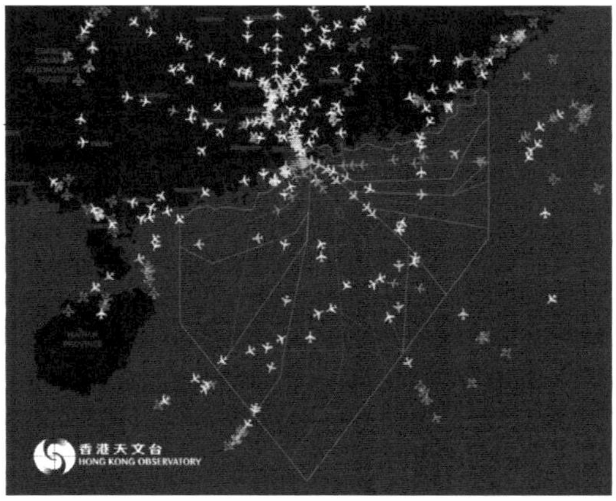

Figure 1. Sample display of the ADS-B aircraft position data as received by HKO. The white lines mark the HKFIR boundary.

The ADS-B data is updated up to every second (or more) and contains various flight parameters such as callsign, latitudinal and longitudinal position, geometric height, flight level, ground speed and true air speed, etc. For the purpose of this study, only the positional data will be used. Quality control procedures are applied to filter out potentially erroneous data (e. g. missing or incomplete data fields, repetitive entries, discontinuous timestamps or trajectory, unphysical values in parameter space).

2.2. Air traffic data

Air traffic flow rate at HKIA may be derived from the actual number of aircraft movement, including arrivals and departures, at each hour. During the study period (and up to the time of writing), HKIA operates two parallel runways (the North and South Runways along the 070-250 direction) in independent segregated mode, supporting a maximum of a total of 68 movements per hour (Figure 2). In reality, the maximum number of arrival and departure flights per hour would be influenced by various operational and regulatory factors including weather [7] and aircraft wake turbulence separation [8].

Figure 2. The existing two runways of HKIA, with the new Third Runway to their north under construction as of the time of writing. (Image taken from Google Earth.)

The observed hourly arrival and departure rates will form the basis for training and validation of the machine learning prediction model.

2.3. Artificial intelligence algorithm

In this study, we apply the open-source machine learning algorithm of XGBoost [9] as implemented on the R software platform [10]. XGBoost, or "extreme gradient boosting", is a computationally efficient artificial intelligence algorithm based on iterative ensembles of regression trees with gradient boosting, which is suitable for supervised learning involving large data sets.

Figure 3 shows the schematic diagram of the model development process. ADS-B and air traffic data from 2017 are used. The training period covers the 3-month period of April to June. The validation period covers July to November. The distribution of aircraft positions as derived from ADS-B data over the previous hour is used as the predictor, while the hourly arrival rate and departure rate over the next hour is used as the predictand. In this way, we generate short-term prediction of air traffic flow rate for HKIA at hourly intervals, each covering the next hour.

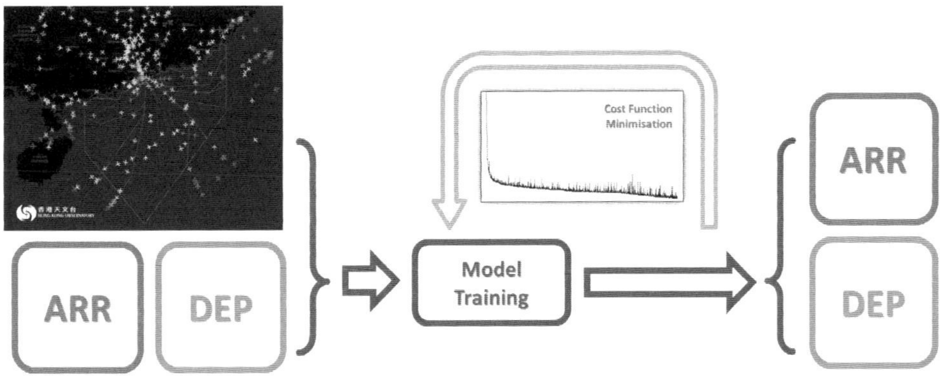

Figure 3. Schematic diagram of the development of the machine learning prediction model.

3. Verification Results

In this section, we present the validation results of the machine learning prediction model. Figure 4 compares the results of the machine learning prediction model against the corresponding actual arrival rate at HKIA. The x-axis shows the predicted values while the y-axis shows the observed values. The black line shows the reference slope of 1:1. In the diagram, the colour of each pixel indicates the number of data pairs falling into that category. The value represented by each colour is explained in the legend to the right.

It can be seen from Figure 4 that the machine learning prediction model results exhibit strong linear correlation with the observed values. The forecast-observation pairs generally clutter along the line of unity throughout the considered range of values. The regions with highest density of data points (including those in yellow, orange and red) also occur closest to the line of unity. Results for departure rate is shown in Figure 5. It can be seen that, similar to Figure 4, predictions from the machine learning model show good correlation with the observations across the whole range of observed values. These results demonstrate that the machine learning prediction model exhibits positive skill for both arrival and departure rate prediction at HKIA.

We may further examine the skill level of the machine learning prediction model by comparing its mean absolute error (MAE) against the "persistence" method. The "persistence" method refers to using the observed value from the same hour of the previous day as the predicted value for upcoming hour. As indicated in Figure 4 and Figure 5, the MAE of the machine learning prediction model are 1.9 and 2.4 respectively for arrival rate and departure rate. The performance of the "persistence" method is shown in Figure 6 and Figure 7 respectively for arrival rate and departure rate. By comparison, the "persistence" method gives MAE of 2.5 for arrival rate and 3.2 for departure rate. This further confirms the skill level of the machine learning prediction model.

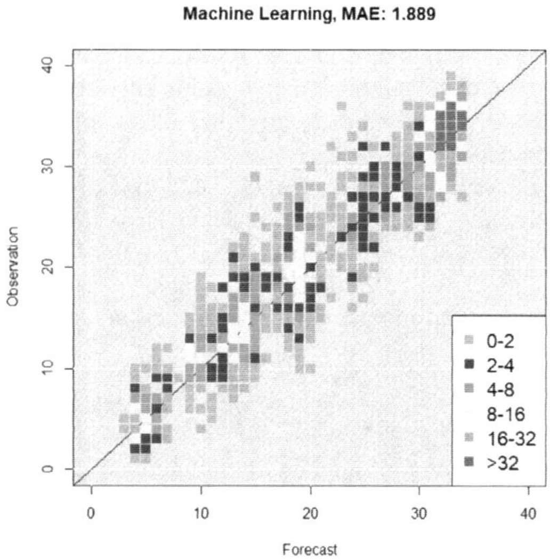

Figure 4. Density plot of the machine learning prediction algorithm results (x-axis) against the corresponding actual arrival rate (y-axis). The colour of each pixel indicates the number of data pairs in that category.

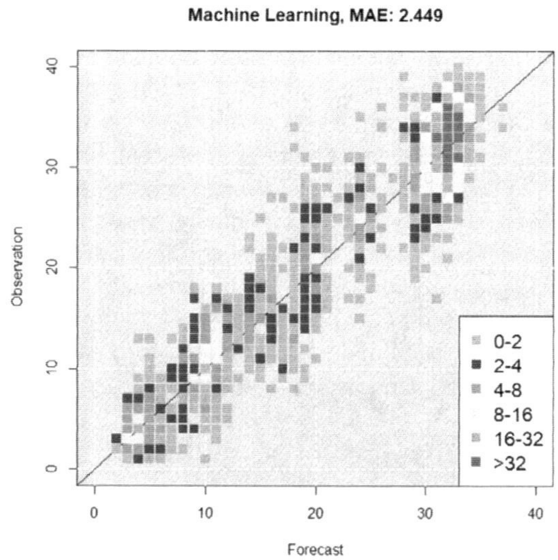

Figure 5. Same as Figure 4 except for departure rate.

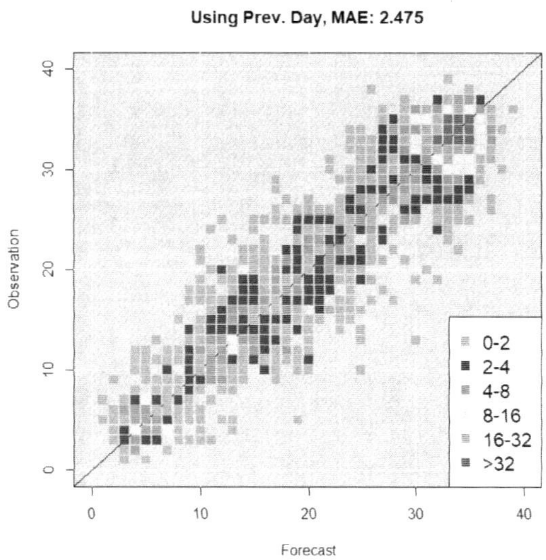

Figure 6. Same as Figure 4 except using the "persistence" method.

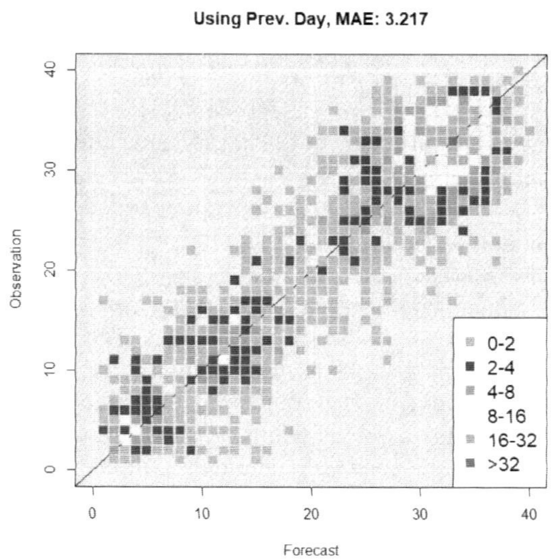

Figure 7. Same as Figure 5 except using the "persistence" method.

4. Discussions and Summary

Air traffic flow rate is a complex inter-disciplinary parameter with considerable operational significance for airport management. This study presents one of the first efforts in applying artificial intelligence techniques in making short-term predictions of this important parameter. By using aircraft position data around HKFIR with high spatio-temporal resolution, the XGBoost-based machine learning prediction model is able to produce skillful predictions for both arrival rate and departure rate over the next hour at HKIA during the study period. Furthermore, the machine learning prediction model also demonstrated superior skill level over the "persistence" method from the previous day in terms of MAE.

Despite the initially promising results, we acknowledge various limitations in the present study. Firstly, the data set should ideally be expanded to cover multiple years in order to better establish robustness of the prediction model. This may be achieved by collecting long-term data to extend the training and validation sets. We also bear in mind that the air traffic system, whether in Hong Kong or other locations around the world, are not static but rather constantly evolving. Regulatory requirements, aircraft performance and handling, as well as air traffic control procedures and practices may change from time to time, to name a few. As such, the prediction model might need periodic updating to keep up with the latest environment, which in turn creates additional challenges for long-term validation. Furthermore, weather is known to exhibit considerable impact on air traffic over a range of timescales. Methods need to be developed for effectively incorporating such effects in the prediction model, as with other non-linear influence such as time-lag or cascading effects well-known to the traffic engineering community [11]. These would be possible areas of focus for future studies.

References

[1] Hong Kong International Airport, 2017: https://www.hongkongairport.com/iwov-resources/file/the-airport/hkia-at-a-glance/facts-figures/2016e.pdf

[2] OAG, 2020: Busiest Routes 2020, http://www.oag.com

[3] K.K. Hon, 2018. Big data opportunities in high-impact public and aviation weather at the Hong Kong Observatory. WMO CBS TECO, 23-26 Mar 2018, Geneva, Switzerland

[4] Hon, K.K., C.W. Ng and P.W. Chan, 2020: Machine learning based multi-index prediction of aviation turbulence over the Asia-Pacific, Machine Learning with Applications, Volume 2, 2020, 100008, https://doi.org/10.1016/j.mlwa.2020.100008

[5] Uzun, M., M.U. Demirezen, G. Inalhan, 2021. Physics Guided Deep Learning for Data-Driven Aircraft Fuel Consumption Modeling, Aerospace, https://doi.org/10.3390/aerospace8020044

[6] Hong Kong Observatory, 2019: A new ADS-B ground station at Tai Mo Shan, https://www.hko.gov.hk/en/aviat/outreach/35th/adsb.htm

[7] Lui, G.N., R.P. Liem and K.K. Hon, 2020: Towards understanding the impact of convective weather on aircraft arrival traffic at the Hong Kong International Airport. IOP Conf. Ser.: Earth Environ. Sci. 569 012067, https://doi.org/10.1088/1755-1315/569/1/012067

[8] Hon, K.K. and P.W. Chan, 2017: Aircraft Wake Vortex Observations in Hong Kong. Journal of Radars, 2017, 6(6): 709-718, https://dx.doi.org/10.12000/JR17072

[9] Chen T, Guestrin C, 2016: XGBoost: a scalable tree boosting system. Proceedings of the 22nd ACM SIGKDD international conference on knowledge discovery and data mining. ACM; p. 785–94

[10] R Core Team (2017). R: A language and environment for statistical computing. R Foundation for Statistical Computing, Vienna, Austria

[11] Zhai X Q, Ou C Q, Xing F, Huang H L, K.K.Hon, Tony Sze and Zhan Z Y, 2019: Hourly associations between weather factors and traffic crashes: non-linear and lag effects. Anal. Methods Accid. Res. https://doi.org/10.1016/j.amar.2019.100109

GPS/BDS displacement monitoring of railway slopes based on coloured noise analysis

Jun Ma[1, 2]

[1]China Railway Siyuan Survey and Design Group Co., LTD, Wuhan 430079, China

[2]Corresponding author's e-mail: 794598508@qq.com

Abstract. To analyse the influence of coloured noise in displacement time series on the global navigation satellite system (GNSS) displacement monitoring of railway slopes, and to compare the GPS, BDS and GPS/BDS displacement monitoring results, in the present research the global positioning system (GPS), the BeiDou navigation satellite system (BDS), and GPS/BDS displacement time series data from the GNSS monitoring points on the slopes on both sides of a tunnel entrance of the Changgan High-speed Railway are taken as an example to analyse the displacement of monitoring points before and after removal of coloured noise under the model of white noise + flicker noise. The experimental results show that the flicker noise causes uncertainty around displacement velocity estimation that exceeds the velocity itself, and there is a difference between the GPS and BDS velocity estimation. Eliminating the flicker noise in the displacement time series reduces the uncertainty of the displacement velocity estimation. Most of the uncertainties are less than 0.01 mm/d. Using GPS/BDS technology can help reduce flicker noise and the difference between GPS and BDS speed estimates. That is, the integrated use of GPS/BDS and the colored noise filtering technology in the displacement time series improves the reliability of the slope displacement monitoring results. In addition, the low slope on the side of the railway had a 4 mm displacement in the northerly direction during the selected period. The slope on both sides near the railway has a displacement of 1 to 2 mm in the easterly direction and a rise of about 4 mm in the vertical direction. The displacement of the slope further and higher from the road-bed is not more than 1 mm in both horizontal and vertical directions.

1. Introduction

Railway lines will inevitably pass through many complex geological areas such as mountains and hills. At the same time, they are also facing more serious safety-related risks of slope deformation and instability, which are directly related to the safety of railway construction and operation. Slope deformation and instability are sudden and difficult to judge and predict in advance, therefore, long-term deformation monitoring and analysis is needed, which is of great significance to ensure the safety of railway operations. Global navigation satellite system (GNSS) automatic monitoring characterised by high accuracy, unlimited by climate conditions, a high degree of automation and no need to see between stations, offers real-time dynamic monitoring and other advantages. Therefore, it is often used to monitor the deformation of slopes in real time and provide necessary data for slope design, construction, maintenance, and reinforcement [1-4].

The slope GNSS deformation monitoring actually uses the GNSS displacement time series of the monitoring point to estimate the displacement velocity of the monitoring point, and then obtain the displacement amount. The GNSS displacement time sequence of the monitoring point is composed of displacement signals and noise. The noise includes white noise and colored noise, and colored noise dominates the noise, which is caused by incompletely eliminated tropospheric delay, satellite ephemeris error, multipath effects, and ignored seasonal signals [5]. Before estimating the displacement velocity, it is necessary to use the prior variance of the noise to construct the covariance matrix of the observation value, and then calculate the displacement velocity of the monitoring point according to the function model of the observation value [6]. The random characteristics of colored noise are different from white noise. However, in the previous GNSS slope deformation monitoring, when estimating deformation parameters, the noise in the displacement time series is treated as white noise and the colored noise is ignored. A covariance matrix that cannot accurately express the random characteristics of the observations is constructed, which is not conducive to obtaining an accurate rate of displacement and total displacement at each monitoring point.

BDS is a high-precision GNSS developed in China. At present, GPS technology is mostly used for GNSS slope deformation monitoring and analysis, while BDS technology is rarely used: under the influence of coloured noise, the monitoring results obtained according to GPS technology may be different from those arising from the use of BDS. Since the comprehensive use of GPS and BDS for deformation monitoring is helpful to reduce the noise in the deformation time series [7], we take the slope at the entrance of a tunnel of Chang-Jiangxi high-speed railway in China as an example and analyses the displacement of slope monitoring points by mainly using GPS/BDS coordinate time series data, taking into account the influence of coloured noise. At the same time, GPS and BDS coordinate time series are used to compare and analyse the influences of different positioning techniques on displacement monitoring results.

2. Experimental data

The slope of Chang-Jiangxi Line is located at the exit from one side of a tunnel in Xiajiang County, Jiangxi Province. The road-bed is in hilly terrain and has a length of about 370 m (Figure 1). The slope on the left-hand side of the tunnel entrance is significantly higher than that on the right. To monitor the slope displacement on both sides of the subgrade in this section in real time, 20 GNSS monitoring stations were installed on the slope, and a solar-powered reference station was established on the roof of a 3-story building with relatively stable geological conditions within 2km [8].

Figure 1. Slopes on both sides of the tunnel portal.

Due to the limitation of space, the observation data from four monitoring points were selected for analysis herein. Among them, monitoring points P1, P3, and P4 were close to the subgrade, with monitoring point P4 located on the side of the low slope of the subgrade, and monitoring point P2 was located on the side of the high slope of the subgrade (Figure 2, where JZ represents the reference station). All monitoring point receivers receive GPS and BDS satellite signals. Second generation

BDS satellite data were used due to limitations imposed by receiver conditions. Integrate GPS and BDS observation data, and use real-time kinematic (RTK) carrier phase difference technology to obtain GPS/BDS coordinate time series of monitoring points in the east (E), north (N) and vertical (U) directions. The sampling interval of the coordinate time series was five minutes, and the time span was from 0:00 on 28 April to 23:00 on 24 May 2020. Taking the coordinates at 0:00 on 28 April as a reference, the displacement of the monitoring point relative to the reference time point at each sampling time point is obtained, and finally GPS/BDS displacement time series in the three directions are formed. In order to compare with the results of GPS and BDS slope monitoring, this paper uses the same method to obtain GPS and BDS displacement time series (Figure 4).

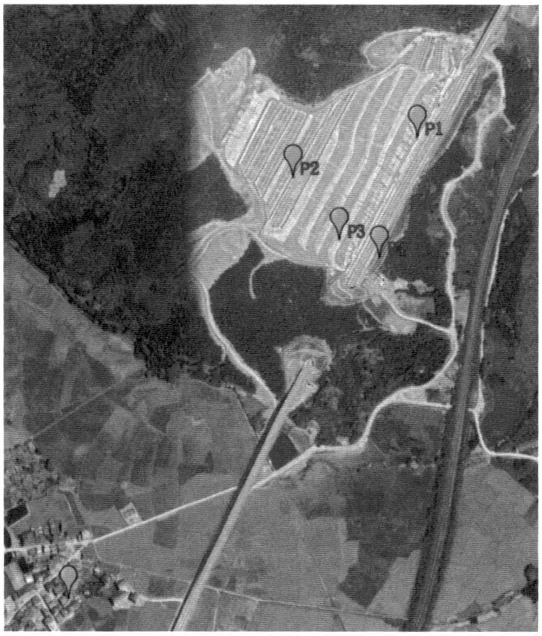

Figure 2. Distribution of monitoring points and reference stations.

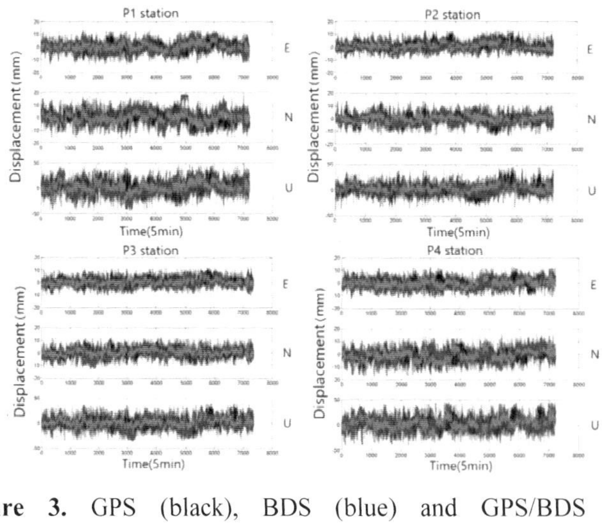

Figure 3. GPS (black), BDS (blue) and GPS/BDS (red) displacement time series data.

3. GPS/BDS displacement monitoring analysis

Since slope monitoring focuses on the displacement of monitoring points, only the linear movement in the displacement time series is considered, and the periodic signals are regarded as coloured noise. It can be seen from Figure 3 that the GPS/BDS displacement time series fluctuates greatly, which makes it impossible to distinguish the displacement trend of the measuring point. Therefore, the displacement speed of the monitoring point must be estimated based on a suitable noise model. Since the best noise combination model of GNSS coordinate time series data is white noise (WN) + flicker noise (FN) [6-9], and flicker noise is colored noise, under the combined model of WN + FN, the commonly used noise estimation software Create and Analyze Time Series (CATS) is used to calculate the noise amplitude (Table 1) and velocity (Table 2) in the displacement time series of each monitoring point [10]. It can be seen (Table 1) that the estimated amplitude of flicker noise is about four to seven times that of white noise, so flicker noise plays a dominant role in the residual of deformation time series data.

Table 1. Amplitude estimation of white noise (mm) and flicker noise (mm/d$^{0.25}$) in the displacement time series of the monitoring station.

	E-direction		N-direction		U-direction	
Station	WN	FN	WN	FN	WN	FN
P1	1.79	7.78	1.46	9.15	5.04	25.00
P2	1.09	6.49	1.46	7.01	3.39	19.63
P3	1.29	6.52	1.48	6.90	4.05	19.77
P4	1.56	7.54	1.77	8.70	4.48	24.31

Table 2. Estimation of velocity considering coloured noise (mm/d).

	E-direction		N-direction		U-direction	
Station	Estimation	Uncertainty	Estimation	Uncertainty	Estimation	Uncertainty
P1	-0.01	0.14	-0.04	0.13	0.17	0.46
P2	-0.03	0.12	0.05	0.13	0.04	0.37
P3	0.06	0.12	0.03	0.13	0.15	0.36
P4	0.07	0.14	0.17	0.16	0.36	0.45

As can be seen from Table 2, the uncertainty in almost all velocity estimates is greater than the velocity itself. This is due to the large amplitude of the coloured noise in the time series [11-13], which almost overwhelms the displacement information, resulting in greater uncertainty and lower reliability of the displacement velocity estimation. Therefore, to improve the accuracy of velocity estimation and reduce the uncertainty therein, the joint wavelet transform and information entropy theory are adopted to eliminate coloured noise in displacement time series data [14]. Figure 4 shows the GPS/BDS displacement time sequence of the monitoring point after the colored noise is removed. It can be seen that, after removing the coloured noise, there is a significant linear trend in the displacement time series, which indicates that the signal-to-noise ratio of the remaining displacement time series is high, and the influence of noise is reduced. To compare and analyse the noise changes in the displacement time series after removing coloured noise, Figure 5 shows the logarithmic power spectrum before and after removing flicker noise from the GPS/BDS displacement time series. The power spectrum of the displacement time series has a certain slope before the flicker noise is eliminated, and the energy is concentrated in the low-frequency region. After removing the coloured noise, the power of the displacement time series decreases, and the decrease in power at low frequencies is significantly greater than that at high frequencies. This is because most of the power of the coloured noise is concentrated in the low-frequency range [15], and the low-frequency power decreases the most after it is removed. It also shows that the joint wavelet transform and information entropy theory can eliminate the coloured noise in the displacement time series.

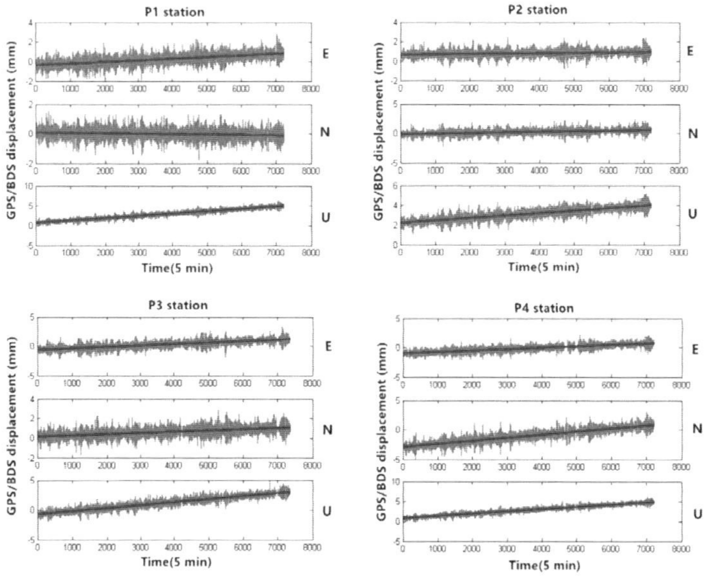

Figure 4. GPS/BDS displacement time series (red) and linear trend (blue) of monitoring points after removing flicker noise.

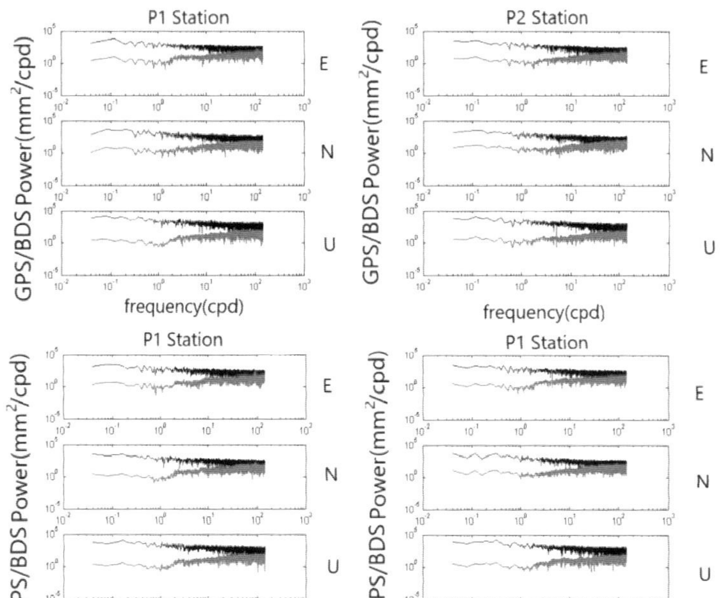

Figure 5. GPS/BDS displacement time series (red) and linear trend (blue) of monitoring points after removing flicker noise.

The same method was used to calculate the filtered noise amplitude (Table 3) and the velocity estimate (Table 4). By comparing Table 1 and 3, the estimated amplitude of flicker noise after filtering is no more than 0.02 mm/d$^{0.25}$, while the estimated amplitude of white noise also decreases to a certain extent. This indicates that the flicker noise in the displacement time series is almost entirely eliminated, and the white noise is weakened to a certain extent, which is consistent with the changes to the power spectrum before and after filtering in Figure 5. By comparing Table 2 and 4, the uncertainty of velocity estimation after filtering is significantly reduced, and most of the uncertainty is less than 0.01 mm/d. This shows that the flicker noise, which plays a dominant role in noise, is removed, leaving low-amplitude white noise.

Table 3. Amplitude estimation of white noise (mm) and flicker noise (mm/d$^{0.25}$) after removing flicker noise.

	E-direction		N-direction		U-direction	
Station	WN	FN	WN	FN	WN	FN
P1	0.44	0.02	0.34	0.01	0.35	0.02
P2	0.33	0.01	0.41	0.02	0.31	0.01
P3	0.46	0.02	0.40	0.02	0.38	0.02
P4	0.38	0.02	0.52	0.02	0.35	0.01

Table 4. Estimation of velocity considering flicker noise (mm/d) after removing flicker noise.

	E-direction		N-direction		U-direction	
Station	Estimation	Uncertainty	Estimation	Uncertainty	Estimation	Uncertainty
P1	0.05	<0.01	-0.01	<0.01	0.17	<0.01
P2	0.01	<0.01	0.03	<0.01	0.07	<0.01
P3	0.07	<0.01	0.04	<0.01	0.15	<0.01
P4	0.07	<0.01	0.15	<0.01	0.16	<0.01

Table 5. Displacement of each monitoring point after flicker noise elimination (mm).

	E-direction		N-direction		U-direction	
Station	Displacement	Uncertainty	Displacement	Uncertainty	Displacement	Uncertainty
P1	1.28	0.04	-0.21	0.02	4.54	0.02
P2	0.32	0.02	0.79	0.02	1.90	0.02
P3	1.92	0.04	0.98	0.02	3.96	0.02
P4	1.85	0.02	4.01	0.04	4.41	0.02

Table 5 lists the displacements at monitoring points in three directions within the selected time range obtained based on the GPS/BDS displacement sequence. It can be seen that in the N-direction, the displacement at point P1 to the south is about 0.2 mm; the displacements at points P2 and P3 are less than 1 mm; and the displacement at point P4 is about 4 mm. In the E-direction, the displacements of points P1, P3, and P4 exceed 1 mm, and the displacements at points P3 and P4 are close to 2 mm, while those at point P2 are about 0.3 mm. In the U-direction, the displacement of points P1, P3, and P4 is about 4 mm, and that of point P2 is about 2 mm. The above analysis shows that, in the selected time range, the low slope on the right-hand side of the railway undergoes greater displacement in the

IOP Conf. Series: Earth and Environmental Science **865** (2021) 012052 doi:10.1088/1755-1315/865/1/012052

northerly direction compared with the high slope, and the slope on both sides near the railway also undergo a certain displacement in the easterly direction. Moreover, the slope on both sides of the railway has an upwards trend. Figure 6 shows the historical temperature record for the slope in April and May: in April, the highest and lowest local temperatures rose from about 15°C to 30 and 20°C, respectively. The average maximum and minimum temperatures in May remain at 30 and 20°C, therefore, slope heave is likely to be caused by the expansion caused by continuous heating of the surface and metal pier [16]. In addition, the slope further from, and higher than, the road-bed undergoes a smaller displacement in the horizontal direction (albeit no more than 1 mm).

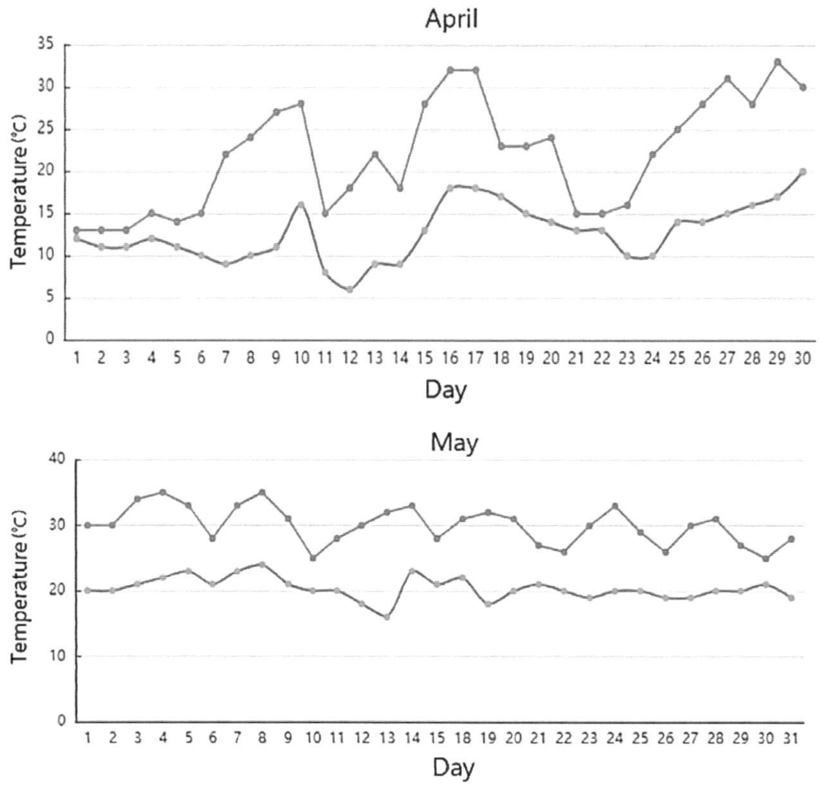

Figure 6. Logarithmic power spectrum of displacement time series in a certain direction of the monitoring point before (black) and after (red) removing flicker noise.

4. Comparison with GPS and BDS displacement analysis

In order to compare and analyze the GPS and BDS slope displacement monitoring results, the same method is used to eliminate the flicker noise in the GPS and BDS displacement time series. The displacement velocity and noise of the monitoring points before and after filtering are calculated. The results are shown in Table 6 and Table 7, where the symbols "I" and "II" represent the results before and after filtering, respectively. Comparing Table 1 and Table 6, it can be seen that the estimated flicker noise amplitude in GPS deformation time series data exceeds that in BDS, and the estimated noise amplitude in GPS/BDS deformation time series data is the smallest. This may be because the system error of GPS is greater than that of BDS, and the integrated application of GPS and BDS for data processing increases the number of visible satellites, decreases dilution of precision (DOP) values [17], and thus improves positioning accuracy.

Table 6. The amplitude estimation of white noise (mm) and flicker noise (mm/$d^{0.25}$) in the GPS and BDS displacement time series of the monitoring station before and after filtering.

Station	Time series	Before and after filtering	E-direction		N-direction		U-direction	
			WN	FN	WN	FN	WN	FN
P1	GPS	I	2.49	9.28	1.69	12.50	6.54	32.78
		II	0.43	0.02	0.34	0.01	0.34	0.01
	BDS	I	2.75	8.83	2.31	11.39	9.56	28.27
		II	0.44	0.02	0.41	0.02	0.45	0.02
P2	GPS	I	1.75	10.58	1.69	13.09	4.45	36.83
		II	0.28	0.01	0.23	0.01	0.23	0.01
	BDS	I	1.61	7.20	2.3	8.19	5.51	18.99
		II	0.35	0.02	0.41	0.02	0.41	0.02
P3	GPS	I	2.13	8.96	2.16	10.77	5.63	32.74
		II	0.38	0.02	0.33	0.01	0.32	0.01
	BDS	I	1.64	8.36	2.21	8.24	6.54	24.14
		II	0.34	0.01	0.40	0.02	0.40	0.02
P4	GPS	I	2.30	9.81	2.25	13.81	5.48	33.91
		II	0.37	0.02	0.28	0.01	0.28	0.01
	BDS	I	2.39	8.84	3.10	9.16	8.55	34.16
		II	0.44	0.02	0.44	0.02	0.38	0.02

Table 7. Estimation of velocity after flicker noise elimination (mm/d).

Station	Time series	Before and after filtering	E-direction		N-direction		U-direction	
			Estimation	Uncertainty	Estimation	Uncertainty	Estimation	Uncertainty
P1	GPS	I	0.03	0.13	0.02	0.17	0.20	0.45
		II	0.06	<0.01	-0.10	<0.01	0.22	<0.01
	BDS	I	-0.03	0.13	0.02	0.17	0.12	0.41
		II	0.06	<0.01	0.06	<0.01	0.20	<0.01
P2	GPS	I	-0.01	0.15	0.02	0.18	0.31	0.51
		II	0.33	0.33	0.33	0.33	0.33	0.33
	BDS	I	0.01	0.10	0.05	0.12	-0.13	0.28
		II	0.04	<0.01	0.06	<0.01	-0.11	<0.01
P3	GPS	I	0.08	0.12	-0.01	0.15	0.29	0.45
		II	0.09	<0.01	0.09	<0.01	0.35	<0.01
	BDS	I	-0.03	0.12	0.12	0.12	-0.11	0.34
		II	0.02	<0.01	0.14	<0.01	-0.12	<0.01
P4	GPS	I	-0.03	0.13	0.13	0.19	0.41	0.47
		II	-0.01	<0.01	0.14	<0.01	0.35	<0.01
	BDS	I	0.10	0.13	0.13	0.13	0.31	0.50
		II	0.12	<0.01	0.16	<0.01	0.11	<0.01

As can be seen from Table 7 that before filtering, for the same direction of the same monitoring point, there are differences in the displacement velocities of the stations obtained according to GPS, BDS, and GPS/BDS displacement time series data. Among them, 58% of the GPS velocity estimates differ from those made using BDS by more than 0.1 mm, and 66% of the velocities are estimated to have been in the opposite direction, for example, for the E-direction displacement time series at station P1, GPS speed and BDS speed have the same magnitude but opposite direction. BDS speed and GPS/BDS speed have the same direction, but the speed is three times that of GPS/BDS. For example, for the U-direction displacement time series at station P2, the GPS velocity is eight times that of GPS/BDS and twice that of BDS, and the direction is opposite that of BDS. Moreover, after filtering, about 30% of GPS velocity estimates differ from BDS by more than 0.1 mm/d, and about 40% of GPS, BDS, and GPS/BDS velocity directions differ. In the E-direction at station P1 station and the E and N-directions at station P3, the estimated directions of movement become consistent after filtering. In addition, about 75% of GPS/BDS velocity estimates are between those made using GPS and BDS. This shows that the difference between GPS and BDS speeds can be reduced by eliminating flicker noise and using GPS/BDS technology.

5. Conclusion

Taking a tunnel entrance slope of Chang-Jiangxi high-speed railway in China as an example, GPS, BDS, and GPS/BDS coordinate time series data from 28 April and 24 May 2020 are respectively used to analyse the slope displacement under the condition of white noise and flicker noise. The experimental results show that:

(1) The amplitude of flicker noise in the displacement time series is four to seven times that of white noise, which leads to an uncertainty in velocity estimation exceeding the velocity itself; the estimated flicker noise amplitude in GPS displacement time series data exceeds that in BDS data, and that in GPS/BDS displacement time series data is the least.

(2) There are differences in magnitude and direction among GPS, BDS, and GPS/BDS displacement velocities. By eliminating flicker noise in the displacement time series, the uncertainty of velocity estimation can be significantly reduced, and most of the uncertainties are less than 0.01 mm/d. The difference between GPS and BDS speed can be reduced by eliminating flicker noise and using GPS/BDS technology.

(3) Within the selected time range, the monitoring point on the low slope on the right-hand side of the tunnel entrance has a 4 mm displacement in the northerly direction; the monitoring points on both sides of the slope near the railway also undergo a displacement of 1 to 2 mm in the easterly direction, and rise about 4 mm in the vertical direction. The further and higher from the road-bed on the left-hand side of the tunnel entrance, undergoes a displacement of less than 1 mm in both the horizontal and vertical directions.

References

[1] Luis A, De L M, Ramos M, et al 2018 Displacements study of an earth fill dam based on high precision geodetic monitoring and numerical modeling *Sensors* **18(5)** 1369
[2] Benoit L, Briole P, Martin O, et al 2015 Monitoring landslide displacements with the geocube wireless network of low-cost gps *Engineering Geology* **195** 111-121
[3] Abolmasov B, Milenković S, Jelisavac B, et al 2015 The analysis of landslide dynamics based on automated GNSS monitoring—a case study *Engineering Geology for Society and Territory-Volume 2 Springer Cham* 143-146
[4] Komac M, Holley R, Mahapatra P, et al 2015 Coupling of GPS/GNSS and radar interferometric data for a 3D surface displacement monitoring of landslides *Landslides* **12(2)** 241-257
[5] Langbein, J, 2008 Noise in GPS displacement measurements from Southern California and Southern Nevada *Journal of Geophysical Research Solid Earth* **113(B5)**
[6] Ma J, Jiang W, Zhou X, et al 2018 Coordinate Time Series of IGS Station in China Based on the Combination of Wavelet Spectral and Variance Component Estimation *Geomatics and*

information science of wuhan university **43(04)** 629-636

[7] Xi R , Chen H , Meng X , et al 2018 Reliable Dynamic Monitoring of Bridges with Integrated GPS and BeiDou *Journal of Surveying Engineering* **144(4)** 04018008.1-04018008.15

[8] Nandakumaran N et al 2018 Multi-GNSS PPP-RTK From Large- to Small-Scale Networks *Sensors* **18(4)** 1078

[9] Ma J, Jiang W, Deng L, et al 2018 Estimation Method and Correlation Analysis for Noise in GPS Coordinate Time Series *Geomatics and information science of wuhan university* **43(10)** 1451-1457

[10] Williams S D P 2003 The effect of coloured noise on the uncertainties of rates estimated from geodetic time series *Journal of Geodesy* **76(9-10)** 483-494

[11] He X, Bos M S, Montillet J P, et al 2019 Investigation of the noise properties at low frequencies in long GNSS time series *Journal of Geodesy* **93** 1271-1282

[12] Jiang W, Ma J, Li Z, et al 2018 Effect of removing the common mode errors on linear regression analysis of noise amplitudes in position time series of a regional GPS network & a case study of GPS stations in Southern California *Advances in Space Research* **61** 2521-2530

[13] Ma J, Li Z, Jiang W, 2021 A new three - dimensional noise modeling method based on singular value decomposition and its application to CMONOC GPS network *Earth and Space Science* **8(2)** e2020EA001250

[14] MA J, CAO C, JIANG W, et al 2020 Method for Removing Colored Noise of GNSS Station Coordinate Time Series Based on Wavelet Packet Coefficient Information Entropy *Geomatics and information science of wuhan university* doi.org/10.13203/j.whugis20190353

[15] Langbein J 2012 Estimating rate uncertainty with maximum likelihood: differences between power-law and flicker–random-walk models *Journal of Geodesy* **86(9)** 775-783

[16] Wang K, Jiang W, Chen H, et al 2018 Analysis of Seasonal Signal in GPS Short-Baseline Time Series Pure and Applied *Geophysics* **175** 3485–3509

[17] Ma X, Yu K, He X, et al 2020 Positioning Performance Comparison Between GPS and BDS With Data Recorded at Four MGEX Stations *IEEE Access* **8** 147422-147438

AUTHOR INDEX

Aidarkozhina, A S ... 289
Cao, Jianping ... 95
Castagna, John ... 122
Chelnokov, G A ... 289
Chen, Aiqiong ... 275
Chen, Jun ... 275
Chen, Leishi ... 348
Chen, Li ... 400
Chen, Maolin ... 108, 221
Chen, Xiaoying ... 262
Chen, Yingzhi ... 54
Chen, Yonghang ... 172
Chong, Jinxin ... 207
Claramunt, Christophe ...
Cui, Chen ... 41
Dan, Linlin ... 81
Deng, Jian ... 22
Deng, Jianbo ... 348
Ding, Hao ... 342
Du, Dongsheng ... 348
Du, Jinyang ... 328
Duan, Shengyang ... 172
Fan, Xiang ... 328
Fan, Xiaoyong ... 89
Fang, Yanguo ... 115
Feng, Changmao ... 122
Feng, Yangjie ... 328
Feng, Yu ... 59, 355
Fu, Lei ... 35
Gao, Fan ... 235
Gao, Shang ... 35
Guan, Ronghao ... 335
Guo, Guo ... 380
Guo, Kai ... 14
Guo, Q ... 289
Han, Meitao ... 95
Han, Qinzhe ... 348
Han, Tao ... 67
He, Chicheng ... 115
Hon, Kai-Kwong ... 407
Hou, Lei ... 370
Hou, Qian ... 207
Hu, Tianyi ... 342
Hu, Yan ... 162, 251
Hu, Zheng ... 380
Jiang, Lijun ... 215
Jiang, Liyong ... 1
Jiang, Renqi ... 122
Jiang, Wenbo ... 235
Jiang, Zelin ... 22
Jin, Shangyong ... 294
Jin, Xiaohong ... 67
Ko, Chuan-Chuan ... 321
Kong, Yawen ... 143
Lavrushin, V Y ... 289
Li, Ang ... 335
Li, Congqing ... 89
Li, Fuxiao ... 14
Li, Laixin ... 74, 268
Li, Lanyu ... 282
Li, Li ... 294
Li, Lingqi ... 1
Li, Maohua ... 115
Li, Shaokai ... 115
Li, Wei ... 400
Li, Xin ... 182
Li, Xueping ... 207
Li, Yuanshan ... 321
Liang, Quan ... 342
Liao, Tongzhou ... 59, 355
Lin, Hong ... 8
Lin, Yilin ... 348
Liu, Chien-Yu ... 321
Liu, Chunmei ... 321
Liu, Chunyu ... 143
Liu, Furong ... 189
Liu, Guiqin ... 315
Liu, Hongwu ... 162, 251
Liu, Huanqian ... 251
Liu, Lina ... 122
Liu, Liping ... 54
Liu, Lixin ... 207
Liu, Qiong ... 172
Liu, Qiuxiang ... 189
Liu, Shuguang ... 143
Liu, Wei ... 41
Liu, Xiangjiang ... 108, 221
Liu, Xiaohao ... 136
Liu, Xingchen ... 335
Liu, Xingyan ... 46
Liu, Xuefeng ... 41, 243
Liu, Yunsheng ... 275
Lu, Anling ... 262
Lu, Jinjin ... 182
Lu, Xiaohui ... 262
Luan, Haijun ... 22

Luo, Qi....315
Luo, Zijun....54
Ma, Dongmin....74, 268
Ma, Jianwu....363
Ma, Jun....182, 412
Ma, Qianling....294
Ma, Xiaolin....74
Ma, Yupo....59, 355
Nawari....300
Pan, Genyun....54
Qiu, Zhangjin....22
Qu, Bo....1
Ren, Chonglei....243
Ren, Jiabao....243
Shao, Kai....74, 268
Shen, Lingzhi....196
Shi, Guoning....59, 355
Shi, Wei....189
Song, Lei....143
Sun, Ming....122
Sun, Qinbang....235
Sun, Yaqin....400
Tan, Chunsen....108
Tang, Jianhua....228
Tang, Jie....162
Tang, Minghui....251
Tang, Yiyang....89
Teng, Yuntian....89
Tian, Maozhong....380
Wang, Changcheng....59, 355
Wang, Difan....342
Wang, Dongdong....268
Wang, Hao....315
Wang, Junmin....35
Wang, Kun....122
Wang, Lei....41, 243
Wang, Li....315
Wang, Ruifang....54
Wang, Shijie....155
Wang, Wukang....315
Wang, Xiaomei....89
Wang, Xinran....81
Wang, Xiyu....221
Wang, Yalin....363
Wang, Yang....196
Wang, Ying....67, 315
Wang, Yu....189
Wang, Zhonghang....115
Wang, Zhuowei....335
Wei, Kejia....182
Wei, Li....81
Wu, Changhuai....35
Wu, Jincai....122
Wu, Kai....1
Wu, Leixiang....335
Wu, Mengfan....67
Wu, Qiwang....189
Wu, Shujuan....136, 228
Wu, Xiaodong....342
Wu, Zengbin....59, 355
Xiao, Jianhua....136, 228
Xie, Yun....196
Xu, Jingyu....162
Xu, Shengli....81
Xu, Shujie....41
Xu, Wen....315
Xu, Zhihao....67
Xue, Huiying....370
Yan, Fangping....189
Yan, Wenqi....155
Yan, Xiaoying....67
Yang, Fei....1
Yang, Gang....268
Yang, Haoqing....262
Yang, Lina....81
Yang, Wenguang....108
Yao, Rong....162
Ye, Xiangyu....189, 342
Ye, Xiaomin....196
Yi, Facheng....328
Yi, Hai....122
Yin, Huijuan....1
Yin, Yanjun....81
You, Aiju....35
Yuan, Xiangyong....321
Zeng, Fanyang....8
Zeng, Yuting....251
Zhan, Zhanyu....59, 355
Zhang, Chong....235
Zhang, Enhui....95
Zhang, Laichen....363
Zhang, Lihua....380
Zhang, Lin....74, 268
Zhang, Lishuo....8
Zhang, Panwei....335
Zhang, Pei....74
Zhang, Xian....400
Zhang, Xinyi....108, 221
Zhang, Yanming....215
Zhang, Yongqi....95
Zhang, Yueqiao....115
Zhang, Yumiao....235
Zhang, Yuxiang....95
Zhang, Ziguang....275

Zhao, Chenhang 172
Zhao, Pengfei 235
Zhao, Yizhan 182
Zhao, Yukun 46
Zheng, Jieqiong 294
Zhong, Guangjian 122
Zhong, Wenhan 67
Zhou, Gang 122
Zhou, Lv 182
Zhu, Meiheng 275
Zulkarnaini 300
Zuo, Zhenyu 59, 355